In the Wake of Tacoma

Suspension Bridges and the Quest for Aerodynamic Stability

RICHARD SCOTT

ASCE PRESS

American Society of Civil Engineers
1801 Alexander Bell Drive
Reston, Virginia 20191–4400

Abstract: The spectacular 1940 collapse of the Tacoma Narrows Bridge fundamentally changed the course of suspension bridge engineering. *In the Wake of Tacoma* surveys the changes the collapse has imposed on the design of suspension bridges to the present day. It describes design trends before the collapse, the collapse itself, and subsequent investigations to determine its cause. It then examines aerodynamic and other design and construction developments and their application in suspension bridges world-wide in the six decades since the collapse. The book contains 30 photographs and six line drawings, and will appeal to anyone with an interest in bridges and engineering history.

Library of Congress Cataloging-in-Publication Data

Scott, Richard, 1956 Feb. 10-
In the wake of Tacoma : suspension bridges and the quest for aerodynamic stability / Richard Scott.
p. cm.
Includes bibliographical references and index.
ISBN 0-7844-0542-5
1. Suspension bridges—Design and construction—History. 2. Bridges—Aerodynamics. 3. Tacoma Narrows Bridge (Tacoma, Wash.)—History. I. Title.

TG400 . S38 2001
624'.5'09--dc21 2001034137

Manufactured in the United States of America.
Second printing.

To my parents, for their curiosity about the world,

and to Renée.

TABLE OF CONTENTS

LIST OF FIGURES

FOREWORD

The collapse of a major structure understandably makes the news headlines. Those that involve significant loss of life, such as the Quebec Bridge, remain as part of a country's history. The Tacoma Narrows Bridge is perhaps unique in that its failure, with no casualties, has remained in the public awareness for some 60 years after the event, and is as well known today as those involving human tragedy on a much greater scale.

Its failure from aerodynamic instability has, of course, had a profound effect on suspension bridge design, and provides the absorbing theme for this book. Richard Scott has used it, legitimately, to present a comprehensive world view of the history of suspension bridges, as wind instability has been a problem dating back nearly two hundred years. The authors' research has continued for over ten years, but this is a particularly appropriate time to publish with the recent completion of exciting projects in Scandinavia and Hong Kong and the record breaking Akashi Straits Bridge in Japan.

The major suspension bridges such as the Brooklyn, George Washington, Golden Gate and Humber are well known, but this book brings to light lesser known projects and their designers in countries such as Norway and China. Newly developing technologies are discussed as applicable to very long spans for the future, where aerodynamic concerns will dominate. The recent vibration problems on the new Millennium footbridge in London are mentioned as a reminder that new systems can still present dynamic surprises and that a knowledge of bridge history and past performance is still important.

While covering technical issues at the expected level, the human interest items also figure prominently. Examples such as the rivalry between renowned engineers Ammann and Steinman, the controversy over the English design of the Severn Bridge compared to the American design of the Verrazano Narrows Bridge, plus the authors' observations on bridge aesthetics should widen the appeal of the book.

To me, the most surprising aspect of this unique book, providing broad coverage of technical, historical, aesthetic and human interest topics, is that it has been written by someone not involved with bridges at all in his career as a planner. I refrain from suggesting that suspension bridges are a hobby for Richard Scott, as I know that they have been more than that; more like a consuming fascination for a period of over 20 years.

I have had the real pleasure of reviewing the manuscript under development at the author's invitation. This association brought to mind my good fortune in having been able to pursue my own interest in suspension bridges at both the professional and personal levels. That interest has been heightened by the author's great enthusiasm for the subject. That enthusiasm is evident in the book, and I am sure the author will feel well rewarded for his years of diligent work if some of that enthusiasm is absorbed by readers of this book.

Roger Dorton
Buckland & Taylor Ltd.

ACKNOWLEDGMENTS

A great many people have freely given both time and information over the many years involved in preparing this book.

First and foremost go sincere thanks to Dr. Roger Dorton, who as early as 1984 read a very early draft of the manuscript and has since meticulously read several more drafts. His insightful comments, patience and stamina in explaining concepts, loans of copious amounts of material, and enthusiasm for the history of bridges have enriched both the book and the experience of putting it together in myriad different ways. He also kindly agreed to write the foreword. For his generosity and efforts I shall be forever grateful.

Dr. Hiroshi Tanaka of the University of Ottawa deserves special mention for his thoughtful and valuable insight in explaining aerodynamic concepts, correcting errors, and for providing a fascinating perspective on the history of bridge aerodynamics. I hope his passion for the subject is somehow communicated in this book.

Dr. Alan Davenport and Dr. J. Peter C. King of the Boundary Layer Wind Tunnel at the University of Western Ontario generously took a day from their busy schedules to explain concepts related to bridge aerodynamics. Their advice produced fundamental changes to the text.

Professor Niels Gimsing of the Technical University of Denmark kindly reviewed the text on the Storebælt East Bridge, and took the time to meet during a September 1998 visit Copenhagen. Similar thanks go to Klaus Ostenfeld and Anton Peterson of COWI Engineers and Planners A/S for meeting in Copenhagen, and for providing additional information on various bridges designed by COWI.

Dr. Allan Larsen of COWI generously provided Figure 18-1 that illustrates his recent findings on the torsional instability of plate girders and the Tacoma Narrows Bridge collapse. Dr. Larsen's offer to review material at short notice, his continued referral of subsequent findings, and his enthusiasm for his chosen field have been greatly appreciated and motivating.

Others who have given generously of their time to comment on portions of the manuscript include Dr. Satoshi Kashima, Executive Director, Honshu-Shikoku Bridge Authority; Mr. Arne Vangsnes, Head, Bridge Division, Norwegian Public Roads Administration; Mr. Robert Passfield; Mr. Lawrence Rubin, formerly of the Mackinac Bridge Authority; Professor Dario A. Gasparini, Case Western Reserve University, Cleveland; Mr. Ian Firth, Flint & Neill Partnership; Mr. Allen Paul, Ove Arup & Associates; Mr. David Goodyear, TY Lin International; Stuart and Helen Scott; and Nora Haslam. The book has been improved immeasurably because of their efforts.

Many people responded generously to requests for information. They include:

In North America, Lona R. Babbington, American Institute of Steel Construction; Frank L. Stahl, Ammann & Whitney; Arvid Grant, Arvid Grant Associates; R. H. Wildt and David R. Young, Bethlehem Steel; Greg Bayol, Caltrans; William P. Miller, Delaware River and Bay Authority; C. Carlton Read, Delaware River Port Authority; Gerwick Engineers; Bruce Selby and Dan Mohn, Golden Gate Bridge, Highway and Transportation District; Wendy Barnable and Steve Snider, Halifax-Dartmouth Bridge Commission; Henry J. Stanton, New York State Bridge Authority; Gladys Para, Peninsula Historical Society; Geoffrey Katz and Diane Koch, Port Authority of New York and New Jersey; Thomas E. Freburger, Maryland Transportation Authority; Modjeski & Masters; George Gesner, Parsons Transportation Group (Steinman); Professor David Billington, Princeton University; Gail Schwartz, Steinman; Harold Clinton, Sverdrup and Parcel; Ingolf Hohn and Catherine Sweeney, Triborough Bridge and Tunnel Authority; Tom Horken, United Infrastructure; Robert H. Krier and Tim Moore, Washington State Department of Transportation; and Margaret Vatcher, Vancouver.

In the United Kingdom, J. B. Robinson and Dr. David Fisher, Freeman Fox & Partners; D.C.C. Davis, Acer/Freeman Fox; Michel Virlogeux, Association Français pour la Construction; M. Parker, Bridon Ropes; J. W. Swift, Cleveland Bridge & Engineering; B. G. Grewar, Forth Road Bridge Joint Board; Keith Seacroft, Hull Daily Mail; M. Stockwell, Humber Bridge Board; Ian Hunter and Cheryl Eaton, Kværner Construction; J. M. Finch, Mott, Hay & Anderson; A. G. Simpson, and R. P. Stanley, Mott MacDonald; and Laurainne Copnall, Trafalgar House.

In Scandinavia, B. Grabonski, Hanne Meister, and Torben Forberg, COWI; Ib Møller, Møller and Grønborg; Inge Pontoppidan, MT Steel; Gitte Graven, Rikke Jensen, Michael Mølmer, Christine Secher, Gitte Tage Olsen, and Jacob Vestergaard, Storebælt A. S.; Sigmund Lønset, R. Rimehaug, K. Gjerding-Smith, and Odd Williamsen, Norwegian Public Roads Administration; Jon Krokeborg, Strait Crossings Secretariat; Gerner Rörso Jörgensen, Kjessler & Mannerstråle AB; and Werner von Olnhausen, P. A. Ortendahl, and Catharina Ottander, Swedish National Road Administration.

In the rest of Europe, Hans Störringer, Munich Reinsurance; J. Chafik, Société Nationale d'études du Détroit de Gibraltar (Morocco); Dr. Massimo Tarquini Guetti, Stretto di Messina; and Nurseli Cezayirlioglu of the Turkish General Directorate of Highways.

In Japan, Kyoko Arai, Tsuyoshi Matsumoto, and Ysuji Saeki, Bridge and Offshore Engineering Association; Y. Yamasaki, Chodai Ltd.; Satoshi Nishimoto and Hajime Shibuya, Hokkaido Development Bureau; Hiroshi Fujii, Koichiro Fumoto, Atsushi Goto, Toyoaki Ito, Katsuya Ogihara, Kazuyoshi Sakai, and Tsutomu Takazawa, Honshu-Shikoku Bridge Authority; Hiroshi Tanaka, Takai Nishijima, and Mitsuaki Iriya, Hitachi Zosen Corp.; Rikiya Iizuka, Infrastructure Development Institute, Japan; Etsure Natori, Ishikawajima-Harima Heavy Industries; Yasufumi Iseki, Kajima Corp.; Gary I. Tsuchida, Kobe Steel; Yoshiharu Shibarmura, Japan Highway Public Corp.; S. Ohte, Japan Society of Civil Engineers; Eric Beyer and I. Miyaki, Matsuo Bridge Co.; Seiji Ohmichi, Mitsubishi Heavy Industries; and George Kuruda and J. Yanai, Nippon Steel.

In China, C.K. Lau, Hong Kong Engineering Development Department; K.T. Cheung and Sam C.K. Wong, Hong Kong Highways Department; Stanley C.Y. Chan, Anthony Kau, and D. English, Hong Kong Transport Department; Joan Sin and Michael W. H. Leung, New Airport Projects Co-ordination Office, Hong Kong; Mai Jie, Lingdingyang Bridge Project Office; and Shi Zhong Zhou of the Jiangsu Province Department of Communications.

The following people kindly provided photographs and/or illustrations: Dr. Alan Larsen, COWI; Dr. Roger Dorton; David Lee Photography, Hull, England; Bob David, Golden Gate Bridge, Highway and Transportation District; Wendy Barnable, Halifax-Dartmouth Bridge Commission; Taku Hanai and Toyoaki Ito, Honshu-Shikoku Bridge Authority; Graham Whitehead and Nicky Watkins, Hyder Consulting; Leslie Ironside; Inge-Lise Leihof and Inge Pontopiddan, MT Steel; Hank Lotoszinski, Mackinac Bridge Authority; Kelly L. Melhem, Maryland Transportation Authority; Linda Norris, Michigan Department of Transportation; George Gesner, Parsons Transportation Group (Steinman); Martin Bodman, Mott MacDonald; Constance Nardella, Port Authority of New York and New Jersey; Merete Anderson, Storebælt BilledArchiv; Massimo Tarquini Guetti, Stretto di Messina; Laura Rosen, Triborough Bridge and Tunnel Authority; Kristin Kinsey, University of Washington Libraries; Robert H. Krier and Randy Cavanagh, Washington State Department of Transportation. Uncredited photographs were taken by the author.

The staff and resources at two Ottawa-based research libraries of Canada's National Research Council—the Canadian Institute of Scientific and Technical Information (CISTI) and the Institute for Research in Construction (IRC) were invaluable; the Engineering Library of the University of Toronto and the Ontario Ministry of Transportation library also provided a wealth of information. My thanks go also to François Lapointe of the National Capital Commission for permitting the flexible working arrangements that were necessary to complete the project in its late stages. Others who helped along the way include Janet Gooderham for translation services; Mara and Rota Bouse and Janice Miller for their generous loans of computers; Sinisha Brdar, Michel Céré, and especially Anne Heaney and Ilia Faghfou for invaluable advice and assistance with illustrative material; Nelson Edwards for indexing assistance; and Paule Ouellet for printing assistance.

The staff at ASCE Press has shown great patience and understanding in first showing interest in this project, and then sheperding it through the myriad stages of acceptance and production: my thanks to Joy Bramble, Joe Norris, Betsy Shepard, Donna Dickert, and Alyson Stein. Suzanne Coladonato deserves special mention for her enthusiasm and encouragement, as does Julie Kostecky, whose diligent editing of the manuscript extracting the author from several literary corners. My thanks also to the anonymous experts who reviewed early versions of several chapters on behalf of ASCE Press, and the anonymous technical reviewer who vetted a final draft of the entire manuscript, again on behalf of ASCE Press.

Finally, a project such as this—which is ultimately a solitary pursuit—cannot be undertaken without the support, patience, and understanding of family and friends. Most especially my thanks to my wife Renée, who has quietly but firmly encouraged me in my passion for bridges over the years and sacrificed much to help see this project through to completion.

An overview of this subject will be influenced by what has, or what has not been, consulted during research. What one person considers an important concept—or bridge—and its relative contribution to suspension bridge aerodynamics or bridge engineering in general may not be shared by another. The task of fairly presenting information is further complicated by the fact that an interpretation of recent history cannot benefit from the perspective of time. Finally, debate still surrounds the subject of suspension bridge aerodynamics; consensus does not exist in all areas and the field remains very much in a state of evolution.

The author's intent has been to present achievements, failures, and significant ideas—not all of which saw the light of day—that in some way have influenced the development of suspension bridge design and construction. The advice and counsel of those consulted has therefore been invaluable in helping to distinguish those ideas and developments that are more obviously important and significant from those that are less so, and from those where consensus may not yet (if ever) exist. This does not, however, in any way implicate those who contributed to the final product in any errors, omissions, or views with which they might not agree. The responsibility for whatever faults may remain lie solely with the author.

Finally, some housekeeping notes: all weights, dimensions, and distances are expressed in the International System of Units (SI). All heights and depths (e.g., towers, foundations) are measured from mean water level; truss or box girder depths are measured from center-to-center of chords. All costs are given in US dollars. Exchange rates may vary significantly over a period of years, and in some cases dramatically even over a short period. As such, costs quoted in the text may not reflect actual costs in local currencies, or the currency (or combinations thereof) for which payments to build a particular bridge were made. Page numbers for citations in the text are provided in the References, and full listings are provided in the Bibliography.

INTRODUCTION

The stage for the events of November 7, 1940 was set innocuously enough. As morning dawned, the winds blowing into the Tacoma Narrows in northwest Washington State began to increase as the storm forecast for that day approached. Presenting a thin profile to the wind—now gusting about 68 km/hr—were the graceful plate girders of the 853-m span Tacoma Narrows Bridge, the world's third longest suspension bridge. Opened four months earlier, this sleek bridge was known to behave in a peculiar manner. Most noticeable was a slow vertical rhythmic motion of its narrow two-lane solid plate girder. While the motion was not considered dangerous—the span had earned the name of "Galloping Gertie" and had become something of a novelty because of its fickle behavior—vehicles were often seen to alternately disappear and reappear in the trough of several gentle vertical waves gripping the main span. Some restraining measures had been added. When the gale of November 7—stronger than usual—blew into the Narrows, some motion would have been expected. But the events of this day would alter the course of suspension bridge engineering forever.

By 9:30 A.M., the characteristic vertical motion was well established but within another half-hour had reached sufficient proportions to prompt the span's closure. Engineers hurriedly sought to devise additional temporary restraining measures for installation following the storm's passage. Then, shortly after 10:00 A.M., a violent twisting motion consumed each half of the main span, alternately wrenching one edge of the flexible plate girder up and then brutally slamming it down. By 11:10 A.M., one of the most famous—and perhaps the most spectacular—of engineering failures ever witnessed was over. Galloping Gertie's slender but tortured main span lay at the bottom of a turbulent Narrows; its battered towers and cables stood as the wounded remnants of the most graceful suspension bridge ever built. The entire spectacle was recorded on some of the most memorable motion picture footage ever captured.

The immediate reaction to the catastrophe was one of profound shock, not just for the span's engineer—Leon Moisseiff, arguably the most respected and prolific suspension bridge engineer of the first half of the 20th century—but for the entire structural engineering profession. While the formative years of suspension bridge development in the early 19th century had been marred by an alarming frequency of failure—much involving loss of human life—the tragedy at the Tacoma Narrows, with no human fatalities whatsoever, would strike at the core of the engineering profession more profoundly than any previous bridge disaster. The collapse of the Tacoma Narrows Bridge almost instantly condemned a generation of suspension bridge engineering practice, but out of the wreckage emerged the new science of bridge aerodynamics, a discipline with profound implications for suspension bridge design.

The collapse also struck a chord with the general public. Faith in the apparent infallibility of engineering was shaken, but the disaster also spoke of a public fascination with this largest but most graceful of bridge forms, the suspension bridge. These finely-spun webs of steel stretched tautly across the sky on a monumental scale rarely fail to evoke emotional response. American engineering and architectural historian Carl Condit (1961, 129) wrote that the suspension bridge expresses a clarity of function so simple that even the most uninitiated can understand its workings; he considered it the most

> impressive type of bridge in appearance. . . . the most popular of structural forms; indeed, to make it the nearest thing to a public monument that contemporary building has produced. . . .
>
> As pure empirical form, there is nothing that can quite match them.

Such spans as San Francisco's Golden Gate Bridge or the Brooklyn Bridge in New York City have become powerful symbols, synonymous with the cities they serve. Few other structures can claim to play such a role in civic or cultural identity.

Perhaps it is the rarity of the suspension form that appeals; generally, only the largest bridges employ it. Few cannot be impressed by the sight of a great suspension bridge looming in the distance, the drama of the approach, and the crossing of the bridge itself. Landscape architect Lawrence Halprin, for example, (1966, 48) has described the north approach to San Francisco's Golden Gate Bridge as "one of the exciting visual events of our time." The crossing of a suspension bridge—the delicate net of suspenders rising higher and higher, the near tower framing its twin standing ephemerally in the distance, and the grand sweep of the cables—focuses the eye upward, offering a sense of freedom and a touching of the sky rarely experienced elsewhere.

The primary structural element of a suspension bridge is the wire cable, whose shape when suspended between two points has been described as "a direct visual expression of the operation of the force of gravity" (Appleton 1990, 16). Suspension bridges, per unit weight, have been lighter than any other type of bridge form and for this reason they have long dominated the genre of long spans. It is from the main cable—and its thousands of pencil-thin but incredibly strong steel wires—that this lightness is derived. Yet a suspension bridge is a juxtaposition, of simplicity in appearance with sophisticated engineering, of visual delicacy with strength. The reliance on tension makes this structural form both beautiful and graceful but at the same time responsive—and at worst vulnerable—to the dynamic forces of wind. These forces would prove the undoing of both Galloping Gertie and an increasingly bold—and perhaps overly-confident—generation of engineering thinking.

The wounds inflicted by Gertie on the engineering profession have for the most part healed, expressed over the past six decades in a cornucopia of ever larger and safer suspension bridges. Yet the repercussions of the collapse continue to profoundly affect practitioners today. The spectacle of November 7, 1940, the developments and personalities which lead up to it, and the continuing struggle to come to grips with aerodynamic phenomena, thus form the subject of this inquiry. Its prime focus is to document the evolution of the changes imposed by the Tacoma Narrows Bridge collapse and their manifestation in the many monumental suspension bridges built in the six decades since; spans such as David Steinman's legacy—the Mackinac Bridge; Othmar Ammann's Verrazano-Narrows Bridge, a span Steinman sought to build for almost half a century; and Gilbert Roberts' Severn Bridge between England and Wales, whose streamlined box girder revolutionized suspension bridge design. Also examined are the more recent contributions of the Orient, and in particular Japan and China, where in the 1980s and 1990 a plethora of huge suspension spans was built. A long but little known-tradition of Scandinavian suspension bridges is also explored, as are spans on the drawing boards and new concepts for ultralong spans that today reside in the imaginations of engineers.

The story is one of developments in bridge aerodynamics, construction advances, aesthetic trends, economic conditions, and engineering personalities. It is a story of how these trends have interacted to guide suspension bridge design and construction, extending from the era that produced the Tacoma Narrows Bridge through the more than six decades that have reflected its legacy. Aerodynamics forms the central theme of the narrative, but this survey also provides a general overview of the development of suspension bridges in the post-Tacoma era. The story is written from the perspective of engineering history in a way, it is hoped, that will appeal to both engineers and non-engineers with an interest in bridges and engineering generally. An overriding emphasis is placed on vehicular spans, with occasional forays into such interesting topics as pipeline suspension bridges, pedestrian suspension bridges—where much innovation has occurred—and the rehabilitation of large suspension spans.

The Tacoma Narrows Bridge collapse affected many prominent engineers responsible for other sleek plate girder suspension bridges, many of which embodied the same principles found in Moisseiff's ill-fated span. If these smaller contemporaries had not stretched the limits as had the bridge over the Tacoma Narrows, there was nonetheless cause for introspection by all; several of these spans had proven susceptible to the same seemingly harmless vertical motion that afflicted the Tacoma Narrows Bridge during its short life. No one, however, had foreseen the onset of torsional motion—the horrifying twisting and wrenching of the slim plate girder—that gripped Galloping Gertie in her final hour. The burning question for a stunned engineering profession was how a bridge designed to withstand the static force imposed by 160-km/hr gales could succumb so completely in a wind almost one-third that velocity. The answer was not a simple one, and an explanation for the collapse involved more than the peculiar circumstances of Tacoma's design and wind environment. In fact, it would be inextricably tied to rapid advances in suspension bridge engineering that emerged in the early part of the 20th century. Moreover, the evidence would reveal that the phenomenon at Tacoma was not new, but forgotten.

John Roebling appreciated the implications that wind held for suspension bridges when he designed New York City's landmark Brooklyn Bridge in the 1860s. Behind him lay a litany of failed suspension spans across Europe and America, some wracked to destruction by the sheer force of wind, others succumbing to strange rhythmic contortions that consumed suspended structures until they literally tore themselves apart. The triumphal 1883 opening of Roebling's 486-m span in New York would instill a new-found confidence in the form among both engineers and the public alike as a solution for long spans. This confidence was not misplaced; Roebling had armored his bridges well to stiffen them against the dynamic forces of wind.

Ironically, in one sense this emerging confidence in the suspension form would ultimately sow the seeds of the Tacoma Narrows Bridge collapse; engineering practice had seemingly shed the consequences of past failures. Roebling's admonitions against unstiffened suspension bridges were well documented in his writings and in his structures; the visually distinct, delicate array of wire stays radiating from his tower tops to the deck—most popularly known in the Brooklyn Bridge for their artistic composition—were borne out of the functional necessity to stiffen the structure. With Roebling's advances, the suspension bridge had become an inherently safe structural form. Yet the lessons he so ardently preached would be either forgotten or taken for granted as suspension bridge engineering advanced at an astonishing rate in the first four decades of the 20th century.

Several influences presaged spans of increasing size. New, complex mathematical theories suggested suspension bridges with deep heavy trusses were over-designed; combined with a shift from heavy rail to lighter vehicular loads, trusses became more flexible and shallower. Towers became more flexible too, with steel the material of choice instead of stone. The great spate of suspension bridges of unprecedented span built during the 1920s and 1930s thus reflected great advances in material and engineering science. With the effects of static wind loads and traffic loads seemingly mastered, aesthetics became an important principle to help guide design; no longer were steel juggernauts as New York's Williamsburg Bridge acceptable. Spans became narrower too, with the Depression not only pinching bridge finances, but also expected traffic growth.

Othmar Ammann's George Washington Bridge, opened in 1931, virtually doubled the span record to 1,067 m in a single bound, and embodied these changes unlike any other bridge. The debate over the appearance of its naked steel skeleton towers was indicative of an increased interest in suspension bridge aesthetics. Its pencil-thin suspended structure was completely devoid of stiffening trusses, yet so wide, massive and heavy that weight alone—and the stiffness of its huge main cables—were thought sufficient to buttress the bridge against static wind loads. This significant and influential characteristic of the George Washington Bridge imbued engineers with the confidence to design suspension bridges with increasingly thin—and graceful—suspended structures. The Golden Gate Bridge, for example, featured the slimmest trusses of any of its long-span contemporaries when completed in 1937. Yet perhaps the most significant departure from tradition was the replacement of the ubiquitous truss with plate girders. These solid but thin steel ribbons exemplified the economic and aesthetic doctrines of the 1930s and reflecting the engineering confidence of the era. Their introduction in American suspension bridges during the second half of the decade would be characterized by tremendously bold leaps in span within just a few years.

Then came the Tacoma Narrows Bridge. With its collapse came the recognition that static wind loads formed only one part of the wind-loading puzzle. This revelation sparked a period of intensive research and sometimes emotionally charged debate within the engineering profession. Much of this debate involved two dominant figures of 20th century bridge engineering—Othmar Ammann and David Steinman—continuing one of the greatest engineering rivalries of all time. The foundations of modern suspension bridge aerodynamics that emerged from the collapse continue to fundamentally shape the design of suspended structures. Bridge engineers continue to extend their knowledge of the dynamic effects of wind and to unravel its intricate and sometimes mysterious relationship with suspension bridges. In fact, the field is still rife with discussion and differing opinions on the exact aerodynamic phenomena at work, how they interact with one another and how suspension bridges respond to them, and the most appropriate methods by which to model various aerodynamic phenomena and structural response. Opinion on the exact cause of the Tacoma Narrows Bridge collapse is even today not unanimously shared.

As the understanding of suspension bridge aerodynamics continues to evolve, the suspension bridge's 75-year supremacy in the genre of long-span construction is now being seriously challenged by the cable-stayed bridge—a structural form virtually unheard of at the time of the Tacoma Narrows Bridge collapse. Other new hybrid forms are emerging for spans over 2 km in length. The future promises an exciting rivalry between—and possible integration of—these two bridge forms, both of which found expression in the work of the master-builder John Roebling almost 150 years ago. Were Roebling alive today, the foundations he laid would surely enthrall him.

Chapter 1

TURBULENT BEGINNINGS

1.1 ROOTS IN ANTIQUITY

The origins of the suspension bridge are manifested in nature. Tension structures have been known from the tropical rainforests of South America and Southeast Asia where vines—such as those in the liana family—have draped themselves between trees, providing passage for tree-dwelling mammals. It is then not surprising that suspension bridges using natural fibers are first known from the tropics, where use, thought to have occurred by the onset of the Bronze Age (circa 3500 B.C.), even predates that of masonry. The earliest suspension bridges are thought to have originated in the eastern Himalaya and, later, along the coast of western South America. It is not known whether or how knowledge migrated, but China, India, and Japan and Borneo in Asia; the west coast of Africa; and, in the new world, Mexico and Peru are all known to have built early suspension bridges.

The birth of the modern suspension bridge is tied to the development of metals, in particular iron and steel. Bridges with planks laid transversely over parallel chains are thought to have first appeared in the Chinese mountain province of Yunnan at the beginning of the Christian era, although iron chains may have been used as early as 200 B.C. during the Han dynasty. These bridges, however large in span, were not without fault, for writings of the period allude to suspended decks of bamboo planking wobbling uncontrollably under moving loads. Not until 1586 was the first known suspension bridge with a form of stiffening completed; in the Ji-Hong Bridge, auxiliary chain stays radiated from the towers to points along the deck—a prototype of today's cable-stayed bridges.

Knowledge of Chinese bridges percolated slowly to the West, largely as a result of travelogues published by Athansius Kircher in 1667. The conceptual origins of the modern suspension bridge have been attributed to the Venetian Faustus Verantius in *Machinae Novae,* first published in 1615. This remarkable work describes two types of tension structure prevalent today. One bears a strong resemblance to contemporary cable-stayed bridges; the other suggests the classic suspension bridge that would emerge in the 19th century. A more recent (early 1990s) study in the Americas of the remains of a remarkable 183-m long multispan 7th century Mayan suspension bridge has revealed the use of concrete footings, rope cables, and a horizontal deck connected to the main cables by suspenders.

Europe's first permanent chain bridge (although more cable-stayed than suspension) was the 21-m span Winch Bridge, completed in 1741 across England's River Tees. In spite of the well-intentioned use of chains to connect the span's quarter-points to the river banks below, one observer remarked that the unstiffened bridge was:

> planked in such a manner that the traveller experiences all the tremulous motion of the chain, and sees himself suspended over a roaring gulph on an agitated, restless gangway to which few strangers dare trust themselves (Hopkins, 1970).

Nonetheless, the span survived more than 60 years until an 1802 collapse which killed at least one person.

1.2 MODERN BEGINNINGS IN AMERICA AND EUROPE

The prototype of the modern suspension bridge emerged in America from the work of James Finley (1756-1828). His first bridge—a 21-m span—was likely completed in 1801. Finley incorporated two critical advances; the first was a horizontal deck connected to the span's two wrought iron link chains by suspenders of varying length. Finley also used a handrail braced with diagonals and longitudinal joists beneath the deck to resist wave-like deformations created by moving traffic. These were important defenses, for traffic loads were large in comparison to deck weight. Although probably not realized at the time, these measures likely helped to guard against the static and dynamic effects of wind, as did generous deck widths of up to 9.1 m. In these ways, Finley's spans would presage the form in use today.

Up to 40 or more spans were attributed to Finley or those who built under license, most before 1815.[1] Finley's concept was largely devoid of theoretical basis, and several spans also collapsed under load. One replacement bridge—a 124-m span built by the nearby wire manufacturer White and Hazard in 1816 across the Schuylkill River at Fairmount near Philadelphia, Pennsylvania—was the first wire suspension bridge in America, and possibly anywhere. Anchorage was crude: one main cable was wrapped around a tree trunk; the other passed through the attic of one of the firm's factory buildings! The bridge later collapsed under snow and ice, but its replacement would be equally significant.

Outside of Finley, few built suspension bridges in America. Finley foresaw spans of 457 m and proposed a bridge with 25 spans of 68 m across the Susquehanna River in Pennsylvania. His work was cited in Thomas Pope's 1811 *A Treatise on Bridge Architecture*, the first American exposé on bridge building. Pope also recognized the problems of deck vibration, which he said could occur from the passage of "even an animal of small weight" (Peters, 1987). France, where the Revolutionary and Napoleonic wars had inhibited information exchange, and England, still at odds with its former colony, were also influenced by Finley's work, via Pope. Yet Finley's approach to stiffening was largely ignored in place of varied but largely unsuccessful cable-stiffening systems. The great 20th century bridge engineer Othmar Ammann would write (1923) that "Had Finley's sound principles of construction been adhered to many of the early suspension bridge failures would have been avoided."

Although America turned its back on suspension bridges, the new form blossomed in Europe, particularly in Great Britain and France. British engineers had experimented as early as 1807; Scotland's 34-m span Galashiels Bridge—completed just months after the bridge over the Schuylkill River at Fairmount—also featured wire suspension cables. Several small cable-stayed spans were also built in Great Britain. In 1821, French architect Bernard Poyet published drawings of a cable-stayed span, followed in 1823 by compatriot Claude L.M.H. Navier. Yet cable-stayed bridges faded after two collapses. England's Dryburgh Abbey Bridge, an 1817 pedestrian span of 79 m across the River Tweed, fell in an 1818 ice storm when several stays broke; oscillations, which had occurred before, destroyed the span. For repairs, the great Robert Stephenson suggested longitudinal stiffening. Then came the 1824 failure of a German span that killed an unknown—but large—number of people. Navier's subsequent (and perhaps unfair) condemnation of cable-stayed bridges signaled their demise for almost 150 years until the emergence of analytical tools to calculate cable loads and techniques to prestress the stays.

Briton Samuel Brown (1776-1852) dominated much of the early British suspension bridge building, but his spans' record was spotty. His 1818 replacement for the Dryburgh Abbey Bridge marked the first ever use of eyebar suspension chains, it also featured stays, as well as a stiffened railing. His record 133-m span Union Bridge (1820) was also the first British span designed to support vehicular loads; 12 eyebar chains, six on each side, supported its unstiffened deck. In spite of long-standing rumors of its early demise, the bridge remains standing as Britain's oldest extant vehicular suspension span (steel cables were added in 1903). Brown's 1821 Trinity Pier—comprising three end-on-end spans of 64 m—at the Firth of Forth featured 1.2-m high parapets of wrought iron, but a storm soon led to the addition of stabilizing underfloor stays.

Brown's most famous—and notorious—bridge was the Brighton Chain Pier (1823), with four continuous spans of 77-m each. In 1833, its third span was partially destroyed under uncertain but wind-related circumstances. In November 1836, the same third span collapsed in spectacular fashion in high winds before a large crowd in broad daylight. Lt. Col. William Reid recorded its last moments in meticulous detail, but the full significance of his sketches and notes would not be appreciated for more than a century. The repaired pier eventually succumbed in an 1896 gale.

In March 1830, Brown's 131-m span Montrose Bridge (1829) collapsed under the sudden movement of 700 spectators during a boat race. Although the total number of dead is not known, this failure ranks as one of the worst bridge disasters of all time. The span was repaired but was then severely damaged in an October 1838 storm. This time the strange rhythmic contortions that had gripped Brown's Brighton Chain Pier were observed. James Rendel's 1840 rebuild featured 3.5-m deep through stiffening trusses, as well as cross-trusses; he wrote that such spans should possess the "greatest possible amount of stiffness" (Petroski, 1994). Ammann (1923) credited Rendel as one of the first to recognize the value of an inherently stiff suspended structure. The rebuilt bridge survived until 1930.

Other English spans failed under load, including Brown's 1829 Broughton Bridge. Its 1831 collapse from marching soldiers setting up rhythmic vibrations resonant with the natural period of vibration of the bridge is said to have led to the rule to break step over bridges, although a French and Austrian span later collapsed for the same reason. The 1845 collapse of a small span in Yarmouth, England under excessive load was also said to have killed many people.

The rise of Thomas Telford (1757-1834) marked a watershed. His 1814 proposal for a 305-m span across the River Mersey at Runcorn, England was the first to feature wrought-iron cables (although Brown later convinced him to switch to eyebar chains); he even built a 15-m long model of the bridge. Telford achieved greater renown at the Menai Straits, where in 1826 he completed a record eyebar span of 177 m. The bridge was immediately hailed an engineering masterpiece. Telford, like many British engineers, was skeptical of theoretical approaches, but the span at Menai Straits was one of the first to benefit from suspension bridge theory when the mathematician Davies Gilbert suggested an increase in cable sag. Telford had considered a trussed deck but decided to only add stiffening later, should it be required. This decision proved portentous. Undulation of the flexible deck would cause minor damage on several occasions, and had occurred even during construction. Barely a week after opening, the span was lucky to remain standing after a gale incited the same unusual rhythmic twisting responsible for the demise of Brown's Brighton Chain Pier.

Subsequent deck repairs and strengthening heralded a decade of relative calm until January 1836, when another storm induced waves of almost 5-m double amplitude.[2] Incredibly, the span suffered only minor damage. Then in January 1839, a storm lashed the bridge with winds of hurricane force. More than 50 m of deck flapped about in the poststorm breezes, and more than one-third of the suspender rods had snapped. The deck was rendered virtually unsalvageable and was replaced with a strengthened version incorporating trussed railings. The bridge performed well for another century and was extensively refurbished in 1939 and 1940 with four high-tensile steel eyebars replacing the 16 original wrought-iron chains. The bridge over the Menai Straits continues to carry traffic, and remains a structure of great engineering significance. Telford's extensively modified 1826 span of 100 m at Conway in north Wales also remains standing.

There were unqualified successes, including W. Tierney Clark's 122-m span Hammersmith Bridge (1827) across the River Thames near London. His 71-m span Marlow (1831) Bridge over the same river is his last extant span, albeit extensively remodeled. Influenced by the behavior of Telford's bridge at Menai Straits, Clark used stiffening trusses, which undoubtedly contributed to the longevity of his spans. Isambard Kingdom Brunel's 206-m span Hungerford pedestrian bridge became Britain's longest single span in 1845, but the pinnacle of the British chain era was Brunel's Clifton Suspension Bridge over the Avon Gorge at Bristol. Although completed in 1864, its genesis extends back to an 1829 competition involving Brown, Brunel, and Telford. Brunel—at age 24—won the commission, but financing and other problems would see 34 years pass before the 214-m span was completed. Ironically, Brunel's death in 1859 provided the impetus to finish the bridge, and the 1862 demolition of the Hungerford Bridge saw its suspension chains re-used for the Clifton Bridge (the chains originally intended for the Clifton Bridge ended up in Brunel's Royal Albert Bridge across the Tamar, opened in 1858 with two spans of 139 m). Situated in a spectacular setting, the Clifton Suspension Bridge has survived to this day, mostly unaltered, through careful maintenance.

The Clifton Suspension Bridge marked the end of the British chain era. Between 1830 and 1850, several collapses and a need for bridges to support an expanding rail network pointed to a need for more rigid, stronger structural forms. Developments on both the continent and in America would ultimately have a more profound and enduring influence on the future of suspension bridges.

French engineers had been insulated from British developments, largely because of France's self-imposed isolation beginning in 1789 during the French Revolution. Knowledge of Finley's achievements was more common given America's sympathy for the French, and France quickly became a leader in this nascent technology. French engineers had strong academic training and a theoretical bent, as opposed to the British, who relied more on experience and testing. British engineers largely ignored developments in France, while French engineers learned from—but did not mimic—British experience. France would build at least 350 suspension bridges up to 1870, but the wire cable, rather than the chain, dominated. Two visits of Navier (1785-1836) to England—in 1821 and 1823—were influential, and his 1823 *Rapport et Mémoire sur les Ponts Suspendus* is recognized as the first systematic theoretical and practical treatise on suspension bridge design, construction, and behavior under load.

The world's first permanent wire suspension bridge, with two spans of 40 m each at St. Antoine in Geneva, Switzerland, was completed in 1823 by French engineer Henri Guilliame Dufour (1787-1875). Both he and Navier believed weight implied greater stability and, unlike Finley, considered deck stiffness to be of secondary importance. Navier also considered a low sag ratio (that is, a flatter cable arc) a more effective stiffener than a truss; thicker cables would handle larger tensile loads. Dufour also wrote that under full dead load "nothing is to be feared from the live forces," especially in longer spans with a greater ratio of dead-to-live load. Yet he also recognized that repeated and regular vibrations could be amplified to the point of alarm (Peters, 1987).

The Seguin brothers (nephews of Joseph Montgolfier, the balloonist)—and Marc Seguin in particular—did much to entrench the wire suspension bridge in France. Marc Seguin's 1824 publication *Des Ponts en Fil de Fer* on wire cable suspension bridges provided a practical counterpoint to Navier's theoretical prescriptions. Seguin considered rigidity essential, and his 1824 span across the Galaure River was the first to use a trussed handrailing for stiffening, aided by underfloor stays. His 1825 Tain-Tournon Bridge was France's first large suspension bridge and the first to use wire (over Navier's objections) in support of vehicular traffic.

Navier's influence waned with the partial collapse in 1826 of his almost-completed 170-m span Pont des Invalides over the River Seine in Paris. The collapse sparked several investigations, most notably by Louis-Joseph Vicat (1786-1861). Vicat found wire bridges to be rife with problems; many had collapsed or were severely damaged from wire failure. In spite of these problems, Vicat's support of wire cables was unequivocal. Many hundreds of wires ensured a span against failure, and the drawing of the wire during its manufacture provided in situ strength testing; by contrast, a single imperfection could bring down a chain bridge. Nonetheless, Vicat did recommend major changes. He preferred a single large cable to the several small ones of the so-called "garland" system; it would act as a whole and be stronger, stiffer, and more corrosion-resistant. His most enduring suggestion was to erect wires in situ rather than on the ground. He proposed to loop individual wires over a wheel attached to an endless hauling rope traveling between anchorages. The method promised greater accuracy and made his large, single cables possible. Vicat claimed major labor and machinery reductions, and he suggested that spans of up to 500 m were now possible. With this, Vicat established the theoretical and practical basis for cable spinning. Although the term is inaccurate, the practice itself would revolutionize suspension bridge construction.

Vicat's 1831 report coincided with the onset of a frenetic period of iron wire suspension bridge construction in France. The Seguins would build 86 documented spans (estimates range as high as 186), but almost all, without exception, retained the intricate garland system rejected by Vicat. Several multiple spans exceeded 305 m, the longest being the André de Cuzbac (1839) over the River Dordogne with five spans of 109 m. The Roche-Bernard Bridge (1840) was France's longest single suspension span at 198 m, and it may also have been the first to use Vicat's largely ignored method of aerial spinning. Only one other span (at Tours) was known to have used it, but both were among the largest of this early period of French suspension bridge building.

Nevertheless, collapses continued from flood, landslide, test loading, and wire corrosion, but unlike the British experience, wind would not figure prominently. Yet one disaster would signal an end to the era. On April 16, 1850, some 500 troops were marching across the 102-m span Basse-Chaine Bridge across the Angers River during a storm. Severe vibrations gripped the span, one cable snapped near the anchorages, and of the 478 soldiers that followed the deck into the river, 226 died in what remains the worst bridge disaster ever. The tragedy stunned French engineers. Its cause: cable wire corrosion in the anchorages. Ironically, Vicat's system of cable anchorage—devised to prevent corrosion—was the main culprit.

Literally hundreds of spans using this system were now considered suspect, but they were difficult to inspect. A moratorium was slapped on new suspension bridge construction in France. Then just 2 years later, the Roche-Bernard Bridge was heavily damaged in an 1852 gale, likely because of a lack of stiffening. Underfloor stays were subsequently added, but these two events brought disfavor to the suspension bridge in France. The form's apparent unsuitability for an emerging network of railways also saw steel cantilevers and arches dominate. The prohibition ended only in 1870—with strict conditions—but the heyday was over. Even then, problems would continue; for example, in 1894, the 184-m span St. Christophe (1847) would go down in a gale. Nonetheless, the legacies of Navier, Vicat, Dufour, and the Seguins remain today.

Other European nations would make significant contributions to the nascent form. One example was Ignace Edler von Mitis' 105-m span Karl's Bridge over the Danube Canal in Vienna. Completed in 1828, it was the first western bridge known to use steel, which it did in its two eyebar chains. Germany completed its first suspension bridge in 1827 and, in 1839, produced the first braced chain span, a form in which vertically arranged parallel chains are connected and strengthened by members arranged in a stiff triangular pattern. This concept was applied to Czechoslovakia's largest suspension bridge, the 132-m Moldau Bridge in Prague, completed in 1842. There were problems, however; in the winter of 1833 and 1834, 12 of the 16 eyebar chains supporting the unstiffened Nassau Bridge (1830) broke during a wind storm, dropping the 75-m span into the Lahn River.

Joseph Chaley's record 265-m span Fribourg Bridge, completed in 1834 over the

Sarine River in Switzerland, brought about major advances.[3] Chaley used many of Vicat's prescriptions, especially in the anchorages and cables. Each of the four record 1,056-wire main cables was assembled into 20 strands on the valley floor to facilitate lifting into the bridge, whereupon they were bound into a single cable. Chaley also inclined his suspenders transversely to help brace against wind (although vertical oscillation did occur in spite of supplemental bracing in the form of Seguin-inspired railings). The Fribourg remained the world's longest span for 15 years and Europe's longest for 67 years until demolished in 1923; the Swiss have yet to exceed it. Chaley used in situ spinning in six of his later bridges and introduced catwalks—continuous working platforms slung beneath the line of the future cable to facilitate wire assembly (the name was derived from the view that only cats would tread across such a precarious platform). Chaley would later introduce endless wires, in which the ends of adjacent wires were spliced together on the bridge. This feature, introduced in the 227-m span Gottern Bridge(1840), dispensed with hundreds of individually anchored cable wires dimensioned to the span length and also increased the strength of the cable. It established a prototype for modern cable construction which persists today.

Portugal's first large suspension bridge was completed in 1842, a 170-m wire cable span across the Douro River. In Budapest, Hungary, the massive 203-m span Széchenyi Bridge featured a 14-m wide deck suspended from four eyebar chains. It was built between 1839 and 1849 to a design of Clark, of Hammersmith Bridge fame. Clark provided longitudinal and cross-trusses for stiffening, but oscillations were observed under heavy load or in strong wind. The bridge was largely replaced (save stone towers and anchorages) in 1914 and 1915 but then destroyed by a retreating German army in 1944. Its 1949 replacement emulated the remodeled span, and a plaque commemorating Clark was installed as recently as 1998.

British engineer Charles Vignoles built another monumental chain span at Kiev over the River Dnieper between 1848 and 1853. Its four 134-m spans and flanking 69-m side spans featured an unusual truss that deepened above the roadway at the main span's quarter-points. The massive 16-m wide deck was hung from four large chains; its $2.5 million cost was also extraordinary for the times. Unfortunately, the span was demolished by Russian troops retreating in the face of Hitler's armies in World War II.

The abrupt end to French suspension bridge building left much unused knowledge and experience for others to adapt and improve. In America, two protagonists—native-born American Charles Ellet Jr. and Prussian immigrant John Augustus Roebling—both engineers of European training, would dominate American activity and catapult the suspension bridge into a new era.

1.3 MATURITY IN AMERICA—ELLET AND ROEBLING

In America, the suspension bridge had languished while British and French engineers embraced it. Yet in the 1840s, the form experienced a resurgence that would see America gain supremacy in the field of long-span suspension bridges for more than a century. Ellet (1810-1862), a flamboyant and inventive American, is credited with importing French expertise to America. Ellet had been exposed to the work of Vicat and the Seguins while studying at the École des Ponts et Chaussées in

Paris in 1830 and 1831. He then toured France, Germany, and England to examine suspension bridges, including Chaley's Fribourg Bridge, then in the early stages of construction. Ellet returned to America in early 1832 and immediately proposed a 183-m suspension span at Washington, D.C., and a huge 366-m span at St. Louis, Missouri. His first bridge—the Fairmount—was a more modest span of 109 m, and was completed in 1842 across the Schuylkill River near the site of White and Hazard's wire span of 1816. Ellet adopted many French features in this bridge, including garland cables assembled from wire on the ground. The Fairmount was also the first American span to use stone towers yet was largely unstiffened, although Ellet claimed its weight and framing provided rigidity. It was replaced in 1874.

Roebling (1806-1869) had immigrated to the United States in 1831 at the age of 25, 5 years after graduating in civil engineering from Berlin's Royal Polytechnic Institute, where he had written his thesis on suspension bridges. Roebling had also read German translations of Navier's work. He was the antithesis of Ellet—scientific, forceful, rigid in his thinking and carrying a stern, intense visage. He cast an intimidating shadow. In 1841, Roebling began to supply wire rope to Pennsylvanian canal companies and, in 1842, patented a process for manufacturing iron wire. Even then he recognized that many suspension bridges—particularly British chain bridges—were "entirely too light" to forestall wind-induced undulation (Petroski, 1994). Roebling followed Ellet's progress at Fairmount closely and, in 1844, began his first suspension bridge, using iron wire cables of his own manufacture. The Pittsburgh aqueduct across the Allegheny River was completed in 1845. With seven spans of 49 m, it was an impressive debut. Its continuous cables contained 1,900 wires each, placed in situ almost identically to Vicat's prescriptions. The span was the world's first suspension aqueduct, the first large American multispan suspension bridge, and the first anywhere to use the rudiments of modern cable spinning.

In 1846, Roebling completed the Smithfield Bridge across the Monongahela River, again in Pittsburgh, Pennsylvania. It was his first road suspension bridge but also his only span to shun cable spinning. Latticed truss railings and wire rope stays radiating from the tower tops helped to stiffen its eight spans of 57 m. The stays would become a Roebling trademark, and reflected his concern over rigidity. Roebling had denigrated Ellet's Fairmount Bridge for its absence of stiffening. He felt French methods ignored the wind problem, and although not fully understanding the problem's mechanics, Roebling buttressed his bridges against it. Ellet, for his part, derided the use of trusses as "a charletan proposal" (Buonopane and Billington, 1993)—a view that would cost him dearly.

Roebling improved his methods and received a patent for cable spinning in 1847. Between 1848 and 1850, he completed four more suspension aqueducts. Only the 1848 Lackawaxen Bridge across the Delaware River—the largest, with four 41-m spans—remains standing as possibly the oldest extant American suspension bridge. Roebling also moved his wire business to Trenton, New Jersey, in 1848. Yet in spite of Roebling's advances, it was Ellet who landed the plum job of spanning the Ohio River at Wheeling, West Virginia. His record-breaking span of 308 m was opened in 1849 and provided America a foothold it would not relinquish for more than 125 years. However, the span's almost complete absence of deck stiffening would ultimately serve to secure Roebling's dominance in the rivalry.

Both engineers had also tendered plans for a road/rail suspension bridge connecting Canada and the United States about 3 km downstream of Niagara Falls. In his 1847 bid, Roebling proclaimed rail suspension bridges with 457-m spans to be feasible and claimed, "There is not one good suspension bridge in Great Britain," a reference to the lack of stiffness in Great Britain's bridges (Steinman and Watson, 1957). These faults, he asserted, could be overcome with trusses and stays. Ellet ended up winning the contract and built a temporary span in 1848, but a dispute over tolls saw Roebling take over the job and begin construction in 1851 of a permanent 250-m span of his own design. These delays had in the meantime allowed Edward Serrell's 317-m span Lewiston and Queenston Suspension Bridge to be opened across the Niagara gorge first in 1851.[4] Then in May 1854, with truss erection about to begin at Niagara, Ellet's Wheeling Bridge went down during a storm in a well-documented collapse. The flexible deck—stiffened only with railings—was utterly destroyed.

Roebling's Niagara Suspension Bridge was opened in 1855. With a 5.5-m deep timber truss connecting upper rail and lower road decks, it was the first large suspension bridge with a deep stiffening truss. The demise of Ellet's Wheeling Bridge had prompted Roebling to add wire stays down from the tower tops and up from the slopes below to the truss to help stiffen it. By 1860, 45 trains crossed the span daily, albeit at a sloth-like 5 km/hr, but the truss and stays combined to limit deflections to 10 cm. Moreover, the use of multiple forms of stiffening—weight, trusses, and stays—heretofore used individually in previous spans, came together for the first time. Improved cable spinning procedures introduced at Niagara would also remain essentially the same for 125 years. Robert Stephenson's written words to Roebling that "If your bridge succeeds mine [the great Britannia Tubular Bridge of 1850] is a magnificent blunder" would become well known (Steinman and Watson, 1957).

With the Niagara's record as the first successful rail suspension bridge, Roebling envisaged much larger spans. He also recognized the potential of new materials:

> Bridges of half a mile span, for common or Railway travel, may be built, using iron for the cables, with entire safety. But by substituting the best quality of steel wire, we may nearly double the span, and afford the same degree of security (ibid.).

With regards to wind, Roebling wrote:

> Weight is a most essential condition where stiffness is a great object, provided it is properly used in connection with other means. If relied upon alone, as was the case in the plan of the Wheeling Bridge (U.S.), it may become the very means of its destruction. That bridge was destroyed by the momentum acquired by its own dead weight, when swayed up and down by the wind (Billington, 1979).

This wind, he continued, when "acting on a suspended floor, devoid of inherent stiffness, will produce a series of undulations, which will be corresponding from the center each way" (Henry and Jerome, 1965). He indicated the "necessity of introducing the principles of the triangle, so as to form stationary points and thus check vibrations and restore balance." The Wheeling Bridge's demise, he indicated, "was clearly owing to a want of stability and not a want of strength . . . over-floor stays, truss railings, underfloor stays or cable stays" (Plowden, 1974) would provide the recipe for stability.[5] Roebling took a poke at his competitors by writing:

> The catalogue of disastrous failures is now large enough to warn against light fabrics, suspended to be blown down, as it were, in defiance of the elements. A number of such fairy creations are still hovering about the country, only waiting for a rough blow to be demolished (Plowden, 1974).

Other suspension bridges would span the Niagara in the latter half of the 19th century, but none as significant as Roebling's. Serrell's 1851 span was nearly wrecked by wind in 1855; underfloor stays were subsequently installed on Roebling's advice using his wire. Then in February 1864, the span went down in a winter gale; ironically the stays—disconnected the previous month to avoid being imperiled by an ice jam—had not been retightened. Its main cables hung as an epitaph of the event for another 34 years until a replacement span was opened in 1899.[6] Samuel Keefer's 386-m span Upper Suspension Bridge then opened in 1869 as the world's longest single span bridge. Just months after a major strengthening, a violent January 1889 storm tore away the entire deck in circumstances that were never fully known; the omission of underfloor stays in the reconstruction may have abetted the span's destruction. The bridge was reopened that spring and lasted another decade. None of the Niagara suspension bridges survive today, and all except Roebling's succumbed to wind.

Many other midsized spans from other engineers were completed during this period, in Nashville, Tennessee; Charleston, West Virginia; and Minneapolis, Minnesota, to name a few. Serrell designed Canada's longest suspension bridge, a 191-m span completed in 1852 at St. John, New Brunswick.[7] But Roebling would dominate. In 1860, he completed Pittsburgh's Sixth Street Bridge over the Allegheny River; its 12.2-m wide deck was suspended over two 105-m spans and flanking spans of 52 m. Here, Roebling for the first time used wrought iron instead of wood trusses and also used iron towers. This span featured his now-characteristic cable stays and introduced son Washington (1837-1926) to the craft.

Roebling began an even larger 373-m iron-and-wood truss-stiffened railway suspension bridge over the Kentucky River in early 1853. With the towers and anchorages completed (and cable wire and stiffening truss fabricated), a lack of finance then brought work to a halt, permanently, in 1860. Roebling's bridge at Cincinnati, Ohio would endure even more trials and tribulations. In an 1846 proposal for this Ohio River crossing, Roebling wrote that "no competent builder will hesitate to resort to spans of 1,500 ft [457 m] and more"—this at a time when the Fribourg Bridge's 265-m span reigned supreme (Steinman and Watson 1957). Work on this record 322-m span began in 1856 and, after interruptions from the 1857 economic panic and the Civil War, was completed in 1867 under Washington Roebling's direction at a cost of $1.8 million. The 3-m deep wrought-iron truss was continuous through the towers over the 493-m of suspended span but was hinged at the center, a practice later found to be undesirable. More than 100 cable stays augmented support. The British journal *Engineering* described the bridge as "undoubtedly the finest work, structurally, and architecturally, of its class" (Hopkins, 1970). The Cincinnati-Covington—now the John A. Roebling—Bridge remained the world's longest span for only 2 years until Keefer's Niagara span of 1869. A drastic late-1890s makeover added two more cables and superimposed a huge truss that marred the bridge's fine lines. Nonetheless, after 140 years, the towers, two original cables, and anchorages remain intact. The bridge is now a National Historic Civil Engineering Landmark.

Roebling's attention was soon focused on what was to become his masterwork—one he would not live to see. New York City was destined to have great bridges; its East and Hudson rivers have attracted the greatest engineers of each generation—Roebling, Gustav Lindenthal, David Steinman, and Ammann, to name a few. The city has on four occasions possessed the world's longest span. But the first, a bridge across the East River between Manhattan and Brooklyn, remains the most cherished.

The construction of the Brooklyn Bridge is an epic tale—of vision, perseverance, courage, personal tragedy, politics, corruption, and engineering triumph. The story has been documented by several authors, including the great 20th century American bridge engineer Steinman and more recently David McCullough, who produced the definitive work on the bridge. The structure has elicited more emotional response than perhaps any other product of civil engineering, and it has long been a favored subject of artists, photographers, and poets. Its gothic stone towers, slender webs of cable stays, and delicate truss have helped the bridge become a symbol of New York, one of constancy as the stone towers that once dominated the city's skyline have been inexorably swallowed by Manhattan's steel and glass towers. In spite of several major overhauls, the bridge's appearance has remained essentially unchanged.

Roebling devised the basics of the bridge in 1865. Two years later, the New York Bridge Company appointed him chief engineer. The 486-m span was half again the length of anything that Roebling—or anyone else—had ever built. With side spans, Roebling's plans called for 1,036 m of suspended bridge. This was ambitious indeed, but in the year of his appointment, he proclaimed spans of 914 m feasible in a report on a crossing of the Hudson River, thanks to the miracle material of steel. Roebling planned to introduce steel wire for the first time into the Brooklyn's main cables—a far-sighted decision. The four 40-cm diameter cables would together consume almost 22,000 km of wire. The 26-m wide deck would be more than twice the original width of the bridge in Cincinnati. Although the truss would be just 4.6-m deep, Roebling wrote that it would:

> support itself without the assistance of the [main] cable, the supporting power of the stays alone will be ample to hold up the floor. If the cables were removed, the bridge would sink in the center, but would not fall (Steinman and Watson, 1957).

Three days after final approvals were obtained, Roebling miscalculated the timing of an incoming ferry and his right foot was pinned against some pilings, crushing several toes. Tetanus set in. After refusing medical treatment and enduring almost 3 weeks of agonizing pain, Roebling died. Washington, then aged 32, assumed control just as an arduous 3 years of caisson sinking began. During this period, fire consumed the Brooklyn caisson, and the bends—later found to be caused by too rapid a decompression when leaving the caisson—killed several workers and crippled Roebling himself in the summer of 1872. Perilous finances, the use of substandard wire by a Roebling competitor, and political maneuvering plagued the work. Roebling's decision to use stiffening trusses of steel rather than iron sparked more controversy, but he did, fortunately, prevail. The key role his wife Emily played in relaying instructions to the bridge site only intensified speculation concerning Roebling's ability to manage the project. Sitting at his window aided by a telescope, Roebling would remain a controversial and enigmatic figure during the final years of construction. An 11th-hour decision to strengthen the span to handle subway trains (which it never did) would also prove contentious but ultimately fortuitous.

After 14 years of construction, $16 million, more than 20 deaths (including John Roebling's), a bankrupted bridge company, and the sacrifice of Washington's health, the span opened on May 24, 1883. Roebling would eventually recover much of his mobility; he died in 1926 at the age of 89 as ground was about to be broken on another epoch-making New York City bridge, Ammann's mighty George Washington Bridge across the Hudson River. Only then would the Brooklyn Bridge's total suspended length—comprising main and side spans—be exceeded. The choice of steel for the cables and truss would serve the Brooklyn Bridge well—it would carry three times the loads for which it was designed for 60 years before Steinman's sensitive upgrade in the early 1950s. Urban historian Lewis Mumford would later describe the bridge as "perhaps the most completely satisfactory structure of any kind that had appeared in America" (Billington, 1977).

The Brooklyn Bridge was the pre-eminent suspension bridge of the 19th century and was designated a National Historic Landmark in 1964. David Plowden, in his superb survey of North American bridge development, aptly remarked (1974) that the Brooklyn Bridge represented in both symbolic and concrete terms "the bridge between the old era and the new—a bridge of stone and of steel." It was the last large suspension bridge to use stone towers, yet its use of steel would presage trends that would propel the suspension bridge into the 20th century.

ENDNOTES

1. Emory Kemp (1993) states that "more than 20" Finley bridges were built.
2. The term double amplitude refers to the total vertical distance between the crest and trough of a wave.
3. The 265 m denotes the distance between towers; the suspended span was 246 m.
4. The 317 m denotes the distance between towers; the suspended span was 259 m.
5. Roebling was not the only advocate of deck stiffening. Steinman and Sara Ruth Watson (1957) cite American John Trautwine as being the one of the first to design deep trusses for a suspension bridge. In 1851, before Roebling developed his Niagara design, Trautwine proposed a massive bridge of four 305-m main and two 152-m end spans across the Delaware River at Philadelphia with 6.1-m deep trusses. This bridge was never built (although a model of it was), nor were any of his other designs.
6. The replacement bridge remained in use until 1962.
7. The 191 m denotes the distance between towers; the suspended span was 177 m.

Chapter 2

A NEW FOUND CONFIDENCE

With the heyday of the railroads in the late 19th and early 20th centuries, the cantilever dominated long-span bridge construction. Such bridges as the twin 521-m spans of John Fowler and Benjamin Baker's 1890 Forth Rail Bridge and Quebec's 549-m span of 1918 symbolized the new age of steel. Yet the railroad era reached its nadir after World War I. Road-only spans, with their lighter live loads, soon came to dominate—and make possible—even longer spans. By 1929, a convergence of new steels, analytical methods, and lighter traffic loads had seen the suspension bridge regain its former hegemony, one that has been maintained to this day.

2.1 EMERGING CONCEPTS

After their checkered 19th century genesis in modern form, suspension bridges did experience a resurgence in Europe. M.F. Arnodin (1845-1924) would build many spans in the late 19th century and has been credited with introducing stiffening trusses to French practice. He is best known for a system substituting vertical suspenders over the main span from the towers to its quarter-points—and over the side spans—with stays radiating from the tower tops. His first span of this type was completed in 1879, and the era would culminate in G.L. le Cocq's 274-m span Aramon Bridge (1900). Arnodin's 163-m span Bonhomme Bridge (1904) stands today, but the most spectacular example was the 1911 Constantine Bridge in Algeria, whose 164-m span over a 174-m deep gorge was then the world's highest.

Arnodin's contemporary Albert Gisclard developed a cable-stayed variant whose rigidity was well-suited for railway use; many were built in France and overseas in the early 20th century, the most notable extant example being the 156-m span La Cassagne Bridge in the Pyrenees in France (1909).[1] Arnodin's transporter suspension bridges, in which traditional suspension cables supported a high-level truss from which suspended cable cars shuttled vehicles, also appeared around the turn of the century. His largest was a 1905 span of 305 m at Runcorn-Widnes in England (coincidentally in the same locale and of the same span as Telford's proposal of 1814); an Arnodin span of 197 m at Newport, Wales (1906) stands today. Arnodin also adapted Bernard Poyet's 1821 concept of fanned stays, most notably in a 1905 240-m transporter bridge at Marseilles. Le Cocq would advance the cable-stayed bridge in the 112-m span Lézardrieux Bridge (1924), as would Spaniard Eduardo Torroja in the first reinforced concrete bridge of the type (a 60-m span completed in 1926). Le Cocq's 1911 book on the theory of several suspension forms is also a notable contribution of the period.

Another form was the braced chain suspension bridge, which transferred stiffening from the suspended structure to the chain itself. Several European examples followed the first of this type, an 1839 German span. An 1860 replacement for Ignace Edler von Mitis' 1828 span in Vienna was also Europe's only rail suspension bridge, and it would last for 25 years. London's 1895 Tower Bridge also used the concept.

Simple chain bridges also emerged. In 1910, the 126-m Kaiser Bridge in Breslau, Germany was opened. The Cologne-Deutz Bridge, a 1915 span of 184 m, would have a remarkable influence internationally and was also the first large self-anchored suspension bridge. Yet the most monumental chain bridge of all was the Elizabeth Bridge, a 290-m eyebar span completed in Budapest, Hungary, in 1903 across the Danube River. This bridge remained Europe's longest suspension bridge for more than a quarter century, but it was destroyed in the dying days of World War II.

Chain bridges would find only sporadic use in Europe, mainly in Germany. In America, the chain captured the imagination of the great Gustav Lindenthal who, like John Roebling, was a product of a European engineering education. Still, Edward Hemberle—and not Lindenthal—would design America's first braced chain bridge, the 244-m span Point Bridge in Pittsburgh, Pennsylvania, over the Monongahela River. This 1877 bridge was destined to remain America's longest eyebar span (it was replaced in 1926). Lindenthal completed his only braced chain bridge in 1884, the Seventh Street Bridge over the Allegheny River in Pittsburgh. This bridge featured twin main spans of 97 m. Few other chain bridges were built in America; an 1891 span of 122 m in St. Louis, Missouri, was in use until the late 1950s, and more recently, a braced cable span of 88 m built in rural Montana in 1950 has come to light. Lindenthal would become the braced chain's main advocate. He proposed an ambitious rail bridge of 536-m span at Quebec in 1910, as well as others in New York City—most notably across the Hudson River. But none would ever be built.

The geographic range of the suspension bridge also expanded; the first modern South American span—the 287-m Occidente Bridge across the Cauca River in Columbia—was completed in 1894. It was one of the lightest spans ever for its length and featured an extremely flat (1:27) cable sag. William Hildenbrand, a protégé of Washington Roebling, completed Mexico's 314-m span Miampimi Bridge, in 1900. In America, many spans were built across the Ohio River after a 25-year hiatus. Herman Laub's 215-m span East Liverpool and E.K. Morse's 244-m span at Rochester both opened in 1897 and were the first suspension bridges to use steel exclusively. Both also featured deep trusses up to 6.1-m deep, as did three other Laub spans built over the Ohio River up to 1916. Laub's work has gone largely unrecognized, but he did much to develop the classic suspension bridge.

New York City would host the next wave of large suspension bridges. The Williamsburg Bridge, spanning the East River north of the Brooklyn Bridge, was completed in the 20th century but is a product of the 19th century. Leiffert L. Buck, assisted by O.F. Nichols, designed a record span of 488 m. Buck advanced the art in some respects; he produced the first large suspension bridge to be built entirely of steel. A capacious 12.2-m deep truss supported four trolley tracks, two elevated lines, and several vehicular and pedestrian lanes over two decks. The wire in its four 47-cm diameter cables could carry 27% more load than could the wire of the Brooklyn

Bridge. Paradoxically, its massive steel towers still reflected the tradition of stone; these and the gargantuan truss would earn the Williamsburg Bridge's its chief—if dubious—renown; the bridge could claim none of the majesty of Roebling's masterpiece. Construction began in 1897, and the bridge opened in December 1903. Lindenthal, who became the city's Bridge Commissioner in 1902, described the towers as the "ugliest possible," and architecture critic Montgomery Schuyler derided the new bridge as a "distinct artistic retrogression" (Reier, 1977). The rather spartan design of this bridge—its cables, for example, were ungalvanized—would combine with long-term neglect to spawn a 15-year, $750 million restoration launched in the 1990s.

Less than a kilometer to the east of the Brooklyn Bridge stands the Manhattan Bridge. Under a design lead by Nichols (Buck's assistant on the Williamsburg Bridge), this 448-m span was completed in 1909. The Manhattan is clearly a bridge of the 20th century, introducing several important innovations in design and construction. It was here that an engineer by the name of Leon S. Moisseiff would make significant contributions. Under Nichols, Moisseiff initiated a trend towards flexibility by being the first to apply the Deflection Theory in the design of a long-span suspension bridge. As such, the Manhattan has been considered to be the first truly modern suspension bridge. Moisseiff was born in Riga, Latvia, in 1872 and emigrated to the United States in 1891 following 2 years at Riga's Baltic Polytechnical Institute. He then graduated from Columbia University in New York with a civil engineering degree in 1895 and was soon employed in the city's Department of Bridges. With a facility for combining mathematics and design innovation, Moisseiff rose rapidly through the ranks and, in 1897, became the Department of Bridges' Engineer of Design.

The significance of the Manhattan Bridge derives from Moisseiff's introduction of a method to accurately assess the interaction between main cable and stiffened suspended structure—the Deflection Theory. Designers of early suspension bridges had little regard for the influence of thc suspended structure on the behavior of the cables. Claude Navier had argued that deck weight and a low cable sag would stiffen the cables; consequently, extra stiffening of the deck was not required. British engineers, too, largely eschewed deck stiffening. Yet the spate of collapses in the first half of the 19th century resulted in a situation, in the words of Othmar Ammann (1923), "that the suspension bridge began to lose favor even before its nature was fully understood." Even Roebling had relied largely on intuition and simple assumptions of bending and tension in his designs.

It is axiomatic that the main cables carry a suspended structure.[2] Briton William Rankine, in the first comprehensive suspension bridge theory of 1858, assumed live loads to be distributed to the main cables uniformly over the length of an undeflecting, rigid truss; the main cable would therefore not deform under live load. This, of course, was a crude assumption, but the spans being considered were small, as were live loads; the lack of economy inherent to deeper stiffening girders failed to materialize because of the short spans over which they would be used. Moreover, a proper accounting of the true truss/cable interaction under the reality of unevenly distributed live loads posed what were then insuperable mathematical difficulties. In any event, Rankine's overly conservative theory saw very little practical application.

The Elastic Theory, which built on Navier's early 20th century theory of arches, was developed primarily by Austrian Josef Melan. It was published in 1888, and was first translated into English in an 1894 report on a crossing of the Hudson River at New York City. The Elastic Theory calculated the bending effect in the truss at a given point using Rankine's assumptions of a parabolic cable from dead load, uniform distribution of live loads across the span, and no cable deflection. Its main difference was that in the determination of uniform loads, it accounted for the elastic stiffness in the main cable under tension and for the stiffness of the suspended structure under bending forces. Yet the theory still overestimated bending moments and shears in the truss from live loads and temperature effects; ergo, it led to unnecessarily deep trusses, as seen in the Williamsburg Bridge. For small spans, this error was not overly significant, and the Elastic Theory became a popular design tool. In large spans, however, the theory's inaccuracy would be profoundly large.

It is somewhat paradoxical that the Elastic Theory also encouraged simpler structures. Stays and suspended side spans transmitted forces between spans, creating complex structures that proved difficult to analyze mathematically; they were shunned, again, as demonstrated in the design of the Williamsburg Bridge. Moreover, deeper stiffening trusses made stays superfluous.

In 1906, Melan published an improved version of the theory that would provide the first accurate means to design suspended trusses. It now accounted for the deflection of the superstructure from actual variations in live loads carried by the suspenders across a span. Moisseiff (1937) would write that, "The introduction of the Deflection Theory led to a correct determination of deflections and made it possible to design the [Manhattan] bridge with relatively shallow trusses." The major requirement of the truss was to distribute local live loads to the main cables and avoid local grade changes. It therefore did not require the rigidity formerly required under the Elastic Theory. The Deflection Theory (so named by Moisseiff) was especially significant for longer spans, in which dead loads were significantly higher in proportion to live loads. Here, the use of the theory made possible even further reductions in cable and stiffening truss sizes.

With the introduction of the Deflection Theory, trusses designed under the Elastic Theory were now considered overproportioned. Estimates would eventually place savings in truss steel from 20% to as high as 65% (Steinman, 1934). Further savings emanated from smaller cables, anchorages, and lighter towers. Improved aesthetics came with thinner trusses, too; the main cables could hang above the top chord of the truss at midspan, and the roadway could be raised to the top chord with little effect on approach span length. Moisseiff's use of the Deflection Theory in the Manhattan Bridge also extended Melan's theory from single spans to three spans, providing a critical breakthrough. Notwithstanding the theory's complexity because of the nonlinear relationship between load and deflection, long spans became both structurally and economically possible.

The Manhattan Bridge would, however, undergo many trials and tribulations before the final design emerged, and Lindenthal even almost succeeded in imposing a braced chain span. The bridge, built between 1901 and 1909, was significant on other fronts. Its featured the first major use of nickel steel—50% stronger than carbon steel—in a suspension bridge. This and the use of the Deflection Theory permitted the

36.6-m wide suspended structure to be stiffened by a truss just 8.2-m deep. The use of a triangular Warren truss, which most effectively accommodated stress reversals, would become standard practice in long span design. The Manhattan Bridge's slender steel towers also flexed with the longitudinal movement of the cables from expansion, contraction, and live loads; this allowed the saddles to be fixed in place. The towers' resulting slenderness marked a watershed in aesthetics, too; Moisseiff (1939) would later write that, "so daring was the appearance of the tower at that time that for many years the author was haunted by its excessive slenderness."[3] The towers' construction also introduced an innovation used to this day—the self-elevating crawler crane—and cable spinning incorporated several advances that would become standard practice. The Manhattan Bridge launched Moisseiff on a long and distinguished career as the most prolific suspension bridge engineer of his generation; he would contribute to virtually every major American suspension span built in the first 4 decades of the 20th century.

The Brooklyn Bridge's 486-m span was first exceeded in 1924 by the 497-m span Bear Mountain Bridge across the Hudson River 65 km north of New York City. The placement of its roadway atop a deep truss opened up views for bridge users and would become common practice. Yet the opening of a record $37 million span 2 years later in Philadelphia, Pennsylvania, would really set the stage for the future.

Carl Condit (1961) described the 533-m span Benjamin Franklin Bridge as "the first distinctly modern suspension bridge built on the grand scale." Designed by Ralph Modjeski, the bridge incorporated elements that would come to characterize the form—two main cables hanging free above the truss, unfettered diagonally braced and flexible towers of cellular construction, and a shallow (8.5-m) but wide (38-m) Warren truss. Moisseiff—having ended his tenure in 1914 as Engineer of Design at New York City's Bureau of Bridges—designed the truss using the Deflection Theory. Cellular construction, whereby the tower legs are built from small units to form a slender but strong honeycombed structure, would become standard long-span practice for the next 4 decades. The only discordant note in the Benjamin Franklin Bridge is found in architect Paul Crét's heavy and monumental stone anchorages, which do not reflect the structural modernity of the bridge.

One other notable span of this period was the Mid-Hudson Bridge, about 45 km north of the Bear Mountain Bridge. Built between 1925 and 1930 to a design of Modjeski, this 456-m span is even today recognized as one of the most graceful suspension bridges ever built. Its diagonally braced towers with open-fluted legs are especially distinctive. The 2-year founding of one of its tower caissons would also provide one of the most horrendous experiences in the annals of bridge building. Deep transverse girders connecting 6.1-m deep Warren through trusses support a two-lane roadway; as was typical of the era, upper lateral bracing was absent.

2.2 A PROLIFERATION OF CONCEPTS

The frenetic period of suspension bridge building in the late 1920s and early 1930s also spawned a proliferation of small- to medium-span suspension bridges. Dozens were built around the world, in France, Norway, Latin America, and elsewhere. Most notable was activity in North America, most especially in the United

States, where innovation in structural forms and cable composition took hold.

One unusual development of this period was the Florianopolis truss. The first large bridge of this type was a Brazilian span (officially the Ponte Hercilio) for which the form is named. Its design involved David B. Steinman (1887-1960), whose extraction from academia by Holton D. Robinson would see him eventually become one of the great American engineers of the 20th century. Unlike Ammann, Moisseiff, Modjeski, and Lindenthal, the other greats of early 20th century American bridge engineering, Steinman was a native-born American. He grew up in New York in relative poverty in the shadows of the Brooklyn Bridge, which he vowed to emulate. Steinman combined his passion for bridges with an astounding proclivity for mathematics, and by 1911, he possessed a doctorate and in that year published his dissertation, *Suspension Bridges and Cantilevers: Their Economic Proportions and Limiting Spans*. In 1910 at age 23, Steinman began a 4-year stint at the University of Idaho as the country's youngest professor. He also translated two of Melan's books; the *Theory of Arches and Suspension Bridges* was published in 1913 and included the Deflection Theory. Steinman's own 1922 *A Practical Treatise on Suspension Bridges: Their Design, Construction, and Erection* became a classic, and a revised edition was issued in 1929. Steinman also simplified the Deflection Theory, making it easier to use. While his long-time rival Ammann would build fewer but larger bridges, Steinman's career would be incredibly prolific and he would design literally hundreds of bridges around the world.

From 1914 to 1918, Steinman assisted Lindenthal out of New York on that city's Hell Gate arch and the Scotioville continuous truss over the Ohio River; in fact, on the Hell Gate Bridge, Steinman temporarily replaced Ammann, who had been conscripted by the Swiss Army in 1914. Yet by the end of 1918, Steinman was out of work, and Ammann was supervising a New Jersey clay mine. Steinman had dabbled in several jobs when Robinson—almost 25 years his senior—approached him in 1920 about the Florianopolis. Steinman would first aid Robinson in the Rondout Bridge, a 1922 span of 215 m north of New York City and the first American suspension bridge with a continuous truss between anchorages. Their design for a conventional 340-m span at Florianopolis carrying road, rail, and an aqueduct would evolve to a unique structure whose eyebar chain also functioned as the top chord of a variable-depth stiffening truss over the central half of the main span. With the maximum truss depth at the span's quarter-points adding strength at the point of maximum deflection, the design offered four times the stiffness of the original plan while using two-thirds the steel. Rocker towers, hinged at their bases, were also used, and the bridge was designed using the Deflection Theory. Completed in 1926, the Ponte Hercilio remains recognized for its inventiveness and is also the world's longest eyebar suspension bridge (although it has since been closed to all traffic for more than two decades).

However innovative, the Florianopolis truss never really caught on. One observer would later write in a review of Steinman's 1929 edition of *A Practical Treatise on Suspension Bridges: Their Design, Construction, and Erection* that the economy of the concept "remains yet to be proven." The reviewer—Ammann (1929)—would become Steinman's greatest rival over the next 30 years. Nevertheless, some Florianopolis spans were built. The largest other example was the Indoroopilly Bridge, an unusual 183-m span completed in 1936 in Brisbane, Australia, whose large concrete towers also contained offices. Its wire cables had been previously used as construction tiebacks for the Sydney Harbour arch, a bridge for which Steinman unsuccessfully proposed a 503-m span Florianopolis road/rail truss in 1923. The span remains the continent's longest suspension bridge. Two Florianopolis-type eyebar suspension bridges of 213 m span opened across the Ohio River in 1928 and 1929 respectively—one at Point Pleasant and a second almost identical span, the St. Mary's Bridge. These bridges also introduced high-strength, heat-treated carbon steel eyebars to America. The Point Pleasant Bridge would later go down in infamy when undetected corrosion in an eyebar precipitated the span's collapse in 1967, killing 46. Steinman would not design another Florianopolis truss again until his 1950 proposal for a 1,524-m span over Italy's Messina Strait. A 1954 Japanese span of 114 m would be the fifth—and last—to use the concept.

The Florianopolis Bridge did establish Steinman's reputation, and his Carquinez Bridge—a California crossing opened in 1927 comprising tandem 351-m cantilever spans—was the first large bridge designed to withstand earthquake loads. Moreover, Steinman's association with Robinson would remain until the latter's death in 1945 at the age of 82.

The introduction of heat-treated wire was another truncated and almost disastrous development, one played out in two spans—Steinman's Mount Hope Bridge in Rhode Island and the Ambassador Bridge across the Detroit River between the United States and Canada. The 366-m span Mount Hope Bridge was Robinson & Steinman's first major suspension bridge on American soil. Steinman's attention to aesthetics was manifested in well-proportioned diagonally braced towers and the span's green patina. The Ambassador Bridge, with unsuspended side spans, was more utilitarian in appearance, but its 564-m span would return the rank of world's longest span to the suspension bridge. Moisseiff, advisor on the superstructure, eliminated overhead laterals and halved the truss depth. Work on both spans began in 1927. Traditional cold-drawn wire had been specified for both bridges, but the higher yield point (the point at which irreversible deformation from stretching occurs) of a newly developed heat-treated wire offered smaller and more economical main cables. With the same contractor at work on both spans, this wire was adopted.

Spinning on both spans was finished at the end of 1928. By mid-January 1929, the Mount Hope Bridge's 5.5-m deep truss was in place. Minor amounts of broken wire had been discovered at the anchorage strand shoes (where the wire loops back into the span), but at the end of February the cables were found to be riddled with breaks and, by late March, were declared unsalvageable. The same decision was made at the Ambassador Bridge, with one-quarter of its main truss in place and just 6 weeks away from opening. The trusses and cables of both spans were dismantled, cables respun with cold-drawn wire, and trusses then reassembled with remarkable speed; more than 730 m of main and side span truss was lifted into the Mount Hope Bridge in just 13 days. Both bridges opened in late 1929, the Ambassador Bridge an incredible 9 months ahead of the original schedule. Not until 1936—after exhaustive investigations—was it determined that heat-treated wire was susceptible to changing construction stresses, notably those transmitted to the cables by traveler cranes used to erect the trusses. This type of wire was never used again.

Smaller spans embodied more successful cable innovations. Several Ohio River spans had their cable strands spun on land adjacent to the bridge in a process resembling early 19th century French methods. The British-designed Chirundi (or Otto Beit) Bridge also used a variant of offsite cable assembly in the form of 40 horizontal layers of wire. Ralph Freeman of Sir Douglas Fox & Partners, designer of the famous steel arch at Sydney, Australia, designed this 1939 span of 320 m across the Zambezi River in southern Africa. Land spinning would remain a novelty, as longer spans made opportunities for its use rare, but the advantages of prefabricating parallel wire cables would re-emerge in the 1960s.

More enduring were prestressed helical wire strand cables, in which each strand comprises layers of wire twisting over its antecedent in an opposite lay. The prestressing of wire rope had been developed by the J.A. Roebling's Sons Company for the footbridges of the George Washington Bridge across the Hudson River at New York, and its successful use there prompted further development in the form of helical wire strands. These prefabricated, prestretched cables were quickly installed, and often the need for catwalks was eliminated. A further European refinement was the locked-coil strand, whose outer layer of Z-shaped wires locked together as loads were applied, increasing resistance to stretching and improving water-tightness. This cable type was first used in modern form in the 315-m Cologne-Mulheim Bridge across the River Rhine in 1929.

The first large use of helical strands was in Steinman's 289-m span Grand'Mère Bridge in Quebec, Canada, completed in 1929. Steinman would also use them in the Waldo-Hancock Bridge, a 244-m span across the Penobscot River in Maine (1931). This bridge marked the first use of Vierendeel trusses—or portal braces—in a steel suspension bridge tower in place of the usual diagonal bracing. It was built in just 16 months and was recognized by the American Institute of Steel Construction (AISC) for its artistic design. Robinson & Steinman's 368-m span St. Johns Bridge in Portland, Oregon (1931), then set a new standard for aesthetics with its Mid-Hudson-like gothic towers; even the railings and light standards were architecturally detailed. The J.A. Roebling's Sons Company erected all 182 helical wire strands in just 23 working days and placed up to 16 strands per day. The 122 strands of the Maysville Bridge, a 323-m span over the Ohio River opened in 1931 and designed by Modjeski, Masters & Chase, were placed in just 9 days. Canada's Isle d'Orléans Bridge, 15 km downstream of the famous Quebec cantilever, was completed in 1935 to a design of the firm of Monsarrat & Pratley. This 323-m span also used helical strand cables, as did the Iowa-Illinois Memorial Bridge, a 226-m Modjeski & Masters-designed span completed over the Mississippi River at Moline, Illinois, that same year.

One other form that enjoyed limited success was the self-anchored suspension bridge. Here, strengthened, deeper girders absorb the cable forces normally taken up by anchor blocks, with cost and aesthetic benefits seen to accrue from the lack of anchorages. Eyebars (which were more commonly used in this span type on account of their compact anchorage) or cables would be anchored directly to the stiffening girders at the ends of the span. The girders—usually solid and not trussed—would also be continuous through the towers to transmit forces directly to the anchorage points. The first modern bridge of this type was completed at Lübeck, Germany, in 1899. It was followed by the Cologne-Deutz Bridge, the 184-m span opened in 1915 over the River Rhine at Cologne, Germany. This influential span was of surprisingly modern appearance. It was emulated in the Kiyosu Bridge, a 91-m span Tokyo bridge with "spine" box girders opened in 1928, and in the 1929 Rio Quequen Bridge, a cable-supported (and gravity-anchored) span of 150 m in Buenos Aires, Argentina.

The most noted imitators of the Cologne-Deutz Bridge were three almost identical spans of up to 135 m completed across the Allegheny River in Pittsburgh, Pennsylvania, between 1926 and 1928 under the direction of V.R. Covell. The 6th, 7th, and 9th Street bridges were America's first self-anchored suspension bridges and constitute half of the entire stock ever built there. They were America's first eyebar suspension bridges since the 19th century, and their 2.75-m deep solid stiffening girders were also unusual for American practice. The design of these spans was greatly influenced by the city's art commission, the effort being rewarded by the AISC. Yet apart from a few small spans (including the Steinman-designed Wabash River Bridge, a 107-m rolled I-beam span in Indiana), the form never really took hold in the United States; for one, its construction required complex and costly scaffolding.

Other more traditional midsized spans of this period include the Rio Higuamo Bridge in the Dominican Republic, a span of 166 m whose creation—from design initiation to finished bridge—occupied just 5 months in 1934. Another Dominican Republic span of 137 m completed a year earlier was built in just 10 weeks. The J.A. Roebling's Sons Company built both spans. One of the more spectacular efforts of this period is the Royal Gorge Bridge, whose gossamer span of 268 m soars a full 321 m above the Arkansas River in Colorado. Completed in 1929, it remains the world's highest bridge. In Canada, Vancouver's Lions Gate Bridge was built in just 17 months. The firm of Monsarrat & Pratley designed this $6 million, 472-m span bridge. When completed in 1938, it was the longest suspension bridge outside of the United States; its helical wire cables remained the longest anywhere until France's Tancarville Bridge was completed 20 years later. Innovation took hold on a small scale as well; a 1934 span of 61 m near Los Angeles, California, featured arc-welded (instead of riveted) towers and stiffening trusses, as did an earlier Swiss pedestrian span of 49 m.

In South America, Columbia produced two spans of more than 180 m in 1926 and 1929. France continued to innovate in small suspension spans with concrete towers, concrete suspended structures, and even concrete-encased cables. Two notable truss-stiffened spans completed in 1932 and 1933, respectively, include the 209-m span Veurey-Voroize sur l'Isère and the Serrières Bridge, a 185-m span opened in 1934 over the Rhône. Still, the spotty reputation of early suspension bridges had yet to be completely shaken. In June 1931, a 76-m span east of Bordeaux undergoing a combined dedication and load test collapsed when a truck drove into and broke a suspender; as many as 15 people died. An even worse American disaster occurred in 1923 when a 91-m span timber suspension bridge in Washington collapsed, killing 40. Yet in spite of these setbacks, the confidence in suspension bridges was such that only this bridge form would now be considered for long spans. In fact, the genre was about to make an enormous leap.

2.3 A GENERATION OF ENORMOUS SPANS

Although the plethora of midsized spans advanced the art considerably, the late 1920s and early 1930s were most notable for major breakthroughs in long-span design. The most startling of these projects was a long-discussed bridge over the Hudson River in New York City. David Billington (1977) has described the George Washington Bridge (*see* Fig. 2-1) as "unique in the history of bridges." It is clearly a bridge of profound engineering significance. In a single stroke, it doubled the longest suspension span and ranks among the most influential bridges of the 20th century.

Billington's attribution goes beyond the massive size of this bridge to controversies that continue to surround its design—the resolution of aesthetic and engineering questions and, perhaps more importantly, their influence on subsequent suspension bridge design in the 1930s. The George Washington Bridge would close and cement the careers of two great engineers—Lindenthal and Ammann. The story of its design and construction is an epic in and of itself which space does not permit telling here in any detail. Yet in spite of its significance and the debate it has engendered, the George Washington Bridge has not garnered the popular attention of Roebling's Brooklyn Bridge or San Francisco's Golden Gate Bridge.

Fig. 2-1. George Washington Bridge, before the addition of a lower deck
PHOTOGRAPH COURTESY OF THE PORT AUTHORITY OF NEW YORK AND NEW JERSEY.

Lindenthal's long affiliation with a Hudson crossing extended from 1885 until his death in 1935 at age 85. He would conjure up several colossal—some would say preposterous—schemes of increasing boldness. His 1921 proposal for a 987-m span at 57th Street is one of the most stupendous bridges ever conceived. Braced eyebar chains hung from 251-m tall masonry-clad steel towers would support a mind-boggling 12-track lower deck and an upper deck of 16 traffic lanes, four trolley/bus lanes, and two walkways spread over a width of 72 m. Gigantic 20-story buildings supported by massive ceremonial arches straddling the roadway capped its anchorages. A $180 million price tag was equally overwhelming.

The year before, Lindenthal had retained an unemployed Swiss engineer—Ammann (1879-1965)—to help craft the design. Ammann had immigrated to America in 1904 after graduating in 1902 with a civil engineering degree from the Swiss Federal Polytechnic Institute. He first worked for a New York consulting engineer and then for the Pennsylvania Steel Company. The young and industrious Ammann assisted C.C. Schneider in the investigations of the tragic 1907 collapse of the cantilever at Quebec. In 1912, he began work as Lindenthal's assistant on the Hell Gate arch bridge, a post that lasted 6 years (save his brief 4-month stint in 1914 with the Swiss Army). Ammann's 1918 paper in the American Society of Civil Engineers' *Transactions* on the project earned him the ASCE's Rowland Prize. By this time, however, there was no room for either him or Steinman. Lindenthal's Hell Gate and Scotioville bridges were now completed, with no immediate prospects in sight.

Ammann was the antithesis of Steinman; the two did not get along. Austere in emotion and precise in demeanor, Ammann had a quietness that was underscored by a firm resolve, and he commanded great authority and respect. In 1920, with prospects for a Hudson crossing revived, Lindenthal plucked Ammann from his managerial post in New Jersey (a job Lindenthal had helped secure for him). Yet by 1923, Ammann had become despondent over Lindenthal's grandiose illusions and the scheme's location. Having failed to convince Lindenthal to alter the proposal, a frustrated Ammann left in March 1923. And so Ammann, who had yet to design a large span of his own, set out to conceive the largest single span ever built to that time. He chose a location further north, at 179th Street. By mid-1925, the newly created Port of New York Authority had embraced his concept for a 1,067-m span, and Ammann was made its chief engineer. It had taken more than 4 decades to increase the longest suspension span by 47 m; now, in the space of 5 years, it would double. Many notable engineers would help craft the design, including Moisseiff, Allston Dana, and Daniel Moran for the foundations. Cass Gilbert, designer of the Woolworth Building, served as architectural consultant. Ammann did retain Lindenthal as a special consultant, and the two would ride in the same automobile to the span's official opening in 1931.

The George Washington's capacity was impressive. Its four-lane upper deck could support four additional lanes over its open median, and a lower deck could handle six lanes or four subway tracks.[4] To carry both decks, Ammann (1933) produced a "very shallow and flexible truss, which not only resulted in far-reaching economy, but also effected a light and graceful appearance." Its 8.8-m depth—1/120 of the main span—would act mostly to eliminate "excessive local deflections" (Ammann, 1953). Incredibly, the span would be built initially with only its upper deck—in essence the future upper level of the truss—in place. Moreover, only cross-girders and stringers would support its concrete roadway. The suspended structure was thus virtually bereft of vertical stiffness. This Ammann (1933) freely admitted: "The system in its initial state, being without lower floor and stiffening trusses is entirely flexible, that is, without any stiffening effect upon the cables." Ammann's comparison with the Manhattan Bridge is stark; in that bridge, the truss constituted 30% of the bridge's weight and almost 1.5 times the weight of the cable system. In the George Washington Bridge, the final truss would comprise just 6% of the superstructure weight and just 20% of the cable weight.

Ammann's confidence in the design was reflected in several features that stiffened the span, especially the four huge cables—grouped in pairs—which would be in place with only the single deck. Their great stiffness and weight would benefit the deck (and vice versa) thanks to the tenets of the Deflection Theory by—for different reasons than postulated by Navier—adding stiffness to the overall system. The huge weight of the deck also provided inertia, another source of stiffness. Other factors included the generous 36-m deck width and the short 186-m and 198-m side spans, whose steeply inclined and taut main cables would enhance rigidity. Only a light wind truss confined to the upper deck was required to resist horizontal displacement from static wind loads; Ammann (1933) even suggested that the "bridge as a whole would be perfectly safe and sufficiently rigid even without a wind truss." The ratio of dead to maximum live load in the span's final form was almost 5:1, an unheard of proportion. Even then, live loads were assumed to be less than the theoretical maximum on account of the improbability of the span and all lanes being fully loaded (a tactic used in the Manhattan and Benjamin Franklin bridges); it was designed to support just 17% of a fully congested load (Doig and Billington, 1994).

The George Washington Bridge represented the ultimate application of the Deflection Theory. A heavy suspended structure without vertical stiffening (apart from that to avoid local grade changes) would not detract from the rigidity of the system writ-large; a truss designed under the Elastic Theory would have been "virtually unbuildable" (Buonopane and Billington, 1993). Ammann speculated this approach saved almost $10 million, through direct steel savings and lower tower, cable, and anchorage costs. It was a far cry from Steinman's view in *A Practical Treatise on Suspension Bridges: Their Design, Construction, and Erection* (1929a) that "the unstiffened suspension bridge is not used for important structures."

There were many other design issues. A debate over eyebars versus wire was resolved in the form of four wire cables of unprecedented size, each 91-cm thick and containing 26,474 wires. More enduring was the controversy over the design of the towers, a debate of interest even today. Concrete encasement and masonry cladding were originally to have enveloped a steel structural skeleton, but Depression Era economics and the striking—and unintended—visage of the naked rigid steel frames lead to the decision to leave them as is. The 184-m tall towers—each comprising more than 18,600 tonnes of steel and with a still-unsurpassed width of 64 m and 17-m thickness—remain unique. Nothing like them has been built since. As Condit (1961) wrote, they "must be seen to be believed." Ammann would never again propose a rigid frame or masonry cladding.

Work on the George Washington Bridge began in 1927. All sorts of innovations characterized its construction, particularly in the spinning of the 172,200 km of cable wire. The bridge was completed in late 1931, 8 months ahead of schedule at a cost of $55 million, $5 million under budget. It would remain Ammann's favorite; French architect Le Corbusier (1947) would consider it "the most beautiful bridge in the world" and "the only seat of grace in the disordered city." The George Washington Bridge cemented Ammann's stature as a leading engineer. He would lead all subsequent suspension bridge design commissions in New York City, as well as that for the world's longest steel arch bridge, a span of 504 m opened the same year between nearby Bayonne, New Jersey and Staten Island.

In October 1981, on its 50th anniversary, the George Washington Bridge was designated a National Historic Civil Engineering Landmark by the ASCE. Yet even almost 50 years after its opening, the design evolution of the George Washington Bridge would still generate debate. A 1977 paper by Billington ascribed to Ammann an apparent lack of concern in the design of the bridge over the poor aerodynamic performance of the early unstiffened British suspension spans. The defense of Ammann, particularly from many of his former colleagues, was vocal. Yet as Henry Petroski (1994) observed:

> By glossing over the root causes in failure of the evolutionary changes that had taken place in suspension bridge design throughout the nineteenth century, Ammann was inadvertently providing a model for twentieth century designers that suppressed rather than highlighted the real problems that they faced in extending the envelope of experience.

The George Washington Bridge heralded a decade of unprecedented activity in America in which five suspension bridges with spans of more than 700 m were built. The decade also reflected a heightened interest in aesthetics, one embodied in Ammann's next span, the Triborough Bridge. The centerpiece of this 5.6-km bridge complex linking Manhattan, the Bronx, and Queens is a 421-m suspension span. Inauspiciously, work had begun the day before the 1929 stock market crash. Sporadic work ground to a halt in 1932 for lack of money, but the newly created Reconstruction Finance Corporation (RFC)—an arm of Roosevelt's depression relief public works program—allowed Robert Moses, head of the newly-chartered Triborough Bridge Authority, to restart the $60 million project in 1934. Ammann was retained early that year, and he reduced what was planned as a 16-lane bridge to one of eight lanes. Architect Aymar Embury II of Chrysler Building fame helped overhaul the towers and anchorages; he would collaborate with Ammann on other spans and publish several articles on suspension bridge aesthetics.

The Triborough Bridge’s eight-lane roadway was perched atop a 2.5-m deep floor beam, with no upper laterals. A 6.1-m truss depth may appear surprising in light of the dictates of the Deflection Theory and Ammann's approach to the George Washington Bridge, but the theory's advantages of flexibility and shallowness waned in smaller spans, as weight and cable stiffness diminished. The Triborough Bridge opened in July 1936. It was the first large span truly designed for the auto age, for rail transit was never contemplated in its design. The suspension bridge portion of the crossing was deemed the most beautiful long-span bridge of 1936 by the AISC, vindicating the efforts of Ammann and Embury to salvage the design.

That year, the focus shifted to San Francisco Bay, where two suspension bridges of record-setting dimensions—the world's longest twin suspension bridge and the world's longest single span—neared completion. The first was a long-discussed crossing between Oakland and San Francisco across 7.25 km of San Francisco Bay; only the midchannel island of Yerba Buena provided respite from its deep waters. The memories of the 1906 earthquake that leveled San Francisco also remained fresh. Discussions heated up after World War I, and by 1928, in spite of a 1921 rejection of any crossing by the War Department, some 38 schemes had been proposed.

Several were Lindenthalian in nature, including a 1926 plan by Charles Evan Fowler for a mammoth 1,478-m span suspension bridge hung from 8 main cables over the deep 3.2-km wide west crossing. The eminent J.A.L. Waddell once described it "the boldest bridge design ever made" (Plowden, 1974). Detailed investigations for the final bridge were lead by C.H. Purcell, with Charles E. Andrew as bridge engineer and Glenn B. Woodruff as engineer of design; the consulting board included Modjeski, the ubiquitous Moisseiff, Charles Derleth, and foundation expert Daniel Moran. The options boiled down to a single span, multiple continuous spans with tie cables connecting the tower tops, and end-on-end suspension bridges with a central anchorage. Models, increasingly used for design and construction, aided in the decision to build twin spans of 704 m centered by a huge central anchorage—at 64-m deep, the deepest suspension bridge pier to that time.

Construction of the double-deck Transbay Bridge began in May 1933. Its upper level would cradle six automobile lanes; on the bottom of a 9.1-m deep Warren truss were three truck lanes and two transit lines—the last to be designed into a large American suspension span for more than 6 decades (the transit lines would be removed in the 1950s). The Transbay Bridge pioneered the lifting of large, preassembled truss units into place from barges, instead of piece-by-piece assembly of individual members from the bridge. Another innovation was the use of an 18-m long model to determine the sequence of truss erection. The entire crossing, including a 427-m span cantilever bridge, was completed 2 months ahead of schedule at a cost of $79 million—and 24 lives—in November 1936; its construction required just 3 years and 7 months, a remarkable achievement. The Transbay Bridge would remain the most impressive bridge assemblage and hold the record for end-on-end suspension bridges for more than half a century. On its 50th anniversary in 1986, the bridge was designated a National Historic Civil Engineering Landmark.

Only the Brooklyn Bridge rivals the second San Francisco span—the Golden Gate Bridge—for its symbolic association with its host city (*see* Fig. 2-2). The story of its genesis and construction is similarly suspenseful. The physical obstacles—open sea, deep water, strong tides, paralyzing fog, storms, complex geology, and earthquake concerns—were matched by those of a political nature. The vision and tenacity of its diminutive engineer Joseph B. Strauss (1870-1938) were in the Roebling tradition. Many have chronicled the story, but some salient points are worth noting.

Strauss first proposed an ungainly 1,222-m span cantilever/suspension hybrid in 1921, but it proved singularly unpopular. By the late 1920s, metallurgy and the Deflection Theory had overtaken the design; moreover, Ammann's George Washington Bridge was proof positive that a pure suspension bridge was now feasible. The next few years were absorbed by political battles, eventually won. The

Fig. 2-2. Golden Gate Bridge, construction of the initial truss configuration
PHOTOGRAPH COURTESY OF THE GOLDEN GATE BRIDGE AND TRANSPORTATION DISTRICT.

design for a longer 1,280-m suspension span was finished in April 1931, but legal wrangling delayed the award of contracts until January 1933. Strauss's role in the design itself has been recently questioned, with evidence suggesting Charles Ellis of Strauss's Chicago office to be largely responsible. Clifford Paine—who replaced Ellis—played a key role, as did the stellar engineering board, including Ammann, Moisseiff, and Derleth. Strauss suffered a breakdown during construction, recalling Washington Roebling's travails half a century earlier. Irving Morrow, the consulting architect responsible for the art deco styling of the span's 227-m tall towers and light standards, as well as the span's international orange patina, also casts a large shadow.

Setbacks would plague the south tower footing 365-m offshore; its 2-year sinking has been described as "one of the most difficult, frustrating and remarkable achievements in bridge construction history" (Seim, 1987). The towers went up with no great difficulty. The 91-cm diameter cables were the largest and—at 2,332 m—the longest to that time, but here the J.A. Roebling's Sons Company improved spinning efficiency by 60% over its performance at the George Washington Bridge. Truss assembly was aided by the first ever use of a safety net that stemmed the otherwise fatal plunge of 19 workers; in fact, no one had died up to just months before the bridge's opening. Then in February 1937, a terrible accident occurred when a scaffold beneath the bridge wrenched loose and plunged into the Golden Gate, tearing 640 m of netting and taking 10 workers to their deaths. This tragedy still ranks as the worst single suspension bridge construction accident ever.

The Golden Gate Bridge was completed in May 1937 at a cost of $35 million, and it garnered the AISC's 1937 long-span award for artistic bridge design. The bridge is renowned as a rare example in which a human creation has intensified the emotive impact of a dramatic natural setting, as Lawrence Halprin (1966) noted. Its main span would remain the world's longest for almost 3 decades (*see* Fig. 17-1) and its towers stand as the tallest for twice that length of time. But the bridge had broken Strauss; he died just a year after its opening at the age of 68. In spite of the recent second-guessing of Strauss's engineering prowess,[5] there is little doubt that without him the Golden Gate Bridge would not carry the same allure that has helped it become a worldwide icon. It stands today as one of the 20th century's great bridges.

The Golden Gate Bridge's saga would not end with its completion. The truss had benefited not only from the Deflection Theory but also from recent theoretical work of Moisseiff and Fred Lienhard, as described in Chapter 3. In a reverse configuration of that employed in the Triborough Bridge, the roadway was perched atop a 2.6-m deep floor beam with no bottom lateral bracing; the truss's 1:168 depth-to-span and 1:47 width-to-span ratios were also slender (*see* Fig. 7-2). Strauss himself (1938) described the stiffening truss as "exceedingly flexible" and the wind truss over the main span as being "likewise very flexible." These features were seen to be benign, because the large, rigid cables would take up a "considerable transfer of wind load" imposed on the truss. Strauss wrote that, "Unquestionably the bridge has ample stiffness in both vertical and horizontal directions. It is believed that the depth of the stiffening truss could have been reduced considerably without detriment."

The flexibility and slenderness of the Golden Gate Bridge would place it in a category of aerodynamic vulnerability whose significance would soon emerge. Yet suspended trusses would not spark this concern; rather, a new, dramatically different form of suspended structure would irrevocably change the future course of suspension bridge engineering—the plate girder.

ENDNOTES

1. It was this span that ended Gisclard's life, along with five others, when a train plunged off the bridge into the gorge below.
2. A free-hanging cable in the form of a catenary was first described mathematically by Johannes Bernoulli in 1691; Niklaus von Fuss first analyzed its shape as a parabola under a load uniformly distributed over the span in 1794. The actual shape of a loaded suspension cable lies somewhere between a catenary and a parabola.
3. In this article, written 30 years after the Manhattan Bridge opened, Moisseiff (1939) now considered the towers to "appear to him rather stubby."
4. The George Washington Bridge's four upper deck lanes were added in 1943; its six-lane (without a subway) lower deck was built from 1958 to 1962 (*see* Chapter 10).
5. In fact, a 1998 editorial in the *Engineering News Record (ENR)* describing the Golden Gate Bridge's design evolution cites both Ellis and Morrow but omits Strauss's name altogether.

Chapter 3

THE PLATE GIRDER SUSPENSION BRIDGE

The 1930s would produce five of the world's longest suspension bridges. The Deflection Theory had made such spans economically possible, but it also paved the way for engineers to introduce a new form of suspended structure, the plate girder. This alternative to the open truss featured a solid—but shallow—profile, and it would revolutionize suspension bridge design in the second half of the decade.

3.1 EUROPEAN ORIGINS

Plate girder suspension bridges first emerged in Germany as self-anchored spans, the most notable early example being the 184-m span Cologne-Deutz Bridge (1915), which used eyebar chains. Similar self-anchored spans in Tokyo, Japan, and V.R. Covell's Pittsburgh, Pennsylvania, triumvirate were stiffened by box spine girders and plate girders respectively. Another eyebar span, the Admiral Scheer Bridge in Berlin, Germany, was a 1927 span of 96 m stiffened by 2.2-m deep girders. The use of plate girders in a self-anchored suspension bridge then achieved a new threshold in the Cologne-Mulheim Bridge, a low-level German span completed across the River Rhine in 1929. This 315-m span was the longest in continental Europe and will only lose its distinction as the longest self-anchored suspension bridge ever built in the near future. The Cologne-Mulheim was also the first self-anchored suspension bridge to use cables instead of eyebars.

A suspension bridge almost didn't come to pass. The winning design in a 1927 competition went to a 333-m span steel arch bridge, but the decision was later overturned in favor of a suspension bridge designed by Karl Möhringer. Its simple rectangular towers were—unusually—hinged at their bases and devoid of ornamentation; only a single horizontal strut was placed above the 22-m wide deck. Continuous steel plate girders passed through narrow slots in the towers and over unsuspended 91-m side spans. The bridge's 56-cm diameter main cables also featured the first major use of locked-coil strands. The Cologne-Mulheim was the first bridge to cross the Rhine in a single span, and it was highly regarded for its clean appearance. Sadly, this bridge suffered the fate of the Cologne-Deutz when retreating German forces dynamited it in 1944.

Another significant self-anchored, plate girder span was the massive King Alexander I Bridge in Belgrade, Serbia, a 261-m span completed over the Save River in late 1934. Its capacious 21-m wide deck was supported by just two 4.3-m deep girders spaced more than 14-m apart. Two large prestressed locked coil cables 66 cm in diameter suspended the roadway. This bridge was also destroyed in World War II; its replacement, a 1958 steel box girder built on the same piers, held the record for this span type for many years.

Another large, partially self-anchored plate girder span was built in Vienna, Austria, over the Danube River from 1934 to 1937. The 241-m Reichsbrucke was built on the widened piers of an 1870s truss bridge; the former bridge was first shifted 26-m upstream to allow construction to begin, no small task given its 336-m length and 4,350-tonne weight. The new bridge, one of the last of the eyebar-supported suspension bridges, featured a 26-m wide deck supported by 1.4-m deep plate girders. Designs for a fully gravity-anchored eyebar span had to be modified when investigations revealed poor soils; the girders were strengthened to provide self-anchorage, although the eyebar chains would also be partially gravity-anchored. In 1937, a 107-m self-anchored suspension span stiffened by slender spine box girders was opened at Chelsea across the River Thames in London, England. Unusually, its towers featured rectangular columns with no connecting upper beam. The designers, the firm of Rendel, Palmer & Tritton, used locked-coil cables.

Traditionally anchored plate girder suspension bridges also emerged in the late 1920s. France produced a number of them. The Ivry Bridge, a 1928 span across the River Seine, preceded the Teil Bridge, a 235-m span over the River Rhône with 2.4-m deep plate girders. The 300-m span Cavaillon Bridge over the Durance—a span framed by enormous solid reinforced concrete towers—featured virtually no deck stiffening at all. It was then France's longest span, but succumbed during World War II when it was destroyed in 1944. The Bonneuil-Matours Bridge (1932), an 82-m span, featured rolled I-girders, and the 1935 Houate Durance Bridge, near Mirabeau, was an elegant but deep plate girder span hung from concrete towers.

It was the building of the German autobahnen in the 1930s that propelled the plate girder to another plane with the Rodenkirchen Bridge, a low-level span built across the Rhine south of Cologne between 1938 and 1941. This 378-m span was then Europe's longest. Over the main and 94-m side spans, Fritz Leonhardt, A. Klönne, and Paul Bonatz (as architect) produced a striking span noted for its unfettered, almost harsh appearance. Austere box-like towers featured only a simple upper crossbeam. The use of gravity anchorages made possible lighter and slender plate girders just 3.3-m deep with a 1:115 depth-to-span ratio. The girders pierced the tower legs to avoid the need for a lower tower beam and, like all of Leonhardt's suspension bridges, were continuous between anchorages. The bridge was a model of aesthetic efficiency but—like the Cologne-Mulheim—capitulated to war. It was first seriously damaged in 1944 from a chance bomb hit in one anchorage, and was then mortally wounded by the retreating German forces in January 1945.

The Rodenkirchen Bridge was Leonhardt's first major suspension span, but its legacy, according to Leonhardt, was transoceanic; he later claimed that its use of plate girders resulted in their "imitation" in the United States (1982). Leonhardt (1910-1999) established his own firm in 1939 and after World War II helped pioneer cable-stayed bridge design. Although not building many suspension spans, Leonhardt would challenge design orthodoxy for almost half a century. He was particularly influenced by Othmar Ammann's George Washington Bridge, which he visited in 1932 during a 2-year stint in America as a graduate student at Purdue University. He would write (1966) that the "slenderness of [the Rodenkirchen's] stiffening girder was influenced by the fact that [the] George Washington Bridge did work well without any stiffening truss, an observation which had great influence on my later ideas."

Norway had also experimented with solid cross girders with little longitudinal stiffening as early as 1913 in a 54-m span and in two other 1918 spans of less than 60 m. The 1930 Bingsfoss and 1931 Framnes spans—130 and 150 m, respectively—both used rolled I-beam cross girders; both were then exceeded by the 168-m span Vrengen (1932), the 160-m span Elverum, and 230-m span Fyksesund (1937) bridges. Other plate girder spans of this era included the 1929 Rio Quequen Bridge in Buenos Aires, Argentina, whose 150-m span was modeled on the Cologne-Deutz Bridge. Most of these bridges, because of their relatively small spans, often wide decks, and, in the case of self-anchored spans, deep and heavy plate girders, were fairly stiff. Low navigation clearances created less wind exposure, too, with the result that most averted wind-related problems in service. In America, however, plate girder suspension bridges evolved along a different path, with very different consequences.

3.2 EARLY APPLICATIONS IN AMERICA

The plate girder found rapid acceptance in American suspension bridges where, notwithstanding Leonhardt's assertions, several spans were underway as the Rodenkirchen was being built. Fewer spans were called on to carry rail, subway, or trolley traffic, and this reduced the need for stiffness against static deformation from live loads. Erection was faster and simpler, as larger girder sections could be lifted in lieu of piece-by-piece truss erection. For engineers and bridge authorities grappling with the harsh economic realities of the Depression, the plate girder was attractive.

This pursuit of economy dovetailed well with the Deflection Theory, which had greatly reduced requirements for vertical stiffness. An extreme example, of course, was the unstiffened suspended structure of Ammann's George Washington Bridge. Leon Moisseiff (1939) wrote that this bridge "brought home the seeming paradox that [Moisseiff's italics] "*the cable is the stiffest member in a suspension bridge*." Simplified approximations of the Deflection Theory soon emerged that avoided its complex calculations. In 1930, Stephen Timoshenko devised a direct method to calculate deflections in the main cables from live loads or temperature change. Shortridge Hardesty and Harold Wessman devised a method well-suited for preliminary design and the analysis of flexible suspended structures such as plate girders. David Steinman's 1934 method aided the design of suspension bridges with continuous spans; others such as Britons R.J. Atkinson (1939) and C.D. Crosthwaite (1947), and perhaps most significantly H.L. Tsien in China (1948), followed.

In 1933, Moisseiff, with Fred Lienhard, a Port of New York Authority engineer, published a paper that was perhaps the most important theoretical advance of the decade. Moisseiff, of course, commanded attention, and *Suspension Bridges under the Action of Lateral Forces*, published in the *ASCE Transactions,* was influential. Moisseiff and Lienhard's Theory of Elastic Distribution extended the tenets of the Deflection Theory to horizontal bending under static wind load. They demonstrated that the stiffness in the main cables—via the suspenders—would absorb up to one-half of the static wind pressure pushing a suspended structure laterally. This energy would then be transmitted to the towers and anchorages. The theory accounted for an equilibrium position of a suspended structure resulting from static wind forces pushing it horizontally and the stiffness of the cables resisting that force. According to the theory, the lower the span's sag ratio, the lower the horizontal displacement. This in turn meant that less lateral stiffness in the suspended structure was required—thus, lighter systems of lateral bracing. Under the theory, vertically slender decks would derive the greatest benefit, as they would be subject to less horizontal wind pressure. Moisseiff and Lienhard even claimed long suspension spans would be two to four times stiffer laterally than beam or arch spans of similar width.

The paper elicited effusive response from such leading engineers as Charles Derleth, Glenn Woodruff, and Charles Ellis. Derleth suggested the cables reduced by more than 80% the wind loads formerly thought to have been carried by the suspended structure of the George Washington Bridge. Narrow long spans of low traffic capacity now became economically feasible by avoiding deep and, in particular, wide trusses. A debate over deck width that had once embroiled the Golden Gate Bridge was now rendered irrelevant; its trusses would have to resist just 50% of the wind loads they faced. The Transbay Bridge could be built with two narrower decks, instead of a single wide deck, reducing tower and anchorage costs. Towers, too, would benefit from lower assumed lateral loads.

An increased confidence in engineering theory and the availability of higher strength steels also allowed engineers to enhance the appearance of their bridges. The material lightness of the plate girder was matched by its streamlined appearance, a trend influencing many areas of industrial design from automobiles to steam locomotives. Plate girders offered a sinuous, clean, and graceful profile unmatched by trusses; their ribbed solid faces could transform an otherwise staid truss design into modern structural sculpture.

This combination of factors brought about a rapid acceptance of the plate girder suspension bridge, and although American engineers did not pioneer the form, they would become its leading exponents. This reliance on analysis, without the living memory of the collapses of the past century, would prove a dangerous combination. Henry Petroski (1994) would write that:

> in a matter of a generation or two of major bridge designers, the design climate changed from one in which the occurrence of failures had to be explicitly addressed by anyone proposing a new suspension bridge to one of such confidence and even hubris that designers took off from the achievements of Roebling and his successors in directions that appear almost at times to have focused more on economy and aesthetics than on structural behavior.

The first notable American plate girder span—the Anthony Wayne Bridge—was completed in 1930 across the Maumee River in Toledo, Ohio. The firm of Waddell & Hardesty, with the aid of Moisseiff and architect Cass Gilbert, designed the 239-m span. Its 3-m deep plate girders supported an 18-m wide deck; both were generously proportioned in relation to the span length. The $2.6 million bridge, with diagonally braced towers and unobtrusive anchorages, earned an honorable mention from the

AISC. A second plate girder span was Pittsburgh's South 10th Street Bridge, a 221-m parallel wire span opened in 1933 and designed by S.A. Shubin. Its 16.5-m wide deck was stiffened by plate girders slightly more than 2-m deep. Attention to aesthetics was reflected in steel towers with single top pierced portal braces—and rewarded with an honorable mention from the AISC.

In 1937, the Dominion Bridge Company completed a 91-m span across Quebec's Gatineau River north of Ottawa in Canada. The clean portal tower braces of the Chute de Brûlé Bridge complemented its plate girders and earned the span accolades from Elizabeth Mock, who in *The Architecture of Bridges* (1949) described it as "probably the finest small suspension bridge in the Americas." Its deep girders, small span, and wide proportions rendered it stiff, as did the similar proportions of a small 1940 plate girder span completed across the Klamath River in northern California. This 110-m span, designed by F.W. Panhorst of the state highways department, was recognized by the AISC as the most artistic in its class that year.

3.3 STEINMAN AND THE THOUSAND ISLANDS AND DEER ISLE BRIDGES

North America's first flexible plate girder suspension span was designed by Steinman, who had already done much to promote aesthetically pleasing suspension bridges. The Thousand Islands Bridge, a five-bridge crossing centered by two slender two-lane suspension bridges, was completed in 1938 between Canada and the United States over the picturesque St. Lawrence River. Typical of Depression-era projects, extremely tight budgets influenced its design.

Early attempts in the 1920s to build a crossing were dogged by criticisms of its remoteness. New York Governor Alfred E. Smith, who felt a bridge should be a public venture, thwarted a privately financed scheme in 1926. In 1930, the state created the St. Lawrence River Bridge Commission to examine crossings to Canada over the St. Lawrence. The firm of Modjeski, Masters & Chase recommended three sites—the Thousand Islands; between Cornwall and Massena; and at Ogdensburg, New York. The report died in the state legislature the following year. The Depression and the availability of public works relief money then rekindled efforts. In 1932, local citizens in both countries retained the firm of Robinson & Steinman; New York relented the following year and established the Thousand Islands Bridge Authority. By 1934, both nations had approved the project, but Public Works Administration (PWA) funding was unobtainable. More than $3 million of private finance was eventually secured in February 1937, and ground was broken on April 30, 1937, the last day permitted under the terms of the federal franchise to the Bridge Authority.

Whereas other estimates had ranged up to $8 million, Steinman's 1936 estimate of $2.2 million had made the project possible. He took advantage of a rock outcrop to reduce the Canadian span from 305 to 228 m; its American counterpart would span 244 m. Plate girder depths of 1.8 and 2 m for the two spans, respectively, produced depth-to-span ratios of 1:125, slender indeed but deeper than the 1:164 proportions of the truss-stiffened Golden Gate Bridge. Construction proceeded quickly, and the bridge was opened in August 1938 after less than 16 months of work and almost 10 weeks earlier than planned. Steinman used prestressed helical strands in the 23-cm diameter cables; over the longer American span, they extend a distance of 647 m. No catwalks were used, and the Canadian strands were strung in bitter winter weather. Stiffening girders, support steel, and roadway concrete were all placed from an overhead cableway strung between the towers, which also ferried workers in small cages to install cable bands, suspenders, and cable wrapping. Frugality was such that the girders supporting the tower erection crane were later placed in the approach spans. One unusual but expressive feature is the exposed cable splays entering the anchorages, involving a new anchoring system developed by Robinson that allowed for the individual adjustment of cable strands. The Canadian anchorage would also serve as the north footing of a 106-m span steel arch.

Steinman had stressed aesthetics, even with the limited budget. The Thousand Islands was unusually graceful for a small high-level suspension bridge; the 46-m clearance under the American span was more than half the 76-m height of its towers. The two-lane roadway was perched high between the slender plate girders to offer views of the Thousand Islands. Steinman's approach spans maintained the theme of slenderness, and the pale green of the ensemble was designed to complement the forest verdure of the bridge's surroundings.

In the first month, the bridge carried 10 times the traffic predicted. Yet not all was well, for here emerged the first manifestations of an aerodynamic problem. Mild vertical undulation had been observed in both spans during the late stages of construction. The undulation was especially pronounced in the larger American span. Motion was first observed during formwork installation for the concrete floor. It was believed that the weight of the concrete floor would quell the motion, but placement of the floor did not resolve the issue. Even moderate breezes would induce undulation—particularly in quartering winds—which at times involved the side spans as well, with one large half-wave raising the main span up to 61 cm while lowering the side spans less so. The main span also tended to sway along its axis, alternatively pushing and pulling against the towers. At times, motion was clearly discernible to onshore observers, and it occasionally included a mild twisting action. About 2 weeks before the span's opening, the onsite engineer summoned Steinman from New York. From his own observations and eyewitness accounts, Steinman hastily devised a temporary remedy of diagonal wire ropes shaped as an inverted V to connect the central cable band to the top flange of the plate girder at adjacent suspender points.

Although longitudinal motion was eliminated, the vertical motion of the main span persisted—if in magnitude diminished. Side span motion was also reduced—to 30.5 cm—but still in the form of a single long half-wave. The cable ties had caused other changes, too. Whereas one long wave had previously engulfed the main span there were now two, with a node centered about the midspan cable tie. Midspan, previously the point of greatest vertical displacement, was now an area of calm. Within the two oscillating segments, maximum vertical motion now occurred at the quarter-points of the main span with a double amplitude of 76 cm. The two segments also oscillated out of phase—while one half of the main span was raised, the other half was lowered, and vice versa.

The reduction in the magnitude of motion, its change of character, and the disappearance of torsional motion indicated a stiffening role for the main cables. Yet Steinman was not satisfied, and more tests were conducted before the span's mid-

August 1938 opening. He settled on a temporary system of wire rope stays, with two stays radiating from the top flange of each plate girder at the towers to suspender cable bands along its corresponding main cable near the quarter-points of the main span. Although not entirely understanding the modus operandi of the motion, Steinman sought to restrict movement indirectly by stiffening the main cables. The stays would also later be found to act against torsional motion.

The last stay, culled from surplus hoisting rope, was installed the morning of the span's opening. These remedies worked to limit vertical motion to no more than a double amplitude of 30 cm, and Steinman used three stays of stronger wire in the final installation of January 1939. The flexible midspan ties were also replaced at this time with rigid double-angled struts. Virtually no perceptible motion was recorded afterward. The more sheltered, smaller, and less afflicted Canadian span also received identical treatment.

Steinman presented his findings to an October, 1938 meeting of the Structural Division of the ASCE in Rochester, New York, and described them in the November 1938 issue of the ASCE monthly *Civil Engineering* (Steinman, 1938). Ammann and Moisseiff attended the meeting, where Steinman predicted that two larger plate girder bridges on which construction had begun—Ammann's Bronx-Whitestone and Moisseiff's Tacoma Narrows—would endure similar vertical oscillation. Steinman's contemporaries, however, were yet to be convinced of the warning signs.

Steinman had also designed a similar span on the east coast of Maine that opened in 1939. His Deer Isle Bridge, a slim two-lane plate girder span of 329 m, was designed less for aesthetic appeal than to meet an extremely frugal budget. Steinman is quoted in his effusive 1959 biography by William Ratigan as saying that, "The PWA gave us only one week to design the bridge. I needed three million dollars, but all they allowed me was $1,300,000. They said 'Give us a bridge that will just stand up'. That's what it was." The austere nature of the design reflected the difficulties in getting the project off the ground at all. In the early 1930s, the Deer Isle-Sedgewick Bridge District was formed to apply for newly available PWA funding. Preliminary plans prepared in 1933 indicated a cost of about $850,000. The state authorized the sale of $490,000 in bonds, and a $315,000 PWA grant brought the Bridge District close to its target. Construction appeared imminent, but then complications set in. By legislation, the budget was fixed, in spite of the preliminary nature of the plans. In fact, the bonds were sold before the Bridge District retained Robinson & Steinman to develop final contract plans. However, not until 1936—3 years after the preliminary estimates—were construction contracts able to be advertised.

Steinman's 244-m span suspension bridge—flanked by 122-m side spans—placed the tower piers on rock from 23- to 38-m deep. Yet the sole bid to build the foundations would have pushed costs to more than $1.1 million, almost $300,000 more than the budget allowed. The Bridge District directed Robinson & Steinman to modify the plans, but a February 1937 bid opening again brought unacceptably high bids. The foundation plans were revised yet a third time, but the April 1937 bid opening produced little improvement; once again, all bids were rejected. The need for drastic change became apparent. Steinman determined that the savings from shallower tower piers would more than offset the higher superstructure costs of a longer span. This led to a 329-m span centering lengthened side spans of 147 m.

Detailed plans were prepared, and the state contributed an additional $170,000 to help out. The fourth bid opening in November 1937—almost a year to the day of the first attempt—bought an $870,000 bridge, almost within the original budget.

The depth problem had been overcome, but the 56-m tall towers now stood on steeply sloping rock; a thin veneer of overburden would do little to prevent the piers from sliding. To overcome these obstacles, the tower cofferdams were assembled on land—their bases contoured to fit the undulations—and then sunk into place.

Economics had greatly influenced the form of the suspended structure. Its 6.6-m width, although adequate for expected traffic, was not overly desirable in terms of width-to-span proportions. The 2-m depth of the plate girders also made the suspended structure exceptionally shallow in relation to its length. It was light, too, weighing just 1,180 tonnes. Yet the Deer Isle Bridge reflected little of the grace of the Thousand Islands Bridge; its deck rose sharply to lessen the height and length of the approach spans; only a short vertical curve marked the grade transition at midspan. The towers, although similar to those of the Thousand Islands Bridge, did not have the same pleasing proportions. The stringent budget had also affected construction methods. Each of the 19 strands in each 19-cm diameter helical strand cable was erected over the tower centerline and was then shifted into the saddles; no catwalk was used. Even accounting for their lightness, extreme diligence was required during plate girder erection to avoid overstressing the light and flexible towers. Yet work proceeded without incident, and the bridge opened in June 1939.

While wrestling with the instability of the Thousand Islands Bridge, Steinman had recognized that the similarly austere Deer Isle Bridge might be prone to harmonic motion. Its cross-section, although similar, was narrower and thinner in relation to span, and the deck camber was also greater than in the Thousand Islands Bridge. The lengthening of the main span from the original scheme had undoubtedly increased its flexibility; static wind forces in the highly exposed and windswept arm of the Atlantic Ocean had already been a concern. As it happens, Steinman would be the first beneficiary of his experience at the Thousand Islands Bridge.

Construction was not far advanced by the time Steinman had devised his remedies for the Thousand Islands Bridge. He now designed a similar system of center cable ties and cable stays that was installed as a preventive measure on the Deer Isle Bridge before its opening. Motion, however, was first observed shortly after installation of the first set of cable stays, but before they had been fully tensioned on account of the need to pour the concrete deck. About half the deck was in place by late May 1939 when the first significant motion developed. All main span stays were in place, but they were still not fully tightened. A strong east wind set off vertical motion with a single amplitude of 18 cm. The center cable tie split the wave pattern as it had at the Thousand Islands Bridge into two oscillating segments split by a midspan node. The cable stays were tightened but not fully, as the concrete deck was not yet fully placed. Three weeks later, after concreting, larger oscillations occurred in the unrestrained side spans. This time, single amplitudes reached 25 cm, while the main span—checked by the cable ties and stays—remained calm. It was evident that the side spans would require restraint, too, and so temporary cable ties were installed there in time for the span's opening.

These defenses still did not quell the Deer Isle Bridge's aberrant behavior. Main and side spans oscillated repeatedly and in various combinations in the weeks following the bridge's opening, causing concern for motorists and engineers alike. In August 1939, the temporary side span stays were replaced with a stronger, permanent installation. Improper fabrication of the main span stays saw minor oscillation continue as the fault was corrected. By late 1940, the new stays were described as "fully effective," with the bridge experiencing "no perceptible oscillation" (Anonymous, 1940b). This situation would prove falsely reassuring.

3.4 AMMANN'S BRONX-WHITESTONE BRIDGE

Steinman's Thousand Islands and Deer Isle bridges introduced flexible plate girders to America in two-lane bridges of relatively modest span. Yet the rapidity of plate girder evolution was reflected in the completion of two suspension bridges on the east and west coasts of the United States that would dwarf Steinman's efforts. The first of these spans—the Bronx-Whitestone Bridge (*see* Fig. 3-1)—had actually opened 2 months before the Deer Isle Bridge, in April 1939. This crossing of the East River east of the newly opened (and already congested) Triborough Bridge would also provide access to the 1939 New York World's Fair. Its 701-m span was the third longest in the world. A plate girder just 3.3-m deep imbued the Bronx-Whitestone Bridge with a depth-to-span ratio of just 1:209; along with great attention to aesthetics, these attributes helped create what many consider the most graceful suspension bridge built to that time.

The bridge formed part of Triborough Bridge Authority czar Robert Moses' vision of a metropolitan network of spacious parkways. Moses entrusted its design to Ammann, the Bridge Authority's chief engineer. Design work began in 1935, a year before the Triborough Bridge opened, but unlike the Triborough, the Bronx-Whitestone Bridge would be financed entirely through the private markets. Ammann used the Triborough team, with Allston Dana as design engineer and Aymar Embury II as consulting architect. Moisseiff acted as the consulting engineer for design. Twin span arch and truss designs were examined, but a suspension bridge was ultimately adopted. One can only speculate as to Moisseiff's influence on Ammann's choice of a plate girder—which may, in fact, have preceded Steinman's Thousand Islands Bridge design—but its use would generate a significant $2 million savings over the cost of a truss. Ammann himself, with the George Washington Bridge, was already convinced of the merits of a flexible deck, as illustrated in his rationale for the use of plate girders in the Bronx-Whitestone Bridge:

> It is now well established that long-span suspension bridges for modern highway traffic may have a relatively flexible stiffening system, and that the degree of flexibility has a material effect upon the economy of the design. In this respect the Bronx-Whitestone Bridge marks another radical departure from past theories and practice. Its stiffening girders have greater flexibility in relation to span length than any other suspension bridge built in recent years, except the George Washington Bridge in its present state with only a single unstiffened highway deck. The latter bridge, however, with a present roadway capacity equivalent to that of the Bronx-Whitestone Bridge, has a

suspended dead weight per lineal foot 2.5 times greater, a center span 56 per cent longer, and side spans somewhat shorter, all of which factors contribute to the greater rigidity of the unstiffened cables (Ammann, 1939a).

Ammann's discussion then took a crucial turn:

Fig. 3-1: Bronx-Whitestone Bridge, in its original configuration
PHOTOGRAPH BY RICHARD AVERILL SMITH, COURTESY OF THE MTA BRIDGES AND TUNNELS SPECIAL ARCHIVE.

> It was the aim . . . on esthetic as well as structural and economical grounds, to restrict the height of the floor structure to a minimum, to avoid trusses, and to keep the top at such an elevation above the floor as not to obstruct the view of the landscape from passing vehicles. A depth of 11 ft [3.3 m] for the stiffening girders was found to be sufficient . . . [which] . . . is only 1/210 of the length of the center span and 1/70 of the [224 m] side span.

These statistics are all the more remarkable as the Bronx-Whitestone Bridge used largely the same loading specifications as the George Washington Bridge. Although its depth-to-span ratio was unprecedented—save the George Washington Bridge—the 25-m deck width translated to 1/34 of the span length, well within acceptable limits.

Ammann's 115-m tall towers constituted another technical and aesthetic achievement. They stood in refutation of his earlier views that had guided the conception of the massive, sturdy frames of the George Washington Bridge. Their weight of only 3,175 tonnes each reflected significant economies. They were the first closed box towers to be braced with single arched braces above and below the roadway; static wind pressure was taken up by strong but clean legs in lieu of multiple portal braces or diagonal bracing. Here was the essence of simplicity, elegance, and slenderness; the Bronx-Whitestone towers are still admired as among the finest examples of this form.[1]

Embury was conscious of the visibility of the entire bridge and extended the design philosophy to the treatment of its individual elements. Most prominent were the anchorages, molded as large, triangular blocks devoid of "architectural embellishments" (Ammann, 1939a). Their front faces were perpendicular to the line of the cables, and four diagonal grooves etched into their sides served to visually emphasize resistance to the pull of the cables. Their upper edges curved gently, following the path of the hidden cable strands and anchorage chains. Embury (1938) admitted that "We are conscious of the fact that this design is an experiment, perhaps a rather daring one, but we hope that its very boldness and stark simplicity will mark it as belonging to a bridge and to nothing else in the world." The design contrasted greatly with the monumental anchorages of Modjeski's Benjamin Franklin Bridge, and Ammann would use essentially the same concept in the Verrazano-Narrows Bridge some 20 years later. The desire for visual simplicity extended to the continuous plate girder approaches, providing an uncommon consistency in overall profile that enhanced the already graceful suspended spans. Aesthetically, the Bronx-Whitestone Bridge constituted a radical departure in suspension bridge design.

The entire bridge, including approaches, cost just $18 million, in spite of deep anchorage and tower piers, whose supporting caissons are embedded up to 50-m deep in poor soils. It was also completed 2 months ahead of schedule in April 1939 after just 23 months of construction. More than 1,800 tonnes of main cable wire were spun in just 2 months; the tower saddles were also the first with a circular—rather than stepped—bore to maintain the shape of the cables. Ammann (1939b) expressed the view shortly before the span's opening that, "if it proves to be amply rigid, as the writer does not doubt, it should set a precedent for spans of that magnitude and similar proportions." His confidence was such that, in the Golden Gate Bridge, he felt "even shallower trusses or solid plate girders would be as adequate."

This belief would soon prove misplaced. At about the time that Steinman encountered problems in the Deer Isle Bridge, Ammann was introduced to vertical oscillation on the Bronx-Whitestone Bridge. The ramifications were much larger, as the bridge was three and a half times longer, six times heavier, and funneled infinitely more traffic than either of Steinman's spans. Paralleling Steinman's experiences, the first vertical motion was felt—rather than observed—by workers installing formwork for the concrete roadway. The motion typically split the main span into two oscillating segments, one-half being raised, while the other was deflected downward. Occasionally, two nodes split the main span into three segments. The main span, rarely, would also oscillate in one segment extending from tower to tower. At no time did the suspended structure oscillate laterally or torsionally. However, as with Steinman's bridges, vertical motion was accompanied by a sway along the span's axis. The vertical waves—with a double amplitude of up to 91 cm—were also observed most frequently when winds blew at an oblique angle to the bridge.

Although there was no evidence that the oscillation posed any structural threat, efforts were made to arrest it prior to the span's April 1939 opening. An interim midspan cable tie similar to that used by Steinman was installed, as were buffer blocks between the plate girders and the towers to resist longitudinal sway. These devices reduced motion, but within months it was apparent that they were insufficient for the task; the midspan cable ties were strengthened, and friction brakes replaced the buffer blocks at the towers. As it happens, this would not be the end of the story.

The Bronx-Whitestone Bridge represented the antithesis of the early 20th century suspension bridge. Yet even it would be overshadowed by a span designed by none other than Leon Moisseiff. In August 1939, Moisseiff had written an article in *ENR* entitled "Growth in Suspension Bridge Knowledge." Moisseiff lauded developments in theory and practice—many of them his own. His most prophetic statement was that wind pressures had only recently been more fully understood—a reference to his and Lienhard's work on lateral wind forces. At this time, Moisseiff's design for a startling suspension bridge was under construction over the Tacoma Narrows in Washington State. It was the first suspension bridge largely of his own creation, for in spite of the impressive array of spans to which he had contributed, Moisseiff had led none of the efforts to design them. Here, on a blustery November day in 1940, Moisseiff's comments would take on a wry and dramatic irony, for events at the Tacoma Narrows would shift the course of suspension bridge evolution—violently and irrevocably.

ENDNOTES

1. Leonhardt, the noted German engineer, proposed similar towers for a 700-m span suspension bridge in Hamburg as an alternative to massive granite Roman arches envisaged by Hitler. This enormous span, with a 47-m wide deck, was never built.

Chapter 4

EPIPHANY AT THE TACOMA NARROWS

4.1 THE RISE AND FALL OF THE TACOMA NARROWS BRIDGE

A connection from the city of Tacoma to the sparsely developed Olympic Peninsula in northwest Washington had been the subject of discussion for many years. The Tacoma Narrows, where Puget Sound narrows to about 1,400 m, was a logical choice for a crossing, but its 37-m depths and strong tidal currents had made a bridge an expensive proposition. A vehicular ferry was established in 1929, but the onset of federal Depression-relief funding and the need for improved access to a large U. S. Navy yard on the Olympic Peninsula soon conspired to breathe new life into a Narrows crossing.

David Steinman's 1930 suspension bridge design for the Tacoma Chamber of Commerce was one of the first proposals. Its two-lane, 12.2-m wide deck occupied just two-thirds of a 18.3-m wide floor truss; this great width acted with a 7.3-m deep stiffening truss to statically resist the Narrows' high winds. Only a decade later would the full significance of these measures become apparent. In 1933, a private venture obtained state approval to build another design, but a lack of federal financing killed it. Then in 1933, a two-lane, 366-m span cantilever bridge from E.M. Chandler received War Department blessing; it, like the Steinman scheme, went nowhere.

Local pressure soon led to the creation of the Washington State Toll Bridge Authority in 1937. It first considered an unusual multispan suspension bridge of three 366-m spans and two 183-m anchor spans with tie cables connecting the tower tops. This design was then jettisoned in favor of a clear span of 792 m flanked by 305-m side spans. A narrow 9.1-m wide deck was unprecedented for a span of this length, but only two lanes were required and made the span affordable; this factor would loom large in light of subsequent events. The State Highways Department, working on behalf of the Toll Bridge Authority, prepared a design under the direction of engineer Clark H. Eldridge. The narrow deck favored a shorter span to resist static wind pressure, whereas great water depths demanded a longer span. A compromise emerged in mid-1938 in the form of a 792-m span flanked by longer 396-m side spans, with a 12-m wide floor truss stiffened by a relatively conservative 6.7-m deep stiffening truss. In June 1938, the Toll Bridge Authority received a PWA grant of \$2.7 million to help fund the span's construction; however, its application for a \$3.3 million loan from the RFC remained unacted on.

Then came a fundamental change in the design. The PWA grant had obliged the Toll Bridge Authority to obtain an independent review of its plans. The Authority called on Leon Moisseiff, arguably the most prolific suspension bridge designer of the era, to evaluate the superstructure design. Moisseiff, with Fred Lienhard, proposed changes that, if approved by the PWA and the Toll Bridge Authority, would radically alter the adopted design. The first inklings of change came in a July 17, 1938 letter from Moisseiff to Lacey V. Murrow—Chief Engineer of the Toll Bridge Authority—outlining preliminary findings. Towers of differing height had been envisaged to accommodate unequal bank elevations, but Moisseiff, with an eye to economy and aesthetics, urged Murrow that "the towers should be of identical design and fabrication. Economic fabrication and good appearance demand it. The symmetry of the bridge should be adhered to" (Board of Engineers, 1941). To this end, Moisseiff raised the west end of the bridge by 6 m. He then suggested the 6.7-m deep truss would provide sufficient stiffness only "at great cost," and that it "may practically be omitted and the desired rigidity . . . be obtained by other means." He recommended a lengthened span of 853 m, with shorter side spans of 335 m. Lower 136-m tall towers were also recommended with bases set 15-m apart instead of the planned 18 m. The main/side span redistribution and the lower sag ratio of 1:12 had in Moisseiff's judgement provided the necessary rigidity—including lateral stiffness against static wind pressure—to facilitate his most crucial change. He now proposed a solid plate girder just 2.4-m deep and half the weight of the truss. Moisseiff's own theories had permitted such vertical and horizontal flexibility, and he now produced a span of remarkable economy and grace.

Tacoma's main span would rank third in length behind the Golden Gate and George Washington bridges, yet its proportions were unprecedented (*see* Fig. 6-2). The bridge's depth-to-span ratio was a startling 1:350, almost a tenth of the 1:40 proportions of the ungainly Williamsburg Bridge. Its ratio of width-to-span was equally eye-catching. Contemporary practice suggested a ratio of 1:30 and in the Bronx-Whitestone Bridge was a comfortable 1:31. Yet Tacoma's 11.9-m wide suspended structure more than doubled the ratio to 1:72, exceeding by a large margin the record 1:47 proportions of the Golden Gate Bridge. Yet in all other respects, the span conformed to contemporary norms for live loads and static wind pressure. The Toll Bridge Authority and the PWA accepted Moisseiff's changes—the PWA candidly stating it assessed the engineer's reputation and not the design. Toll Bridge Authority staff executed the detailed design under Eldridge in July and August of 1938, and this work was referred to Moisseiff's office in New York City for approval.

The Toll Bridge Authority now faced a PWA deadline for tendering construction contracts. Expecting RFC loan approval, the Toll Bridge Authority advertised bids in August 1938. The Authority's own advisory board of engineers—chaired by Charles Andrew, formerly Bridge Engineer at the Transbay—reviewed Moisseiff's design. In a report dated August 31, 1938, 2 days before the receipt of bids, Andrew stated that, "We have full confidence in Mr. Moisseiff and consider him to be among the highest authorities in suspension bridge design" (Board of Engineers, 1941). Concern was confined to the narrowness of the span, not its shallowness:

> It might seem to those who are not experienced in suspension bridge design that the proposed 2800-ft [853 m] span with a distance between stiffening trusses of 39' [11.9 m] and a corresponding width of span ratio of 72, being

> without precedent, is somewhat excessive. In our opinion this feature of the design should give no concern (Board of Engineers, 1941).

Andrew also considered the Deflection Theory to render width-to-span ratios and the degree of vertical and lateral deflection not so much as a structural concern as one of motorist comfort. He cited the low sag ratio as further stiffening the main cables, which, in accordance with Moisseiff's and Lienhard's Theory of Elastic Distribution, benefited the suspended structure. Andrew wrote:

> In a long narrow span the force of the wind is to a large extent transferred to the cables and through them to the towers rather than through the stiffening trusses. The greater the total dead load stress in the cables the more difficult it becomes to deflect them laterally and the less the side deflection of the bridge (Board of Engineers, 1941).

In fact, the main cables of the Tacoma Narrows Bridge were estimated to take 80% of the static wind load. Andrew considered Moisseiff's projected 2.8-m lateral displacement at midspan in 96-km/hr winds to be virtually unnoticeable (in 161-km/hr winds, it would be 6.1 m), and he pointed out that a truss would be prone to the same displacement. The Toll Bridge Authority's advisory board endorsed the bridge's vertical slenderness by stating that, "This feature of stiffening truss design is strikingly demonstrated in the George Washington Bridge where no stiffening truss is used." Yet the advisory board's support was somewhat ambiguous, as it concluded that,

> We believe the present span could be materially increased if it were necessary, keeping the same width without any detrimental effect. In consequence, we have no concern as to the general features of the proposed design of the superstructure (Board of Engineers, 1941).

With still no word on the RFC loan, the Toll Bridge Authority faced another setback when the only bid received in early September was $750,000 more than budgeted for. Its advisory board was called on to help lower costs. Although the board modified the open-bottomed cellular tower caissons,[1] a December 31 PWA deadline hampered efforts to fully review the superstructure design. The board did, however, modify the floor system and road deck. PWA blessing was received, and new bids were called for the end of September.

The RFC delay was perhaps partially explained by the first real doubts over Moisseiff's design. It had retained T.L. Condron, a septuagenarian engineer, to review Moisseiff's plans. In a late September 1938 report, Condron expressed concerns, couched with some deference to the legendary Moisseiff:

> In view of Mr. Moisseiff's ability and reputation, I hesitate to make any criticism of the structural design, but from a practical standpoint, I would feel that the width of this Bridge relative to the length of spans was open to criticism, particularly, since it was without precedent (Board of Engineers,

1941).

Although his examination was cursory, Condron had sought clarification from Moisseiff and had traveled to Berkeley, California, to meet R.E. Davis. In 1933, Davis had, by way of models, verified the vertical deflections predicted by the Deflection Theory. His tests had also suggested that Moisseiff and Lienhard's Theory of Elastic Distribution accurately predicted lateral displacement under static wind forces. Condron reported Davis to be "reasonably confident that the lateral deflections of the [Tacoma Narrows Bridge] . . . would be in no way objectionable to users of the bridge" (Board of Engineers, 1941). These findings suggested few issues. Moisseiff himself had responded to a Condron inquiry in mid-September. After comparing calculated lateral deflections of the Tacoma Narrows Bridge design with those of the Golden Gate Bridge, he wrote:

> To obtain a reasonable degree of lateral stiffness for the bridge flatter cable curves than usual have been adopted. . . . thus increasing their resistance to distortion. . . .
> . . . For a bridge of such horizontal slenderness the obtained stiffness is rather satisfactory. I see no structural objection to the magnitude of the horizontal deflection (Board of Engineers, 1941).

Condron considered Moisseiff's use of "rather satisfactory" unconvincing, even to the point that he felt Moisseiff somewhat unsure. Condron suggested it "advisable to widen the super-structure to 52 ft [16 m] . . . This width would give a ratio of 1:54 approx. . . . The cost would be increased considerably, but the additional cost would certainly be justified." He concluded by saying the bridge:

> is probably technically sound notwithstanding its narrowness, but there are several reasons why it would be of material advantage if the bridge could be widened at a reasonable increase in the cost, and therefore, I recommend that serious consideration be given to the possible increase in the width of this structure, before the contract is let or work begun (Board of Engineers, 1941).

Condron recognized that a required widening of the anchorages would be a major obstacle, and he suggested that at the very least a widened roadway of 9.1 m be considered within the existing cross-section. Yet Condron's report was submitted after contract bids had been called, and Moisseiff's design emerged unaltered. The $3.3 million RFC loan was approved in October, and a $5.95 million contract awarded in late November saw foundation work begin in earnest in January 1939.[2]

The Tacoma Narrows Bridge's false-bottomed caissons would be sunk through 37 m of water—twice the depth achieved before—in currents of up to 13 km/hr and a 5-m tidal range. Not even the infamous south tower of the Golden Gate Bridge had encountered such arduous conditions. Dozens of 550-tonne concrete blocks were first dumped with difficulty because of the currents and steep slopes near the Tacoma pier; from these, wire ropes guided the descent of each caisson. Contractors worked around the clock to lodge the caissons in overburden before spring tides peaked in June 1939;

the last pier touched down with just 8 hours to spare. Overburden was then excavated through 32 dredging wells. Yet more difficulties still loomed: currents upended several concrete barges, and in early April 1939, a crane boom buckled and dropped a load of steel into one caisson, injuring two and delaying work for 10 days.

Poor rock then forced the Tacoma pier down an additional 7.3 m before resting at a depth of 68 m, a record for open-bottomed caissons and one of the deepest bridge piers ever; its west twin was sunk to a depth of 54 m. A 7.6-m thick bottom seal was then cast in the deepest use of tremie[3] concreting to that time. The piers were completed in September 1939, but it perhaps came as no surprise when the contractors later sued the Toll Bridge Authority (successfully) over what they argued was faulty tidal information. Chandler also (unsuccessfully) sought damages for the Toll Bridge Authority's alleged use of information gleaned from his defunct 1933 cantilever and a later Pierce County toll scheme.

The tower piers consumed 40% of the budget and served to highlight not only their magnitude but the remarkable economy of the $3.5 million superstructure. Bethlehem Steel, the steel subcontractor, moved quickly. Tower assembly began in July 1939. Each tower leg comprised just seven cells in cross-section; their solidity was soon demonstrated when in early November the almost-completed towers held firm in a large earthquake centered just 65 km away. Other notable features included the use of welded—rather than cast—steel cable saddles for only the third time anywhere; those at the Tacoma Narrows were the largest yet.[4] Welded cable bands were also used for the first time. By the end of 1939—just 12 months after work had begun—spinning of the 6,308 wires into each 44-cm diameter cable had begun. After just 2 months of winter work, J.A. Roebling's Sons had laid all 38 strands of 332 wires. Lifting of the plate girders required just 6 weeks to complete, with an average of 61 m added each working day. By May, a thin ribbon of steel stretched across the Narrows and, in early June, paving was finished.

After a record construction period of just 19 months, the bridge was opened on July 1, 1940 at a cost of just less than $6.6 million. Visually, the Tacoma Narrows Bridge was stunning; its slenderness elicited much comment (*see* Fig. 4-1). The pamphlet marking the 5 days of opening celebrations was full of superlatives; most prophetic was a passage that only months later could have served as the span's elegy:

> Dead materials in themselves, the steel and concrete of this bridge have come to life when combined under the magic touch of the bridge-makers. It is vibrant with energy transmitted to them. Stress counteracts stress. Weight balances weight. Force meets force. It breathes as though alive, expanding and contracting with the changes in the seasons (National Youth Administration, 1940).

And here lay the seeds of unease. It had been surmised that the slender proportions of the Tacoma Narrows Bridge might render comparisons with other suspension bridges under horizontal wind and traffic loads inappropriate. F.B. Farquharson, a professor of civil engineering at the University of Washington, was retained to develop a dynamic model of the entire bridge. In contrast to the static model used to aid in deck erection, this 1/100 scale model could be mechanically

Fig. 4-1: Tacoma Narrows Bridge, 1940
PHOTOGRAPH (NEGATIVE MSSUA690) COURTESY OF THE SPECIAL COLLECTIONS DIVISION, UNIVERSITY OF WASHINGTON LIBRARIES.

induced into periodic horizontal motion through the sequential activation of 100 electromagnets spread across its 16.5-m length. In late 1939, as the model was being readied, reports of mild but persistent vertical motion in the Bronx-Whitestone and Deer Isle bridges convinced the Toll Bridge Authority to investigate further, before the plate girders at the Tacoma Narrows were erected.

Suspicions were confirmed when, during the final stages of work, an unusual rhythmic vertical motion began to grip the main span in only moderate winds. First noticed by workers concreting the deck, these gentle but perceptible undulations were sufficient to induce both bridgeworker nausea and engineering concern. Farquharson (1949) later wrote that, "the realization that a potentially dangerous phenomenon was developing became inescapable." The somewhat primitive mechanics of the model were then modified to impart vertical forces, and Farquharson achieved remarkable success in modeling the span's vertical undulations, inducing waves with up to nine segments in the main span. It was apparent, as Andrew (1949) would write, that, "Energy was being fed into [the] structure by some form of pulsating force derived from the wind, separate and apart from the normal wind pressure forces considered in structural design." Farquharson (1940) summarized the studies in the July 1940 issue of *Civil Engineering* and opined that the narrowness of the Tacoma Narrows Bridge might be responsible. He also recognized uncertainties introduced by recent design trends:

> During the last decade startling developments in the art of suspension bridge design have pushed the frontiers of achievement far beyond the dreams of yesterday. Three trends are notable in modern design—a growing ratio of span length to width, an increasing sag ratio, and an ever-decreasing rigidity in the stiffening trusses. . . .
> Prior to the building of the Golden Gate Bridge, a ratio of span to width of 40 was considered daring. In that structure it jumped to 48, and in the Tacoma Narrows Bridge it has reached 72. As might well be expected, such a drastic change has introduced complications that render prediction exceedingly precarious and necessitate greatly extended theoretical investigations.

Farquharson was exploring new frontiers—there were no precedents for emulating a motion whose origin was not understood. Ameliorative strategies included damping out the undulations or dissipating their energy, or reducing the energy absorbed by the bridge. This second approach required modifying the form of the plate girders, and so damping measures were devised before the span's July 1 opening. Diagonal wire rope ties connecting the midspan cable band to the plate girder were installed in early June. At the end of the month, hydraulic buffers (the Bronx-Whitestone Bridge's friction brakes had worn out quickly) were installed at the towers to help restrain the plate girder. Stays extending from the towers to the deck were found to be somewhat beneficial, as were those from the deck/tower juncture to the cables, but both remedies were rejected because of the difficulty of maintaining adequate tension in the stays.

In spite of these measures, the vertical motion continued and became the subject of much local interest. Motorists ventured onto the Tacoma Narrows Bridge to observe vehicles ahead of them slowly disappear in the trough of a wave. There was little perceived danger—indeed the bridge earned the popular if dubious nickname of Galloping Gertie (an appellation once said to have been bestowed upon Charles Ellet's Wheeling Bridge [Hopkins, 1970]). A local bank even planted a billboard along one of the span's approaches proclaiming it "as safe as the Tacoma Bridge." The bridge soon proved itself financially, too, and not just on account of pleasure seekers; within a month, commuting tolls were reduced by 25%. Yet the span's behavior continued to concern Toll Bridge Authority engineers, as reports of nausea emerged with increasing regularity; some motorists even avoided the span altogether.

At the end of July, it was decided to devise additional restraining measures. To enable engineers to measure the magnitude and character of vertical motion, markers were attached to the lampposts from the Tacoma shore to mid-span, and a movie camera was placed atop the toll building. All through August and intermittently to mid-October, film was obtained, during which vertical waves of up to almost 1.2-m double amplitude and up to 25 cycles per minute were observed. A single wave with a double amplitude of up to 61 cm was most common, followed by a double-noded pattern symmetrical to either side of midspan with a double amplitude of up to 81 cm. Up to nine nodes were also observed, either symmetric or antisymmetric to midspan, but there was little evidence of the side sway that had caused Condron concern. Interestingly, winds of less than 3 km/hr generated perceptible motion, and winds of just 5 to 6.5 km/hr would generate double amplitudes of up to 76 cm. Paradoxically, in winds of up to 56 km/hr, little or no motion was apparent. No pattern could explain the onset of oscillation, which occurred just 10% to 15% of the time, and traffic remained unimpeded in all motion. Although 12 different wave patterns were observed, a frequency of 12 waves per minute was increasingly common, a concern because it matched the natural frequency of the towers.[5] To everyone's later regret, accurate readings of wind speed and direction were not obtained to allow detailed comparisons with the types of observed motion.

Mindful of stronger autumn winds, engineers added holddown cables to restrain the east (Tacoma) side span, attaching them on October 4 some 106 m from the anchorage to 45-tonne concrete blocks buried in the slopes. The same operation was repeated 82 m into the west side span 3 days later, with wires anchored in wooden piles driven offshore. These measures were considered interim until a permanent solution could be devised. Interim they were—the south tie snapped in a storm just 1 week after installation. It was re-installed 3 days later.

About 2 weeks before the holddown cables were installed, it was decided to perform static tests in the University of Washington's 2.4- by 3.6-m wind tunnel to further examine the theory of autorotation (or galloping, a self-sustaining vertical motion). This, along with random wind gusts abetted by the Narrows' steep terrain, was considered a possible cause of Gertie's vertical motion. Gusts coinciding with the natural frequency of the bridge were soon eliminated from consideration because such periodicity was thought to be extremely rare. Galloping appeared more promising. The phenomenon had been known to induce oscillations of up to 6.1 m in ice-coated high-tension wires; it occurs when lift forces from random gusts exceed drag resistance.[6] Early tests suggested that lift forces were influenced by deck cross-section and the angle of attack of the wind. With oscillation confined to a vertical mode, it was decided to reduce lift through subtle changes in form. Tests on a 1:20 scale static (i.e., fixed to supports in the wind tunnel) "sectional" model representing 49 m of plate girder suggested two possible remedies. The first deflected the wind using fairings or deflector vanes attached to the girder fascia. A second involved punching holes in the girder above and below the deck to reduce wind pressures. This approach reduced lift but was less favored because of its irreversible nature. By contrast, all variants of a deflector vane eliminated lift in the model tests. Military use of the wind tunnel pre-empted further testing until November 2, but a simple 23-cm tall curved fairing was devised, supported 1.9 m from the girder face by two struts and centered about deck level. Engineers prepared to order materials.

On the night of November 6, an expected storm began to assert its influence on the bridge. The first indications of concern came from a Toll Bridge Authority employee stationed overnight on a scow beneath the west side span. Vibration of the temporary holddown cables had awakened him at about 3:30 A.M. on the morning of November 7. A daylight inspection revealed a minor loosening of a base connection, but the side spans remained relatively placid, and by 5:00 A.M. their motion had for the most part subsided. By 8:00 A.M., the wind had increased to 61 km/hr. Eldridge drove across the span around 8:30 A.M. and described the bridge as acting:

> in the customary manner, the east side span being practically quiet, the main span oscillating in a four noded manner and the west span oscillating considerably from the temporary hold down to the tower. All of these motions, however, were considerably less than had occurred many times before so I came to the office at about nine o'clock (Board of Engineers, 1941).

The day before, Eldridge, Andrew, Farquharson, and Bethlehem Steel project engineer L.R. Durkee had decided to install the curved fairings on the span's south side as soon as possible. Upon reaching the office, Eldridge prepared preliminary drawings as a basis for contractor estimates. Then around 9:30 A.M., a contractor's employee noticed the center ties on the north cable "alternately tightening and loosening causing considerable snapping at this point" (Board of Engineers, 1941). His motion picture film would provide crucial evidence in light of subsequent events, as would similar observations of a U.S. Coast Guard captain whose cutter passed beneath the span around 10:00 A.M. Farquharson arrived at the bridge about 9:45 A.M. and—the story goes—then rushed to a local camera shop to borrow a movie camera. He later reported:

> A series of observations and photographs taken just before 10:00 a.m. showed the bridge to be moving with the usual vertical motion with a frequency of 36 cycles per minute. . . . Thru several observations this condition did not change. At the same time, it was observed that the center line of the bridge was deflected to the north. Observations suggested a wind of 42 mph [68 km/hr] striking the bridge obliquely. The lateral deflection of the center of the bridge was not measured but did not appear excessive, perhaps four times the width of the yellow center line (about 2 ft. [61 cm]) (Board of Engineers, 1941).

This wind was possibly the strongest encountered to-date; the oscillation frequency was considerably higher than previously experienced and generated undulations with waves of about 90-cm double amplitude in a nine-noded pattern. Yet there appeared little cause for concern other than to close the bridge and wait out the storm. Then, shortly after 10:00 A.M., to astonished onlookers and aghast engineers,

Gertie's behavior took on a totally unexpected and alarming turn, graphically described by Farquharson (*see* Fig. 4-2):

> While checking the frequency from a position on the roadway to the north of the toll plaza a few minutes later, a violent change in motion was noted. This change appeared to take place without any intermediate stages and with such extreme violence that the span appeared to be about to roll completely over. The most startling condition arose out of the fact that from a line of sight very nearly parallel to the bridge the upper side of the roadway was visible while what appeared to be a nearly perpendicular view of the bottom of the roadway was offered on the Tacoma side. The motion, which a moment before had involved a number of waves (nine or ten), had shifted almost instantly to two (Board of Engineers, 1941).

This phenomenon had not been encountered in Farquharson's wind tunnel tests at all. The wave frequency had changed dramatically, too, from 36 per minute to 14, and the main cables appeared to vibrate out of phase. The torsional motion was divided into

Fig. 4-2: Torsional oscillation, Tacoma Narrows Bridge
PHOTOGRAPH (NEGATIVE FARQUHARSON 4) COURTESY OF THE SPECIAL COLLECTIONS DIVISION, UNIVERSITY OF WASHINGTON LIBRARIES.

two segments, with a node at midspan. One side of the deck would heave up at a quarter-point, while the same side at the opposite quarter point would pitch violently down. Each cycle took about just 4 seconds.

One of the last to venture onto the bridge before it was closed was a newspaper reporter, now immobilized about a quarter of the way across the writhing main span. The roadway was twisting almost 45° from the horizontal, with one side lurching 8.5 m above the other. The reporter's vehicle slid across the roadway; a door popped open, and a figure tumbled out and scrambled to the curb. During a temporary lull in the motion, the reporter stumbled to the relative safety of the east tower and then off the gently (and vertically) rolling side span. Left behind in the vehicle was a recalcitrant dog. A truck moving material to the west side of the bridge was also stranded; its two occupants escaped just before it was tossed onto its side. By now, most Toll Bridge Authority engineers had arrived. Eldridge, having earlier met Farquharson on the bridge and observing the apparent success of the wire guys in restraining the side spans, now began making arrangements with Bethlehem Steel to install cables from the tower piers to points along the main span once the winds eased. Farquharson and others were still on the main span, experiencing firsthand its contortions. He wrote:

> On reaching the axis of the bridge just in front of the toll house, it was apparent that a motion of catastrophic proportions had developed, the greater part of which was confined to the main span. As quickly as possible, a station was reached just outside the towers, from which a photographic record of the main span was obtained. A considerable amount of shock motion of small amplitude was evident in the side span, enough to make operation of the motion picture camera difficult. It was quite impossible to operate a camera on the main span, so violent was the motion. . . .
>
> At this time, not more than 8 or 10 minutes after the start of this violent motion, a considerable amount of damage had been done to the concrete in the sidewalks and curbs and several of the lamp posts in the vicinity of the Tacoma tower were already loose on their bases. At this time, one lamp post was seen to fall on the Gig Harbor side of the span . . . (Board of Engineers, 1941).

At brief intervals, the midspan node disappeared and a single wave engulfed the main span. Farquharson took advantage of the fact that the torsional motion was centered along the span's axis. On the spectacular film footage of the span's last moments, Farquharson can be seen taking the sensible(!) precaution of staggering down the center line while the deck on either side of him pitches violently up and down at horrifyingly steep angles. The scientific curiosity of the few who had accompanied Farquharson deferred to a sense of impending disaster, and those observers rushed off the span. Farquharson continued out to the abandoned vehicle during a short lull in the motion in an attempt to rescue the stranded dog. The motion's quick return prevented him from reaching the car, which, according to Farquharson, "began to shift about in a most alarming manner" (Board of Engineers, 1941). At this time, Farquharson noted more lampposts falling off the west part of the main span, where a more pronounced torsional motion was evident. Motion between the deck and plate girders also suggested to him a progressive failure of the girders' lateral bracing. Farquharson hurried off the bridge for more film, but on returning to the Tacoma tower, he noted a change in vibration frequency from 14 to 12 per minute, as well as a new and larger horizontal oscillation.

It became increasingly apparent, even to the indefatigable Farquharson, that the bridge could not endure this unrelenting punishment much longer. The end began over the west half of the main span when a plate girder buckled. Shots sounding like gunfire were heard as several suspenders snapped from their sockets. At 10:30 A.M., a floor panel broke apart and fell into the Narrows. This event had a marked effect on Farquharson, for on the film he can be seen running to the security of the east tower. The coup de grâce was delivered at about 11:00 A.M., when a 183-m section of suspended structure peeled away near the west quarter-point of the main span, overturning as it fell before crashing into the Narrows more than 60 m below; this event was captured by a movie camera now set up on shore (*see* Fig. 4-3). The span then appeared to steady itself, but the oscillations quickly reasserted themselves. Without the restraint of the still-intact holddown cables, the slightly sagging side

Fig. 4-3: Collapse of the main span, Tacoma Narrows Bridge
PHOTOGRAPH (NEGATIVE FARQUHARSON 12), COURTESY OF THE SPECIAL COLLECTIONS DIVISION, UNIVERSITY OF WASHINGTON LIBRARIES.

spans—which to this point had remained relatively stable—were now incited, with the east side span under a small amount of torsional motion. What remained of the main span progressively disintegrated, throwing Farquharson to the deck twice during his retreat over the east side span while still taking observations.

By 11:10 A.M., the last of the main span had plunged into the water, taking with it both abandoned vehicles and one dog. The side spans rose and then, without the counterbalancing weight of the main span, sank some 18 m (60 ft) before recoiling upward to rest at a final sag of about 9 m. The rump of the main span wallowed gently like a beheaded monster, with shell-shocked engineers watching helplessly and half expecting the towers—now bent more than 3.6-m shoreward—to go down as well. The most spectacular bridge collapse ever was over. The Tacoma Narrows Bridge had lasted just 4 months, but with its demise went an entire era of engineering philosophy.

4.2 THE AFTERMATH OF THE COLLAPSE

By some miracle, no human lives were lost, but a Toll Bridge Authority engineer did have to be restrained from leaping into the Narrows when the span went down. The bank billboard was also hastily removed later that same day. But the biggest casualty was a generation of engineering practice. The questions before an incredulous engineering profession were troubling: how could a span designed to withstand 161-km/hr winds and a static horizontal wind pressure of 146 kg/m^2 succumb under a wind of less than half that velocity imposing a static force one-sixth the design limit? Just before the collapse lateral sway was just 61 cm, yet the bridge could be safely deflected up to 6.1 m out of alignment. More fundamentally, how could horizontal wind forces be translated into dynamic vertical and torsional motion? A baffled Moisseiff was quoted to be "completely at a loss to explain the collapse" (Robson, 1994). The disaster had exposed a profound ignorance of the dynamic effects of wind on bridges. For decades, engineers had successfully accounted for traffic and static wind loads. Stiffness in torsion had received scant consideration, and only then for transient asymmetric traffic loads that would be countered by deck and cable stiffness. It now appeared that the dynamic action of wind was a fundamental yet unknown and enigmatic phenomenon influencing suspension bridge behavior.

The collapse was remarkably well-documented on motion picture film by Farquharson, other cameras set up on shore, and the original installation atop the span's toll booth. This spectacular footage would prove of incalculable scientific value to subsequent investigations. The first effort to discern the cause of the failure was initiated by the PWA, much of whose $2.9 million investment now lay at the bottom of the Tacoma Narrows. The Board of Engineers (known as the Carmody Board, after PWA head John M. Carmody) established in the wake of the collapse included Othmar Ammann—who in 1939 at age 60 had retired from the Triborough Bridge Authority to set up his own firm—and Glenn B. Woodruff, Engineer of Design for the Transbay. Its third member was Theodore von Kármán (1881-1963), director of the Guggenheim Aeronautical Laboratory at the California Institute of Technology in Pasadena. A Hungarian with a doctorate from the University of

Göttingen in Berlin, Germany, von Kármán was a leading airfoil aerodynamicist and had pioneered the study of wake turbulence since 1911. He took on the California Institute of Technology post in 1929 and would spend the rest of a brilliant career in the United States.

The colorful von Kármán was the subject of an autobiography (co-written with Lee Edson) published posthumously in 1967. In it, he recalled his interest being piqued on hearing a news story the day after the collapse that quoted the state governor affirming the soundness of the original bridge and suggesting a similar replacement. That night, von Kármán took home a small jury-rigged model of the bridge. By adjusting the speed of a household electric fan, he induced severe oscillation, confirming his suspicions that vortex shedding—a phenomenon studied as early as 1908—might be a factor in the collapse. Von Kármán wired the governor that night and wrote both Farquharson and *Engineering News Record.* His letter to *ENR*, published 2 weeks after the collapse, offered a phenomenological and mathematical explanation for it. Von Kármán described how the lateral oscillation of a truss (which he appears to have used generically in reference to the Tacoma Narrows Bridge) under static wind load would induce a slight twist, exposing it to lift forces concentrated midway between the centerline and leading edge of the cross-section. The stiffness of the truss and main cables would resist these lift forces to restore the truss to equilibrium. Although the restoring forces varied directly with the angle of twist, the "aerodynamic moment increased with the square of the wind velocity. Hence, there must exist a critical value of the wind velocity, for which the structure becomes unstable" (von Kármán, 1940). In other words, the restoring forces could exacerbate the twisting. Von Kármán was in fact describing flutter, a major concern in aircraft wing design (this phenomenon is described more fully in Chapter 5). For a flexible suspended structure in which the main cables provide the primary resistance to twisting, von Kármán showed how to predict wind speeds that marked the onset of torsional oscillation. His estimate for the Tacoma Narrows Bridge, although assuming a waveform without a midspan node, was just 16 km/hr higher than the wind that brought it down. A month later, he found himself on the Carmody Board.

The Toll Bridge Authority, meanwhile, was faced with more immediate concerns. It had to determine what parts of the bridge, if any, could be salvaged, and then negotiate a settlement with the span's 22 insurers. The insurers had established a Narrows Bridge Loss Committee to determine the extent of loss; the Loss Committee in turn retained Clifford Paine, Joseph Strauss's second-in-command on the Golden Gate Bridge, to establish an engineering rationale for a settlement.

The bridge had been insured for $5.2 million, less than the $6.6 million it cost to build. Rumors had circulated before the collapse, suggesting that the Toll Bridge Authority intended to reinsure the bridge in mid-November at a lower value to reduce premiums, but these would now remain conjecture. In late November, the Toll Bridge Authority appointed its own three-man board to establish an insured value. It included Francis Donaldson, of Grand Coulee Dam fame; L. J. Sverdrup, a noted St. Louis-based engineer; and Russell G. Cone, the engineer responsible for the operation of the newly completed Golden Gate Bridge.

Steinman was conspicuous by his absence from the formal investigations, despite his three recent plate girder spans and an interest in aeronautics—he had taught

America's first university course on the subject at City College in New York during World War I. Some have speculated his omission from the Carmody Board was a result of Ammann's influence. No other engineer—save Moisseiff—approached these two in experience, particularly in plate girder suspension bridges in which both had wrestled with harmonic oscillation. An intense, bitter, emotionally charged, and highly public debate between Ammann and Steinman would emerge over design remedies. Yet the collapse provided only one chapter in the history of wind-induced oscillation in suspension bridges, for the autopsy following the failure of the Tacoma Narrows Bridge uncovered hauntingly similar—but forgotten—reminders from previous eras.

4.3 REMINDERS OF HISTORY

Farquharson (1949) would write that, "the collapse . . . came as such a shock to the engineering profession that it is surprising to most to learn that failure under the action of wind was not without precedent in the history of suspension bridges." In fact, several narrow 19th century spans had succumbed to torsional oscillation in ways that bore more than a passing resemblance to the Tacoma Narrows Bridge; these failures now acquired great currency. J.K. Finch of Columbia University first summarized them in an *ENR* article 4 months after the collapse. In 1941, A.A. Jakkula of the Texas Agricultural and Mechanical College published a comprehensive bibliography of suspension bridges up to 1940; begun before the collapse, this work would now provide an unexpectedly rich source of information.

Early British engineers had largely ignored deck stiffening in favor of improved eyebar chain cables. The 1818 collapse of the Dryburgh Abbey involved a form of unrecorded oscillation that wrenched the inclined chain stays from the deck; rudimentary handrails had offered little protection. The 1834 collapse of Germany's Nassau Bridge from chain breakage was thought to have emanated from stresses imposed by the oscillation of its unstiffened deck. But the loss of Samuel Brown's Brighton Chain Pier in 1836 would provide the first clear evidence of resonance. A storm had set the seaward three spans into gentle motion, but the third span generated the most interest. Lt. Col. William Reid witnessed the collapse, describing (and illustrating) it in great detail:

> For a considerable time the undulations of all the spans seemed nearly equal. The gale became a storm about eleven o'clock in the forenoon, and by noon it blew very hard. . . . soon after mid-day the lateral oscillations of the third span increased to a degree to make it doubtful whether the work could withstand the storm; and soon afterwards the oscillating motion across the roadway seemed to the eye to be lost in the undulating one, which in the third span was much greater than in the other three; . . . but there was also an oscillating motion of the great chains across the work, though the one seemed to destroy the other, as they did not both (at least as far as could be seen) take place in a marked manner at the same time . . .
>
> At last the railing on the east side was seen to be breaking away, falling into the sea; and immediately the undulations increased; and when the railing on

> this side was nearly all gone, the undulations were quite as great as presented in the drawing (Lt. Col. W. Reid, quoted in Russell, 1841).

The partial 1838 collapse of Brown's rebuilt Montrose Bridge (just 8 years after its notorious 1830 failure under a load of spectators) generated reports of oscillation that split the main span into two segments before much of it went down. James Rendel, who handled the bridge's second reconstruction, differentiated between longitudinal motion and transverse sway, noteworthy observations for the time. Brown could now lay claim to the dubious distinction of having two of his bridges heavily damaged by wind-induced oscillation.

Thomas Telford's masterpiece across the Menai Strait provided the most compelling evidence of aerodynamic instability. The storm of March 1826, barely a month after the span's opening, left no doubts as to the phenomenon at work:

> The motion which had been anticipated was that of simple undulation flowing in waves at right angles to the length of the bridge; . . .
> The movement of the undulating wave, however, was oblique . . . with the general direction of the bridge. In other words, when the summit of the wave was at a given point on the windward side of the bridge, it was not opposite this point on the leeward side, but in relation to the flow of the wave, considerably behind it . . .
> The motion was observed to be greatest about half way between the pyramids [towers] and the center of the bridge. The wave increased in its progress from the pyramid till it attained its maximum altitude near the first quarter, and at the same instant the extreme depression was near the third quarter. The wave then gradually diminished to the center of the bridge and afterwards increased to the third quarter, when it attained its greatest height at the same time that the first quarter was most depressed. The platform and main chains were equally subjected to this undulatory motion (W.A. Provis, 1842, quoted in Finch, 1941).

The similarities between this motion and that which occurred in the Tacoma Narrows Bridge more than a century later are extraordinary. There is no doubt that the Menai was wracked by torsional motion, its main span split by a midspan node in an oscillation similar to that which gripped the Tacoma Narrows Bridge in its final hour. Minor damage inflicted on the Menai Bridge a decade later during a violent storm caused the deck to oscillate with even greater double amplitudes estimated by the span's gatekeeper to approach 4.9 m. There was no direct reference to a recurrence of torsional motion, but the waves peaked at the span's quarter-points; such observers as Yves Rocard (1957) and Allan Larsen (1999) concluded it did occur. The storm of 1839 also inflicted heavy damage; significantly, the quarter-points of the main span displayed evidence of severe abuse, suggesting similarities to the torsional motion of 1826. No fewer than four European suspension bridges (three of these in Great Britain) were destroyed or severely damaged in strong winds in the 1830s; most succumbed to some form of harmonic oscillation. The 1840s was relatively quiescent, but in 1852, France's Roche-Bernard Bridge met its demise in winds of hurricane force that tore the roadway from its suspenders. Evidence of harmonic oscillation, however, could not be distilled from accounts of the collapse.

Then came the spectacular 1854 collapse of Ellet's Wheeling Bridge. The final minutes of this 308-m span—then the world's longest—in a May 1854 hurricane were recorded in meticulous detail in the *Wheeling Intelligencer*:

> For a few minutes we watched it [the deck] with breathless anxiety, lunging, like a ship in the storm; at one time it rose to nearly the height of the towers, then fell, and twisted and writhed, and was dashed almost bottom upward. At last there seemed to be a determined twist along the entire span, almost one-half of the flooring being nearly reversed, and down went the immense structure from its dizzy height to the stream below, with an appalling crash and roar . . .
> The great body of the flooring and the suspenders . . . was swayed to and fro, like the motion of a pendulum. The cables on the south side were finally blown off the apex of the eastern tower, retaining their position on the opposite side of the river. This destroyed the equilibrium of the swinging body; and each vibration giving it increased momentum, the cables, which sustained the whole structure, were unable to resist a force operating on them in so many different directions, and were literally twisted and wrenched from their fastenings (Podolny and Scalzi, 1986).

The similarity of this description to the last moments of the Tacoma Narrows Bridge is remarkable, and the story's raconteur inadvertently provided a penetrating insight into the self-exciting mechanism of the failure. Steinman and Sara Ruth Watson (1957) would write that the newsman "wrote better than he knew" when stating that an engineering solution to the problem awaited "further developments." The 1864 collapse of Edward Serrell's 317-m span Lewiston and Queenston Suspension Bridge over the Niagara 10 years later (abetted—ironically—by the temporary removal of storm guys) had, apparently, been preceded by oscillation of unrecorded form. Then came the 1889 failure of L.L. Buck's upgrade to Samuel Keefer's 386-m pedestrian span, again at Niagara, marking the third collapse of an American span exceeding 300 m. There were no witnesses to the early morning failure, but the last person across the span around midnight stated that the deck rose and fell 6.1 m and that, "The bridge rocked like a boat in a heavy sea and at times it seemed to tip up almost on its very edge" (Finch, 1941). By daybreak, the span had been destroyed by winds of up to 120 km/hr.

Several engineers possessed at least a rudimentary understanding of the problem. As early as 1823, Marc Seguin described British engineer Marc Brunel (father of I.K. Brunel) as recognizing the need "to prevent the bridges from entering into vibration [during windstorms] and to oppose the cumulation of motion that follows" (Kranakis, 1997). John Scott Russell, in an 1839 discourse to the Royal Scottish Society of Arts,[7] examined the Brighton Chain Pier collapse and suspension bridge vibrations in general. He noted the presence of nodes in the wave pattern and proposed the attachment of stays to the deck between the nodes to damp out harmonic motion. John Roebling wrote in his report on the Niagara Suspension Bridge that, "although the weight of a floor is a very essential element of resistance to high winds it should not be left to itself to work its own destruction. Weight should be, simply, an attending element to a still more important condition viz: stiffness" (Billington, 1979).

Many of these suspension bridge failures had left ample evidence of harmonic motion. Roebling more than any other of his era appreciated the nature of the phenomenon—if not scientifically but intuitively—and reinforced his spans with trusses and tower stays. The introduction of the Elastic Theory in the late 19th century emphasized deep trusses to support the heavier and faster-moving live loads of trolleys and subways; stays also disappeared because of calculation difficulties associated with their indeterminate stresses. Such stronger and deeper trusses, unconsciously, solved the aerodynamic stability problem, but the renaissance in bridge engineering wrought by the Deflection Theory in even longer and lighter road-only spans saw a return to flexible suspension bridges. Twentieth century engineers had no direct memory of aerodynamic failure; the hard-won lessons and intuition of previous generations were largely lost.

4.4 BACK TO THE FUTURE

Not all bridge engineers had ignored the dynamics of wind. In a series of eight articles published in *ENR* during 1934 and 1935, W.W. Pagon applied aerodynamic research and testing techniques from the aircraft industry to structural engineering design; shape, he stressed, was important. Farquharson (1949) would later describe this work as "must reading." Yet Pagon's work had little impact, in spite of increasing evidence of aerodynamic instability. Nor was the phenomenon just confined to North America. Norway, a prolific builder of thin, narrow, flexible suspension bridges, saw its manifestations, too. That nation's last truss-stiffened span was completed in 1923; between 1927 and 1937, 22 spans were built using rolled I-beams—first introduced in 1913. Many were suspended by below-deck cross-beams, with little longitudinal stiffening. It emerged (via Steinman almost a decade after the Tacoma Narrows Bridge collapse) that several spans ranging from 70 to 230 m displayed problems predating any experienced in North America. On the 150-m span Framnes Bridge, storm-induced motion snapped several suspenders and buckled a stiffening girder at quarter-span before the pouring of the concrete floor, although the bridge behaved after opening in 1931. The Vrengen Bridge, a thinner and more flexible span of 168 m opened in 1932, was prone to mild vertical oscillation split into two segments that broke two center span suspenders in 1934. Mild oscillations continued afterward.

The most severe problems were reserved for Norway's longest span, the Fyksesund Bridge. This slender, flexible two-lane span of 230 m was designed by Arne Selberg (1910-1988) and opened in 1937. Stiffening from rolled I-beam girders just 45-cm deep created a startling depth-to-span ratio of 1:511—much thinner than the Tacoma Narrows Bridge—although its width-to-span ratio was about twice as conservative and the side spans were unsuspended. On two occasions soon after opening, vertical oscillations gripped the bridge, splitting the main span into two to four segments with waves with a double amplitude of up to 160 cm. Although the girder suffered no damage, longitudinal deck motion incited by two-segment waves did damage the towers. Sectional model tests in 1944 at the Norwegian Institute of Technology (NTH) replicated the oscillations; main span stays linking the girders to the tower legs below deck level were installed in 1945 after delays imposed by World War II and appeared to stem further motion. The continuous concrete deck slab common in Norwegian spans was also thought to have prevented torsional motion.

The Fyksesund Bridge episode spurred Selberg to study suspension bridge aerodynamics, and his work over the next 2 decades would see him emerge as a leading authority in the field. French engineers, too, had to grapple with aerodynamic instability. In 1935, the 275-m span Aramon Bridge over the Rhône had oscillated in strong winds during lifting of its slender 2-m deep suspended structure; connection of the main cables to the deck at midspan appeared to stem further motion (Carriere *et al*, 1966).

In North America, Steinman cited rumors of vertical waves of up to 1.8-m double amplitude in the George Washington Bridge; by coincidence, it shared the 1:350 depth-to-span ratio of the Tacoma Narrows Bridge. Limited vertical and torsional motion had occurred (Board of Engineers, 1941) but in Ammann's (1953) words was "entirely harmless." Its suspended structure, on a per-unit-length-of-span basis, was more than five times as heavy as the Tacoma Narrows Bridge and twice as wide in proportion to span; its wide central gap, as yet unoccupied by roadway, allowed wind passage. Moreover, its use of cross-girders and stringers in place of a stiffening truss or plate girder produced a very shallow profile that reduced wind resistance. Ammann (1953) would later "confess that [at the time of the George Washington's design] there was no doubt in my mind that a span of 3,500 feet [1067 m] with an ultimate suspended dead weight of . . . 15 to 20 times the maximum wind force which could possibly be exerted vertically on its floor structure, could not be moved by wind."

Even truss-stiffened spans were not immune; the Golden Gate Bridge's shallow truss and the absence of lower lateral bracing soon suggested inadequacies. A February 1938 storm with gusts of up to 121 km/hr pushed the newly opened span up to 3-m off the horizontal; vertical waves with a frequency of 20 to 30 per minute and up to 61-cm double amplitude were also observed. The bridge emerged unscathed, but in February 1941, gusts of up to 96 km/hr from the southwest induced vertical motion of 61-cm double amplitude with a frequency of more than seven per minute. Despite the new-found concern for aerodynamic instability, no corrective measures were considered, perhaps because the bridge again suffered no damage. War-delayed sensors added in 1945 and 1946 augmented instrumentation installed in early 1942 to measure wind and seismic motion as a part of later investigations into suspension bridge aerodynamics.

Steinman's 1938 paper to the ASCE describing remedies for the Thousand Islands and Deer Isle bridges had by now acquired greater relevance. He wrote that he had contacted Moisseiff to render assistance at the Tacoma Narrows Bridge but was rebuffed. Later, Steinman took exception to the type of midspan cable stays used at the Tacoma Narrows:

> They did not know that the wire-rope mid-span stays on the Thousand Islands Bridge were only an emergency, a temporary installation that we replaced a few months later by the permanent stays made of rigid structural-steel angle

> members. Even so, the mid-span stays at Tacoma, copied from me, although inadequate, were the only thing that kept the bridge from going into destructive oscillations during the four months of its life . . .
> If they had let me help them, I could have saved the bridge, as I have saved other bridges. I could have made the Tacoma span safe for a very small expenditure. But my volunteer offer went begging (Ratigan, 1959).

It was speculation as to whether Steinman's remedies would have prevented Gertie's demise. Nonetheless, he had reduced oscillation in his own plate girder spans, and his firm earned a patent on the diagonal cable stays in 1940. Yet the Deer Isle Bridge would soon suggest that remedial measures were no substitute for inherent stability. Steinman had augmented its stay system with "transverse diagonal stays" extending from one side of the girder to the opposite main cable. These helped to limit oscillations to 20 cm. Then, in December 1942, high winds thrashed the span with vertical waves peaking at a horrific double amplitude of 7.6-m. No torsional motion developed, and miraculously the span survived relatively intact. Steinman immediately set out to strengthen the stay system, but no sooner had work begun when a storm of even greater intensity touched off vertical oscillations of such force that many stays were ripped out. The bridge was seriously damaged, but the stay system likely saved it. Steinman heaped particular praise upon the transverse stays and later suggested they could have helped to save the Tacoma Narrows Bridge. The system was subsequently buttressed with the aid of Roebling engineers into a profuse array of stays, braces, and cross stays. As it turns out, the Deer Isle Bridge would still not be fully tamed.

The persistence of motion in Ammann's much larger and heavier Bronx-Whitestone Bridge (now the longest extant plate girder span), however small, prompted further study—at the time not so much to safeguard the bridge but to mollify public concern. Targets strung out across the deck from the towers to the quarter-points of the main span in early 1940 allowed measurements to establish the forms of oscillation affecting the bridge; they revealed vertical motion of up to 35-cm double amplitude. A 1/100 model of the bridge was also built at Princeton University to help evaluate the effectiveness of the friction brakes and center cable ties and to aid in the development of further restraining devices should they be required. E.K. Timby—who in the late 1920s had assisted Steinman in model studies of the Mount Hope Bridge—devised a model that, like Farquharson's, involved the release of loads along the sides of the plate girder in various combinations to simulate the effect of wind forces. In spite of the crude method and the use of a single cable to reduce costs and complexity, Timby duplicated observed motions of the bridge with remarkable accuracy. Stronger friction brakes at the towers and a second set of cable ties were recommended; their installation in the summer of 1940 apparently restricted vertical motion to a few centimeters. Motorists, too, appeared happy, but Timby, like Steinman, sought additional restraint in the form of wire stays. Timby differed in reversing their direction, and settled on single stays radiating from the tower tops to the sides of the plate girder over the main and side spans. By December 1940, less than a month after the Tacoma Narrows Bridge collapse (and perhaps because of it), the tower stays were being installed.

Two back-to-back articles appeared in a December 1940 issue of *ENR* describing Ammann's stays in the Bronx-Whitestone Bridge and those used by Steinman in the Thousand Islands and Deer Isle bridges (Anonymous, 1940a; 1940b). Ammann asserted comparisons between the Bronx-Whitestone and Tacoma Narrows bridges inappropriate; the suspended structure of the Bronx-Whitestone Bridge, he noted, was more than twice as heavy, almost twice as wide, and four times as rigid as that of the Tacoma Narrows Bridge. Moreover, its lateral bracing was 10 times as rigid, and its span was 152-m shorter. Yet the real issue was the contrasting approach to the use of stays, sparking an acrimonious debate between Ammann and Steinman conducted entirely through the engineering press; not once did they face one another. Steinman's exclusion from the investigations made the engineering press his medium, and he would be prolific in its use. He fired the first salvo, arguing that tower stays would have little effect in damping motion in the Bronx-Whitestone Bridge; in fact, he suggested the opposite:

> This system of stays helps to bring main span, towers, and side-spans into unified harmonic motion, instead of checking such motion . . .
> . . . [and] directly helps to bring the torsional flexibility of the towers into play with any tendency of the two sides of the span to get out of phase in their undulations. The stays thus help to build up a twisting undulation, instead of checking it (Steinman, 1941a).

Steinman also speculated that seasonal temperature changes would alternately over- and understress the tower stays, necessitating "continual attention and adjustment." If left slack, he wrote, they would be useless. Worse, he argued, their attachment to the points of maximum inflection in the side spans would, rather than restrain motion, allow them to participate in it and bring main and side spans and towers into a closer—and potentially mutually reinforcing—resonance. He argued that his stays would inflict less distortion by distributing tensile forces over several points along the cables and plate girder. *ENR* submitted these claims to Ammann before publishing Steinman's letter. In an acrid response published with Steinman's letter in late February 1941, Ammann (1941) wrote:

> It must be clear that such general unqualified assertions as are made in the above letter are valueless unless they are substantiated by most careful analysis, . . .
> Some time ago Dr. Steinman made unsuccessful attempts to sell to the Triborough Bridge Authority his services and the use of his patented stay ropes which he endeavors to advertise as being superior to anything else. This may explain the motives behind his criticism of the Whitestone Bridge installation.

Ammann asserted that temperature-induced stress was not a concern. With respect to Steinman's concerns over harmonic motion, he wrote of "such a complex problem that no one, not even the most learned physicist, could make a reliable analysis without experimental investigation. Dr. Steinman's medley of arguments is pure guesswork expressed in impressive sounding scientific words." Ammann cited model tests and the experience with the Bronx-Whitestone Bridge itself to counter Steinman's claims. He wrote that calculations had revealed Steinman's cable stays would not have decreased undulations appreciably; such stays had been discarded even before the Princeton model tests began. He then suggested that Steinman's stays would impose stresses on the cable bands from variations in temperature, wind, and live load. "It would be of interest to know," he conjectured, "whether the Steinman patented stays applied to the two Thousand Island Bridges and in the Deer Isle Bridge are under all conditions safe against cable slipping." Ammann quoted calculations from the Bronx-Whitestone Bridge suggesting stresses up to three times the safe level; he believed the stay connections would require "constant watching."

Steinman (1941b) renewed the debate in May 1941. He described his initial letter as "a sincere invitation for open discussion of the subject by the profession." His crucial concern, the potential of Ammann's tower stays to "bring main span, towers, and side spans into unified harmonic motion, instead of checking such motion," should, he wrote, be "obvious to any suspension bridge engineer, even without the aid of laboratory models." He interpreted Ammann's early rejection of cable stays as an "admission" that they were not studied adequately, and he suggested that "by his [Ammann's] own criteria of safe design, the mid-span stay connections on the Whitestone Bridge call for review or for constant watching." Slippage of the midspan cable band, Steinman reminded readers, was now recognized as a contributing factor in the Tacoma Narrows Bridge collapse. He concluded by saying that, "I am not advertising any system of stays. I sincerely hope there never again will be a need for them. I am merely recording a plea for more candor, more professional courtesy, and more open-minded consideration among bridge engineers."

Studies would reveal a reduced incidence of oscillation on the Bronx-Whitestone Bridge but not in its magnitude. Motion of up to 76-cm double amplitude was largely confined to winds between 40 and 65 km/hr, whereas winds twice that speed (peaking at 130 km/hr in September 1944) generated vertical waves half the maximum experienced amplitudes. The need to increase the span's stiffness was now considered serious enough to warrant radical surgery. The initial scheme was to clip a 7.6-m deep truss to the inside of the plate girder. This proposal was complex, and it was instead decided to add a 4.3-m deep truss to the top of the 3.3-m deep plate girder to form a 7.6-m stiffening "truss" with a very deep bottom chord—the existing plate girder. The depth-to-span ratio of the Bronx-Whitestone Bridge would increase from a slender 1:209 to a conservative 1:92. Its sidewalks would also be lost to two more traffic lanes. E. Lionel Pavlo, who had worked for Steinman and Moisseiff in the 1930s and was now with Madigan-Hyland—designers of the overhaul—would consider it to provide more a psychological benefit than fulfil a structural need.

By the time plans were completed in 1941, war had intervened, and work on the retrofit only began in early 1946. It was an exacting job; truss fabrication and erection required great precision, as the exact profile of the plate girder, including irregularities, had to be matched. The $1.3 million system endured its first test in March 1947 when winds of up to 65 km/hr struck the bridge at an angle that previously would have triggered maximum motion; vertical movement exceeded no more than 23 cm. Although an apparent technical success, the transformation—in spite of a curving of the top chord to follow the arc of the main cables over midspan and Ammann's assertions (1946) that the truss would "not be sufficiently conspicuous to mar the [bridge's] graceful appearance"—greatly impaired the outline of what many regarded the most elegant suspension bridge of the era. Initial optimism notwithstanding, the Bronx-Whitestone Bridge would prove a difficult beast to subdue; 5 decades later, engineers still wrestled with her instability, as they would with Steinman's smaller plate girder spans.

Interestingly, German and French plate girder spans were generally not afflicted by instability. They were proportionately wider in relation to their span and featured low navigational clearances; this latter factor, because of surface shear of wind, made these spans less exposed to high winds.

4.5 THE AUTOPSY RESULTS

In late March 1941, Ammann, von Kármán, and Woodruff submitted their report. It was made public in early May, and was extensively covered in the same issue of *ENR* that carried Steinman's May 1941 rebuttal to Ammann's criticism of cable stays. The Carmody Board considered the crucial event in the collapse to be the sudden change from a vertical to a torsional mode of oscillation; this it traced to a slippage of the midspan cable band on the north cable. Once loosened, the band repeatedly scraped more than a meter of cable, breaking more than 350 wires. The sliding split the cable into two unequal segments; the extreme flexibility of the plate girder was such that it reflected this imbalance, and thus was borne torsional oscillation of a "first antisymmetric mode," in which the deck twisted in opposite directions on either side of a single node at midspan.[8] The concrete roadway yielded to the twisting as it increased in magnitude; this in turn overstressed the plate girders, whose buckling produced crushing impacts on the suspenders, one of which was wrenched out of its deck socket and flung upward. Adjacent suspenders were further overstressed and snapped one by one as the span progressively disintegrated into the Narrows.

Ironically, the midspan cable ties—designed to help prevent motion—rendered the Tacoma Narrows Bridge vulnerable. The Deer Isle Bridge had demonstrated insufficient tensioning would not adequately resist vertical motion. For some reason, the ties in the Tacoma Narrows Bridge were never fully tightened. The vertical motion on the day of the collapse did not, in the Carmody Board's estimation, threaten the bridge. Yet the motion alternately slackened one undertensioned cable tie while overstressing the other. One tie finally snapped, and the forces grew to exceed the frictional resistance of the cable band. The Carmody Board also recognized that the extreme flexibility of the plate girders in the Tacoma Narrows Bridge could not resist the torsional forces that, once begun, eventually destroyed it.

The Carmody Board concluded that the bridge could stabilize itself under vertical motion; its towers were also stronger in torsion than in bending along the axis of the span. This explained (Anonymous, 1941) why "torsional oscillations did not occur at lower velocities and why they were not caused by the random wind effects." However, the distortion of the bridge's wind truss, with a lateral moment of inertia 35 times that of the plate girders' vertically, indicated the tremendous energy absorbed by the structure. The Carmody Board determined that aerodynamic instability was not threatening so long as a span could damp the motion. Yet the flexibility of the Tacoma Narrows Bridge was such that its damping ability deteriorated in winds of about 65 km/hr and, according to tests by Farquharson, was 1/30 that of its contemporaries.

The Carmody Board asserted that in every other respect the bridge was well designed and built; all static forces imposed by wind and traffic were more than adequately provided for. When the last of the main span fell, the tower tops were thought to have been wrenched shoreward up to 8.2 m, causing 12 times the stress for which they were designed; 2-cm thick plates on their shoreward faces buckled, and rivets on their channel sides sheared. Twisting bent the upper portal struts, and the channel-side tower bases were lifted 13 mm. Moreover, the 18-m sag of the side spans exceeded their maximum design deflection by a factor of six. The Carmody Board (1941) remarked:

> That under these severe stresses the main carrying members, in particular the towers, did not actually fail attests to the excellence of their design, quality of materials and workmanship, and to the ample margin of safety which the structure would have had except for the severe dynamic motions.

Yet the Carmody Board concluded that designers erred in creating:

> an extraordinarily small width of structure as compared to its span. Their expectation that the bridge would have adequate lateral rigidity under wind pressure was unquestionably justified, but the combination of this narrow width with great vertical flexibility, proved to be responsible for the excessive torsion under aerodynamic action which led to the failure of the floor. This experience reveals the danger of excessive narrowness.

In fact, the Tacoma Narrows Bridge was several times more torsionally flexible than any other large span. The Carmody Board revealed a unique mix of circumstances, events, and inherent vulnerability that triggered Gertie's demise. It felt the center cable ties might have resisted torsional forces, but it stated:

> It must not be inferred . . . that the failure of the bridge would have been averted if the details of the center ties had been made stronger, or if center ties had not been installed. The evidence is conclusive that very high torsional forces existed and that their magnitude increased rapidly with the wind velocity. Even assuming that the bridge, without the slipping of the cable band, had weathered this particular storm, it is now apparent that it would have failed under wind velocities far below that of 100 m.p.h. [161 km/hr] for which it was designed.

As Steinman (1946a) later wrote, the ties "permitted—not caused—the catastrophic oscillations that wrecked the structure." Although the side span holddown cables substantially reduced motion over the side spans, they had little effect on the behavior of the main span. The hydraulic buffers at the towers were similarly felt to have been ineffective—in fact, they were practically useless. Seals on one buffer had been damaged during sandblasting of the bridge in preparation for its painting, allowing sand to enter the cylinders. Ironically, they had been considered a great improvement over those used in the Bronx-Whitestone Bridge. The Carmody Board (1941) concluded that the effect of low cable sag on vertical stiffness had been "greatly overestimated" and that the light plate girder had practically no stiffening effect on the main cables. Tower stays might have enhanced deck stiffness, the Carmody Board said, but more study was required. It also felt the long side spans of the Tacoma Narrows Bridge, as compared to the Bronx-Whitestone Bridge, had worked to its disadvantage and that fairings, although reducing static wind forces by a third, would have contributed little to vertical rigidity and might have even increased torsional instability.

The Carmody Board did feel that weight played a beneficial role, even if it might reduce structural damping (that is, the ability of the structural characteristics of a bridge to damp out oscillation). Ammann's unstiffened but heavy George Washington Bridge was estimated to deflect less than half—both vertically and torsionally—than the Golden Gate Bridge and significantly less than even the Transbay. To match the vertical rigidity of the Golden Gate and Bronx-Whitestone bridges, the suspended structure of the Tacoma Narrows Bridge would have to be two and a half times as heavy than it was. To achieve the Golden Gate Bridge's torsional rigidity, the bridge would have required a 16-m wide deck and a 7.3-m deep stiffening truss twice the weight of the former plate girder suspended structure. The Carmody Board also abruptly dismissed Steinman's cable stays by stating that his three smaller spans could "give no clue to the behavior of a suspension bridge 3½ times longer and 6 times heavier." In fact, Steinman's bridges were not even assessed by the Carmody Board. If anything, the Carmody Board's findings suggested that depth- and width-to-span ratios, although important, were insufficient to explain or predict a span's aerodynamic behavior. It concluded by stating:

> The excessive vertical and torsional oscillations [of the Tacoma Narrows Bridge] were made possible by the extraordinary degree of flexibility of the structure and of its relatively small capacity to absorb dynamic forces. It was not realized that the aerodynamic forces which had proven disastrous in the past to much lighter and shorter flexible suspension bridges would affect a structure of such a magnitude . . .

ENR captured the essence of the Carmody Board's findings by writing:

> The fundamental weakness of the Tacoma Narrows Bridge was its great flexibility, vertically and in torsion. The comparatively narrow width gave adequate lateral rigidity but combined with the extreme vertical flexibility, the narrowness made the structure extremely sensitive to torsional motions created by aerodynamic torsional forces . . .
>
> The resistance of a suspension bridge to vibrations from dynamic forces is furnished by its rigidity, inertia and structural damping. The Tacoma Narrows Bridge was deficient in all three respects . . . (Anonymous, 1941).

It was perhaps no surprise that Steinman, in an August 1941 letter to *ENR*, criticized the Carmody Board's findings. Although regarding its report a "valuable contribution on the subject of aerodynamic oscillations of bridges," he stated:[9]

> the body of the report appears to contain a number of questionable statements and conclusions . . .
> It goes out of its way to build up a case for bridges of greater width, using various comparisons in an attempt to support the pre-conceived conclusion that wider bridges offer greater rigidity against aerodynamic effects (Steinman, 1941c).

Steinman set out to prove mathematically that critical velocity and, by extension, rigidity, were not proportional to deck width. He wrote that the Carmody Board significantly overstated the vertical and torsional stiffness of the Bronx-Whitestone and Golden Gate bridges. Moreover, although the Carmody Board had shed no conclusive light on the tower/cable stay debate, Steinman was astonished that it felt the study, design, and installation of wire rope stays in the Tacoma Narrows Bridge would have required months. He wrote, "In the case of the Thousand Islands Bridge the emergency system of end-span stays was decided upon, ordered, and installed—all in *one week* [Steinman's italics] before the bridge was opened . . ." He then (1941c) insinuated that the designers:

> of the Whitestone and Tacoma bridges evidently profited by the prompt publication of timely information, for they copied the mid-span stays as described by me, but without a word or credit or acknowledgement. In fact they appeared to keep their duplicate installation a secret, for I knew nothing of it for two years until after the failure of the Tacoma span. The engineers of both bridges, moreover, made the mistake of copying our initial *temporary* installation, with wire ropes [Steinman's italics]. In our permanent installation, completed and installed at leisure five months later in January, 1939, we knew enough to make our mid-span stays of rigid members.

In the same letter, Steinman criticized the Carmody Board's dismissal of his plate girder spans by stating that:

> I submit that suspension bridges of 750, 800, and 1,080 ft. [229, 244, and 329 m] spans, exposed to actual wind conditions, can give better and more realistic information than small-scale models exposed to electro-magnets or to dropping weights. If the mid-span stays at Tacoma had been more adequately proportioned, as in our *permanent* [Steinman's italics] installation at Thousand Islands, and if they had been supplemented by the necessary end stays, I am confident that the Tacoma span could have been saved.

Steinman concluded by suggesting the Carmody Board did not go far enough:

> [The report] leaves many questions unanswered. It does not tell what

> combinations of cross-sections produce aerodynamic instability, how aerodynamic instability can be reasonably predicted or readily tested, nor how it can be prevented. There are ways of designing sections to avoid or eliminate aerodynamic instability, and that is perhaps better engineering than merely placing all of the emphasis on adding rigidity to provide increased resistance to a dangerous inherent characteristic. It is more scientific to eliminate the cause than to build up the structure to resist the effect.

This emotionalism was characteristic of Steinman and in a conservative profession was generally not well received.[10] Yet his last comment would prove more prophetic than he could have ever imagined, long after his decade-long pursuit of independent research into bridge aerodynamics.

4.6 THE CARMODY BOARD AND THE DYNAMIC ROLE OF WIND

Although the Carmody Board had described a sequence of events leading to the failure of the Tacoma Narrows Bridge, fundamental questions concerning the aerodynamics of suspension bridges remained unresolved. The Carmody Board had performed wind tunnel tests on sectional models and a 6-m full model of the Tacoma Narrows Bridge at the California Institute of Technology's Guggenheim Aeronautical Laboratory under Louis G. Dunn to investigate aerodynamic damping. W.D. Rannie developed preliminary equations of dynamic motion but he considered the problem to be "even more complex than those of the deflection theory for static loads" (Board of Engineers, 1941). The Carmody Board also concluded that amplitude was independent of wind speed and that a critical velocity at which torsional motion began depended on oscillation frequency and could be raised—notwithstanding Steinman's assertions to the contrary—by increasing deck width.

Yet the Carmody Board wrestled, with some dissension, over aerodynamic phenomena. Wind tunnel tests on sectional models of the Tacoma Narrows Bridge at the California Institute of Technology and of the Golden Gate Bridge at Stanford University were inconclusive in terms of vertical instability. However, before the Carmody Board's first meeting in Seattle, Washington, a model of the Tacoma Narrows Bridge tested in the California Institute of Technology wind tunnel at von Kármán's instigation had "vibrated violently as its oscillation synchronized with the frequency of the vortex shedding" (von Kármán and Edson, 1967). In the Carmody Board's 1941 report, von Kármán discusses the role of vortices. He wrote of the findings of model tests that suggested "The mechanism of [Tacoma's] torsional instability is probably connected with the formation of alternating eddies created by the incipient torsional motion . . ." (Board of Engineers, 1941). This statement was a prescient one, as later research would reveal, yet von Kármán failed to convince Ammann and Woodruff of any possible role of vortex shedding in the Tacoma Narrows Bridge collapse. The final report states that, "It is very improbable that resonance with alternating vortices plays an important role in the oscillations of suspension bridges."

After commenting on no strong relationship between wind speed and oscillation frequency (contradicting von Kármán's conclusion 10 pages earlier), the Carmody

Board expressed perhaps more than it realized when it wrote that, "It seems that it is more correct to say that the vortex formation and frequency is determined by the oscillation of the structure than that the oscillatory motion is induced by the vortex formation." Paradoxically, the Carmody Board concluded that, "There is . . . no convincing evidence that the vertical oscillations were caused by so-called aerodynamic instability," and instead said they could be "explained satisfactorily as forced vibration excited by random action of turbulent wind."

The Carmody Board suggested that intermittent gusts—or "natural turbulence"—struck the bridge at different angles (horizontally and vertically) and with changing velocities. This in turn imposed vertical forces of randomly varying amounts and directions, which were felt to have the same net effect as a large number of separate but superimposed periodic impulses. Interestingly, this was the phenomenon Farquharson had rejected 2 months before the collapse. The Carmody Board did not consider this action in itself to have been a precursor to Gertie's demise. However, the vertical forces were felt to incite periodic deflections whose frequency matched the natural period of the bridge. It was surmised that the vertical oscillations on November 7 roughly corresponded to one of the span's natural frequencies. Vertical deflections, the Carmody Board wrote, "will be amplified, especially if the structure is not sufficiently damped." The slippage of the midspan cable band appeared to explain the rest.

In a concession to von Kármán, the Carmody Board acknowledged that, "Further experiments and analytical studies are desirable to investigate the action of aerodynamic forces on suspension bridges." Wind tunnel tests would be conducted at the University of Washington and the California Institute of Technology before a new bridge was built. Nonetheless, in his 1967 autobiography, von Kármán was critical of his engineering colleagues, and in particular Ammann:

> I hadn't reckoned on the depth and long standing of the prejudices of the bridge engineers. Their thinking was still largely influenced by consideration of static forces, like weight and pressure which create no motion instead of dynamic forces, which produce motion or change of motion . . .
> Bridge engineers, excellent though they were, couldn't see how a science applied to a small unstable thing like an airplane wing could also be applied to a huge, solid, nonflying structure like a bridge.

Von Kármán (with Edson, 1967) would parallel the Tacoma Narrows Bridge's malaise to "the structural instability of a badly designed airplane wing under certain flying conditions." Blake Mills, a mechanical engineer at the Massachusetts Institute of Technology, had suggested a similar line of thinking to that of von Kármán's in two letters to *ENR* published in December 1940 and January 1941. He suggested galloping and vortex shedding could evolve into flutter, concerns he felt had been neglected in suspension bridge design.[11]

The Carmody Board's conclusions—and to a lesser extent von Kármán's—would eventually be proven off the mark. If anything, the Carmody Board did, perhaps unintentionally, suggest that a number of wind mechanisms could have been responsible for inciting motion in the Tacoma Narrows Bridge. An understanding of aerodynamic instability in suspension bridges would emerge only after subsequent research that largely vindicated von Kármán. Yet regardless of the disagreements following the release of the Carmody Board report, change was in the wind. Farquharson wrote in July 1941 that the failure of the Tacoma Narrows Bridge "brought to light a disturbing and dangerous tendency in suspension bridge design that has been developing slowly for 25 years without having been properly appraised and recognized." Steinman (1945c) would write:

> improvements in analysis, in economy, and in aesthetics—all combined to accelerate the trend toward increased slenderness of proportions. Engineers began to forget or to discount the principles that Roebling's genius had evolved and demonstrated. With more mathematics than the pioneer master builder had available in his time, but with less native intuition and constructive perception, a later generation of designers carried the reverse trend to a climax until they were awakened by the impact of catastrophe—the destruction of the Tacoma Narrows span.

More than half a century later, engineering historian Henry Petroski (1994) suggested engineers had forgotten to use "avoidance of failure as a guiding principle of design"—an approach Roebling had used with success. Engineers instead extrapolated the successes of the Tacoma Narrows Bridge's immediate predecessors to new and uncharted territory, with unexpected and shocking consequences.

Moisseiff was not vilified, as Thomas Bouch had been 6 decades before in the terrible collapse of Scotland's Tay Bridge. Steinman and Watson (1957) would write that:

> "The span failure is not to be blamed on him; the entire profession shares in the responsibility. It is simply that the profession had neglected to combine, and apply in time, the knowledge of aerodynamics and of dynamic vibrations with its rapidly advancing knowledge of structural design."

Much of the subsequent 6 decades of suspension bridge design has been dedicated to this application. The dynamic interaction between moving air and structures is a fickle phenomenon, and an understanding of its various guises is still incomplete. Wind is turbulent and approaches from different angles and aspects; these characteristics—along with the largely unstreamlined cross-sections wind strikes—set bridges apart from airfoils and complicate theory and testing. Nevertheless, engineers have, over the decades, sought to make their bridges torsionally stiff, aerodynamically streamlined, or both. The Tacoma Narrows Bridge had the benefits of neither; it has been speculated that had it possessed one of these properties, it likely would not have gone down (Scanlan, 1979). Other bridges were blessed with an unintended combination of aerodynamic profile, stiffness, and damping properties to inhibit aerodynamic instability. Some suspension bridges that might have been aerodynamically suspect avoided extreme wind exposure—in magnitude or direction—because of the peculiarities of their location and, in some cases, low navigation clearances. Yet even had the Tacoma Narrows Bridge not collapsed, its behavior had caused a crisis of confidence and an irreversible change in the future course of suspension bridge engineering.

ENDNOTES

1. The firm of Moran, Proctor, & Freeman, famous for its work at the Transbay Bridge, had redesigned the foundations originally specified by state engineers.
2. The grant was soon increased to almost $2.9 million and the loan to just more than $3.5 million, bringing the total cost to $6.4 million.
3. Tremie concreting refers to the placement of concrete underwater, which eventually displaces the water.
4. Canada's Isle d'Orléans (1935) and Lions Gate (1938) bridges and Robinson & Steinman's Wabash River Bridge (1939) were the prior applications.
5. This figure had been obtained during the November 1939 earthquake just prior to the towers' completion.
6. Jacob P. Den Hartog provided the first mathematical description of the phenomenon in 1930.
7. Russell's 1839 discourse, entitled "On the Vibration of Suspension Bridges and other Structures; and the Means of Preventing Injury from this Cause" was published in 1841 in the *Transactions* of the Royal Scottish Society of Arts and reprinted in the Carmody Board report.
8. The first symmetric mode describes an oscillation of one-half amplitude, with equal magnitude and direction on either side of midspan (which experiences a vertical displacement). The first antisymmetric mode features a single full wave form of double amplitude, split into equal but mirror images by a midspan node at rest. Second, third, and subsequent symmetric and antisymmetric modes are characterized by an increasing number of vibrating segments separated by nodes across the span. The relatively benign vertical oscillation of the Tacoma Narrows Bridge featured multiple modes. In antisymmetric modes, the tower tops will be pulled in the same direction; in symmetric modes, the tower tops move in opposite directions.
9. Steinman only highlighted the contributions of W.D. Rannie and Louis Dunn of the Guggenheim Aeronautical Laboratory of the California Institute of Technology as having particular value.
10. Steinman's promotional flair included a regular mailing of press clippings that referenced him or his work to an audience of 600 (Rubin, 1985).
11. William Addis (1994) reports the discovery of a letter received by Farquharson from Mills less than 2 weeks before the collapse noting the parallels between galloping transmission wires and the motion of the Tacoma Narrows Bridge. In a reply penned just 4 days before the collapse, Farquharson professed some ignorance of recent work into the phenomenon, which is somewhat surprising given the prior study of autorotation (galloping) with static sectional models in the University of Washington wind tunnel.

Chapter 5

A DIFFICULT REBIRTH

5.1 INSURANCE BATTLES, REPLACING GERTIE, THE DISMANTLING

The Carmody Board's exoneration of the Toll Bridge Authority in the Tacoma Narrows Bridge collapse provided fleeting solace, for after a reinstatement of the former ferry service, the Authority was left with the problem of the tattered remains of the bridge. Whether any parts could be salvaged was not at all clear, yet reconstruction hinged largely on the ability to reuse the main cables and towers. Both had been manhandled. The cables had been wracked by vertical displacements of more than 30 m; where one cable had been wrenched out of the cable bent saddle, a broken light standard now pierced 8 cm into the cable. Wire breakage was also feared to lurk elsewhere than at the midspan's north cable tie. The shoreward deflection of the towers was a concern, too. To relieve the towers of high stresses, removal of the dangling side spans—a continuing danger with continued mild oscillations in moderate winds—had begun within a week of the collapse. The towers then returned to less than 91 cm off the vertical.

The year 1941 also brought a lawsuit from the superstructure contractor claiming $620,000 in unpaid bills. The Toll Bridge Authority had yet to receive a final portion of its PWA grant but had been advised by its insurers that payment could be construed as a carte blanche acceptance of the contractor's work, even if the contractor was in some way responsible for the collapse. The courts were unsympathetic to the plight of the Toll Bridge Authority and in April 1941, after the Carmody Board had absolved workmanship of any role in the bridge failure, they directed it to pay the full amount. The Toll Bridge Authority appealed, claiming the state government responsible for the shortfall, but the courts rejected this rather specious argument in June.

The Toll Bridge Authority was also pitted in a battle with the bridge's insurers, which had given it until March 7, 1941 to file claims. The Authority had become convinced of the need for a new bridge long before this deadline; its own board of inquiry had already painted a bleak picture in a February 1941 report. The towers, cables, and west side span cable bent were considered write-offs, as was some anchorage concrete, whose removal was required to facilitate the spinning of new cables. A limited amount of tower steel might be re-used, pending dismantling.

In the midst of this gloomy prognosis was the relatively intact condition of the tower piers. Their re-use would substantially reduce reconstruction costs, as would that of much of the anchorages. Leon Moisseiff's reduction in tower width had not altered the piers, and it was felt they could easily accommodate towers up to a third wider than the originals. Yet thought of a replacement bridge was largely speculation until tower conditions could be verified. Unfortunately, dismantling of the side spans proved a delicate operation and was not completed before the claim deadline. On March 1, the Toll Bridge Authority officially declared the bridge a total loss, save for tower piers and portions of the anchorages. Its claim of $5.2 million was 100% of the insured value but included dismantling costs, thought to be $525,000. Ironically, the salvage value of the $6.4 million bridge was estimated to be less than $275,000.

The insurance syndicate's response to the Toll Bridge Authority's claim was predictable. Based on work of its Narrows Bridge Loss Committee under Clifford Paine (who, beginning in February 1941 was assisted by others, including Holton Robinson, David Steinman's business partner)[1], the syndicate insisted the existing bridge could be restored for $1.8 million. It considered the towers and cables repairable, and a deep stiffening truss would replace the plate girders. Although the Toll Bridge Authority reached a $240,000 settlement with two of its 22 underwriters in April 1941, it became increasingly apparent that the courts would be called on to determine the value of the Toll Bridge Authority's claim.

Until the insurance issue was resolved, any bridge was fantasy. Fearing years of litigation, the mayor of Tacoma threatened to organize a new bridge authority. He intimated that three backers—insurance companies, incredibly enough—would invest the required $10 to $12 million. This scheme never materialized, but it may have galvanized the state government into action. In July 1941, it in effect prejudged the outcome of the insurance dispute by giving the Toll Bridge Authority $75,000 to initiate plans for a new bridge. By this time, Paine's committee had devised a rebuild of the existing bridge. Rehabilitated towers—including a replacement of their bottom six shoreside tiers and the top portal strut—and respliced cables would support a 7.6-m deep through truss with a 14.8-m wide floor truss. A 7.9-m wide two-lane roadway of open grating placed on the bottom lateral bracing would allow wind passage and make the truss no heavier—but 18 times stiffer—than its predecessor. The committee confirmed its earlier estimate of $1.8 million for the rebuild. Conversely in July, the Toll Bridge Authority's insurance board declared the towers a write-off. Then in August, as a September court date loomed, the insurance syndicate capitulated and offered the Toll Bridge Authority $4 million. The offer was accepted.[2]

Although substantial, the settlement still left the Toll Bridge Authority with a shortfall. Yet a new bridge was needed. Increased navy yard activity had generated three times the traffic carried by the ferries; this volume was 60% higher than predicted (which, ironically, might have funded a more substantial bridge in the first place). Now, pending assurances that the recent past would not be repeated, it was felt a bridge might appeal to investors. Design was initiated. The first issue: was it cheaper to use the existing piers or to build an entirely new bridge? Eight concepts were developed in the summer of 1941; one, a 488-m span cantilever, carried four lanes and a single rail line—the latter to serve the navy yard. Although this design offered rigidity, prodigious 55-m water depths at the piers and the extra steel required to support train loads rendered the $14 million cost unpalatable. In July 1941, Dexter R. Smith, formerly of the Oregon State Highway Department, was appointed design engineer under Charles Andrew. By October, it was decided to use the existing tower piers in an estimated $7 million span. In place of sinuous plate girders would be a deep, open truss.

That month, the Toll Bridge Authority applied for a permit from the U.S. Corps of

Army Engineers to allow construction. Glenn B. Woodruff and John Parcel of Sverdrup & Parcel were also retained that autumn to advise in the design's development. Yet the progress was illusory. By December, America was embroiled in World War II, and in April 1942, military priorities sounded the death knell of any replacement span. Material—especially steel—was required for the war effort. Financing a now $8 million bridge would be difficult. The war even restricted the availability of engineers and material with which to undertake aerodynamic research and the design of the new span, and both would proceed at a slow pace.

Ironically, material shortages now made remnants of the original bridge marketable commodities. The Toll Bridge Authority had planned to drop the towers and cables into the Narrows' deep waters; galvanizing was thought to render the cable wire unusable for other purposes and the cost of dismantling the towers to likely exceed their salvage value. Yet the military objected to the loss of almost 7,000 tonnes of steel; side span steel—removed in early 1941—had already found its way into the war effort. The Toll Bridge Authority then reversed itself; the first steel—6 tonnes of shock absorber and 7.25 tonnes of cable saddle—sold for less than $600 in January 1942. Some 5.2 km of suspender wire and 900 tonnes of other steel fetched almost $63,000. Yet a sole July 1942 bid to remove towers and cables would cost $450,000—more than twice their estimated salvage value. The job was rebid—successfully—and unwrapping of the main cables began in late August.

The war now had such a pervasive effect that new steel could not even be used to build the equipment required to dismantle the bridge. All materials and equipment were jury-rigged from scrap, much from the remains of the suspended structure. The removal of the main cables began on the second anniversary of the collapse, without the use of catwalks. Barely a month later, after more than half the wire had been removed, a fierce gale struck the bridge, whipping loosely connected cable wire more than 10 m out of alignment. Worse still, a slack strand in the south cable became entangled on the other side of its main cable for a distance exceeding 100 m. The strand resisted efforts to remove it, and the addition of oil as a lubricant only served to envelop the wires in a gooey paste. Friction was now so great that on occasion wires snapped. Three strands were even cut up and dropped into the Narrows before kerosene was discovered to dissolve the paste and overcome the friction. By then winter had set in, prolonging removal of the remaining wire a further 3 months. Dismantling of the towers was, by contrast, uneventful. Steel was lowered to barges by booms built almost entirely from salvaged bridge steel. By mid-May 1943, only the pier heads of the tower foundations remained in an exercise that cost the Toll Bridge Authority almost $650,000 to secure $300,000 of salvaged steel

5.2 A LITANY OF THEORIES—THE CABLE TRUSS AND OTHERS

The Carmody Board had dominated the immediate aftermath of the Tacoma Narrows Bridge collapse, but other engineers were also rethinking suspension bridge design. Various concepts emerged, some speculative, others reflecting a more scientific basis. One of the first came from American engineer C.A.P. Turner (1940), who suggested Steinman-like cable stays to pull the main cables into a circular, rather than parabolic, arc in a "restrained cable structure." With a "box stiffening girder" suspended structure fixed to the main cables at midspan, Turner claimed the concept would reduce bending by five-sixths and approach cantilevers in stiffness against live and wind loads. He also suggested splitting the deck and hanging each half from the outsides of the towers; beams spaced 9.1-m apart would connect the halves, suspend the span, and brace it against lateral wind loads. Both the box strut and the intriguing idea of split decks would re-emerge, but only many years later.

Another unusual proposal from F.D. Chapin (1941) involved reducing inherent natural frequencies through the use of asymmetrical spans and towers of unequal height and width. Chapin also surmised that had the Tacoma Narrows Bridge been loaded with heavy vehicles on the day of its collapse, its vibration period could have been changed and the structure saved. Steinman (1941d) would suggest much the same thing (except loading only the windward edge of the span) as an emergency measure just months later as part of a list of preventive, remedial, and emergency measures published by *ENR* in December 1941. Steinman's preventive measures included the avoidance of plate girders and/or flexible cross-sections in design, as well as transverse stays and horizontal fairings. Two other preventive measures identified by Steinman would find their way into his and other post-war spans—open grid steel floors and/or a gap between the roadway outer edge and the truss to allow wind passage. Remedial measures to retrofit existing plate girder spans included horizontal fairings, midspan braces, and diagonal and transverse stays. Other emergency measures required to quell an episode of instability included the draping of stabilizing horizontal plates in the water from the suspended structure, and the ripping up of sidewalks to balance air pressures. Steinman applied for patents on the gaps, fairings, and transverse diagonal stays; the latter he had first installed on the Deer Isle Bridge.

In the same letter, Steinman illustrated the outer edge roadway gap with a cross-sectional view of his 1930 Tacoma Narrows Bridge design. This measure, of course, had not been devised for the function he now suggested it would perform. Yet Steinman now stated that with this gap alone, his design for the Tacoma Narrows Bridge "would have been assured of aerodynamic stability." The same feature had been used in the Chirundi Bridge across the Zambezi River in Africa, and Ralph Freeman, the span's designer, now (1942) brought it to the attention of American engineers. Again it is doubtful that the full significance of this feature was known at the time it was conceived.

Another offering came from Australian engineer C.O. Harrison (1941), who recommended placing the road deck on top of the plate girders and stiffening them with transverse bracing and trusses between and below the lower girder flanges respectively. Harold Thomas of Pittsburgh's Carnegie Institute of Technology suggested in April 1941 that stiffening girders—even when streamlined—were more apt to form undesirable eddies than were open trusses. He argued that the contemplated addition of fairings at the Tacoma Narrows Bridge would have made little difference, only slightly increasing the speed at which vertical oscillation would have begun and even slightly decreasing the onset velocity for torsional vibration.

Many concepts would remain just that, but other avenues of research would prove unexpectedly fruitful. One discovery in particular would be significant. An article in the March 1941 issue of *Schwiezer Bauzeitung* by F. Stüssi and J. Ackeret of the Swiss Federal Institute of Technology in Zurich contained what F.B. Farquharson described as "interesting suggestions" concerning the cause of the Tacoma Narrows Bridge collapse (Farquharson, 1952a). Farquharson wrote Ackeret to learn more. Ackeret replied in November 1941, apologizing for the delay on account of the need for additional tests. The wait was well worth it. The oscillation of the Tacoma Narrows Bridge had been simulated in a small sectional model using water as the triggering medium. To quell the oscillations, an associate suggested a slot be placed down the middle of the deck. A slot placed between and equal in width to each of two flanking H-girder sections produced spectacular results. Farquharson described, "The effect of opening the slit is striking, and it is clear that, within the range of wind velocity covered, this device completely stabilized the section so far as torsional motion was concerned." He did caution that other factors, such as the model's low ratio of depth to width, contributed to the results. Although military demands prevented further Swiss research, Farquharson initiated tests with slotted decks combined with plate girders. The results were mixed, but the porous slots (which were also suggested in the Carmody Board's report) would emerge as a popular feature of aerodynamically stable trusses. The full significance of Stüssi and Ackeret's arrangement of parallel twin suspended structures for truly long spans would only be recognized some 35 years later, and may profoundly influence 21st century suspension bridge design.

One other development was the cable truss, a concept first discussed by J.A. Ostashevskii in 1940 in the Soviet engineering press. Longitudinally triangulated suspenders would support a stiffened or unstiffened deck, but in either case the structure acted as a rigid truss. J. Gollnow in Germany suggested a similar idea, but American engineers would bring the concept to reality. In 1941, George Maney of the Northwestern Technological Institute in Illinois suggested the superimposition of longitudinally inclined suspenders arranged in an X pattern over a traditional vertical suspension system. For dead loads, the system acted as a suspension bridge; under live loading, it acted as a truss. Maney (1941) described, "The cable, instead of acting as merely a flexible member from which a stiffening truss is hung, acts as a chord of a truss." A thin deck would function as the lower chord of the "truss", the main cables remaining high above the deck at midspan would serve as its upper chord, and the inclined suspender network would link and stiffen the two. Steinman's Florianopolis truss was similar, but the cable truss eliminated rigid stiffening members altogether.

Maney found that the grafting of a diagonal suspender network on a rudimentary model of the Tacoma Narrows Bridge completely damped out oscillation. Removal of the vertical suspenders had little effect, so the diagonal suspenders not only stiffened the deck but suspended it. Tests also suggested the concept's use in the Transbay Bridge in the form of a three-span continuous cable truss would have reduced deflections and avoided its expensive central anchorage. It was thus considered a potential substitute for multispan suspension bridges. Cable truss mathematics was relatively simple, and Maney considered the form competitive in cost with conventional suspension bridges as well.

Interest in the cable truss soon emerged. The U.S. Army was attracted by its light weight and saw potential for use in its Far East campaign. Studies of spans up to 180 m capable of carrying 80-tonne tanks began in April 1944 under William Grove, using a two-dimensional, 1.8-m long model of a 183-m span. These studies suggested the use of tie cables along the deck over the entire span, which were then incorporated into a 3-

m long three-dimensional model of a 61-m span. The U.S. Army also retained J. A. Roebling's Sons in June 1944 to build and study an 18.3-m long model of a 183-m span—including its dynamic response under wind—at the firm's Trenton, New Jersey, plant. The company was already exploring cable stiffening systems; the Tacoma Narrows Bridge collapse had encouraged it to build on the catwalk stiffening methods instituted in the George Washington and Golden Gate bridges. Its supply of wire for Steinman's Deer Isle Bridge—which now almost resembled a cable truss—had also sparked an interest in alternative suspension forms. In 1943, J.A. Roebling's Sons built a small prototype footbridge between two buildings at its Trenton plant. By May 1945, it had tested 50 different patterns of diagonals. The research confirmed Maney's conclusions, and set the stage for the design of a full-scale, single span vehicular bridge of 140 m at the U.S. Military Academy in West Point, New York, just north of the Bear Mountain Bridge. Construction contracts were awarded in the summer of 1945 but were terminated with the end of the war in August. Yet the breakthroughs—most notably by Charles C. Sunderland and Blair Birdsall of J.A. Roebling's Sons and Norman Sollenberger of Princeton University—would eventually pay dividends.

A related idea was the use of inclined suspenders in a traditional suspension bridge. Harrison had mentioned a variation of this concept used in bowstring arches in his September 1941 letter to *ENR*, but the subject was now treated comprehensively by staff from Modjeski & Masters in an October 1941 *ENR* article. Although untarnished by the Tacoma Narrows Bridge collapse, the firm was not unaffected. Four months before the collapse, it had, with Moisseiff, completed preliminary plans for tandem 1,402- and 869-m span plate girder suspension bridges across Michigan's Straits of Mackinac. Modjeski & Masters now declared that plate girders should not be rejected—yet. It pointed out that John Roebling's trusses were weaker than recent plate girder suspension spans and that those of the Golden Gate Bridge were not immune to harmonic oscillation. The authors (1941) stated that a:

> Lack of stiffness . . . is not confined to girder-stiffened structures. . . An arbitrary retreat from their [plate girder] use would indicate temerity out of keeping with modern engineering and design, but it will be necessary to retreat temporarily, until the general problem is solved . . .

In existing plate girder spans, the authors recommended reducing wind exposure instead of adding strength and bulk. Vented roadway curbs and road decks of open steel grid (the latter were used in the Mackinac design to quell uplift—not harmonic—forces from wind) would reduce horizontal wind forces by up to 75%. Most intriguing was a plan illustrating longitudinally inclined suspenders forming a series of sharply-peaked triangles with the deck as their base. Inspired by Roebling's tower stays and armed with studies on a 3.5-m long single-cable model, Modjeski & Masters indicated this system damped out harmonic vibration eight times faster than one with vertical suspenders. This, they said, would permit a shallow stiffening truss while maintaining stiffness. The first seed of a concept to appear 2 decades later was now sown. The authors suggested further studies.

One of the more prescient comments of the period came from Homer M. Hadley of the Portland Cement Association. In an August 1943 letter to *ENR*, Hadley suggested the aerodynamic form of the deck, rather than its depth, to be of greater importance. He stated that, "How wind passes around or through the deck structure and what forces it imparts . . . is really the essence of the problem." He cited Farquharson's work with semicircular fairings on the Tacoma Narrows Bridge model that withstood the equivalent of 201-km/hr winds as evidence of:

> the extraordinary possibilities for stability, at present undeveloped, that exist in shallow, streamlined decks . . . It may even be [that] the depth to span ratio of the Tacoma Narrows Bridge that failed may, in a deck of proper aerodynamic form, presently come to be regarded as very conservative (Hadley, 1943).

Even Hadley could not foresee just how significant his statements would prove to be.

5.3 THE ADVISORY BOARD ON THE INVESTIGATION OF SUSPENSION BRIDGES

In spite of the research to-date, it was clear that a broadly based theoretical understanding of suspension bridge aerodynamics was required. Indeed, the Carmody Board had identified such a need. In late 1941 and early 1942, a committee of the ASCE and an informal group of noted engineers urged the U.S. Public Roads Administration (PRA) to initiate a major research program; both Othmar Ammann and Moisseiff attended meetings of these two groups. Administration head Thomas H. Macdonald responded in May 1942 by creating the Advisory Committee (later Board) on the Investigation of Suspension Bridges. Its research program would ultimately extend more than a decade.

The Advisory Board was charged with conducting research to better understand the dynamic effects of wind on suspension bridges, to develop a theory explaining the relationship between suspended structure characteristics and vibration, and to recommend design measures for future spans—including the new Tacoma Narrows Bridge, which would be designed between 1943 and 1945—and for retrofitting existing spans. Its almost 30 members included Ammann, Farquharson, and Theodore von Kármán, as well as Andrew and Woodruff for the Washington State Toll Bridge Authority. Others included Frank Masters for the Mackinac Straits Bridge Authority; Sunderland of J.A. Roebling's Sons; Canadian P. L. Pratley, designer of the Lions Gate Bridge; Elmer K. Timby, responsible for the Bronx-Whitestone Bridge model tests; and F.H. Frankland, author of the 1934 book *Suspension Bridges of Short Span.* Moisseiff was appointed research consultant. His presence was a reminder to all of just how lucky they were, and that the Tacoma Narrows Bridge collapse was more an indictment of the profession than of just one engineer. Andrew, who would head the design of the new bridge, would write (1949) that in spite of economic constraints:

> the first structure did meet all the requirements of modern suspension design practice . . .
> . . . [Moisseiff] adopted the principle—which had been widely advanced by several prominent bridge designers—that deep stiffening trusses were unnecessary . . .

Moisseiff was to chair one of five subcommittees but he never really contributed. His health was frail—he had suffered a heart attack in 1935—and he died on September 3, 1943 at the age of 70. His legacy of theory and bridges had profoundly influenced a generation of engineers and still does. In recognition of Moisseiff's contributions, the ASCE established an annual award in his name in 1947 for outstanding papers on structural design. Perhaps unfairly, the only black mark on a distinguished career would be his most visible legacy. Yet suspension bridge design would not be what it is today without him—for his Deflection Theory work, the many groundbreaking bridges to which he contributed, and unfortunately the failure of his bridge over the Tacoma Narrows.

The Advisory Board conducted its research in cooperation with the Washington Toll Bridge Authority. The war had virtually eliminated all but the most essential bridge construction, and members could devote more energies to the Advisory Board's activities, although Farquharson (1952a) would lament that equipment and especially personnel shortages would delay the "discovery of principles now understood as basic." Its inaugural meeting in September 1942 also initiated the study and monitoring of existing bridges—although the war would delay instrumentation of the Golden Gate Bridge—and the development of mathematical theory and analysis.

The most pressing need was for a large wind tunnel. The 1:200 sectional models used by von Kármán at the California Institute of Technology for the Carmody Board were inadequate for quantitative results and overestimated damping. Studies in the summer of 1941 also suggested vortex shedding could explain the vertical motion of the Tacoma Narrows Bridge. To explore such questions and to accurately model aerodynamic response, large-scale aerolastic models of entire suspension bridges were viewed as a necessity.

A contract for a 12-m deep by 40-m long wind tunnel at the University of Washington had already been let in October 1941, and Farquharson oversaw its construction in the first half of 1942. That summer, 10 mechanics and engineering students built a 1:50 scale, 30.5-m long steel model of the original Tacoma span with scaled elastic and inertial characteristics. Three months of testing beginning in November 1942 proved the tunnel's merit; Farquharson duplicated all behaviors of the original span, including the torsional motion that brought it down. Its many vertical modes were thought to emanate from changes in wind velocity that would aerodynamically damp out one mode of vertical motion while inciting another. During these tests, the plate girder was given semicircular fairings modeled with paper. Although stable in winds up to 188 km/hr, in winds exceeding this velocity Farquharson (1952b) observed "a rapid augmentation of amplitude which was almost explosive in nature." He had to stop the test to prevent the model from destroying itself. This violent manifestation was, in fact, a form of torsional flutter instability, although Farquharson (1952a) later admitted the phenomenon went unrecognized at the time.

Evidence of a catastrophic flutter, this time involving simultaneous vertical and torsional motion, emerged in late 1943 during tests on a full aerolastic model of a preliminary truss design for the new bridge, but it, too, was not then fully appreciated. This model, comprising a solid, unperforated roadway and a single lateral system, oscillated in a single torsional mode with a frequency slightly less than the expected natural period of vibration; its center of twist was also on the upstream side of the deck—in some cases, even upstream of the truss itself—suggesting a coupling of vertical and torsional modes. The phenomenon was later observed in sectional model tests, but again its import was not recognized for several months (Farquharson, 1952a, 1952b).

Although the full model proved its validity in intensive testing from 1943 to 1945 and in sporadic testing until 1950, it was recognized that such expensive design tools could not always be justified. Tests with 1.5-m long, 1:50 scale sectional models of the original and new spans were conducted to determine their ability to simulate the behavior of the prototype; in total, some 200 cross-sections were tested. The Toll Bridge Authority and the Public Roads Administration sponsored parallel research throughout 1942 and 1943; this work, performed by Louis Dunn at the California Institute of Technology wind tunnel (and overseen by von Kármán), was coordinated with the Advisory Board. The Advisory Board studies examined the aerodynamics of the original bridge and a preliminary truss design for the new one using sectional models that—as it happens—simulated full model results. The effects of slots, cantilevered sidewalks, and handrailings on a new truss design were all tested. Once the basic truss design was finalized, a full model was then built to fine-tune the results.

During this period, Farquharson and Dunn in particular established a solid foundation for wind tunnel techniques. Sectional models are used to this day, as are full-bridge aerolastic models. Unlike the full models, sectional models are rigid. They can be placed on carefully tuned springs to represent the inertia and wind sensitivity of the real bridge or on fixed supports, where forces are measured. (Chapter 18 discusses these methods in more detail.) Full aerolastic models simulate, in miniature, a span's shape, mass, and stiffness, as well as its vibration amplitude in a scaled relationship. Also important are the bridge's static deformation and the accurate modeling of tower stiffness in bending. Variances from a geometrically linear relationship between the model and the prototype must be carefully accounted for. The Reynolds number (the ratio of inertial to viscous forces from air acting on a body), Froude's Law (the ratio of inertial force to gravitational force), and other factors help bridge aerodynamicists understand the nature of the similitude between the model and the prototype bridge. Modeling the natural frequencies of a span and the net of structural and aerodynamic damping—that is, the ability of the bridge to dissipate energy—is critical. Structural damping—the decay in successive oscillations with wind force removed—is measured by a term known as the logarithmic decrement.[3] Aerodynamic damping involves the action of wind forces to reduce or damp out oscillation. Modeling materials have included wood; aluminum is common for deck plates, and more recent materials include carbon fiber composites.

The Advisory Board's investigations culminated in many interim and five final reports issued between 1949 and 1954. These reports examined the aerodynamic behavior of the Tacoma Narrows Bridge and its replacement. They also tested new formulas against dynamic model results to predict oscillation modes, frequencies, and damping of new designs. In plate girder spans, the ratio of width to depth was found to be significant; susceptibility to catastrophic torsional motion would affect shallow, wider girders, and catastrophic vertical motion could grip deeper cross-sections, as it had at the Deer Isle Bridge (Vincent, 1965). Trusses featured a coupled mode dominated by torsion that was incited at higher velocities than were plate girders. This behavior had been reflected in sectional and full-model tests of the new Tacoma Narrows Bridge and in sectional models of the Golden Gate Bridge. Even small changes in wind angle from the horizontal had drastically changed structural response, no small concern as instrumentation at the Golden Gate and Transbay bridges had revealed winds of up to 8° from the horizontal. Sectional model tests on the Golden Gate Bridge greatly influenced the new Tacoma Narrows Bridge design, from which two planes of lateral bracing were recommended that would be up to 40 times stiffer in torsion than a single system (Vincent, 1951). Although the bending stiffness of the towers was not considered significant, their torsional resistance was found to influence oscillation character and natural frequencies (Smith and Vincent, 1950).

The development of formulae to explain the aerodynamic response of suspended structures was a key area of research. George Vincent, Senior Bridge Engineer for the PRA, played a key role in devising practical applications for design purposes. Friedrich Bleich examined the question between May 1943 and June 1947 as a consultant to the Advisory Board in a venture sponsored by PRA and the AISC. It was Bleich who recognized the phenomenon that Farquharson had observed in the models of the new Tacoma Narrows Bridge as a "classic" coupled flutter similar to that in airfoils.[4] As a result of this revelation, Bleich sought to develop a theory of vertical, torsional, and coupled torsional and vertical oscillation in bridges based on airfoil flutter. He modified the theory—with the aid of findings from wind tunnel tests—to account for the role of the blunt form of a truss in reducing the intensity of flutter response, which was otherwise dominated by the airfoil-like flat plate characteristics of the deck. Bleich's work—first submitted to the Advisory Board in 1945 and published in the ASCE *Proceedings* in 1948—spurred additional wind tunnel tests at the University of Washington and ultimately opened up a whole new field of study. Wind tunnel tests on thin plates suggested that higher wind velocities increased the frequency of vertical oscillation while decreasing that of torsional oscillation. When the two frequencies corresponded, a flutter critical velocity was reached, as manifested in a potentially catastrophic coupled oscillation. This is the phenomenon Farquharson encountered; in fact, the girder with semicircular fairings was gripped with catastrophic motion in just one or two vibration cycles after the coupling of vertical and torsional modes.

As the plate girder bore no resemblance to an airfoil, Bleich's work could not be used to explain the Tacoma Narrows Bridge collapse. But his work had great practical benefit for truss design. It demonstrated that top and bottom lateral systems would increase torsional frequencies and thus separate them from vertical frequencies. This would reduce the chances for convergence of vertical and torsional oscillation, which in turn raised the flutter critical velocity, as well as the noncatastrophic critical velocity for torsional oscillation generated by vortex shedding alone. Vortex shedding and flutter (described more fully later in this chapter) were now confirmed as dominating factors in the aerodynamic behavior of suspension bridges. Yet the field remained so uncertain that Bleich *et al* (1950) opened the landmark *Mathematical Theory of Vibration in Suspension Bridges* by stating that, "In certain instances there is evidence that we are not quite sure just what we are searching for."

Bleich died in 1950, the year the report to which he was prime contributor was published. In it, Bleich also incorporated aerodynamic tweakings, such as tower stays and midspan cable ties, into Deflection Theory calculations. Yet uncertainty of mathematical explanation remains to this day. Bleich's efforts to modify airfoil theory would only partially succeed. Robert Scanlan and Ali Sabzevari (1967) would criticize the parallel between airfoil and bluff body aerodynamics—to which many subscribed well into the 1960s—as "doomed to very limited—and even very questionable—applicability." Alan Davenport (1966) would also question its relevancy early on. Even today, the use of models to explain and predict the aerodynamic behavior of suspension bridges remains essential. This quest has seen researchers in these two areas of analysis cooperate—and compete (*see* Chapter 18).

The Advisory Board expressed ambivalence concerning cable and tower stays, considering them to be of some value but problematic in terms of their effect on a span's elastic properties. Although transverse diagonal stays at the quarter-points of the main span and midpoints of the side spans were found to raise critical velocities by 30% to 60%, neither Ammann nor Steinman came out ahead on this issue. Sheer weight was also not felt to add greatly to structural damping (the Carmody Board had felt weight could actually reduce it). Interestingly, plate girders were not completely rejected; a suitable depth-to-width ratio with some undetermined mix of slots and fairings might raise their critical velocity. Yet it was a tepid endorsement; plate girder cross-sections with a depth-to-width ratio greater than 0.2 were considered inherently unstable. The Advisory Board cited the need for more research into structural damping and the characteristics of natural wind, as well as the development of *a priori* knowledge of the aerodynamic characteristics of varying cross-sections, the role of stiffness and weight, and other items such as stays. Farquharson (1954) in particular emphasized structural damping, which was seen as a tool to increase critical velocities:

> Literally no information is available regarding the structural damping existing on any suspension bridge. The total problem of the stability of such structures cannot be considered as fully mastered until this gap is filled, at least for a few representative bridges. This also is a costly venture, but it would seem folly to extend the span of suspension bridges much beyond present practice until such information is available.

In its studies, the Advisory Board had assumed zero structural damping to err on the side of conservatism. The smooth wind flow in the wind tunnel was also considered to be more "severe" (Andrew, 1945) than natural winds of unknown variability in direction, speed, and angle of attack over the suspended spans. Thus, the model of the original bridge oscillated in more modes and at greater amplitudes than the prototype. Vincent would attempt to ascertain the damping characteristics of two small spans of 50 and 83 m in Washington using mathematical prediction and by inciting motion on the bridges themselves, but his more ambitious proposal for inciting motion on the new Tacoma Narrows Bridge was never carried out. Norwegian Arne Selberg would, however, focus some attention on the problem.

In spite of these gaps, the Advisory Board suggested three means to avoid instability. Shaping could be used to alter wind character. Eddy frequencies could be changed by adding more subtle alterations to suspended structure form or by raising a span's natural frequency via increased stiffness (e.g., by bolting the main cables to the truss at midspan). Increased damping capacity was seen as a particularly effective tool against the onset of catastrophic torsional flutter; the use of midspan cable ties could counter asymmetric torsion by, for example, bringing tower stiffness in torsion into play. The Advisory Board's recommended design procedure involved first the calculation of the natural periods of vibration of a cross-section (by inciting vibration of a model in still air), followed by wind tunnel testing of sectional models to determine the nature of the aerodynamic response. Adjustments could then be made to eliminate the instabilities.

For all the Advisory Board's efforts, an explanation of the mechanics behind the Tacoma Narrows Bridge collapse did not really emerge. The bridge's vertical motion had been suggested to emanate from vortex shedding of the leeward girder. Von Kármán and Dunn did in fact provide a hint of a possible mechanism for the torsional motion of the Tacoma Narrows Bridge when testing H-shaped and truss sections in the California Institute of Technology wind tunnel to determine aerodynamic damping forces. Although not seeking a cause for the collapse, they did observe that "an [H] section with a girder on the downstream side only, exhibited stable characteristics, whereas a section only with the girder on the upstream side was unstable" (von Kármán and Dunn, 1952). They then observed that with both up- and downwind girders in place, the section remained unstable, suggesting the downstream girder to have little effect on stability. They wrote that "These observations tend to support the theory that the torsional motions are caused by the vortex shedding at the leading edge." Still, the exact mechanism at work would not yet be uncovered. With truss designs intended for actual construction dominating the attention of investigators, the plate girder became more of a research curiosity.

5.4 STEINMAN REBUTS AND THE DEVELOPMENT OF STIFFNESS INDEXES

The large research efforts of the Advisory Board from 1943 to 1954 generated a great deal of knowledge. Yet full aerolastic models and large wind tunnels were expensive, prohibitively so for most situations. This brought about the development of simplified methods that could economically predict the behavior of new designs using proxies for measuring aerodynamic performance.

Both Steinman and Ammann would devise such tools. Yet Steinman continued to reject most of the Carmody Board and Advisory Board findings, and would vigorously advocate an alternative theoretical explanation with a complexity rivaling Bleich's work. Steinman's exclusion from the Advisory Board may have been because of Ammann's influence, but Steinman would prove an effective force on his own.[5] His own sojourns into the causes of wind-induced bridge instabilities were disseminated through letters, articles, and papers in the engineering press. But his first major paper, submitted to the ASCE in 1941, was not immediately published. Some have conjectured that Moisseiff might have been unduly criticized, although Steinman exonerated him in his own writings after Moisseiff's death. "The Rigidity and Aerodynamic Stability of Suspension Bridges" was published, finally, in the ASCE *Proceedings* in November 1943.

Steinman's interest had been spurred not just by the Tacoma Narrows Bridge collapse. The overturning of a two-span overhead continuous truss over the Mississippi

River at Chester, Illinois, in July 1944 suggested large static lift forces could occur in even moderate winds; his own research indicated winds blowing just slightly off the horizontal could generate static lift forces three to four times those of a horizontal nature.

It was here that Steinman diverged from explanations of flutter or vortex shedding. His premise was that a static instability, rather than a dynamic instability, was at work. Steinman argued that winds blowing slightly upward would create an area of suction or "negative pressure" just beneath and behind the leading edge girder. This phenomenon was related to Bernoulli's principle, whereby faster velocities are characterized by lower pressure. Plate girders split wind unevenly; more flowed below, thus leading to higher wind velocities, lower air pressure, and a downward force. Behind deeper girders, suction forces would extend across enough of the cross-section to produce a dominating downward force, thus creating vertical instability. In girders of moderate depth, negative pressures would lie beneath just downstream of the front girder. In such cases, uplift from the restored wind dominated beneath the leeward side of the cross-section; these girder profiles were "torsionally unstable." With a stable cross-section, positive structural damping would cancel out the upward motion imposed by the wind on the leeward side. Conversely, in an unstable, flexible girder, negative damping would occur because the axis of twist would lie beyond the deck centerline, creating an unbalanced and self-amplifying twist.

Steinman could not secure the use of a wind tunnel and relied on rudimentary cardboard sectional models just 38-cm long. Open grid decks and open railings were modeled with screen wire, and wind was provided courtesy of a small electric fan; Steinman claimed the entire outfit to cost no more than $500. Models, however, could not provide generalizations, and Steinman emphasized mathematics—as would Bleich in his concurrent work—to predict oscillation frequencies and characteristics.

Steinman's November 1943 paper was greeted with a barrage of criticism from mainstream investigators, including Farquharson, Vincent, and Frederick Smith. Louis Balog, who had performed calculations for the Carmody Board, was scathing in his view that Steinman's "equations are useless for the solution of the practical design problem" (Balog, 1945). Yves Rocard (1957) would later describe Steinman's views as "based on hardly verifiable arguments . . . [and] frankly impossible."

The ASCE monthly *Civil Engineering* published Steinman's theories in simplified form in a series of five articles entitled "Design of Bridges Against Wind" from October 1945 to February 1946—after the publishing of his rebuttal to the hailstorm of adverse opinion to his 1943 paper. Here, Steinman discussed history, theories, and remedies. He rejected rigidity alone to "resist the effect" but instead sought to eliminate the causes of instability. His remedies included floor perforations, horizontal fins, stays, weight, and rigidity—Roebling, through intuition, had adopted the latter remedy almost a century before.

Steinman also introduced the concept of "phase difference"—the alteration of pre-existing motion of the suspended structure by the wind as it crosses the deck. As late as 1954, he argued the concept had been overlooked by mainstream investigations, but that he considered it essential to the explanation of the "otherwise mystifying variation of aerodynamic response at different wind velocities" (Steinman, 1954b). Steinman asserted that because of phase difference, even stable bridge sections would be unstable in a small range of wind speeds. However, he wrote, "a basically unstable section will have an upper critical range that is unlimited, and therefore potentially catastrophic" (1945d). Steinman viewed the relationship of wind speed to a girder's natural vibratory period, width, and shape as a critical determinant of instability.

The Advisory Board had concluded that Steinman's theories were not "in full agreement with the observed behavior of bridges and models" (Vincent, 1952); it also cited their inability to explain oscillation decay when winds increased beyond a certain velocity. The Advisory Board was ambivalent on another matter as well—the virtues of vertical stiffness alone. It cited the oscillation of a full-length aerolastic model based on the original Tacoma Narrows Bridge with deeper 3.6-m girders, as well as the vertical motion of the Beauharnois Bridge, a stiff 177-m plate girder span in the Canadian province of Quebec. In contrast, the more flexible Anthony Wayne Bridge in Toledo, Ohio showed no signs of oscillation. The Advisory Board also noted the "insignificant" motion of the vertically unstiffened deck of the George Washington Bridge. It considered stiffness but one factor of aerodynamic stability; more research, it said, was needed (Vincent, 1952). Yet several engineers, including Steinman, Ammann, Balog, and Shortridge Hardesty, would advocate just such a relationship between stiffness and wind-induced motion in "stiffness indexes." These indices were devised to predict the oscillation potential in new designs based on empirical assessments of the relationship between aerodynamic behavior and the design characteristics of existing bridges.

Steinman's index involved a "stiffening ratio," which corresponded to structural damping. A "rigidity coefficient" incorporated deck width, weight, wind speed, and girder depth. Steinman identified eight spans built since 1930 with low coefficients; all had oscillated in some way. Plate girders would be prone below a coefficient of 15; trusses, below a coefficient 6. The aerodynamic effect of various stiffening features could be measured mathematically, such as deep trusses or girders, cable or other stays, continuous suspended structures, unsuspended side spans, and stiff towers. For torsional stability, cross-stays, top and bottom lateral bracing, and other cross-sectional features were suggested. Steinman stressed deep trusses or girders, recommending a depth-to-span ratio of no less than 1:100 in spans from 305 to 610 m and a conservative 1:80 for spans exceeding 1,220 m. Weight alone was no panacea; he considered lighter, deep trusses more stable, an important consideration that helped maintain the benefits of the Deflection Theory. Rocard (1957) would later describe these criteria as "thoroughly justified," notwithstanding the faulty theory on which they were based.

Steinman felt that the vertical instability of the Tacoma Narrows Bridge would not have led to collapse, but that in torsion, the bridge was prone to undampable oscillation and inevitable failure. He reiterated his view that the rigid center ties in the Thousand Islands and Deer Isle spans would have armored the Tacoma Narrows Bridge against torsional motion by keeping the main cables and girder in phase. He considered the Deer Isle Bridge, with no upper critical limit for torsional and vertical motion as originally designed, even more aerodynamically susceptible than the Tacoma Narrows Bridge. Cross-stays as used in the Deer Isle Bridge, he argued, would have increased the torsional stiffness of the Tacoma Narrows Bridge by a factor of 50—tower and cable stays would have wrought a 10-fold increase. Even light systems of top and bottom lateral bracing would have "assured torsional stability at a wind of 100 miles an hour [160 km/hr]" and increased structural damping by a factor of 50 (Steinman,

1946a). A doubled girder depth combined with this bracing, he claimed, would have quadrupled stiffness. He further argued that friction damping (such as used in the Bronx-Whitestone Bridge) was "helpful in resisting the initiation of oscillations, but [has] little effect after any substantial amplitude is obtained" (1946a). He suggested that open grid sidewalks at the Tacoma Narrows Bridge alone could have increased vertical stability by 84%, and he later (1953) wrote that structural continuity would have enhanced the Tacoma Narrows Bridge's stiffness by 24%.

Like others, Steinman felt that a perforation of the Tacoma Narrows Bridge's plate girder would have increased instability while weakening it and that the low cable sag increased instability. He also felt that fairings would have enhanced torsional stability, but at a cost of vertical instability in lower wind speeds and with higher amplitudes. Steinman suggested a deeper girder would have improved the vertical stability of the Tacoma Narrows Bridge by creating an optimal ratio of depth to width. Most provocative was his assertion (1946b) that, "Contrary to popular misconception, the narrowness of the Tacoma Bridge was not its cause of failure." Citing his own formulae, Steinman suggested a wider deck to be more vulnerable to harmonic motion of a magnitude more difficult to remedy.

Steinman also objected to Farquharson's August 1946 writ large indictment of the plate girder in *Civil Engineering*. Farquharson had found deck openings in plate girder suspended structures to add little value—some patterns even abetted oscillation—whereas open decks atop trusses improved stability. Steinman soon rebutted:

> There is no magic aerodynamic virtue inherent in the use of trusses instead of girders . . . The advantages of using trusses are in their usual greater depth, stiffness, and structural damping, not in any mysterious aerodynamic superiority. . .
> . . . An aerodynamically equivalent truss section of equal flexibility and low damping would have been equally vulnerable (Steinman, 1946c).

Steinman felt deck slots on the inside girder faces—either open or roadway grating—could dispel or significantly reduce suction forces inside the leading edge, although dynamic model tests of the Advisory Board would contradict this claim. Steinman also stated that Advisory Board tests of the deep truss for the new Tacoma Narrows Bridge without wind slots had revealed catastrophic oscillation:

> The provision of suitably located openings in the roadway has been persistently advocated by the writer in the face of skepticism for the past six years. These roadway vents, and not the substitution of trusses for girders, should be credited for yielding an aerodynamically stable section for the new Tacoma span (Steinman, 1946c).

In touting the economics and aesthetics of plate girders, Steinman concluded that "It would be unfortunate to scrap these advantages and to limit bridge designers in the future by perpetuating a misleading and unfounded conclusion."

Farquharson, meanwhile, had spent part of the latter half of 1946 in Great Britain observing wind tunnel tests in connection with a proposed large span across the River

Severn. His (1947a) response in the February 1947 issue of *Civil Engineering* was blunt. Farquharson considered Steinman's claims to be a "dangerous oversimplification" based on rudimentary and inappropriate theories and testing methods. He wrote that, "it would be most unscientific to remove the solid plate girder from the dog house where it at the moment most properly belongs." Farquharson did not dispute the aesthetics and economy of plate girder suspension bridges, only their ability to avoid self-destruction. Although he acknowledged that lift played a role in instability, he considered other forces more relevant, a likely reference to flutter and vortex shedding.

In the January and February 1947 issues of *Civil Engineering,* Steinman presented methods to determine the aerodynamic behavior of truss sections. His static wind tunnel tests measured air pressure differences at points across the top and bottom of simple sectional models just inches long held in a fixed position at various angles to the wind. The net pressure differences (or, alternatively, static lift and torque forces) were plugged into various formulae to produce a series of graphs describing the proclivity towards vertical and torsional instability. Steinman (1947a) suggested these made:

> The aerodynamic stability or instability of bridge sections . . . entirely predictable, both in direction and intensity, for all wind velocities and for any angle of attack . . . All that is required is a simple test, requiring only a few minutes time, on an elementary, small-scale model of the cross-section.

In a direct jab at the Advisory Board, Steinman (1947b) wrote that the formulae would answer "all questions on oscillation modes and frequencies more accurately than the most elaborate and spectacular large-scale model tests." Steinman had also reported torsional stability (excepting very low wind speeds) when a broad open section separated two plate girder sections placed side by side. When he had closed this gap in earlier studies, the section was "violently unstable" (Steinman, 1946a). This concept of parallel suspended structures would re-emerge 30 years later as a potential breakthrough for ultralong span suspension bridges.

By this time, Steinman had secured the support of Hunter Rouse of the University of Iowa and F.J. Maher of the Virginia Polytechnic Institute, which allowed him use of the small wind tunnels at both institutions. Steinman (1947a) used his latest writings as a vehicle to launch:

> an appeal, to those who have test facilities, for scientific cooperation. The writer needs more basic tests on various standard bridge cross-sections, and on suggested modifications of these cross-sections, to complete the derived data and conclusions in a form readily usable by the profession.

That summer, Steinman received five offers, including increased funding of the Maher experiments from the Virginia Polytechnic Institute; this Steinman eagerly accepted. In August 1947, he reported in *Civil Engineering* (1947c) that he had developed plate girder cross-sections "of assured aerodynamic stability against vertical, torsional, and coupled oscillations at all wind velocities and at all angles of incidence." These claims partly blunted Farquharson's criticisms, one in July 1947 discarding a theory "similar to that proposed by Dr. Steinman [that] was abandoned in favor of a vortex excitation" (Farquharson, 1947b). Steinman described additional tests in the December 1947 issue of *Civil Engineering* on plate girders with various arrangements of slots, open grid roadways, and fins. He later described a central wind vent combined with fins that provided "a complete cure or prevention of both vertical and torsional instability" (Steinman, 1954a). Most remarkable was his claim that a simple flat plate lacking any vertical stiffening could, if appropriately suspended, be aerodynamically stable.

Apart from disagreement over cause and effect, Farquharson (1947b) questioned Steinman's use of rudimentary sectional models, which, he claimed, could not be extrapolated to characterize the behavior of an entire structure. Steinman (1947c) rebutted in an August 1947 letter to *Civil Engineering* that, "the public and the profession are told that the stability of the new Tacoma design is credited to the wonderful discovery of the magic virtues of the truss type of construction. Years of expensive and mystifying research must be justified."

One interesting sideline to the debate was Steinman's use of analog computers. As one of the first engineers to do so, he is said to have reduced 16 days of hand calculation to 45 minutes for each of 16 graphs in a 1950 ASCE *Transactions* paper "Aerodynamic Theory of Bridge Oscillations."[6] An additional benefit was greater accuracy. This paper had initially been rejected by the ASCE but would end up garnering the ASCE's highest award, the Norman Medal. In the paper, Steinman attempted to incorporate Bleich's flutter work and produced methods applicable to truss and plate girder sections. Steinman's use of tilted models and cross-sections with curved decks (the latter to represent dynamic motion) in static wind tunnel tests to derive measurements of forces in motion was ingenious. Steinman introduced this concept in a 1947 paper that described a series of stable and unstable cross-sections (Steinman, 1947e). Yet in both papers, Steinman remained focused on static instability, and although many respondents praised the work, Farquharson (1950) wrote that Steinman's "dependence on static section model tests . . . does not hold up," a view echoed by Vincent in his discussion of the paper. Steinman (1950) would rebut that "proud reference to the [aerodynamic] theory has become a cult, and any challenge of departure is attacked as heresy."

It is somewhat ironic that for Steinman's efforts to exonerate the plate girder, his later large spans would use deep trusses. It is doubtful—and of this Steinman was surely aware—that few, if any, would invest in a large plate girder suspension bridge, no matter how compelling his arguments might appear. In his first post-Tacoma suspension bridge commission—a 1949 design for the Kingston-Rhinecliff Bridge across the Hudson River north of New York City—Steinman adopted a 518-m span with a 6.7-m deep truss braced with top and bottom laterals. Its depth-to-span ratio of 1:77 was even more conservative than the 1:100 maxim Steinman advocated in 1946. Paradoxically, no wind slots, open grid road decks, or outside vents were provided, although the outer stringers beneath an 11-m wide deck were of porous open web design. The bridge was never built; the preliminary design was abandoned in favor of a Steinman-designed 2,400-m long deck truss completed in 1951.

Ammann, too, had devised a stiffness index as early as 1943, building on his work with the Carmody Board. Ammann had used the index to help devise the truss superimposed on the Bronx-Whitestone Bridge. His suspicion of the value of wind tunnel tests was seemingly confirmed by the results of sectional model tests at the

University of Washington on the Bronx-Whitestone Bridge in 1951 and 1952 by the Advisory Board. (Other spans tested included the George Washington Bridge, the Golden Gate Bridge, and the Paseo Bridge, a small Kansas City span then under design.) This work in smooth air flow had suggested the Bronx-Whitestone Bridge—in both pre- and post-truss configurations—to be prone to catastrophic torsional oscillation in winds of 47 to 79 km/hr. Yet the bridge had easily weathered such winds with just minor vertical oscillation; maximum single amplitudes of 46 cm before the addition of the truss had been reduced to 23 cm following its addition. Vincent (1954) ascribed the variance to "non-uniformity and unsteadiness in the natural wind," a prescient comment. Ammann's skepticism would stay with him, for he would minimize wind tunnel use in his post-Tacoma Narrows Bridge designs and put more stock in his particular variant of stiffness index. Unlike Steinman, Ammann based his index on the measurement of static vertical deflections. His practitioner's bent (and perhaps a jab at Steinman) emerged when he wrote in 1953 that:

> so many treatises of largely academic character and unqualified conclusions on the problem of wind action on suspension bridges, and on ways and means to prevent oscillatory motions, have appeared in print, that a clarification of the present status of this problem from the point of view of the practicing engineer appears timely.

In the same 1953 article, Ammann wrote that he considered the dynamic effects of wind on suspension bridges to be "inadequately explored and will probably never be fully explored." Yet he did offer a practical guide, and he did not differ from Steinman in arguing that with proper design, deck width was not a factor in vertical or torsional instability. Ammann considered the form of a suspended structure and its resistance to deformation as key measures of aerodynamic stability. He used a proxy: the amount of vertical deflection at the quarter-points of the main span under a static load confined to one-half of the main span and an equal upward force over its opposite half. The resulting profile described a first antisymmetric mode (i.e., with a single midspan node). Ammann devised an equation that placed all eight spans that failed under wind below a figure of 200; all had featured a single plane of lateral bracing. Other spans were clustered into categories corresponding to the severity of oscillation experienced; those with no history of motion scored greater than 800. In these figures, Ammann also incorporated the relative contributions of cables, weight, stiffness, and stays.

Ammann also distinguished between short and long spans, an area he felt neglected. He considered the depth-to-width relationship (2.3 to 8.5 m) of the recently-completed but unstable plate girder span at Beauharnois even "more unfavorable" than that of the Tacoma Narrows Bridge. Ammann mildly rebuked Steinman, writing (1953) that, "Greater caution against excessive flexibility would have seemed called for in the case of the Deer Isle and Thousand Island Bridges whose spans and weight are of an order far closer to that of the American bridges which failed in the 19th Century." Ammann—like Steinman—considered the most economical and effective means of enhancing vertical stiffness was to increase truss depth. In spans greater than 762 m, he viewed a double lateral system as the most economical means to enhance torsional rigidity; Ammann felt it would have prevented the Tacoma Narrows Bridge's torsional oscillation, if not its benign vertical motion. Interestingly, Ammann, like Steinman, felt plate girders should not be rejected out-of-hand; such spans as the Anthony Wayne Bridge in Toledo, Ohio had not oscillated and ranked highly on his index. Even the Bronx-Whitestone Bridge in its original form was, according to the index, stiffer than the Golden Gate Bridge. In fact, the Cologne-Mulheim Bridge ranked the stiffest of them all, thanks to 6.1-m deep girders and its self-anchored design. Economics dictated trusses—and not plate girders—for spans greater than 152 m.

Ammann's stiffness index was used in the design of several bridges well into the 1960s—and not just his. It would also be improved, in 1956, to account for deck width and later for torsion; the latter helped counter criticism that vertical stiffness was no panacea for aerodynamic instability. Both his and Steinman's indices were useful for preliminary design, particularly in the era before computers. Today, the approach is viewed as outmoded. Nonetheless, Ammann's and Steinman's indices were important means by which to translate the observed past performance of bridges into a useful design aid for new ones.

5.5 AERODYNAMIC PHENOMENA AND SUSPENSION BRIDGES

Notwithstanding Steinman's claims to the contrary, the Advisory Board now provided a basic understanding of the complex phenomenon of wind-induced motion. This knowledge has governed the aerodynamics of suspension bridge design to this day. It posits that horizontal winds can induce three types of oscillation: vertical, torsional, and coupled. The Carmody Board had examined three possible causes of these oscillations. It favored buffeting—the effects of turbulence from random changes in wind direction, aspect, and speed. Von Kármán vortices—or vortex shedding—involve motion created by eddies brought on by and acting on the lee of an obstruction. These can also precipitate a third type—flutter—which involves self-induced harmonic motion in winds of higher velocity; this is what von Kármán had initially advocated as the cause of Gertic's demise.

Buffeting involves oscillation incited by random gusts occurring for prolonged periods. Buffeting can even occur in a bridge well-designed to suppress flutter, and it reflects the fact that wind is naturally turbulent; adjacent terrain, for example, can greatly influence wind behavior. A bridge's response to buffeting is difficult to analyze because of the randomness of the wind, but it is related to the shape of the suspended structure and its natural period of vibration. The turbulence associated with buffeting also reduces the stabilizing "reattachment" of wind to the suspended structure as it crosses it. The magnitude of oscillation can be lateral, vertical, or torsional in nature, but it is generally not significant. Buffeting can create discomfort for bridge users, and steel fatigue can be a concern if it occurs regularly for long periods.

Vortex shedding is different. Winds passing over a streamlined body such as a wing generally hug its surface and reform in its wake. If there is some separation of the wind from the surface, it may reattach before passing off the body in a relatively smooth flow. By contrast, bluff bodies, such as the plate girders of the Tacoma Narrows Bridge, disrupt this pattern, and separation occurs without reattachment. Two thin, unstable layers interact to generate whorls—or vortices—that are "shed" alternately in rows off the upper and lower downwind sides of the obstruction. Each vortex is connected to succeeding and trailing vortices to create a vortex stream; the lower vortices are positioned midway between the upper ones. The vortices in each row also rotate in opposite directions as they flow off the leeward edge of the obstruction.

This Kármán vortex street—in recognition of von Kármán's epochal work in the field—represents the imposition of alternating cyclical and pulsating vertical forces on the upper and bottom trailing edges of a suspended structure, perpendicular to the horizontal wind direction. The frequency of vortex shedding and these forces is proportional to wind speed, a relationship described by the Strouhal number. If the eddies spin off a flexible suspended structure with a frequency roughly corresponding to (or a multiple of) the span's natural period of vibration in vertical bending or torsion, they can touch off and even amplify vertical or torsional resonance. In this case, an eddy, with a pressure pushing down, moves across the top of the deck as the deck moves down, amplifying its downward movement. The next eddy then moves off the bottom, pushing the deck up as the span's natural period of vibration also brings it up. The nature of the eddies varies with wind velocity and the shape, depth, and width of the suspended structure, which in turn influences the extent to which winds reattach before being shed.

Vortex-induced resonance generally occurs in a small range of low wind speeds, below that for buffeting or flutter. It requires sustained winds to occur. The relatively benign vertical motions in the Tacoma Narrows Bridge have long been considered to be of this type (but *see* Chapter 18), as were those on the Bronx-Whitestone Bridge and the early motion of the Golden Gate Bridge. Suspension bridge towers can also be susceptible. Yet vertical or torsional motion is generally self-limiting, because the motion of the suspended structure disrupts the vortex street and alters the wind forces acting on it; these wind forces, through aerodynamic damping, will "unlock" any synchronicity, and the motion will diminish or subside. The structural damping of the bridge (e.g., through friction between riveted or bolted connections and hysteresis—the span's elastic properties), inertia, and cross-section also affects overall damping. The structural and user effects of vortex shedding are similar to those of buffeting, yet open, sieve-like trusses are less prone to the phenomenon as they break up the wind and prevent regular patterns of vortices from forming.

The third type—flutter—is the most threatening. Flutter is a dynamic instability in winds at and above a critical velocity. It is generally not a problem for bridges in light winds. In higher winds, however, the ability of a suspended structure to damp out the effects of vortex generation can diminish. Flutter is well-known in aeronautics. In streamlined bodies, such as airfoils, flutter occurs from a "coupling" of simultaneous vertical and torsional motion whose frequencies are in phase with one another. This type of flutter is a two degree-of-freedom or classic (or coupled) flutter, which is "hinged" behind the leading edge a quarter of the way across the chord. It is difficult to incite in streamlined bridges except at very high wind speeds. In a bluff body such as a plate girder—or some trusses—the situation is different and more complex (Billah and Scanlan, 1991). Such suspended structures can be more susceptible to severe vertical motion or a torsional (single degree-of-freedom) flutter response, in which the frequency of eddy shedding and the natural period of the bridge have a proportional relationship. The wind flow is more separated from the body in motion and in fact resembles so-called "stall flutter" in engine or turbine blades, in which flow separation exceeds the minimum angle at which stall will occur. In this single degree-of-freedom flutter, the axis of twist follows the centerline of the bridge. Structural damping can have a mitigating effect, but the Tacoma Narrows Bridge's solid plate girder—extremely flexible in torsion—has been considered a prime candidate for this phenomenon. The wild oscillation of the Deer Isle Bridge in 1942 has also been considered to be a form of single degree-of-freedom vertical flutter.

Coupled flutter is of most concern for suspension bridges, although it is torsion that usually dominates the vibration pattern. The mechanism is elegant. When even a small amount of vortex-induced torsional motion is initiated, flow separation increases after wind passes over the leading edge of a suspended structure. This enhanced separation is considered a key factor in the creation of destabilizing wind forces (Scanlan and Vellozzi, 1980). The vortex closest to the deck during flow separation can create suction forces that vary over time between vortices and in position as the vortex crosses the deck. This moving suction can further twist the deck and intensify the lift force acting on it. The elastic response of the suspended structure to resist the force can be either in phase or out of phase with the shed vortices. If in phase (known as negative damping), the elastic response reverses the motion of the deck, and the wind then acts on it in a manner that reinforces the elastic response. A "lock-on" then occurs, whereby the oscillation itself begins to control the frequency of vortex shedding in sync, of course, with its own motion. The vortex pattern is now transformed into a "flutter wake." The suspended structure re-absorbs the resulting forces in phase with its own motion; the forces are now motion—not wind—induced. At this point, the forces overcome the structural and aerodynamic damping of the bridge, and the "self-excited" motion amplifies. Flutter can occur suddenly and increase to violent proportions in a remarkably short period without any increase in wind velocity.[7]

Coupled flutter, once initiated, is almost immune to the effects of structural damping. Knowledge of the vibrational phase relationships of a cross-section at different wind velocities is essential to raise its critical velocity well above maximum probable winds. For a nonbluff body, such as a truss, or a streamlined section, such as some box girders, a high flutter critical velocity is generally achieved by separating the natural frequency—as measured in still air—of its vertical motion from that of its torsional motion to avoid coupling and instability. (Some investigators prefer to use the term coupled-mode instability when referring to trusses, because the extent of flow separation can be greater than in classic flutter, a reflection of the fact that truss sections in particular are different from airfoils.) Design efforts are made to maintain this separation into high ranges of wind speeds, so that the natural decrease of torsional frequency in higher winds approaches the vertical natural frequency at an impossibly high critical velocity for the bridge site. Normally, stability performance is measured over a period that allows motion to incite itself; it may range from 1 to 10 minutes. For bodies prone to single degree-of-freedom flutter (e.g., shallow H sections in bending or in torsion), the separation of vibration modes is less relevant.

A single degree-of-freedom torsional flutter has emerged as the dominant explanation for the demise of the Tacoma Narrows Bridge, but this view is by no means unanimous. The Carmody Board's "random action of turbulent wind" was somewhat ambiguous as an explanation, and the Advisory Board did not provide a definitive explanation of the collapse either. In deference to the Carmody Board, turbulence has since emerged as a prime factor in determining the aerodynamic stability of bridges, but not perhaps in the way the Board intended. Misinterpretations of von Kármán's advocacy of the vortex street have also muddied the waters. Popular and even scientific accounts (von Kármán's autobiography, as well as university physics texts) have cited vortex shedding as the Tacoma Narrows Bridge's executioner. An example: "The wind produced a fluctuating resultant force in resonance with a natural frequency of the structure. This caused a steady increase in amplitude until the bridge was destroyed" (Billah and Scanlan, 1991). Such explanations provide, at best, incomplete and, at worst, incorrect explanations of the collapse.

Even as late as 1991—more than 50 years after the Tacoma Narrows Bridge failure—K.Yusuf Billah and Robert Scanlan sought to bring a mathematical rationale of flutter as an explanation for the collapse into a broader focus. Their work was a byproduct of 1971 research by Scanlan and J.J. Tomko that sought to disassociate flutter in airfoils from that in suspended structures. In the 1991 article, Billah and Scanlan argued that forced resonance did play a role in the Tacoma Narrows Bridge collapse, but the wind, acting as an external force alone, was not the source of impulses responsible for its demise. They concluded that the bridge had responded to vortex shedding—harmlessly—by frequent but mild vertical oscillation during gentle winds. Yet the 68-km/hr winds on the day of the collapse were much too high to generate vortices of a frequency that could have induced the span’s torsional oscillation. In fact, they argued that the vortex frequency would have to be almost five times the oscillation frequency recorded by Farquharson during the catastrophic torsional motion (Larsen [1999] would echo this conclusion). Something other than vortex shedding was inciting the plate girder to oscillate.

The switch from vertical to torsional oscillation abetted by the slippage of one of the bridge's midspan cable bands disrupted the vortex street. It permitted the absorption of additional wind energy that would not normally have occurred at this wind speed, allowing a synchronicity of motion and wind to occur. The vortex street was now dominated by a motion-induced flutter wake with completely different characteristics; this wake was a consequence, and not a cause, of the oscillation frequency. This explanation is compelling, as Farquharson's wind tunnel studies after the collapse would reveal that the Tacoma Narrows Bridge was able to damp out all but one type of motion, an initially small torsional oscillation that built upon itself until destruction; significantly, this was the frequency he had observed in the actual bridge's death throes. Farquharson, who did not pursue flutter initially, later captured its essence when describing fluid flow tests under von Kármán that suggested, "the vortex was discharged at a rate which was some integral multiple of a natural frequency of the bridge. After the oscillation was definitely established, the section appeared to control the vortex which then discharged at the same rate as the oscillation" (Farquharson, 1947b). Ironically, because the plate girder was virtually pilloried, investigators largely shunned it while focusing on trussed suspended structures, and so a definitive explanation of the collapse mechanism was not vigorously pursued. The fact that researchers would continue from time to time to devise explanations of the Tacoma Narrows Bridge collapse suggested uncertainty still remained. This quest for a definitive understanding of the failure has continued to this day (*see* Chapter 18).

Regardless, flutter remains a key concern of modern suspension bridge aerodynamics, even if other phenomena are at work. Moreover, bridge aerodynamicists would come to realize more than 100 years after the fact that torsional flutter was the culprit behind the 1826 damage to Telford's Menai Bridge, and that coupled flutter destroyed Samuel Brown's Brighton Chain Pier a decade later. Farquharson would view the old reports (as described by W.A. Provis) of greater damage to the Menai's leeward handrail during the 1839 storm as telling, for they mirrored his own tests that revealed the center of rotation in coupled flutter to be slightly on the upwind side of the span's axis. Forgotten history now touched engineers directly.

The investigations of the Advisory Board sparked continued study not only in America but in other countries—most notably by Kit Scruton and Dennis Walshe in Great Britain, Atsushi Hirai in Japan, Arne Selberg and Erik Hjorth-Hansen in Norway, S.O. Asplund in Sweden, Martin Jensen in Denmark, Yves Rocard in France, and Alan Davenport in Canada. These engineers and others would make key contributions during the next 2 decades. Research continues; the influence of the Tacoma Narrows Bridge lingers in a phenomenon still imperfectly understood. But with the end of World War II, engineers began to branch from theory and return to the building of suspension bridges.

ENDNOTES

1. Robinson, then aged 78, is said to have clambered across the main cables to obtain wire samples from midspan.
2. One quirky footnote to the whole insurance affair involved a local insurance agent by the name of Hallett French. French, who had sold the Toll Bridge Authority $800,000 of insurance for the Tacoma Narrows Bridge without notifying his employer, retained the premium for himself. He was possibly even more aghast than the span's engineers on hearing news of the collapse. Had rumors of the Toll Bridge Authority's re-insurance plans proven true and the bridge stood for another week, his crime might have gone unnoticed. As it was, in 1941, French pleaded guilty and was sentenced to 15 years imprisonment (Anonymous, 1990).
3. The decrease in energy is measured by comparing the square of amplitude difference between successive oscillations.
4. Theodore Theodorsen developed this theory as applied to thin airfoils in 1934 (Frandsen, 1966; Ito and Nakamura, 1982).
5. William Ratigan (1959) intimates that Frankland and Ammann handpicked the members.
6. The paper was first published in the 1949 ASCE *Proceedings*.
7. A related phenomenon is divergence, in which pitching forces manifested in twisting at a critical wind velocity (slightly higher than with flutter) overcome the stiffness of the bridge and buckle the suspended structure (Richardson, 1984; Davenport *et al*, 1969). Hiroshi Tanaka (Personal communication) describes it as a negative aerodynamic stiffness (and flutter as a negative aerodynamic damping).

Chapter 6

RETRENCHMENT IN AMERICA

The aftermath of the Tacoma Narrows Bridge collapse coincided with the onset of World War II. Plans for several large suspension spans were put in abeyance. Ten years would pass before the first of the large postwar spans—Gertie's replacement—was completed in 1950. The hiatus did allow the intensive period of research to both generate an understanding of aerodynamic phenomena and translate this knowledge into design treatments. This work would profoundly influence the design of all subsequent suspension bridges.

6.1 WAR-RELATED SPANS

World War II had not brought suspension bridge building to a complete halt. Remote parts of North and Central America provided venues to ease the flow of war materiel or to cement political alliances. Even the war zones beckoned; one American span of 128 m built from Bailey Bridge components across the Shwili River on the Stillwell Highway near the China-Burma border was the largest single military span of the war. El Salvador's Cuscatalan Bridge across the Lempa River would provide a key link in the Pan-American Highway. This 250-m span was designed by U.S. Steel and was built between 1940 and 1942; its remoteness required the construction of a temporary village holding up to 500 mainly American workers and their families.

The area's geology was both a help and a burden. Helical strand cables were anchored in tunnels driven 21 m into soft volcanic rock, but requirements for seismic resistance dictated the use of heavily reinforced piers. Aesthetics figured strongly in this flagship span, with steel towers reminiscent of the Mid-Hudson Bridge supported by concrete piers extending to deck level—this to withstand floodwaters of up to 12 m. A 3.5-m deep through truss carried a two-lane roadway on a single lower lateral bracing system. Construction was unorthodox; each unsuspended side span was assembled on the ground and hoisted in a single unit up to 40-m long; only then was tower steel erected. Interestingly, vibration of the open strands occurred over the side spans during construction and, to a lesser extent, once the truss was in place; the use of clamps inhibited further motion. A Victoria, Australia, span of 168-m completed in 1941 avoided slippage on clay soils by the placement of an anchor pier at the end of the unsuspended side spans to take up vertical loads; its main cables terminated in a 272-tonne concrete block 270 m further back of the anchor piers. This rather utilitarian bridge was replaced in 1969.

The largest suspension bridge of the war was built across the Peace River in northern British Columbia, Canada. This 283-m span, along with a 165.5-m suspension span across the Liard River, formed critical links in the Alaska Highway, built at a dizzying pace to move military materiel to the Pacific theater of war. Ice sheets up to 1.2-m thick, deep water, and the need to minimize steel use made a suspension bridge attractive to the span's designers, engineers of the U.S. Public Roads Administration. The site's remoteness and short construction seasons were also major challenges; the nearest rail access was 65-km away, and design and fabrication would take place almost 5,000 km away at the J.A. Roebling's Sons plant in New Jersey. In spite of these obstacles, the bridge would be completed in just 8 months. Excavation of the tower piers to a depth of 15 m began in December 1942 in conditions of extreme cold. Concreting was performed in temperatures plunging as low as -40° C and in weak daylight that persisted just 4 to 5 hours per day. The harsh conditions could be turned to advantage, however, as the thick ice would form a superb work platform from which to erect the 76-m tall towers.

The foundations and towers had to be completed before the spring ice breakup. It would be a suspenseful race, for only in late March were they finished. Cold had delayed the ice breakup, and J.A. Roebling's Sons moved quickly. Less than 4 days later the south tower stood finished, whereupon the erection crane was dragged across the river on rapidly deteriorating ice. Yet 36 hours were then lost trying to find the steel for the north tower, which when wayward somewhere en route. This tower was erected in just 4 days; 5 days later, trucks could no longer cross the ice. Before the end of May, an access trestle twice capitulated to river ice, but all remaining steel made it to the isolated north shore. All 24 helical strands in each cable—the largest open-strand (i.e., unbound) helical cables to that time—were erected in 8 days. The need for speed also produced a unique anchorage of interlocking precast concrete cells filled with gravel. A 4-m deep through truss supported the two-lane deck to lower the approach embankments, an important factor as the span rose 17-m between anchorages. Truss erection maintained the frenetic pace and was completed over the main and 142-m side spans during 20 days in July 1943, the bridge itself serving as a full-scale laboratory to monitor deflections during erection. The concrete roadway was poured in just 3 days, bringing to conclusion 8 months of work under a schedule described as "nearly impossible" (Birdsall, 1944). The cost—just $1.5 million.

Both the Peace and Liard bridges attracted the interest of the Advisory Board on the Investigation of Suspension Bridges because of their mild vertical oscillation in high winds. Both ranked high on Othmar Ammann's stiffness index, again suggesting stiffness was no panacea in and of itself. Then came the 1945 discovery in both spans of minor wire breakage induced by aerodynamic vibration of the open strands and temperatures as low as -57° C. Interestingly, turbulence in the lee of the windward strands, which remained unaffected, was thought responsible. Wood shims to separate and brace the strands solved the problem. Most ominously, the Peace River came to be less benign than its name implied, for problems of increasing severity would plague the bridge for the next decade and eventually signal its demise. A routine inspection in 1947 first uncovered a disturbing situation. Floodwater currents were undermining one pier, causing its tower to lean by 35 cm. To avoid further damage from spring ice breakups, a steel cofferdam was driven through more than 2 m of ice the following winter, the last piles being placed within days of a thaw that would have left the weakened pier exposed. Concrete was later poured for reinforcement.

This work provided only temporary relief, for within a year, more than 3,600 tonnes of boulders were dumped as additional protection. Strong currents continued to eat away at the soft shales of the north bank, and in 1952, some 25,000 concrete blocks were dumped in the river, followed by further stabilization in 1955. Yet the river finally prevailed on October 16, 1957. Aroused in the early morning by the sound of popping steel connections, locals converged on the bridge to find its north anchorage slowly sliding toward the river. Traffic was stopped while engineers of the Canadian Army worked frantically to stem the slide. By midmorning, hundreds had been drawn to the bridge for what was now a foregone conclusion. By noon, a slide of 3.7 m had slackened the main cables, tilting the cable bent alarmingly and slowly crushing the backstay span. At 12:39 P.M., a final lurch tore the suspenders from the north side span, dropping it into the river. The main span now sagged grotesquely, and the north tower tilted by 3 m; the bridge was now unsalvageable. An emergency ferry and then a temporarily decked-over nearby rail bridge sufficed until a five-span deck truss was opened in 1960 next to the original bridge. It is ironic that the first major post-Tacoma suspension bridge would lead a truncated life, albeit in unrelated circumstances. The Liard River Bridge survives, but it suffered from major cracking in its anchorages and a flexible deck that oscillated in strong winds and deflected noticeably under heavy trucks. Upgrading in the 1970s remedied these latter ills.

Further south in Canada, a small suspension bridge southwest of Montreal, Quebec was opened in 1947. The 177-m span Beauharnois Bridge featured a 2.6-m deep plate girder and unsuspended side spans. It was this span that caught the attention of the Advisory Board when in 1950 it oscillated vertically with a double amplitude of up to 33 cm in oblique winds—a surprising phenomenon given its relatively high stiffness. Temporary cable stays were added that year, but when the entire suspended structure was replaced between 1987 and 1989 because of unrelated deterioration, an orthotropic truss—with stays—was installed.[1]

The return of the suspension bridge to the United States began on a modest note in 1947 with a 128-m span over the Suwanne River in Florida. Its designer, E.S. Fraser of the state roads department, designed conservatively; the stiffening truss was 2.3-m deep, with 61-cm strips of open steel grating straddling the 8-m wide roadway and a 1.2-m wide swath of open steel grating comprising its median. J.A. Roebling's Sons erected the towers in less than half a day; another 17 days saw wire rope cables, suspenders, trusses, floor beams, and road deck in place. The bridge cost just $275,000.

Latin America produced a number of spans. A 136-m Chilean span completed in the late 1940s featured—unusually for New World practice—concrete towers. American Bridge built three spans in Columbia: a 263-m span across the Magdalena River on the Bogota-Amero Highway and other spans of 210 and 111 m. It also completed the Puente Duarte across the Rio Ozama in 1956 in the capital of the Dominican Republic, Santo Domingo. Rather spindly steel towers support this 175-m span, four-lane truss-stiffened suspension bridge. Although its $5 million cost was significant for its size, neither the Puente Duarte nor other early post-Tacoma spans provided a major test for the aerodynamic theories and remedies emerging from the Tacoma Narrows Bridge collapse. This situation, however, would soon change.

6.2 THE REBIRTH OF THE TACOMA NARROWS BRIDGE

The most immediate results of the investigations following the collapse of the Tacoma Narrows Bridge were found in the major spans of the initial post-war period. These truss-stiffened bridges marked a retrenchment from the elegance and grace that characterized the plate girder spans of the 1930s; with the failure of the Tacoma Narrows Bridge, this genre disappeared virtually overnight. As Elizabeth Mock (1949) remarked in *The Architecture of Bridges*, from an architectural and aesthetic perspective, "The immediate response was unfortunate, for apprehensive engineers and public officials hurriedly retreated to the apparent safety of deep, ungainly stiffening trusses—the obvious, orthodox antidote to excessive flexibility in suspension bridges."

This return to conservatism was evident in the first large post-war span, a replacement for Gertie herself (*see* Fig. 6-1). Built in just 30 months between 1948 and 1950 under the direction of Charles Andrew, this bridge incorporated many of the features recommended in the wake of the Tacoma Narrows Bridge collapse. Andrew (1949) described its design as "without precedent" and "one which, in a major part, fell outside the structural field and into the aeronautical engineering field." The design of the bridge had been pretty well fixed by February 1946 after a sloth-like pace imposed by war demands. Andrew's (1949) design philosophy was to "set up a condition in which the aerodynamic forces would tend to fight or nullify each other and, at the same time, reduce drag." The open configuration of a truss with smaller steel members offered a means by which to do this. Still, as it was not possible to completely eliminate the aerodynamic forces acting on the bridge, efforts were made to dissipate excess energy through stiffness, weight, and mechanical devices.

The 10-m deep truss—more than four times the depth of the plate girders of the former span—is the most visible outgrowth of the research (*see* Fig. 6-2). The new bridge's depth-to-span ratio of almost 1:85 was the deepest since the Manhattan Bridge. Its four lanes—twice the number in the original span—sat on a 18.3-m wide floor truss that provided a secure 1:47 width-to-span ratio; 60-cm air gaps separated the road deck from the truss edges. The reconstituted Tacoma Narrows Bridge was also the first large span to use top and bottom lateral bracing to ensure a high degree of torsional stability; wind tunnel tests suggested this measure to increase the span's natural torsional frequency by a factor of almost 2.8, providing even greater separation from an unchanged vertical frequency. Yet Andrew's design engineers—Dexter Smith, N.E. Olson, and E.H. Thomas—would incorporate other pioneering features.

Tests were performed on the 1:50 scale aerolastic model of the full bridge during the war in the new University of Washington wind tunnel. Yet a lingering suspicion remained that downward winds coming off nearby bluffs could deflect off the water and angle up into the bridge, inducing torsional oscillation. This concern was confirmed when wind tunnel tests of sectional models with a solid deck even suggested the potential for catastrophic motion. After experimenting with various alterations to the truss leading edge, engineers devised the idea of placing slots between and bordering the traffic lanes. These slots were intended to impede the development of pressure concentrations and at the same time break up the wind.

Fig. 6-1: Tacoma Narrows Bridge, 1950
PHOTOGRAPH COURTESY OF R.A. DORTON.

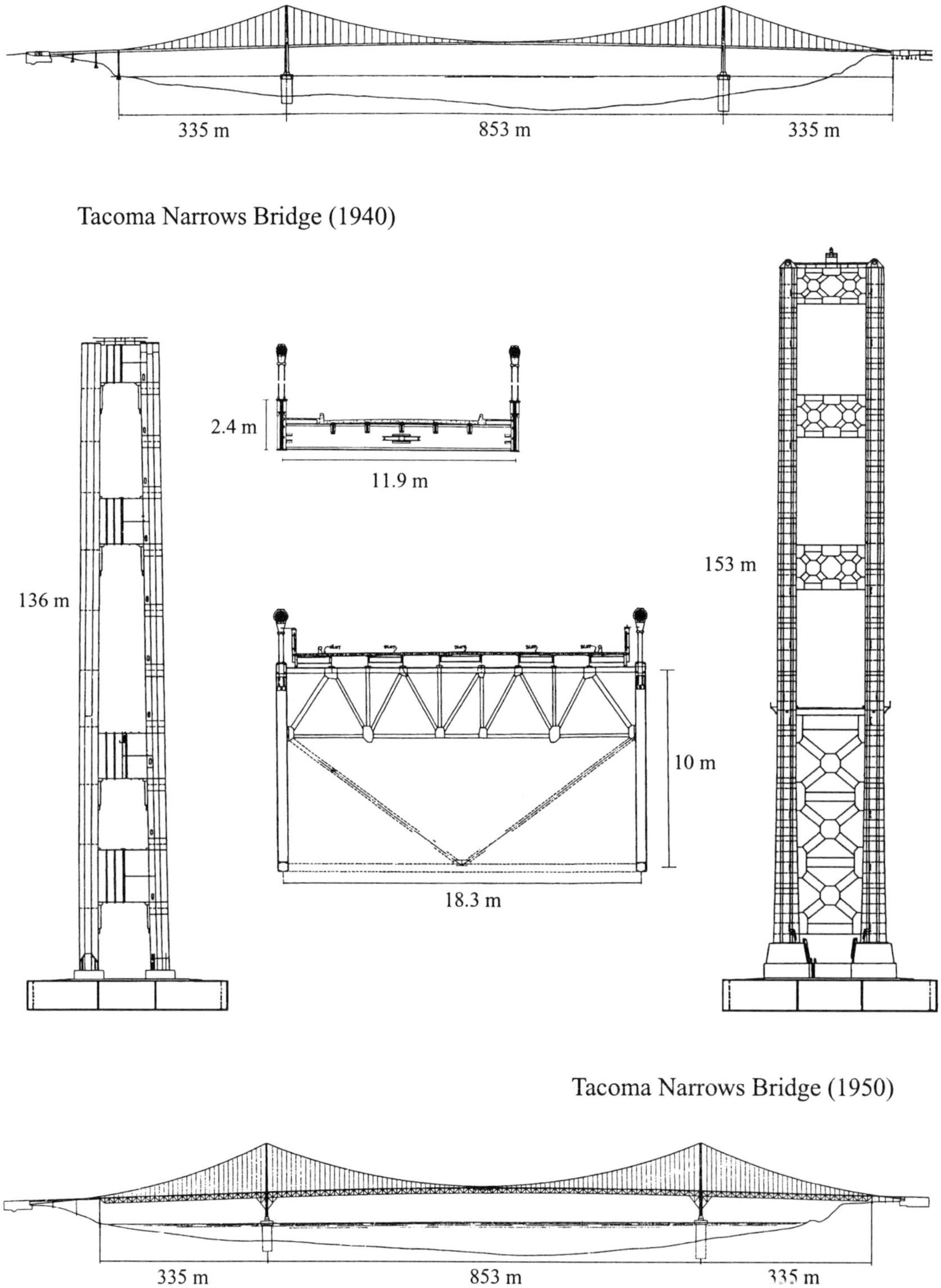

Fig. 6-2. Comparison, first and second Tacoma Narrows bridges
MODIFIED FROM ORIGINAL DRAWINGS COURTESY OF THE WASHINGTON STATE DEPARTMENT OF TRANSPORTATION

Differing combinations of slots tested on sectional models revealed great variation in their effect on aerodynamic stability. The adopted configuration involved three 84-cm wide slots of open steel grating separating all four 2.75-m wide lanes, with 46-cm slots at the curbs. No vertical motion could be induced in wind tunnel tests on sectional models, and torsional motion was limited to what would be less than 8 cm on the actual bridge. When tested with a single lateral design, the slots reduced a susceptibility to catastrophic flutter to a benign, self-limiting oscillation.

Other motion-damping apparatus included large hydraulic buffers at the towers. Struts also braced the bottom of the truss to the tower legs just above the pedestals to act as shock absorbers in the unlikely event of large wind-induced oscillation. They would dissipate energy that might otherwise induce oscillation in the main or side spans (ironically, an inspection soon after the span's opening revealed the buffers had seized from salt-air corrosion). The main cables were also rigidly attached to the truss at midspan via six trussed hydraulic cylinders—a first for a long-span suspension bridge—and open trusses instead of solid girder floor beams supported the floor slab. Finally, the ends of the side spans were firmly anchored to ground. Together, these measures made the new bridge some 58 times stiffer than the original. An increased sag ratio from the 1:12 of the old span to 1:10 was also chosen, the expense of taller towers bringing reduced cable stresses and consequent reductions in cable wire and anchorage concrete.

One other challenge involved the remnants of the original bridge. Could the tower piers support a heavier four-lane bridge, and could they survive earthquake forces? Seventeen earthquakes had occurred since 1939 and two 1946 tremors had rated 5 on the Richter scale. The collapse of the original bridge had also stressed the piers. Fortunately, studies indicated the piers had not shifted and could easily support the new span; in fact, the original superstructure contributed no more than 11% to bedrock pressures emanating from the foundations, and so the 50% to 60% heavier superstructure of the new span proved insignificant. Recycling the tower piers and anchorages (to which additional bulk was added) saved up to $3.5 million.

The new 153-m tall towers—17-m higher and 3-m wider than their predecessors—would be multiply braced with parallel legs comprised of just five cells in cross-section. The wider tower base actually improved the distribution of loads on the foundations, making Leon Moisseiff's decision to retain the original piers when he lessened the width of the original span a fortuitous one. The pedestals were raised to 12-m above high tide—5-m higher than in the original span—to eliminate corrosion from salt spray that had plagued the lower sections of tower steel in the former bridge. The three supports for the west approach span had been virtually undamaged by the collapse and, with minor modifications, were re-used. Lightweight concrete also increased deck weight by only 1.6 times more than the original span, despite a doubling of traffic capacity. The anchorages, however, presented a problem. The initial plan was to retain the original eyebars by slanting the main cables inward as they splayed out in the anchor chamber. In the end, 4 m of concrete along the anchor block sides had to be blasted away to remove the old eyebars. New concrete enclosed more widely spaced eyebars, and an additional 6.1 m thick wall of concrete was added at the back face of the anchorages to resist the increased pull of the larger 53-cm diameter cables. Ultimately, about 65% of the original concrete was salvaged.

By April 1946, final designs had been approved; the bridge was expected to cost \$8.5 million and take 22 months to build. Yet post-war shortages made it impossible for contractors to secure sufficient steel, and only in August 1947 was the project tendered. Bids were opened in October, but the delay was costly; low bids of \$8.3 million from Bethlehem Steel for the towers and trusses and \$2.9 million from J.A. Roebling's Sons for the main cables totaled \$11.2 million, one-third above the Toll Bridge Authority's estimate. Both contractors agreed to keep their bids alive while the Toll Bridge Authority tried to raise money at a time when—ironically—it was also having difficulty arranging insurance for the bridge. A December 1947 bond sale of \$14 million attracted no investors. A March 15 financing deadline now loomed along with a possibility of a retendering. In desperation, the Toll Bridge Authority turned to the RFC, but three days before the deadline a private syndicate purchased the bonds. Contracts were awarded at the end of March 1948, and work began almost immediately.

Several innovations sped construction. A steel frame impregnated with adjustable screws allowed the pedestal tops supporting the tower steel to be leveled quickly, replacing weeks of meticulous grinding. Tower assembly incorporated a novel triangular outrigger rising 15 m from the piers to help statically brace the towers before the main cables did the job permanently; this steel later found its way into the stiffening truss. By April 1949, the east tower stood complete save its top portal strut, and a cable saddle had been lifted onto a temporary platform atop its south leg. Equipment was ready for transfer to begin erection of the west tower. Then, just before noon on April 13, an earthquake measuring 7.2 on the Richter scale—the largest to hit the Pacific Northwest in a generation—struck. The oscillation of the east tower toppled the 19- tonne saddle off its platform, sending it on a 155-m plunge into 43 m of water. En route, it tore through—and sunk—a work barge along with a large amount of equipment. Major delays and remanufacture of the saddle in eastern steel mills loomed if it could not be recovered undamaged. As it turned out, finding the saddle proved the largest challenge, but it was intact, save for a small and easily repairable bend in one corner of its base. Ten days later, the saddle was rehoisted to the tower top.

Hardly had this problem been resolved when in June fire broke out at the base of the west tower, destroying a timber fender. Luckily, the tower suffered only minor damage, but the flames, fed by oil and gasoline, reached scaffolding some 137 m above the Narrows; hoisting engines and derrick booms to be used in cable spinning were extensively damaged. The wreckage of the first bridge also forced the use of a hauling line between towers to raise the first catwalk wires, avoiding immersion and possible entanglement. A severe fall and winter of 1949 and 1950 bore out concerns over the rescheduling of cable spinning occasioned by the delayed construction start. Rain and snow, often accompanied by high winds, fell almost continuously for 6 weeks. This hampered all work but was especially harsh on the spinning of the 458 wire strands into each 8,702-wire main cable. Frozen moisture on freshly spun wire prevented adjustment and banding into strands until the wire could be thawed out. This proved no minor irritation, as it set J.A. Roebling's Sons back almost 9 weeks. Fortunately, a benign spring permitted truss erection to progress swiftly to the span's October 1950 opening. Although the \$13.6 million span shunned the sleek profile of its predecessor, it is surprisingly graceful nonetheless, thanks in large part to its slender, well-proportioned towers.

The new Tacoma Narrows Bridge was the most technically advanced suspension bridge of its day. Its designers may even be faulted for being overly cautious, for many of the span's motion damping devices will likely never be tested, as even Andrew admitted. Yet the climate surrounding the bridge's design was one of uncertainty; no one wanted to ever again craft a design with an even remote possibility of emulating Gertie's behavior. Interestingly, the opportunity to generate confidence in the design soon presented itself. On several occasions between January and March 1951, winds of up to 120 km/hr struck the bridge for periods of up to 30 minutes. Public curiosity in the performance of the new span during these storms was quite high given the disastrous pedigree on which it was built. Although the truss deflected laterally up to 1.2 m, no vertical or torsional oscillation ever made an appearance. The bridge stands today as an epoch-making structure.

6.3 OTHER EARLY POST-TACOMA EFFORTS

Evidence that a stiffening truss was not a panacea in itself was soon revealed at the Golden Gate Bridge. The span had emerged unscathed from its 1938 and 1941 wind encounters, but concern remained. Research by Frederich Bleich and tests on sectional models of the Golden Gate Bridge in connection with the new Tacoma Narrows Bridge design had indicated the span to be susceptible to serious torsional oscillation in winds of 65 to 80 km/hr and in gentler winds approaching at a slightly upward angle. These tests had, in fact, encouraged the use of slots in Gertie's replacement. Three storms at the Golden Gate Bridge between February 1949 and June 1950 then generated vertical motion with a double amplitude of up to 1.1 m. The last storm also—for the first time—generated torsional motion, with eight cycles per minute observed. Fortunately, none of these events inflicted any damage.

Then on December 1, 1951, a major storm slammed in to the Golden Gate. Over a period of 4 hours, gusts exceeding 110 km/hr struck the span obliquely from the southwest. As the storm strengthened, the vertical wave frequency increased from eight to 10 per minute. Torsional motion was then incited in what appeared to be a coupled flutter oscillation of almost 17 cycles per minute.[2] The southeast quarter-point of the main span was observed to twist with a double amplitude of up to 3.3 m; some light standards even rubbed paint off the main cables, and the truss was blown more than 3.5 m off its center line. The bridge was closed for the first time ever. Amazingly, other than severe damage to connections at the towers—whose allowance for limited longitudinal movement of the main span truss had been exceeded—injury was surprisingly light. The motion was found to be remarkably similar to that predicted in sectional model tests; it was now clear that the span's single plane of lateral bracing was inadequate.

In May 1952, Clifford Paine summoned Andrew and Ammann—the latter engineer then aged 67—to help fortify the bridge. They concluded that the rugged north shore had disrupted the angles of attack of wind direction for which the bridge was most susceptible and thus prevented more incidences of torsional motion. Their January 1953 report recommended the addition of bottom lateral bracing, a measure that would raise the span's critical velocity from 80 to 120 km/hr and make it almost three times stiffer in torsion. F.B. Farquharson tested the design in the University of Washington wind tunnel. Center cable clamps as adopted in the new Tacoma Narrows Bridge could also be added as insurance, but open steel grating sidewalks did not perform well in the tests. Installation of the bottom lateral bracing was finished in late 1955 after 18 months of work at a cost of $3.5 million. The main cables easily accommodated the 4,500 tonnes of new steel, which increased axial cable stresses by just 5.5%. The truss has behaved satisfactorily since, with no additional restraining measures.

During this period, Ammann's mighty George Washington Bridge revealed some aerodynamic foibles. In 1951, coupled vertical and torsional motion occurred in waves up to 1.8-m double amplitude. The motion was harmless but highlighted the fact that the span's great weight alone was insufficient to quell its onset. When Hurricane Hazel slammed into New York in October 1954, the deck was again induced into harmonic motion. Winds of up to 182 km/hr blowing at a 45° angle to the bridge translated into a head-on velocity of 128 km/hr. This condition was cited in a somewhat controversial assessment in the late 1960s by Daniel Dicker, who, in calculating the span's critical velocity to be a shade higher, suggested that had such winds persisted the bridge would have been "closer to failure, for a short time . . . than is generally realized" (Anonymous, 1969).

Gertie's replacement heralded a new generation of large spans characterized by rigid and relatively deep trusses. It was soon followed, in 1951, by the Delaware Memorial Bridge on the U.S. eastern seaboard over the Delaware River some 50 km below Ralph Modjeski's Benjamin Franklin Bridge at Philadelphia, Pennsylvania (*see* Fig. 10-1). Efforts to bring about a crossing had begun in 1917 when a bridge was proposed at Wilmington, Delaware, just upstream of the present span. Ferries sprang up along the Delaware River south of Philadelphia in the early 1920s, and in 1926, a bridge and a tunnel at Wilmington competed for attention. The tunnel was squelched in 1937 by Delaware legislators who opposed its construction with private finance.

Meanwhile, a 1932 plan for a bridge or tunnel south of Philadelphia faltered. Delaware's efforts had been further aggravated by a long-festering dispute with New Jersey over jurisdiction of the river itself. The issue was finally resolved in Delaware's favor in 1934, but not before landing in the U.S. Supreme Court. In 1939, the first state-sanctioned studies were launched and were completed in 1941.[3] War thwarted further progress, but in April 1945, a bridge or tunnel 5 km south of Wilmington was authorized with the stipulation that costs not exceed $25 million. A four-lane, 655-m span suspension bridge with 229-m side spans was subsequently adopted that would be the sixth longest in the world. Other approvals were obtained in 1946 and 1947, most notably from the War Department, which was concerned with navy shipyards upstream of the proposed span.

In the meantime, the firm of Howard, Needles, Tammen & Bergendoff—with Ammann—resolved details of the span's design. (In 1946, Ammann had formed a partnership with an expert in reinforced concrete by the name of Charles S. Whitney; the firm continues today.) Not surprisingly, aerodynamic stability was a priority. Ammann's stiffness index was used to help shape the truss, which features top and bottom lateral bracing over its 6.1-m depth. Static load tests for torsional rigidity on a

1:12 scale sectional model at Princeton University under E.K. Timby suggested that the use of bottom laterals would increase torsional stiffness by a factor of 20.[4] With a depth-to-span ratio just shy of 1:110, the Delaware Memorial Bridge is proportionately slimmer than the second Tacoma Narrows Bridge, but the latter's array of motion-damping paraphernalia is absent. Ammann, for example, was not an adherent of open grid decks and slots. He shunned them in his own bridges, citing their discomfort for motorists, and had criticized the use of slots in the new Tacoma Narrows Bridge in a discussion of Andrew's 1949 paper on the bridge's design. Only vented curbs and the placement of the deck slightly below the level of the top chord of the truss, which also acts as a traffic barrier, mark any unusual design treatments.

With the emergence of more detailed designs and inflationary pressures, the former $25 million price tag had now become $40 million. Construction bids in January 1948 on the tower and anchorage foundations also proved substantially higher than expected. The state authorized the raising of an additional $15 million, but plans for a $60 million six-lane span were now permanently shelved. In fact, this issue almost scuttled the project completely. Only last-minute interventions of the consulting engineers that proved a four-lane span could be built within the budget convinced the authorities to proceed. Construction finally began in July 1948.

With bedrock depths of more than 67 m at the towers, it was decided to toe the tower piers into shallower clays using open dredged caissons. The 29- by 67-m dimensions of the Delaware anchorage caisson made it the largest built to that time; it was sunk to a depth of 28 m. The shallower but slightly larger New Jersey cofferdam involved the then-largest continuous underwater (or tremie) pour of concrete when an almost 10-m thick base was poured without interruption for more than a week in 1949. The two main cables, each with 8,284 wires spun from 19 strands of 436 wires, were completed in early 1951 by U.S. Steel subsidiary American Bridge, although spinning operations were dogged by high winds. Then in March 1951, during truss erection, a storm-tossed ocean freighter rammed into the New Jersey tower pier, inflicting severe—but fortunately only cosmetic—damage that was limited to its masonry casing. A month later, after the stiffening truss was in place, cost overruns then threatened to delay the bridge's opening. A wage dispute also brought labor strife that ended only with a court order directing workers back to the bridge in June.

The Delaware Memorial Bridge was opened in August 1951 at a cost of $45.7 million and six lives. The turbulence surrounding the late stages of its construction contrasts with the simplicity and elegance of the suspension span itself. Its 134-m tall portal-braced towers are notable for their clean lines and simple, almost austere, appearance, a consequence of minimal detailing by A. Gordon Lorimer, the consulting architect. The free hang of main cables above the truss at midspan adds to the span's simple lines, although the deep approach span trusses and large anchorages contrast with—and almost dominate—the shallow profile of the suspension span. Considering the era of conservatism from which its design emerged, the Delaware Memorial Bridge is a noteworthy aesthetic achievement. The efforts of its designers were rewarded by the AISC, which cited the Delaware Memorial Bridge as the most beautiful long-span bridge of 1951.

Several smaller American suspension bridges soon followed the Delaware Memorial Bridge. The first of these, the Chesapeake Bay Bridge (now the Preston

Lane, Jr. Memorial Bridge), connects the two halves of Maryland southeast of Washington D.C. It is perhaps best known for its overall length of 6.5 km and the smorgasbord of bridge types used in its design. The 488-m span suspension bridge that forms the crossing's centerpiece did, nonetheless, incorporate several advances in construction technology (*see* Fig. 10-3).

Five decades of proposals preceded construction. A late 1920s plan for a $10 to $12 million bridge foundered because of the Depression. Bridge plans were authorized in the late 1930s, but this effort, too, was stymied, this time by World War II. In 1947, the firm of J.E. Greiner was retained to design a bridge. Ole Singstad, previously involved in the building of New York City's Holland Tunnel under the Hudson River, investigated a tunnel here, too, but it was found to cost almost $10 million more than J.E. Greiner's $35 million two-lane bridge. Approval for the bridge was obtained in 1948, but as contracts were tendered in 1949, an eleventh-hour lawsuit sought to force a call for tunnel bids as well (presumably to protect Baltimore's shipping access). Its rejection paved the way for construction to begin.

The project faced many hurdles, including turbulent tides, high winds, inclement weather, and poor soils. Deep bedrock forced designers to support all piers—including towers and anchorages—on steel piles driven up to depths of 62 m into the Chesapeake Bay muck. In fact, the almost 16,000 tonnes of steel piling was twice the amount of steel used in the superstructure. One tower pier also used a "bell-bottomed" caisson—for only the third time in bridge construction. Each anchorage rests on a large artificial island, and each is supported by a cofferdam which in turn is propped up by 600 piles. Almost totally submerged, the anchorages are revealed only where the main cables exit the anchor chambers. The islands also protrude toward the cable bents to protect the gently rising cables from ship collision.

The Chesapeake Bay's reputation for capricious weather held true. Within a week in November 1950, two storms—one a hurricane—battered the unfinished piers with more than 110-km/hr winds and high tides. One approach pier was destroyed, steel forms for deeper piers severely damaged, the east approach causeway eroded, and a bargeload of piles lost (the piles were later located 11 km from the bridge, but the barge was never found). A month later, part of a steel form in an approach pier collapsed during concreting. Then in December 1951, a derrick boat was wrenched from its mooring and on its wayward journey severely damaged several approach span piers. Six days later, tugs lost a deck truss approach span in strong winds; after striking and damaging another assembled truss, the span floated free, grounding on the coast downstream. The falsework holding up the truss happened to be its future neighbor; both units were disassembled and rebuilt by splicing in new parts.

The span's superstructure involved extensive pre-assembly. The 36-cm diameter main cables support 890 m of suspended span between cable bents, but extend 1,220-m between anchorages and were, at the time, the world's longest helical strand cables. Their erection involved only a third of the time estimated for spinning, as most work, rather than performed high above Chesapeake Bay, was carried out in the factory. The stiffening trusses were assembled on shore in unusually long units of four panels, again reducing work at the bridge site. Approach span trusses were assembled on falsework and barged into place.

The $45 million bridge was opened in July 1952. Perhaps unfairly, the

Chesapeake Bay Bridge has not generally been recognized for its technical innovations and construction methods. Its kaleidoscope of span types and great length have brought it more recognition, some of which has not been complimentary. Christopher Tunnard and Boris Pushkarev (1963), in *Man-Made America: Chaos or Control?,* may have been targeting the Chesapeake Bay Bridge in stating that, "a long bridge consisting of a hodgepodge of every structural system in the book—girder, truss, cantilever, suspension span—may be cheaper than to stick to the dignity of one structural system throughout. But the result is visually disastrous." Even the suspension bridge, taken alone, is not a great visual success. The sweep of the main cables to the bottom chord of the truss reflects an earlier era, as does, superficially, the through truss. Its flat vertical curvature and high passage through narrow 108-m tall towers (adorned with no less than seven diagonal braces) contribute to a somewhat stiff, utilitarian appearance. Aerodynamic considerations did influence the design of the narrow two-lane truss; top and bottom laterals were used for torsional resistance—which is unusual for a suspended through truss—and the center of each lane is also formed from open steel grating to allow wind passage.

One unique bridge of this period spans the Missouri River in Kansas City. Completed in 1954, the Paseo Bridge is one of a handful of American self-anchored suspension bridges and, until San Francisco's East Bay replacement span is completed, the nation's sole example since World War II. In fact, its 188-m span made it the longest of its type in the world—all three longer European spans were destroyed in the war. It would retain this title for almost 4 decades. Designed by Jacob Karot of Howard, Needles, Tammen & Bergendoff, the Paseo Bridge was the first American post-Tacoma span stiffened by narrow, spine box girders. Its relatively small span, a generous 3-m girder depth, and a wide four-lane roadway guaranteed torsional stiffness, enhanced by a necessarily heavy suspended structure to obviate the need for anchorages. Its two 30-cm diameter main cables were formed from 37 preformed helical strands, and the use of low-alloy (in place of silicon) steel in the towers and spine box girders generated weight savings of 10%. Interestingly, open steel grating was used in the median and flanking sidewalks to enhance aerodynamic stability, although wind tunnel tests performed on a sectional model of the span by the Advisory Board on the Investigation of Suspension Bridges suggested a lowering of critical velocity in comparison with no slots.

Construction of the $7.7 million bridge began in October 1952. There were few hitches until early 1954, when the cantilevered box sections were to be bolted together at midspan. But the sections were found to be so misaligned that the bolt holes could not be matched up. The remedy involved moving an armada of 16 heavily laden trucks weighing in at 160 tonnes onto each 94-m side span; this raised the center span and allowed bridgeworkers to connect up the two halves. The bridge was finished without further incident and opened in August 1954 after less than 2 years of construction. In spite of the large proportions of its stiffening girder, the Paseo Bridge possesses a grace and simplicity uncommon to small suspension spans. Its otherwise squat, 41-m tall towers feature slender steel legs. Unusually, the continuous girder floats through the towers above a lower portal strut. Such features helped the Paseo Bridge earn the AISC award for the most beautiful long-span bridge of 1954. It would also be only one of two American post-Tacoma suspension bridges to be stiffened by box girders in the 20th century. Self-anchored spans generally would remain rare—the nascent cable-stayed bridge would not only meet, but exceed, their efficiency for such span lengths.

The first large Canadian suspension bridge since Vancouver's 1938 Lions Gate Bridge was opened the following year in the Atlantic port of Halifax, Nova Scotia. Two wooden rail trestles in the late 19th century had in fact crossed the harbor to neighboring Dartmouth, but both had quickly succumbed to the elements. A 1920 proposal for a low-level span languished. The same fate awaited a 1929 plan for a $3 million 293-m span bridge north of downtown, which was killed by a change in government and the Depression. In 1948, P.L. Pratley was retained to study bridge and tunnel options. Two years later, a bridge commission was created, and in 1951, plans were successfully initiated to replace the congested ferries. In designing the 441-m span Angus L. Macdonald Bridge, Pratley virtually mirrored his previous design for the slightly larger Lions Gate. Its 94-m tall towers, for example, are scaled-down versions of the diagonally braced pylons of its forerunner. Similarly, its two-lane roadway—in a departure from prevailing post-Tacoma practice—was, like the Lions Gate, placed on a 4.6-m deep through truss lacking upper lateral bracing. The truss itself was hung from 61-strand helical wire cables extending 955 m.

The span's construction was not without its perils, for when the stiffening truss was almost fully erected in September 1954, Hurricane Edna ripped through Halifax. Gusts of up to 138 km/hr battered the temporarily connected truss sections, deflecting them laterally as much as 3.7 m out of line. Vertical motion was so violent that it snapped temporary connections, causing some unattached truss sections to dangle precariously over the harbor. Amazingly, damage was only slight, and the span—flanked by 160.5-m side spans—was completed in June of the following year without further incident.

These efforts of the early years following the Tacoma Narrows Bridge collapse set the stage for the first long spans to rival the behemoths of the 1930s. Here, the challenge of aerodynamic stability would truly test emerging theories.

ENDNOTES

1. For a discussion of orthotropic decks, *see* Chapter 8.
2. Robert Scanlan and Joseph Vellozzi (1980) ascribed the cause to buffeting.
3. Louis Balog (1945), in a discussion of David Steinman's 1943 aerodynamics paper, mentions a 914-m span proposed in 1939 for the Delaware River whose floor system sat atop the plate girders, instead of the mid-girder "H" configuration used in the original Tacoma Narrows Bridge and other spans of its era. Balog cites Ammann's use of a similar suspended structure for an unbuilt 1939 proposal for a twin-span suspension bridge between Manhattan's Battery Park and Brooklyn.
4. This figure accounts for the truss, unsuspended; its magnitude in terms of stiffening the entire suspension bridge would be lower. Timby, who had worked for predecessor firm Ash-Howard-Needles & Tammen sporadically since 1938, left Princeton University to join Howard, Needles, Tammen & Bergendoff permanently in 1950. He remained a partner with the firm until 1968.

Chapter 7

THE LEGACIES OF TWO GIANTS

The 1950s would both close and herald work on America's two largest post-war suspension bridges. These spans would also close the careers of two masters of 20th century American bridge engineering. David Steinman would fling his masterpiece across the turbulent waters of northern Michigan; Othmar Ammann would remain in the city that established his reputation, New York. Both spans today stand among the world's great bridges.

7.1 STEINMAN'S TRIUMPH AT MACKINAC

The Straits of Mackinac provide the major outlet for Lake Michigan to the lower Great Lakes. The immensity of the 8-km gap, combined with its great depths and strong currents, long stymied efforts to link the upper and lower peninsulas of Michigan. Enormous ice sheets grind through the Straits in bitter winters, and the area is subject to fierce and sudden storms. Widespread belief that the Straits would thwart attempts to be bridged proved almost as difficult to overcome as the huge physical and technical obstacles. The 70-year campaign for a crossing here would touch almost every eminent American bridge engineer of the first half of the 20th century.

A Western Union telegraph cable had been laid across the Straits as early as 1873, and it was followed in 1881 by the rail ferries. The completion of the Brooklyn Bridge 2 years later inspired a local newspaper advertisement depicting the bridge as "A Glimpse of the Future—Proposed Bridge Across the Straits of Mackinac." With rail access, tourism soon flourished, and in 1887, the Vanderbilt railroad empire built the renowned Grand Hotel on nearby Mackinac Island. Renewed—but short-lived—interest in a bridge emerged. The issue languished until growing summer automobile traffic prompted a 1921 proposal for a "floating" tunnel anchored to bedrock. A state-run automobile ferry began in 1923, and its popularity prompted state interest in a bridge 5 years later. Studies suggested a bridge could be built for $30 million, thought low enough to entice private investment. Financing was even partially arranged when the scheme was dropped.

The Depression imposed a surprisingly short-lived effect on ferry traffic, and in 1933, the state was again studying a crossing. Federal depression relief dollars were funding such projects as the Transbay Bridge; such prospects encouraged the state to create a Mackinac Straits Bridge Authority in 1934. Soon after, Steinman—who back in 1914 under Gustav Lindenthal's employ had examined a Straits crossing—met with Michigan Governor William Comstock to discuss the project. Charles Evans

Fowler was also on the hustings; he had a penchant for spectacular bridge schemes and proposed a four-lane, double-track road/rail span extending north from Cheboygan and touching down on Bois Blanc, Round, and Mackinac islands en route to St. Ignace.[1] The plan was viewed with some skepticism, but in April 1934, the Bridge Authority retained Fowler as its interim chief engineer. The centerpiece of the scheme was a 396-m span cantilever bridge, but almost 10 km of causeway were included in a circuitous 39-km route that considerably exceeded the 8-km width of the Straits. In August 1934, the Bridge Authority requested the RFC finance 70% of the scheme's $35 million cost and sought a PWA grant for the remaining 30%. The bid was unsuccessful, and interest from the railroads was tepid. Moreover, local concern over the convoluted route had surfaced. In July 1935, the PWA rejected the application, citing grossly underestimated costs.

The Bridge Authority must have had its own misgivings, for on the day of the PWA decision it abandoned both the proposal and Fowler. It now adopted a more direct line recommended by the Bridge Authority's secretary, James H. Cissel of the University of Michigan. Developed with Francis C. McMath, the bridge headed east from Mackinaw City, turned north over the Graham Shoals, and then angled northwest toward St. Ignace. Like Fowler, Cissel proposed a cantilever over the main shipping channel, but he increased the span to 518 m. Several continuous cantilevers with alternating 183- and 274-m spans flanked the main span, bringing the cost of the two-lane and single-track bridge to $32.4 million. The Bridge Authority filed a second application for PWA funding in September 1935, 2 months after the Fowler proposal had been rejected.

At that time, the Straits drew the eye of the Corps of Army Engineers, a consequence of political pressure on President Franklin D. Roosevelt. Its studies—released before a PWA decision on the Cissel scheme—offered no specific proposal, but they did assert the technical and economic feasibility of a bridge. By this time, too, the George Washington and Golden Gate bridges had weakened technical arguments against a crossing. Yet not even the purported support of President Roosevelt was sufficient; in December 1935, the PWA rejected the Cissel plan.

The Bridge Authority remained undaunted and in November 1938 retained the firm of Modjeski & Masters to prepare new plans. (Ralph Modjeski had retired in 1937 and would die in 1940.) Frank Masters in turn consulted Leon Moisseiff, fresh from Ammann's Bronx-Whitestone Bridge and the design of the Tacoma Narrows Bridge, which was then about to enter construction. Moisseiff examined several unusual alternatives, including multiple continuous suspension bridges with spans of up to 1,097 m braced by tie cables up to 4,572-m long. His handiwork was unmistakable when plans for a three-lane bridge on a direct line between Mackinaw City and St. Ignace were presented to the Bridge Authority in June 1940. The plans called for tandem plate girder suspension bridges of 1,402 and 899 m span sharing a common anchorage; they would rank as the world's longest and fourth longest spans. Side spans of 488 m flanked the larger span; in the smaller bridge, which spanned a deep secondary gorge, 320-m side spans were used. The tallest of the towers would rise 160 m above the Straits. Both bridges would exceed the span of any plate girder suspension span yet built, demonstrating the remarkably rapid confidence in, and embracing of, this form.

Economy of design was foremost. Masters had rejected a single-track railroad, citing increased costs of 20%. A cantilever had been considered for the subsidiary span, but this, too, was costly. A fourth lane added $16 million and was also rejected. The scheme, as presented, would cost $26.7 million. Most intriguing was the design collaboration with Moisseiff. Even the portal-braced towers were similar to those of the Tacoma Narrows Bridge, as were similarly proportioned side spans. Most striking was the 3-m deep plate girder, which in the larger span produced a 1:460 depth-to-span ratio, far shallower than even the stunning 1:350 proportions of the Tacoma Narrows Bridge. Minor vertical oscillation had by this time been observed on the Tacoma Narrows Bridge prior to its completion, and Masters recognized the potential for such oscillation at Mackinac:

> on account of the great length of the larger suspended span and the resulting slenderness ratio of width to length, in order to determine if a span of these proportions might be subject to oscillations or periods of harmonic vibration due to external wind or live load forces in such a magnitude as to effect the driving conditions, a small scale model of the span should be provided to dampen or control such movements . . . (Modjeski & Masters, 1940).

Masters had designed the three-lane road deck of open steel grating to reduce weight, snow and ice blockage, and uplift from wind. The 10-m wide roadway was cradled within a 15.2-m wide girder whose 1:92 width-to-span ratio would have given Mackinac the narrowest proportions of the large suspension bridges. Whether oscillation would have been a problem, and whether the open grating would have inhibited it, remains an unanswered question.

An autumn 1941 construction start was proposed, pending RFC financing and the preparation of detailed designs. A large proportion of the cost would be borne by the foundations—the seven main piers would cost as much as their $9 million superstructures and require sinking to depths of 61 m in one short construction season. The bitter cold would impose winter shutdowns and prolong construction until late 1946 or early 1947.

The opening of the Tacoma Narrows Bridge in July 1940 followed Masters's report by less than a month. Four months later, the twisted wreckage of Moisseiff's masterpiece lay on the floor of the Narrows. The similarity between the two designs and Moisseiff's association with them evoked inevitable comparisons. Yet the factors leading to Gertie's collapse were as yet unconfirmed. As it happens, the threat of World War II forced the Bridge Authority to postpone construction. Had war been averted, the project might have proceeded, although in light of the Tacoma investigations, it is almost certain that the plate girder would have been abandoned; in fact, Masters would represent the Bridge Authority on the 1943 through 1954 Advisory Board investigations. The scheme's abandonment was not so regrettable in visual terms; a single suspension bridge would have been preferable to asymmetric twin spans. Yet not all was lost. In 1941, the Bridge Authority began construction of the 1,750-m causeway recommended by Masters through the shallows south from St. Ignace. The 1,280-m length built at a cost of $700,000 would eventually reduce costs by more than $3 million. Soundings and borings would also aid later design efforts.

With the end of war in 1945, the state invested heavily in an improved ferry service. But its 1947 abolition of the Bridge Authority inflamed bridge supporters. A state-wide advocacy group was formed in 1949, and Steinman was asked to comment on the feasibility of a bridge. During the war-imposed hiatus in bridge building, Steinman had waged his aerodynamics campaign and published two books: *Bridges and Their Builders*, with Sara Ruth Watson in 1941, and in 1945, the year of partner Holton Robinson's death, *The Builders of the Bridge, The Story of John Roebling and his Son.*[2] Steinman had no doubt that a bridge could be built at Mackinac, but he could not speculate on its cost. Other, less sanguine speculation also emerged that year when a 1944 study by Kenneth Landes of the University of Michigan came to light questioning bedrock conditions and warning of a "possible disaster of shocking proportions" (Rubin, 1985). Landes would later recant, but his words had an impact.

A recalcitrant state government finally yielded and in 1950 created a new Bridge Authority. Ivan Crawford of the University of Michigan recommended three specialists in long-span design to advise the nascent authority. In July, Ammann, Glenn Woodruff, and Steinman held an inaugural meeting on Mackinac Island, marking the first time Ammann and Steinman had worked together professionally since the Hell Gate Bridge over 30 years earlier. All three were masters of the idiom and had contributed to the development of bridge aerodynamics in the aftermath of the Tacoma Narrows Bridge failure. Their preliminary report in January 1951 rejected a $141 million tunnel but considered a bridge feasible. It called for a four-lane, truss-stiffened suspension bridge following the 1940 alignment. Only the Golden Gate Bridge would exceed the 1,097-m span, whose 457-m side spans would also be record-setting. Two 65-cm diameter cables would hang from 172-m tall towers and support a record 14-m deep truss. The estimated $76 million cost was almost three times that of Masters' scheme, although a rejected rail line would have added another $60 million. The engineering studies also laid to rest foundation concerns when borings indicated rock could support four times the weight ever imposed by a bridge.

In spite of the unanimity of the three engineers, the long-festering animosity between Ammann and Steinman boiled over during the 7-month investigation as it became evident that one of the two, and not Woodruff, was favored to design the bridge. Yet the Bridge Authority was prohibited from retaining any of the three engineers to do so, nor did it have the power to arrange financing to build a bridge. Furthermore, the Korean conflict had sent steel prices spiraling upward; steel could only be obtained via special permit and appeared unavailable until at least late 1952. To add insult to injury, in January 1952, an unsupportive state government commissioned a new $4.5 million ferry. The state did grant the Bridge Authority financing powers in April 1952, and it could also now hire any engineer it wanted. Yet the state's refusal to provide bond guarantees for the span's financing created a quandry for the Bridge Authority during this period of post-war recession. Nevertheless, steel shortages had eased by early 1952, and application was made to the RFC to purchase $86 million in bonds.

The Bridge Authority hoped for a construction start in early 1953, but it still lacked a detailed design and an engineer to prepare one. An October 1952 meeting of the Bridge Authority's engineering committee to recommend an engineer saw that Steinman had been inadvertently advised not to attend; the other two did, and

Ammann, in addition to bluntly dismissing Woodruff's capabilities, is said to have derisively referred to Steinman's models as "toys" (Rubin, 1985). After much deliberation, Ammann was endorsed in a split decision. Ammann, however, required an initial payment of $1.25 million, a condition the Bridge Authority could not hope to meet. Steinman, learning of the impasse through a subordinate's chance contact with a Bridge Authority staff member, offered to accept deferred payments for his services.[3] In November, an informal agreement was reached, and in January 1953, Steinman officially landed the prestigious task of designing the bridge. Woodruff became Steinman's consultant. Ammann's intransigence cost him Mackinac, but he would ultimately have the final word in his rivalry with Steinman.

The Bridge Authority was not yet out of the woods. The RFC had rejected the Authority's funding application; the agency that had funded so many bridges in the 1930s was now being phased out. Moreover, higher interest costs had by December 1952 elevated the cost of the now 1,158-m span bridge to a prohibitive $135 million. Steinman modified the design to shave $35 million—$20 million from design changes, the rest from construction methods. Aerodynamic tinkering reduced the truss depth to 11.6 m, saving $5.5 million, whereas lengthened 549-m side spans saved another $1 million. Backstay spans pushed the anchorages to shallower rock, saving $2 million; sloped lower back faces saved another $2 million and distributed forces more efficiently. Foundations were topped off 2.4-m below the water, saving $7 million while reducing ice exposure. The use of piles in place of caissons for the foundations in the subsidiary gorge saved another $1 million.

Steinman renegotiated provisional construction contracts and, by March 1953, had reduced American Bridge's superstructure contract from $56 million to $41.5 million. Still, finance remained elusive, and there was no assurance that work would begin that year. To attract investors, the state assumed annual operating costs of up to $417,000, equal to yearly losses on the ferries. This had little effect, for a financing effort in the summer of 1953 failed. By autumn, the recession began to wane, and in December, just prior to expiry of the state's funding legislation and after a last-minute legal challenge, $99.8 million in bonds were finally sold.[4] Construction would begin in the spring of 1954. Waits of up to 20 hours in 27-km long traffic queues for a 50-minute ferry journey would end; Michigan would finally get her bridge.

The Straits' hostile environment would provide a challenging setting for Steinman's aerodynamic design concepts. Yet Mackinac is perhaps the safest and one of the most beautiful suspension bridges ever built (*see* Fig. 7-1). Its 1,158-m span ranked second to the Golden Gate Bridge when completed in 1957, yet the 2,543 m distance between anchorages, 2,255 m of suspended span, and 2,647 m of main cable established records that endured for more than 4 decades; Mackinac's 549-m side spans remain the second longest in the world (*see* Fig. 17-1). Excepting the towers—which were riveted—more than 1 million high-strength bolts were used for field connections, a first for a large span. Finally, in place of Masters' subsidiary suspension span is a 634-m continuous truss—then the world's longest—where one cofferdam reached a then-record depth of 31 m.

The Mackinac Bridge refutes the aesthetic dictum of short side spans advocated by Elizabeth Mock and noted German engineer Fritz Leonhardt; here, long side spans emphasize the great distance to be spanned.[5] Mackinac's 168-m tall towers are also

Fig. 7-1. Mackinac Bridge
PHOTOGRAPH COURTESY OF THE PARSONS CORPORATION AND THE MACKINAC BRIDGE AUTHORITY.

noted for the quality of their design and remain architecturally relevant today (*see* Fig. 13-3). Their four arched X-pierced portal struts represent the cleanest expression of a motif often used by Steinman. The off-center placement of the tower legs on small circular concrete pedestals and the low elevation of the truss at the towers also emphasize their height. The deep truss has evoked negative comment, again from Leonhardt (1982). Yet unlike many of Ammann's bridges, Mackinac derives its impression of strength through its stiffening truss, not its towers. The continuity of truss depth over almost 6 kilometers of bridge also lends unity to the entire ensemble; slender approach span piers enhance the boldness of the suspension span. Mackinac's two-tone color scheme is also noteworthy. Its "evergreen" trusses and "ivory" towers were inspired by—of all things—an advertisement depicting the future bridge as a backdrop in the July 1954 issue of *Fortune* magazine. No color scheme had yet been adopted, but the artist's intuition struck a resonant chord with the Bridge Authority and Steinman—and, apparently, the AISC, which bestowed its 1958 Artistic Award for long-span bridges on the Mackinac.

If the aesthetics of Mackinac's trusses are open to debate, their aerodynamic stability is not. No suspension bridge has ever approached Mackinac in this regard. Its generously deep 1:100 depth-to-span ratio was the most visible manifestation of

Steinman's design treatments of the 20.7-m wide truss (*see* Fig. 7-2). The extension of the truss 3 m beyond the roadway edges raised its critical velocity to an almost impossibly high 1,017 km/hr. The central two lanes of the suspension span were also built of open steel grating, which Steinman claimed would raise critical velocity to infinity—an improbable claim perhaps; 91-cm wide swaths of open grating also flanked the outer lanes. Finally, the cable and truss were bolted to one another at midspan to discourage relative motion.

Interestingly, Steinman would turn to old adversary F.B. Farquharson to conduct wind tunnel tests to verify the design.[6] Farquharson's tests took place at the University of Washington in 1954, and revealed a remarkable degree of stability. Sectional models representing 76 m of truss resisted all attempts to induce oscillation—hence, Steinman's extrapolated claims. Stability remained unimpaired even with the two central lanes covered to simulate icy conditions. The span's generous static wind load capacity also allowed for up to 6.1 m of horizontal movement in winds 2.5 times the area's highest recorded velocity of 125 km/hr. The Mackinac truss is a significant achievement that would in some ways bridge the gap between the two engineers. Nonetheless, fundamental disagreements remained, for if Farquharson had verified Mackinac's stability in aerodynamic terms, Steinman had fashioned them using his outmoded theories based on avoiding static, rather than dynamic, instability.

One eleventh-hour hitch: the discovery of a rock hump beneath the north tower pier threatened a complete redesign. The solution, to simply move all 34 piers northward by 9 m, overcame the last hurdle to construction. Work on the six main foundations began in the spring of 1954 and involved what was then the largest peacetime assembly of waterborne construction equipment. These six piers alone consumed more than 280,000 m^3 of concrete. The tower piers are footed to depths of 64 m—then only second to the Transbay—and weigh not much less than the anchorages. Three-quarters of the almost 1 million tonnes of concrete and steel in all 34 picrs are underwater, providing a perhaps overly generous safety margin of 20 against ice pressure.

The 35- by 41-m anchorage piers used cofferdams up to 28-m deep and together weigh 326,000 tonnes. Prepakt intrusion concreting—in which grout is injected under pressure into submerged aggregate—was used. The method was not without teething problems; bulges in the south anchorage from grouting pressure buckled portions of its cofferdam. For other reasons, the tower piers proved even more problematic. The lower 15 m of the 35-m diameter steel cylindrical caissons were first built onshore. But the north caisson took weeks—instead of days—to float out of the assembly area because of unexpectedly hard soil. Dynamite helped to deepen the exit channel, but one blast tore a large gash through the caisson cutting edge, requiring several weeks of repair. When finally floated, the caisson immediately ran aground, ergo more days of dredging. Engineers later laid tracks and skidded the 363-tonne south tower caisson over the same distance in 4 hours!

The north tower caisson arrived onsite late, in September; the south caisson followed later that month. Horrible weather hampered sinking, as did large boulders beneath the north caisson. Another scare occurred with the sudden tilting of the south cable bent caisson at a depth of more than 36 m. Construction accidents then took

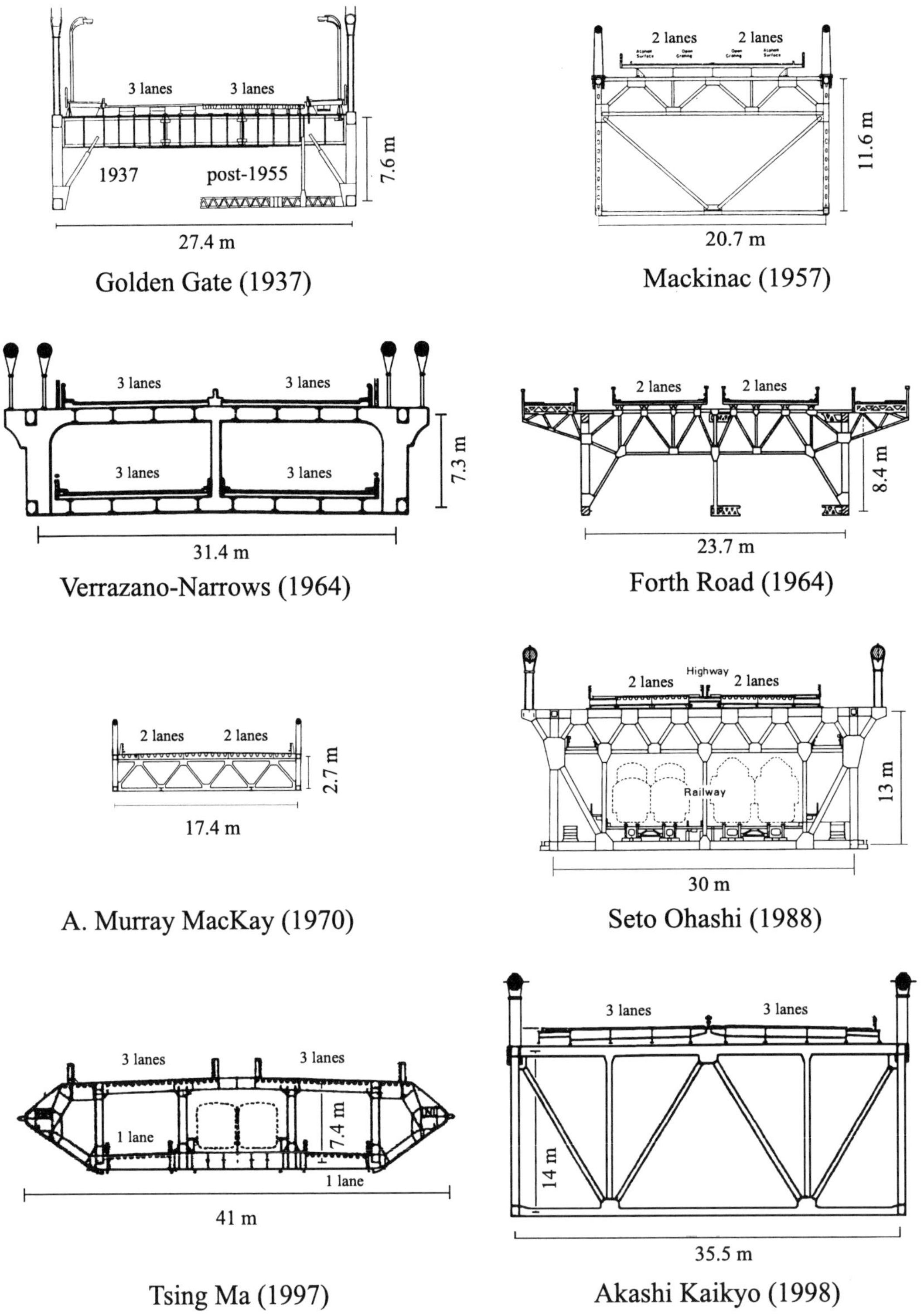

Fig. 7-2. Evolution of post-Tacoma Narrows Bridge truss design
MODIFIED FROM ORIGINAL DRAWINGS COURTESY OF THE GOLDEN GATE BRIDGE, HIGHWAY AND TRANSPORTATION DISTRICT, PARSONS CORP. (STEINMAN), MTA BRIDGES AND TUNNELS SPECIAL ARCHIVE, THOMAS TELFORD LTD., R.A. DORTON, HONSHU-SHIKOKU BRIDGE AUTHORITY, AND HYDER CONSULTING.

three lives within a period of 40 days. Yet mild weather permitted the sinking of both tower caissons until mid-January 1955, when they had almost reached bedrock. When work restarted that spring, a record for underwater concreting was set in May. In all, Prepakt concreting was estimated to have saved a year of foundation work.

By late 1955, the two 5,900-tonne towers and anchorages were ready for cable spinning, and the approach span piers stood completed. The pre-assembled 143-m long, 680-tonne backstay spans were then floated into place in the largest operation of its kind to that time, one estimated to lop several months off the schedule. Their placement occurred just days after a November 1955 storm nearly toppled one span from its onshore scaffolding and battered the half-riveted towers with 116-km/hr gusts. An early winter would delay catwalk erection until the spring of 1956, when tragedy struck. Wire mesh flooring was being installed high up near the north tower when a holdback line snapped. Two bridgeworkers plummeted to their deaths; two others rode a piece of wire mesh down the side span catwalk cables, stopping just before the north cable bent. Steinman's choice of long side spans with their gently sloping cables probably saved their lives. Despite generally poor weather, J.A. Roebling's Sons spun the 12,580 wires in each 62-cm diameter main cable in record time in just 78 working days between mid-July and mid-October 1956. Over the backstay spans, an additional 240 wires were spun to help keep each cable bent vertical from unequal cable forces. Approach span trusswork, however, was stopped just 99 m—one span—short of both anchorages. The truss was lifted into place the following year, the lifting sequence aided by an early use of computers.

The Mackinac Bridge opened in November 1957, just 3.5 years after work began, a remarkable achievement considering its size, the harsh environment, and the short working seasons. For Steinman, Mackinac represented a personal triumph; the unprecedented aerodynamic stability of his first major suspension bridge in 20 years appeared to vindicate a solitary and exhaustive research effort extending more than 15 years. The incentives to create a bridge of such impressive strength were many. Most alluring for Steinman was a chance to lay the still-lingering ghost of the Tacoma Narrows Bridge to rest through the use of his own research findings. Mackinac's span more than doubled anything Steinman had previously built, and although the aerostatic theories supporting its design were debatable, his design has stood the test of time. Mackinac stands as a testament to Steinman's genius and perseverance, and it remains one of the world's great engineering achievements.

7.2 STEINMAN'S CLOSURE—THE ST. LAWRENCE AND ELSEWHERE

In late 1956, as the Mackinac Bridge neared completion, Steinman was retained to design the Karradah (or 14th of July) Bridge over the Tigris River in the Iraqi capital of Baghdad. Aesthetics and poor foundation conditions saw Steinman adopt a 168-m self-anchored suspension bridge with 84-m side spans over four other designs. Interestingly, spine box girders stiffened the span. Simple portal-braced towers atop tall concrete pedestals were adorned with small spires above each leg, typical of Steinman embellishment. Construction did not begin until May 1961, following Steinman's death in August 1960. Long-time associate R.M. Boynton headed up the $3 million project, which opened in 1964. Sadly, the bridge was destroyed in the

1990 Gulf War. The 14th of July Bridge was not, however the sole self-anchored Iraqi span of this period, for British engineers would design a 75-m span completed in 1958 across the Euphrates River midway between Baghdad and Basrah.

Woodruff, who had served as Steinman's consultant at Mackinac, would aid in the design of another suspension bridge, this time in association with Canadian engineers T. Lamb, McManus & Associates. A 1956 decision to replace a ferry across the Peace River in northwest Alberta prompted the consideration of several bridge types. A four-span cantilever was a finalist, but steel shortages and an aversion to its four river piers tipped the scales in favor of a 274-m suspension span with 137-m side spans. In the event, the two river-based tower foundations would prove especially trying, but the $4.5 million all-bolted bridge was completed without further difficulty in the fall of 1960.

The Dunvegan Bridge is unusual in several respects. Its roadway rises sharply, with a height difference of more than 9 m between towers. A complex pattern of double intersecting diagonals characterizes the 4.6-m deep, single-lateral, half-through stiffening truss, analyzed using Steinman's stiffness index. Trussed laterals and open steel grating along the road edges aid aerodynamic stability (and the latter, snow removal) and made it unnecessary to attach the cables to the truss at midspan. The towers—braced by multiple-portal struts above and slender diagonals below deck—are of fine outline, in spite of their different geometries. Both sit atop thin, tall concrete pedestals and, unusually, are pinned at their bases to deal with extreme temperature variations. Helical wire strands form the main cables, which on the north shore are anchored well back of the main bridge over several backstay spans. In spite of the span's marked asymmetry, the Dunvegan Bridge is visually appealing, its rustic patina complementing the open scenery of the majestic Peace River valley.

The late 1950s also brought Steinman back to the St. Lawrence, where by 1960 two new suspension bridges would link Canada and the United States. Bridges downstream of the Thousand Islands Bridge had long been discussed. An 1882 rail bridge charter at Morrisburg, Ontario was never acted on, and the first rail span—completed at Cornwall, Ontario—in 1899 saw 14 die when two spans collapsed during construction. Work began in 1895 on a rail bridge at Brockville, Ontario, but was then abandoned. A high-level road bridge was proposed in 1915, again at Brockville, and then in 1929. Both nations endorsed the latter plan, but New York did not. A 1930 report advocating three separate bridges along the length of the river went nowhere, but a 1934 plan for a 256-m span cantilever at Ogdensburg, New York was, in fact, almost built. Canada resisted the exclusive use of American workers required to secure $2.8 million of RFC financing, and a 2-year stalemate saw the project die in 1936, a year before Steinman's Thousand Islands Bridge opened. Approvals were again secured for an Ogdensburg span, but war then intervened. Political roadblocks dogged early post-war efforts, but by 1952, approvals were finally in place for a bridge at Ogdensburg.

The firm of Modjeski & Masters, with Canadian P.L. Pratley, investigated alignments and bridge types for the 2.1-km crossing. A 1953 route, however, was rejected because of disruptive approaches. An alignment to the east was adopted in late 1955, and in 1956, funds were authorized to build the bridge. Construction began the following year, with American Bridge handling the superstructure of a 351-m span suspension bridge flanked by side spans of 152 m. The Ogdensburg Bridge is probably the last sizable suspension bridge to use a through truss, although its lack of vertical members provides a relatively transparent appearance. The two-lane deck is comprised of an open steel grid to reduce weight; this renders the floor more permeable to wind and reduces snow blockage. Diagonally braced towers feature gently raked legs, and the anchorages are unobtrusively tucked beneath the deep approach trusses. The $20 million span was opened in the fall of 1960; towers, cables, and truss were erected within the space of just 5 months in 1959, a year before the approach spans were completed.

Modjeski & Masters was simultaneously involved in a span more reminiscent of the 1930s. The Iowa-Illinois Memorial Bridge across the Mississippi River near Davenport, Iowa, stands just 23 m from the two-lane 1935 span. In consequence, the new bridge—completed in 1959—is a virtual carbon copy of the first, making a simple job for the designer and for contractor Bethlehem Steel, both of whom had been responsible for the original span. Its 226-m main and 113-m side spans are suspended from helical strand cable draped over diagonally braced towers, and the 4.3-m deep through truss features a single system of lateral bracing.

A second St. Lawrence span—the Seaway International Bridge—was also built to cross new shipping channels at Cornwall. A suspension bridge was an eleventh-hour decision following the abandonment of a circuitous low-level crossing. Pratley was again involved in early studies, but Steinman undertook the final design. The bridge has a 274-m main span and is flanked by 137-m side spans; its towers vaguely resemble those Steinman used at the Thousand Islands Bridge. Problems dogged foundation work, but the towers, cables, and through truss were erected by American Bridge in just 6 months beginning in May 1958. The span—remarkably—was in use that December. Unusually long side spans and the tucking of the anchorages beneath slender approach spans are visually appealing aspects of this bridge.

Steinman contributed to the early planning of two large spans on the other side of the Atlantic—at Istanbul, Turkey, across the Bosporus Strait and at Lisbon, Portugal, across the River Tagus. Yet he would not live to see ground broken on either of them. His death on August 21, 1960 at the age of 73 left a legacy of more than 400 bridges in all parts of the world except Africa. Steinman's greatest bridge—Mackinac—would be one of his last. His greatest ambition would remain unfulfilled, the spanning of the Narrows at the entrance to New York Harbor. This prize would go to Ammann.

7.3 PRELUDE TO THE VERRAZANO-NARROWS—THE WALT WHITMAN AND THROGS NECK BRIDGES

Ammann had not remained idle following his withdrawal from Mackinac. He would become a key figure in the design of the Walt Whitman Bridge, a 610-m span completed in 1957 in Philadelphia, Pennsylvania. The bridge underwent a remarkably short gestation. The need for additional capacity over the Delaware River first emerged in a 1941 plan to convert the outer trolley lines of the Benjamin Franklin Bridge to vehicular use. War temporarily stopped this $3 million conversion, undertaken in the late 1940s to a design of Masters. In early 1948, before work was finished, a new bridge or tunnel was mooted, but financing was thought to delay construction for 8 to 10 years. Six months later, a six-lane, $43 million suspension bridge with a span of 625 m was recommended as part of a port development program. In October 1950, an engineering board comprising Ammann, Masters, and Ralph Smillie was struck to determine where and what type of crossing should be built. Its choice of a bridge about 5.5-km south of the Benjamin Franklin Bridge soon ran afoul of U.S. Navy objections over possible interference with its nearby airfield. A tunnel, however, was found to cost more than $107 million, far greater than the $84 million cost of a bridge. This convinced the governors of Pennsylvania and New Jersey to in turn convince the U.S. Navy to retract its objections, which it did. The two states sanctioned the bridge in July 1952, and the Secretary of the Army issued a permit in February of the following year.

The firm of Ammann & Whitney was retained to design the suspension bridge—with Modjeski & Masters designing the foundations—in September 1952, by which time a seventh reversible-flow lane was added. The final design of the 610-m span—with 235-m side spans—was sufficiently advanced to allow foundation work to begin in August 1953. Contractor Merritt-Chapman & Scott, soon to commence work on the foundations of the Mackinac Bridge, sunk the anchorage foundations to depths of up to 18 m. The front half of each 73- by 44-m anchorage is supported by caissons, the rear half, by 202 piles. Another contractor sunk the tower piers through offshore sand islands, the Philadelphia pier reaching a depth of 32 m. Its shallower New Jersey twin, however, proved more troublesome, with the last 3.6 m to a bed of compacted sand and gravel requiring a month of work to overcome.

The 113-m tall towers are squatter versions of those used in the Delaware Memorial Bridge. Each leg consists of 12 cells enclosing a large central shaft, designers eschewing cruciform shaping in favor of a rectangular form considered to be more efficient. To make up time lost from the foundation delays, high-strength bolts replaced rivets in the upper reaches of the New Jersey tower, a first for an American suspension bridge tower. Each 59-cm diameter cable comprises 11,396 wires. Spinning began in late 1955 and introduced a new position—the tattletale. Emergency switches distributed along the Walt Whitman's catwalks to stop the spinning wheels in the event of an accident were often prone to high jinks; the flip of a switch would bring a superintendent—by no coincidence—from the other end of the span. Upon his arrival, the wheel would just as mysteriously start up again. This game continued for some time until American Bridge rigged up a panel in the superintendent's shack that identified the switch—and the gang responsible for pulling it. The tattletale would become a fixture on subsequent American spans.

Piece-by-piece lifting of the stiffening truss occupied 10 months. Its upper and lower chords were formed from closed hollow boxes, a first for a major suspension span. As in the Delaware Memorial Bridge, Ammann had used his stiffness index in the preliminary analysis of the truss but wind tunnel tests were shunned; they were considered costly, time-consuming, and to offer no "practical value" (Brumer and Hanson, 1956). The 5.5-m truss depth creates a 1:111 depth-to-span ratio, slightly less conservative than that of its contemporaries. Only minor venting of the deck was provided, in the form of 38-cm strips of steel grating flanking the roadway, open grid service walks, and vented curbs. As in the Delaware Memorial Bridge, the top chords project above the deck and double as parapets; the main cables are also not rigidly connected to the truss at midspan.

The Walt Whitman Bridge opened in May 1957 and, for a brief 6 months, was the world's seventh longest suspension span until the Mackinac opened later that year. Other than the extensive use of bolting and closed box members, the span was typical of early post-war American design, but its clean appearance did earn it the AISC's long-span award for 1957.

In New York City, fast-paced post-war growth soon spurred large efforts to deal with traffic congestion. The Throgs Neck—continuing the eastward creep of suspension bridges across the East River—was one of three huge bridge projects initiated in the late 1950s as a result of a 1955 study undertaken jointly by the Triborough and Port of New York authorities. The span would help relieve the traffic crush on the Triborough and Bronx-Whitestone bridges, which together carried almost 80 million vehicles in 1960, but it would also generate much animosity.

Initial plans prepared in 1955 by Ammann & Whitney called for a 945 m span flanked by 274 m side spans. The north anchorage would be footed on a peninsula housing the New York State Maritime College; the south approaches would extend through developed areas in Queens. The span's towers would resemble those later adopted for the Verrazano-Narrows Bridge, save for the lower strut, which was hidden within the truss. In the end, the plans would change drastically. The south end of the bridge would be swung to the east, over the water, to avoid landfall. Span lengths would also be considerably reduced, moving the anchorages into open water. As a result, although the river is about 1,100-m wide, the bridge extends for a distance of more than 4 km. Even so, the bridge's approaches would remove more than 420 homes, most of which were relocated by the Triborough Bridge and Tunnel Authority to two golf courses it had acquired. Critics skeptically viewed the Throgs Neck as yet more evidence of Robert Moses' highway juggernaut steamrolling through New York's fine-grained neighborhoods.

Milton Brumer, an Ammann associate since 1944 who had contributed to the Walt Whitman and a Port Authority colleague before that, headed up the suspension bridge design under Ammann, now well into his 70s. Some 3,200 m of viaduct were designed by the firm of E. Lionel Pavlo, whose founder had helped devise the remodeled Bronx-Whitestone Bridge in the 1940s while at the firm of Madigan-Hyland. The south (Queens) overwater approach curves for 1,250 m to skirt nearby residential areas before meeting the 549-m span suspension span with 169-m side spans. A further 1,950 m of viaduct span the northern peninsula before curving overwater to the Bronx.

Construction of the six-lane bridge began in October 1957. The foundations proved challenging and required 2 years to complete. Each concrete anchorage caisson was built on a temporary sand island and sunk by excavating up to 77 open wells; the Queens caisson—at 44 by 69 m—was the largest to that time. Its sinking was frustrated by a deep layer of hard material that yielded only to high-pressure water jets; it would take more than a year, ending in July 1959, to foot the caisson at a depth of 41 m. The steel tower caissons were sunk to depths of up to 45 m by pouring concrete into spaces between dredging wells, which were then excavated and their sides raised as sinking progressed. Only the lower portions of the wells were concreted; water permanently fills their upper halves.

Bethlehem Steel assembled the 110-m tall towers beginning in the summer of 1959, whereas rival American Bridge would later spin the 10,952 wires in each 58-cm diameter cable and erect the truss. All went without a hitch. Wire in each cable was laid simultaneously from opposite ends of the bridge, and with eight wires laid in one circuit, the cables were quickly completed by late spring 1960. The 5.2-m deep truss was of vintage Ammann design with no steel deck grating, although small gaps between the road and truss edges were provided; the main cables also hung free of the truss at midspan.

The $92 million bridge was opened in January 1961 after just more than 3 years of construction. Its squat towers took the simplicity of the Walt Whitman Bridge a step further, with a featureless top portal strut flush with thick rectangular legs and the lower strut hidden, as in the initial 1955 proposals, within the truss. The anchorages are also imposing and extend more than a third the length of the side spans. The Throgs Neck Bridge has not earned praise from pundits in aesthetics; Christopher Tunnard and Boris Pushkarev (1963) criticized the span for the open water placement of its anchorages and the lack of symmetry with the river it crosses.

If the Throgs Neck Bridge emanated strength, Ammann's final—and biggest—bridge would embody grace at an unprecedented scale. The Verrazano-Narrows Bridge would also mark the final chapter in Ammann's rivalry with Steinman and become one of the world's great—and most expensive—spans.

7.4 AMMANN'S FINAL LEGACY—THE VERRAZANO-NARROWS

The post-war growth that spawned the Throgs Neck Bridge also brought about a long-sought crossing of the entrance to New York's harbor—the Narrows. A bridging of this 1.6-km wide channel would not only open up Staten Island to development, but it would divert traffic from Manhattan's congested streets. The Narrows was an alluring prize; the history of this bridge is fraught with controversy, one dominated by a protracted 35-year struggle between the two 20th century giants of American bridge engineering: Ammann and Steinman.

Most early proposals went under, rather than over, the Narrows. The Baltimore & Ohio Railroad first proposed a tunnel in 1888. A crossing was then suggested in the early 1900s, and in 1910, Charles Worthington proposed a $15 million, 762-m span arch bridge.[7] In 1921, the state approved a rail tunnel almost directly under the line of the present bridge, but a 1923 construction start was soon mired in controversy over the nature of connections to the city's subway system. When it was revealed that the $30 million estimate had more than doubled work was halted; further disagreements between the city and state saw the scheme—and the $4 already million spent—abandoned. Soon after, in November 1926, the visionary Fowler announced a 1,387-m span suspension bridge with unloaded side spans. Influenced by his 1921 proposal at Detroit, Michigan, Fowler's Alexander Hamilton Bridge was, like the tunnel, destined for oblivion.

In 1926, Steinman, with associate Robinson, unveiled plans for the Liberty Bridge, a mammoth project that would consume Steinman for the next 3 decades. Steinman felt a monumental bridge should span the "Nation's Gateway." That it was—its 1,372-m span was more than 305 m longer than Ammann's George

Washington Bridge, then about to begin construction. Its 244-m tall towers supported an eight-lane roadway suspended 72 m above the Narrows from two record 1.2-m diameter cables, each containing 48,000 wires.

In an address to the AISC, Steinman (1929) took aim at Ammann's defense of plans to cloak the steel towers of the George Washington Bridge. He wrote that, "Some designers . . . are resorting to the dubious architectural expedient of building huge bridge towers of steel for strength and then masking them with concrete and stone for appearance." Steinman described this as "subterfuge and evasion." In the Liberty Bridge, he stated, "No attempt will be made to mask the towers in concrete or stone; instead every effort will be made to develop the highest artistic possibility of steel tower design." Steinman remarked that steel towers were visually impressive "not by ornamentations, but by the development of structural forms that will be inherently beautiful in their simplicity." Ironically, his Liberty Bridge towers were neither simple nor unornamented; large Gothic flèches, intricate portal struts, an above-roadway arcade—even an observation platform at the top of the Brooklyn tower—adorned them. Steinman poetically described carillons that would "peal out anthems of Liberty to those who enter through this Nation's gateway" and a beacon that would provide a "symbol of Liberty to guide, welcome and inspire those who have crossed the seas to come to our shores." More prosaic aspects of the design saw the roadway placed on the bottom of a shallow, gossamer stiffening truss devoid of upper lateral bracing.

Steinman had helped to create the Interborough Bridge Corporation to promote the ambitious and expensive $60 million scheme. But the powerful War Department denied permission for fear of potential blockage of the Narrows in wartime. A tunnel now received renewed impetus. In 1929, the city authorized the design of a four-lane tunnel under Ole Singstad, of Holland Tunnel fame. Singstad cited the challenges of great but unknown rock depths beneath the Brooklyn shore, the War Department, and a need for a 1,524-m span, which he considered beyond the capacity of steels then available. The tunnel project received War Department blessing in 1930, but by then the Depression had dried up the funds to build it. A tunnel received state authorization in 1937, and then in 1942, design studies were relaunched under Singstad, now chief engineer of the city's Tunnel Authority. His May 1945 report recommended a $73 million twin-tube tunnel.

The high costs of a tunnel suggested, to some, that a bridge could offer greater capacity, be less consuming in time and money to build, and be simpler to maintain. Steinman now resurrected the Liberty Bridge. In 1945, he and Singstad outlined their proposals at separate meetings of the Staten Island chapter of the ASCE. The Liberty Bridge now featured a lengthened span of 1,408 m, with provision for up to 10 lanes. Its $40 million cost was almost half that of Singstad's four-lane tunnel, and the chapter unanimously endorsed the bridge. Steinman also addressed the Staten Island Institute of Arts and Sciences that December. Oddly, the accompanying illustrations depict the 1920s design, with two added tower struts. The absence of substantial change is peculiar given Steinman's own design developments and those that followed the collapse of the Tacoma Narrows Bridge. One tower variant retains more restrained ornamentation; the transverse bracing is simplified and suggests an embryonic form later found in the Mackinac Bridge. The stiffening truss remains slender, but the roadway sits on the top chord of a truss lacking any bottom lateral bracing.[8]

Steinman's efforts now began to bear fruit. In 1946, the Triborough and New York Tunnel authorities merged; Moses now employed Singstad. Later that year, the newly-amalgamated Triborough Bridge and Tunnel Authority sought to build a bridge, and when it applied for War Department approval in 1948—which it obtained in May 1949—the Liberty Bridge formed the basis of its plans. Steinman remarked that, "I expect the Liberty Bridge to be built and hope to be identified with it. Afterward, I shall retire" (Pfeiffer, 1978). The bridge was even featured on a first-day cover for a 1952 American postage stamp commemorating the centenary of the ASCE. Steinman's hands are shown drafting the tower flèches—again—of the 1926 scheme. The stamp itself featured Ammann's George Washington Bridge—an omen, perhaps—for Steinman's dream was to slowly slip away.

The alignment, for one, was controversial. The Brooklyn approaches would displace more than 1,100 families. In response, the Triborough Authority shifted the bridge south, where it would be anchored in military reservations flanking the Narrows. Then in 1951, fears of Soviet attack spurred a brief flirtation with a tunnel. Yet perhaps Steinman's biggest obstacle was Moses, the hard-nosed and irascible head of the Triborough Authority who ruled over New York's transportation scene with an iron will. Moses disliked Steinman, his outspokenness and self-promotion. Moreover, Moses had a long affiliation with Ammann extending back to the Triborough Bridge; he wanted Ammann for the Narrows. Steinman was nosed out officially in January 1955 with the Triborough and Port of New York authorities' joint release of plans for an Ammann-conceived, 12-lane, double-deck suspension span, part of the program that produced the Throgs Neck Bridge and a lower deck for the George Washington Bridge. In March 1955, state approval for the new scheme was obtained.[9]

The Liberty Bridge was now replaced by a behemoth costing an estimated $240 million—more than twice the cost of the Mackinac Bridge. Although Ammann had dabbled in a Narrows design as early as 1936 for the Port Authority, he had devised the basics of the present scheme in 1954. A record main span of 1,341 m would be flanked by 427 m side spans and the four paired main cables secured by large rectangular, stepped-back anchorages (the latter would later be altered significantly). The 213-m tall steel towers resembled—in outline if not in structural concept—solid versions of the pylons Ammann used in the George Washington Bridge. Although the entire truss would be in place when the Narrows Bridge opened, only the six-lane lower deck of the 31-m wide, two-level truss would be initially opened. The six-lane upper level—carrying passenger cars only—would open when traffic growth warranted the addition of ramps and approach roads. Two years of arduous negotiations with the U.S. Army culminated in a June 1957 agreement that would cost the Triborough Authority more than $26 million to replace military facilities. The main span was also reduced to 1,298 m, with 370-m side spans. Unfortunately, the earlier alignment shift, although sparing one part of Brooklyn, would ravage another and uproot more than 7,000 people and 800 buildings. Opposition intensified. Urban historian Lewis Mumford (1963) described the bridge as a "menacing error" and "a terrifying project." The state twice altered the Triborough Authority's plans, but Governor Averill Harrimann, prodded by Moses, overruled. Gubernatorial candidate Nelson Rockefeller then promised to move the Brooklyn approaches, but Moses convinced the newly-elected governor that the additional $50 million cost would kill the project.

The controversy had by early 1958 created the climate for an alternative scheme across Upper New York Bay just off the tip of Manhattan. By connecting Brooklyn to New Jersey via Governor's Island, this 3-km route would avoid Staten Island altogether. New Jersey cities even helped finance studies, by none other than Steinman. This eleventh-hour effort was folly, but it provided Steinman a last, if desperate, chance of ousting Ammann. The Brooklyn opposition to the Narrows Bridge had forced hearings into the Upper Bay scheme, and Steinman hurriedly completed preliminary plans in December 1958. He proposed two variants of a 1,402-m suspension span with two levels—a $371 million, 12-lane bridge, and a second version with eight upper-level lanes and space for four lanes and two transit lines on a lower level; this second option would cost an astounding $531 million. Moses wrote the scheme off as a financial fantasy, even though estimates for the Narrows Bridge had now reached $325 million. Its lower navigational clearance was also criticized. Most of all, the visual impact of Steinman's bridge and its 244-m tall towers on the Manhattan skyline and the nearby Statue of Liberty would have been disastrous. Ironically, this issue had the potential to resurrect the rancorous debate that had ultimately thwarted one of Moses' pre-war projects, the Ammann-designed twin suspension bridge spanning the East River between Battery Park in Manhattan and Brooklyn (where a tunnel was eventually built between 1940 and 1950).

Steinman's report had no effect, for on the last day of 1958, final approval of the Brooklyn approaches to the Narrows Bridge was secured. The first residents left the following year, and construction began in January 1960. In keeping with tradition, even the August 1959 sod-turning—prudently enacted on Staten Island—evoked passion, this time over the name of the bridge. Florentine explorer Giovanni da Verrazano, who in 1524 was the first European to sail into New York Harbor, emerged as a front runner. Critics pointed to his prior exploits—plundering ships in the Caribbean—and branded the bridge the Guinea Gangplank. Months of debate ended with a state-legislated compromise of Verrazano-Narrows. Further haggling over the spelling was then only resolved with the dispatch of a delegation to Naples.

Meanwhile, Ammann's staff painstakingly worked out the details of the design in an effort extending almost 10 years from 1954, aided by electronic computers. Five years of preliminary design saw more than 20 tower and anchorage locations examined. The 1954 truss had outcompeted two options with eight-lane single decks; in the first option the deck sat atop a 12.8-m deep truss and in the second it was situated midway up a 15.8-m deep truss atop two levels of lateral bracing. Two 12-lane, double-deck variants with 7.9- and 9.1-m truss depths were also studied. Although eight lanes were considered adequate, four more lanes would add 50% more capacity while increasing costs by just 10%. A twin-deck, 12-lane truss was adopted, the final design using a single central post separating the two sets of lanes on the lower level (*see* Fig. 7-2). Unlike the initial 1955 scheme, both levels would carry mixed traffic. The 31.4-m wide truss is just 7.3-m deep; its 1:180 depth-to-span ratio remains the sleekest of any large truss-stiffened, post-Tacoma span. Ammann used his stiffness index in assessing the design, its vertical index now augmented by a torsional index devised during the design of the bridge itself. Sectional models were also subjected to wind tunnel testing. There is no doubt that Ammann benefited from the tremendous capacity and weight of the suspended structure to pare down truss depth, as he had in the George Washington Bridge; at their maximum, live loads would never exceed 13% of the weight of the suspended structure. Yet unlike the George Washington Bridge, the truss would be erected to help stiffen the bridge from day one, with the upper deck opened first.

One significant but often overlooked feature of the Verrazano-Narrows Bridge is its truss framing concept. Ammann designed the truss as a rigid rectangular tube, reducing depths while stiffening against torsion from wind and live loads. Steel stringers supporting the concrete deck are longitudinally continuous over each suspended span and framed into the top lateral, adding rigidity to the entire structure. Long-time Ammann associates Edward Cohen and Frank Stahl (1978) suggested this concept "may have been [Ammann's] most important contribution to bridge design." Ammann also placed 2-m gaps outside the road deck, although he eschewed steel grating. He swung the cables free of midspan, making the Verrazano-Narrows Bridge one of a handful of post-Tacoma spans—mostly his own—to do so. Some 50% of the truss used high-strength, low-alloy steels; vertical members are also confined to every other panel, improving economy and appearance.

Only superlatives describe the Verrazano-Narrows Bridge (*see* Fig. 7-3). Its record 1,298-m main span eclipsed the Golden Gate Bridge by just 18 m (*see* Fig. 17-1), but on a per-unit-length basis, its suspended structure is 75% heavier and has double the traffic capacity. Yet with a proportionately slimmer truss, the Verrazano-Narrows Bridge is 70% more resistant in torsion than even the strengthened truss of the Golden Gate Bridge. It's 211-m tall towers stand 4-cm farther apart at their summits than at their bases on account of the earth's curvature, a consideration for design calculations. In summer, the truss hangs 3.6-m lower than in winter because of expansion of the main cables. Each tower weighs 24,500 tonnes—more than four Mackinac Bridge towers—and comprises 10,000 cells, 1.5 million bolts, and twice that number of rivets; high-strength, low-alloy steels account for 70% of their weight. Each of the four 91-cm diameter main cables contains 26,108 wires; all 230,000 km of wire exceeded that in the George Washington Bridge by a third and could encircle the globe more than five times. In total, the 142,000 tonnes of superstructure steel in the suspension bridge alone could build four Empire State Buildings.

In spite of its tremendous size and weight, the Verrazano-Narrows Bridge is not founded on rock. Singstad's concerns over the Brooklyn foundations rang true; no bedrock could be found, even at depths exceeding 100 m. A strategy was first devised to plunk the enormous 39- by 70-m tower caissons atop concrete-filled tubular steel piles driven to rock. This scheme was abandoned when a deep compacted band of glacial clay was found to provide direct support for the caissons themselves. Sinking began in January 1960 on temporary sand islands; the Brooklyn caisson obliterated a reef that had served as a prison during the American Civil War, then as a naval magazine, and, in World War I, as an ammunition dump. Circular dredging wells similar to those used at the Throgs Neck Bridge were then excavated in a controlled sequence to ensure level sinking as concrete was added in 3-m increments on top.

Fig. 7-3. Verrazano-Narrows Bridge
PHOTOGRAPH COURTESY OF THE MTA BRIDGES AND TUNNELS SPECIAL ARCHIVE.

Controlled sinking became something of a euphemism on at least one occasion when at a depth of 48 m the Brooklyn caisson suddenly lurched downward 1.2 m in just 3 seconds. Soon after, it dropped another 2 m in 5 seconds. This second event fortuitously placed the caisson close to a planned depth of 52 m. Its Staten Island twin would eventually rest some 32 m below ground level. The dredging wells were then plugged with concrete, and 30 months of foundation work at the towers ended with the addition of 10-m tall granite-faced concrete pedestals.

Each Verrazano-Narrows Bridge anchorage weighs 330,000 tonnes; the larger and deeper Brooklyn anchorage covers an area of 105 by 70 m, roughly the size of an American football field. Notwithstanding its smaller size, the Staten Island anchorage would be the most troublesome. A 380,000-m^3 hole was threatened several times by large washouts, and this situation was exacerbated by a 2-month concrete truck drivers' strike that at one time was feared to delay bridge completion by an entire year. Luck intervened, and in September 1961, Bethlehem Steel began the $23.4 million Staten Island tower; the Brooklyn tower was started in early 1962 under a separate contract. The massive legs are up to almost 15-m thick at their bases and taper along the line of the span to a still considerable 10.7 m at their apex. Their dimensions are such that the 9-m long, 161-tonne cable saddles are completely hidden from sight. For the almost 2,000 anchorage eyebars, an economical high-strength manganese vanadium alloy steel replaced traditional forged and heat-treated eyebars.

The huge job of manufacturing the cable wire for the bridge consumed much of America's wire manufacturing capacity. In fact, the $57 million main cable contract, secured by American Bridge in February 1960, was only the second awarded (after the tower foundations). This would allow spinning to start by early 1963, 3 years later; a 13-tonne, 3-m long test section of cable was even assembled to aid in the determination of cable band diameters. Spinning the 61 strands of 428 wires into each main cable began in April 1963. Almost immediately, a storm with 130-km/hr winds forced a 2-day shutdown. Workers returned to see the first four strands in the most southerly cable badly tangled and some tramway frames so twisted that they had to be replaced. After the unraveling, a fast pace soon developed. Eight spinning wheels—two per cable—were used, and in August, the cables hung complete after just 5.5 months of work. Spinning efficiency was twice that at the George Washington Bridge. Nonetheless, the attachment of cable bands and suspenders offered more high drama when a 1-tonne casting plummeted from the span, narrowly missing a hydrofoil and just minutes before the aircraft carrier U.S.S. Wasp passed over the spot where it fell. The factory producing the castings was on strike, so crews scoured the Narrows, finding the casting—intact—several days later in more than 30 m of water.

Truss steel was first assembled into 30-m long truss units typically weighing 352 tonnes. More than 42,600 tonnes of structural steel were then raised, beginning with a 163-tonne midspan lift witnessed by the 84-year-old Ammann in October 1963. The great weight and length of the bridge required lifting from center span to the towers—and simultaneously out from the anchorages—to distribute loads evenly. Eleven units were assembled on land beneath the Staten Island side span, and under the Brooklyn side span, units were slid on large tracks over adjacent roads. Operations did not always proceed smoothly. A dearth of safety nets beneath the growing truss fueled a 5-day strike in December 1963; the nets went in and ultimately saved three lives. On another occasion, a truss unit was left dangling overnight after a hoisting engine failed. But the most spectacular episode occurred when the cables restraining a side span truss unit to the Brooklyn tower snapped as lifting began; the truss careened at low altitude over the Brooklyn shore, miraculously hitting nothing. By the end of May 1964, the truss was in place, and secondary steel and roadway concreting began. When work was finished, the cables hung 8.5 m below their free hanging position.

The Verrazano-Narrows Bridge was opened 6 weeks ahead of schedule in November 1964 at a cost of $305 million, $20 million under budget. Its completion in just less than 5 years was a remarkable achievement, one reflecting advances in design and construction technology. With 70% more steel and three times the concrete, the bridge was built in about the same time it took to build the George Washington Bridge. Ammann, for a second time, would see a suspension bridge of his own design become the world's longest span.

Ammann had sought to create an "enormous object drawn as faintly as possible," one with an "ultimate simplicity of form and a clear expression of function" (Furnas *et al*, 1966). This philosophy belies the Verrazano-Narrows Bridge's tremendous strength and enormous size. Aymar Embury II, who first collaborated with Ammann on the Triborough Bridge in the 1930s, sat on the span's advisory board of architects. His influence is apparent in the large but simply shaped triangular anchorages for which, unlike the curves of the Bronx-Whitestone Bridge, a straight back slope is used. Strakes follow—rather than punctuate, as in the Bronx-Whitestone Bridge—the line of the strands. The uninterrupted line of the main cables, unfettered by a midspan tie, is visually effective. A double fluting of the outer edges of the tower legs and the elegant and deep curved portal strut combine to create the most refined of Ammann's tower designs (*see* Fig. 13-3).

Leonhardt did criticize the design by suggesting the great depth of the portal strut made the towers appear top-heavy. He was quoted as having even offered an alternative bridge design, and that Ammann "simply pulled out the design of the George Washington Bridge . . . out of his files, and that was it" (Anonymous, 1979). He also later (Leonhardt, 1982) described the Verrazano-Narrows Bridge as having "no innovations," although he elsewhere inferred that Ammann borrowed some of his own ideas for its design. Frank Stahl of Ammann & Whitney (1979) rebutted, writing he was "amazed" by Leonhardt's remarks, in particular the lack of recognition of the Verrazano-Narrows Bridge's innovative torsion-tube truss.

The AISC recognized the Verrazano-Narrows Bridge with a special award of merit in 1964 for its "outstanding contribution to technology and esthetics" (although Virginia's 27-km Chesapeake Bay Bridge-Tunnel took the AISC's top prize that year). The Verrazano-Narrows Bridge represented the zenith of the truss-stiffened suspension bridge in the period following the collapse of the Tacoma Narrows Bridge. Yet for all its stature as a gateway to a great metropolis, the Verrazano-Narrows Bridge has never achieved the landmark status of the Golden Gate Bridge. It has not the heroes, history, nor spectacular setting associated with its western counterpart. Its silver-gray patina is mundane in comparison to the Golden Gate Bridge's famous international orange. Moreover, the pedestrian access that so personalizes the experience of the Golden Gate Bridge is lacking in the Verrazano-Narrows Bridge; this reinforces the impression of aloofness, although the bridge is perhaps most popularly known for hosting on its upper level the first stage of the New York marathon, run annually every November. The Verrazano-Narrows Bridge was symptomatic of the automobile's domination that had immersed the Throgs Neck in controversy. The span's lower deck, if anything, confirmed this phenomenon. Although the lower deck had been concreted over the suspended spans in 1964, approach bridges were not built until the need arose, expected in 1980. Yet by 1968, traffic had reached levels predicted for 1978, and work began that year on the $20 million project. In June 1969, the lower deck opened, allowing the bridge to carry a phenomenal 48 million vehicles per year.

Ammann lived for almost another year after the opening of the Verrazano-Narrows Bridge, passing away on September 22, 1965 at the age of 86. His reputation was one only Steinman could rival; Ammann would also be the first civil engineer to receive America's National Medal of Science, bestowed upon him in November 1964 by U.S. President Lyndon B. Johnson. Yet Ammann never considered his largest bridge his greatest. The George Washington Bridge was Ammann's shining achievement; the Verrazano-Narrows Bridge in many ways extrapolated what the George Washington Bridge had proven possible. Nonetheless, New York's largest span remains at the pinnacle of American suspension bridge engineering; it was bigger, and heavier, than any other span built to that time. The nation is not likely to build a larger suspension bridge for some time, if ever.

ENDNOTES

1. David Plowden (1974) also refers to a Fowler "super-arch" at Mackinac.
2. A revised edition of *Bridges and Their Builders* was published in 1957.
3. Steinman would eventually defer payment for some $350,000 of design work.
4. The delays increased American Bridge's contract to $44.5 million. Merritt-Chapman & Scott's substructure contract went from $24.5 to $25.7 million.
5. Carl Condit (1961) held the opposite view, citing the truncated side spans of the George Washington Bridge as detracting from the gentle sweep of the main cables over the main span.
6. Farquharson would also carry out wind tunnel tests on sectional models of a 489-m span truss-stiffened suspension bridge with roadway slots intended for Kelowna, British Columbia. A different type of bridge was eventually built.
7. Worthington had a penchant for arch bridges. In 1910, he proposed a fantastic 549-m span steel arch for the rail bridge at Quebec.
8. Most curious is a simple undated sketch of a variant under the Robinson & Steinman moniker that features sleek portal-braced towers and what appears to be a slender plate girder suspended structure; its resemblance to Ammann's Bronx-Whitestone Bridge is uncanny.
9. This 1955 report also proposed a 1,220-m span two-level bridge across the Hudson River at 125th Street strongly resembling the design later to emerge at the Narrows. The Triborough Authority assumed control of the Narrows' crossing in late 1959, 10 years earlier than planned, when it repaid $30 million borrowed from the Port Authority.

Chapter 8

EUROPEAN RECONSTRUCTION AND REVOLUTION

In Europe, postwar bridge-building faced a huge challenge. Tens of thousands of bridges had been destroyed. This, combined with a scarcity of steel, provided the impetus for new, more economical structures and faster fabrication and erection methods. These circumstances would spur the development of new concepts that would come to transform the classic suspension bridge.

8.1 THE CONTINENT REBUILDS

In newly created West Germany alone, some 15,000 bridges had been destroyed in the war, with almost all crossings of the Rhine and other major rivers lost. Rebuilding the nation's infrastructure created a huge demand for limited amounts of steel. To cope, West German engineers popularized two concepts widely used today—cable-stayed bridges and orthotropic plate decks. Early post-war German suspension bridges would contribute to both concepts.

The first post-war German suspension bridge was a rolled I-beam span built from 1947 to 1949 across the Mosel River at Wehlen. In appearance, this two-lane bridge was almost a miniature of the Rodenkirchen Bridge—perhaps not surprisingly, for its engineer was Fritz Leonhardt. The Mosel Bridge's large width in relation to its 132-m span and its structural continuity through the 37-m side spans guaranteed the aerodynamic stability of the solid 1-m deep girders. Much of the bridge was recycled: the towers stood on the piers of the former bridge, the main cables used locked-coil strands salvaged from the wrecked Cologne-Mulheim Bridge across the Rhine, and the suspenders comprised hoisting ropes formerly used in coal mining. Leonhardt also used cables to prestress the anchor blocks into the ground. The entire bridge—save footings and towers—was replaced in the early-1990s by superimposing a new suspension bridge with an orthotropic deck over the original. To ensure continuous operation, the new bridge has been designed so that suspenders and even main cable strands can be replaced without interruption to traffic.

The first large effort involved replacing the 315-m span Cologne-Mulheim Bridge over the River Rhine south of Cologne. This bridge had been destroyed in 1944. The new bridge—built from 1949 to 1951—used the piers of the 1929 plate girder span. A competition attracted many innovative designs, including a cable-stayed bridge. One concept developed by Franz Dischinger (1887-1953), a thin-shell expert with Dyckerhoff & Widmann, very much resembled M.F. Arnodin's system of the late 19th and early 20th centuries. In this hybrid suspension system, cable stays radiated from the towers to the quarter-points of a 315-m main span and over all 91 m of unsuspended side span, whereas the central half of the main span was suspended in the traditional manner. As an option, Dischinger also offered up a 410-m span. His designs used neither trusses nor plate girders but cross-frames with gently sloped edges and bottoms; their depths ranged from just 2.4 to 3.5 m.

Dischinger had proposed this system as early as 1938 to reduce deflections in an unbuilt 750-m span rail bridge across the River Elbe at Hamburg. He considered the system well-suited for long-span rail bridges and devised several hypothetical schemes, one based on the 1,280-m span Golden Gate Bridge, another involving tandem twin spans of up to 675 m for Scotland's Firth of Forth, and a very long span of 2,000 m. Dischinger published his findings in 1949, obtained a patent, and proposed the concept for a number of bridges. Yet this hybrid concept was never adopted, partly because of its complexity. Ironically, Dischinger's 1953 hybrid alternative to a 160-m span design for a Swedish suspension bridge sparked the revelation that stays alone would suffice; the Strömsund Bridge, completed in 1955, would be the first modern cable-stayed bridge. Dischinger's hybrid was proposed—again unsuccessfully—in 1966 for a span near Dusseldorf, but the form has recently re-emerged in connection with several challenging long spans in East Asia.

In the end, convention ruled for the Cologne-Mulheim Bridge, and a suspension bridge with the similarly terse visage of the former span was adopted. Nonetheless Leonhardt, the span's engineer, made significant changes. Gravity anchorages replaced self-anchorage and the 85-m side spans were not only suspended but continuous with the main span—as in the first bridge. The new 50-m tall towers were built of high-tensile steel and used single-celled stiffened legs. The anchor blocks were prestressed into the ground, and the cable strands were threaded through steel pipes to a more efficient anchorage at the rear of the anchor blocks; this presaged a concept used almost 3 decades later in America.

Leonhardt also used plate girders in the four-lane suspended structure, whose generous 27-m width enhanced the span's torsional stability. The use of a lightweight orthotropic deck—comprising steel plates with ribbing welded to their undersides in place of a heavy reinforced concrete deck—made it possible to almost halve the girder depth of the old span to 3.3 m.[1] In the Cologne-Mulheim Bridge, the deck was welded to the tops of the plate girders to form their top flange *and* support the roadway; only a thin layer of asphalt formed the paved surface. The orthotropic deck minimized bridge weight, cost, and steel usage and, in the Cologne-Mulheim, made smaller cables and anchorages possible. The many design changes and the extensive use of welding reduced steel tonnage from 12,800 tonnes in the original span to 5,800 tonnes—a drop of 55%—which is all the more remarkable given the new span's 250% increase in loading capacity. Equally striking was the speed of construction; onsite steel assembly required only 12 months before the span's September 1951 opening.

Three years later, the 285-m span Duisburg-Homberg Bridge was completed over the River Rhine. Although visually similar to the Cologne-Mulheim Bridge, this self-anchored span is not a true suspension bridge but a bridle chord span, in which the main cables are discontinuous at midspan. The design, by the German firm Demag, was selected out of 35 entries in a design competition. Certain features are notable, such as two parallel 4-m deep, 2.1-m wide steel spine box girders which are spaced

15.5-m apart. Only a thin steel orthotropic deck that for the first time used trapezoidal deck plate stiffeners connected these spine girders (Troitsky, 1967). Extensive welding also contributed to this span's clean appearance. Its erection greatly influenced the development of cable-stayed bridges too; temporary stays supported a cantilevering of the girders from the towers, after which the main cables were installed. Subsequent spans used stays alone as a permanent fixture.

The largest German suspension bridge replacement involved the 1952 to 1954 rebuilding of the 378-m span Rodenkirchen Bridge across the River Rhine south of Cologne. Retreating German forces had brought down its suspended structure in January 1945, and the failure of a main cable had twisted the towers and severely cracked their foundations. Remarkably, towers and anchorages remained standing and could be reused, although the 1,540-tonne west tower foundation required major repairs in an exacting lifting and shimming operation. Stiffening for the new span was obtained from 3.3-m deep stiffening girders with welded and riveted field connections. Its 25-m wide roadway also featured the first composite suspension bridge deck, in which concrete works structurally with the steel girders. The new high-tensile steel suspended structure—continuous over the 567 m between anchorages—was two-thirds the weight of the one it replaced and saved $1.5 million, reducing the total cost to $4.1 million. Hellmut Homberg, the span's designer, would rise to prominence in the 1960s as a pioneer of single-plane cable-stayed bridges.

Almost four decades later, the Rodenkirchen Bridge would again break new ground in an early 1990s twinning. The cheapest means to increase capacity was to replace the bridge—then in need of major repairs—with a cable-stayed span. Yet the 1954 structure was a beloved local landmark. A second deck or deck brackets were expensive, and the approaches to a parallel bridge would desecrate parkland. A compromise was reached by adding a third cable and tower leg on one side of the original bridge; the resulting 53-m deck width would be the largest of any suspension bridge yet.

Work began in May 1991. The new single-celled tower leg is less than half the weight of each nine-celled leg in the original span but is 50% stronger. The new helical locked-coil strand cable is also considerably smaller than the original, thanks in part to an orthotropic deck 10 times as light as the original composite deck. Still, connecting the new deck to the old to form a single bridge while up to 80,000 vehicles crossed the bridge daily was an exacting task.

The challenge lay in equalizing loads on the three main cables. Engineers first hung the new cable lower than its final position. Temporary hinges then connected the two decks over the 567 m between anchorages. Traffic was shifted from the old to new deck in a complicated sequence to permit the original span's concrete deck to be replaced with a lighter orthotropic deck; here, debris from the demolished deck was even placed at strategic points to balance dead loads on the two decks. Once that work was complete, both decks were bolted permanently together and the cable anchors jacked back almost a meter to equalize stresses in the main cables. The $116 million job was completed in late 1994.

One unusual 1946 proposal for a temporary replacement of the famous twin arches across the River Rhine at Dusseldorf deserves mention. Carl Bohny, who from 1927 to 1929 worked for David Steinman (and who penned a 1934 book on suspension bridges) proposed a continuous suspension bridge of four 95-m spans with horizontal cable ties. The bridge was estimated to cost less than $100,000, and although one of the first bridges authorized by German and British military authorities, it was never built. An even more innovative span took its place, one of the three famous Dusseldorf cable-stayed bridges.

France faced similar post-war bridge-building challenges, but other approaches were developed here in a number of small but handsome two-lane suspension bridges that would eventually culminate in the Tancarville Bridge, France's largest. The first was the Rognonas Bridge, a low-level 255-m span with 80-m side spans completed in 1949 over the Durance River. Its heavy half-through truss contrasted with a finely etched diagonal grillage in the top portal strut of its concrete towers that, for all its elegance, was taxing to build. In 1950, an almost identical replacement for the 235-m span, plate girder-stiffened Teil Bridge was opened; Allied bombs had destroyed the original bridge in 1944. A pleasing high-level span was also completed in 1950 across the Aulne River. The 272-m span Terenez Bridge featured elegant towers of reinforced concrete with two struts above the roadway. Like the Rognonas Bridge, a midlevel deck sat on a truss braced only with bottom laterals. Short unsuspended side spans added to the drama of this bridge's rugged setting. The Ancenis Bridge, a 238-m span completed in 1952 over the River Loire, is similar to the Rognonas, especially in its tower design. Another handsome bridge is the Viviers sur le Rhône, a 165-m trussed span whose deck—again—was situated at midtruss level. Graceful concrete towers with a single curved portal stand atop masonry piers that rise to deck level. A second high-level French suspension bridge is the Roche-Bernard Bridge, a slender two-lane span built between 1957 and 1960 over the Vilaine River. Its 245-m span—flanked by 81-m side spans—was suspended between reinforced concrete towers whose solid walls below road level and a heavy top portal beam presaged those used in the Tancarville Bridge. Here, the deck was perched atop a 4-m deep deck truss with lower wind bracing. This fine bridge was replaced in 1995 by a concrete box girder span.

Other smaller suspension bridges used solid plate girders, such as the 121-m span Pont de l'Isère (1954), whose deck was supported by four parallel girders. Its predecessor suspension span had stood from 1833 until 1940, when destroyed in World War II. Other, smaller suspension bridges also emerged. The Canal de Donzère acquired four of them—two spans of 114 and 153 m and two self-anchored spans of 90 and 96 m. The Valence, built in the late 1940s, featured a single tower centering two 112-m spans. Others included the 133-m span Elbeuf Bridge and other bridges across the River Loire and elsewhere. One span over the River Loire—the Varades Bridge—is exceptionally handsome, with simple portal towers and a sleek low-level truss.

Concrete had yet to form the towers of a large suspension bridge—until 1955, when the French commenced the spanning of the River Seine near Tancarville 30 km east of the English Channel port of Le Havre. Here, concrete would make new inroads. This, the first suspension bridge outside of America to exceed 500 m in span, would also launch an intensive period of European activity in the field.

A crossing of the River Seine near Le Havre was first contemplated after the 1870 to 1871 Franco-Prussian War, when a rail link was proposed to enhance area defenses. An 1883 concession for a rail tunnel went nowhere, and a viaduct 15 km to the east proved unpopular with shipping interests in upstream Rouen, which favored a tunnel. An 1895 "ferry" supported by rails laid on the riverbed went nowhere; 2 years later, Arnodin proposed a tramway, but its rejection led him to offer a transporter bridge in 1902. More bridge and tunnel proposals emerged, but even the 59-m vertical clearance of a high-level bridge proposed in 1910 with spans of between 420 and 450 m failed to appease Rouen maritime interests.

This pattern of fits and starts continued after World War I. Costs killed a 1931 tunnel proposal. The idea of a road suspension bridge then emerged in 1926, and by 1932, vehicular ferries were operating. A suspension bridge was again suggested in 1934. Tancarville emerged as a favored location, where the Seine narrows to 600 m and an escarpment on the north shore avoids long approaches; this evolved in 1935 to a 608-m span with a 15-m wide deck supported on the lower lateral bracing of a through truss. Then in 1937, navigation authorities demanded the 50-m clearance be raised by 20 m. This and other changes made costs prohibitive, and although a 1940 law authorizing the crossing produced a 56-m compromise, financing—and then World War II—ground the project to a halt. Studies did continue during the war, and approach embankments were even started. Yet at war's end, money and materials were in short supply; priority was assigned to the replacement of destroyed bridges, not the construction of new ones. In 1945, bridge supporters approached Steinman to seek American assistance in financing, materials, and construction expertise, particularly with respect to assembly of the main cables. Although the contacts continued into 1946, they failed to bear fruit.

By May 1951, a concession had been granted to the Le Havre Chamber of Commerce for the 1935 bridge alignment, and that November, a design competition was launched. Bidders could choose from a 608-m suspension span or a design of their own making. Twenty-three proposals were received—12 of French origin, nine from West Germany, and one each from Great Britain and America. Several were highly innovative. French prestressed concrete pioneer Eugene Freyssinet proposed a daring cable-stayed bridge of 610 m span. Leonhardt proposed an unusual truss-stiffened suspension span of 608 m; its four-lane, 16-m wide deck featured streamlined rounded edges and was supported by a 5-m deep truss of triangular cross-section formed from steel pipes. A second West German proposal for a suspension bridge featured a steel box girder just 2-m deep with rounded edges. In December 1953, a more conventional 607-m span truss-stiffened suspension bridge with 176-m side spans was adopted. The winning French consortium completed basic designs in April 1954 on this less radical, but still innovative bridge, which would be the longest span of any type outside America. Its 123-m tall towers would also be the tallest yet in concrete; their below-deck solid concrete walls made them particularly distinctive. The 1,060-m long helical strand cables would be second only to those of the Chesapeake Bay Bridge and would use anchorages of new design. An $11 million April 1955 contract award launched the span's detailed design, and construction began in October 1956.

Topography influenced the design markedly. The main span rises by 3-m between its south and north towers. Tunnel anchorages extend more than 80 m into the heavily fissured limestone of the north escarpment, their termini connected by a large concrete beam. Prestressed cables extending from the beam to the main cables near the tunnel entrances transmit cable forces to the intervening rock. On the flat south shore, circumstances were entirely different. Early plans had called for a largely buried mass, followed in 1952 by a triangular anchorage similar to those found in Othmar Ammann's Bronx-Whitestone Bridge. In the end, an imposing gravity block was chosen whose 35,000-tonne mass approaches that of the Arc de Triomphe. Only three caissons support it: two small hinged footings under the front corners and a third caisson across the rear of the anchorage.

As Europe's first large span following the Tacoma Narrows Bridge collapse, the Tancarville Bridge came under rigorous aerodynamic scrutiny. Solutions adopted in America—deep trusses, air gaps, and open grating—were all considered but not deemed fully satisfactory. Tests of a sectional model in a purpose-built wind tunnel at the École Normale in Paris also benefited from the aerodynamic theory of Sorbonne professor Yves Rocard and an early use of computers. Three trusses were tested: one lacking bottom lateral bracing, one with flexible lateral bracing, and a torsionally rigid truss with welded bottom lateral bracing. Not surprisingly, the critical velocity of the third variant proved superior. The truss was also made continuous through the towers; it is fixed only at the north anchorage by a strong 6-m deep cross girder, with all expansion taken up at the south abutment. An impressive 712-km/hr critical flutter velocity for the 16-m wide, four-lane truss was achieved—four times the highest winds recorded in the area. A 6-m long midspan cable clamp, stiff concrete towers, and a composite concrete/steel plate deck system also helped to stiffen the truss. The only design maxim reflecting American practice was the 6-m truss depth, about 1/100th of the main span. The limitation of vertical members to every other panel also enhanced the truss appearance.

Construction also spawned innovation, mainly to make up time. Only 18 months into the contract work was already 8 months behind schedule because of design and foundation delays and the severe winter of early 1957. With a need to finish the bridge in 44 months, contractors pulled out all stops—three work shifts, cold weather concreting, additional labor, and a doubling of equipment. Concreting of the south tower began before its supporting caisson was even fully sunk; jacks placed beneath the tower kept it true during the final stages of caisson sinking. Tower formwork—including a crane, footbridge, and the formwork itself—was raised as one unit. Cable assembly also broke with tradition. All 56 helical strands in each cable (with four more over each side span) were draped from a tower-supported cableway without any catwalk. In cross-section, each cable is a 56-cm diameter hexagon in an open configuration. Piece-by-piece riveted truss assembly from an aerial cableway took place along six fronts. The suspender lengths were adjustable, a great benefit for accommodating truss and main cable distortions during lifting of the truss. The deck was also placed before the truss was fully erected and riveted. Opened in July 1959, the Tancarville Bridge was built without a single loss of life and remains the world's longest span with helical cables. Its tolls would fund studies during the 1970s for a 610-m stayed span that would emerge as the Pont de Normandie, a record-shattering cable-stayed span of 856 m completed in 1995. Freyssinet's ambitious dreams for spanning the Seine would more than come true.

France's second largest suspension span—the Aquitaine Bridge—benefited from the experience gained in the Tancarville Bridge. A second road bridge across the Garonne River in the southern city of Bordeaux had been discussed as early as 1850. Arnodin's 1910 proposal for a record 400-m span transporter bridge was killed by World War I, and only in 1954 was the original stone bridge widened. Yet a year earlier, a bypass 5 km to the north had been identified where the river is about 500 m wide. Preliminary designs were approved in June 1955, and although a symbolic groundbreaking had been held in 1960, construction only began in 1963. This 394-m span is smaller than the bridge at Tancarville, yet the topography is remarkably similar. In fact, a steep east bank and a flat west approach requiring more than 1,000 m of approach viaduct make the Aquitaine Bridge longer than its predecessor.

The 21-m wide, four-lane truss clears the river by 53 m. Other than an absence of secondary transverse diagonal bracing connecting the upper and bottom laterals, the 5-m deep truss is of conventional design. The smaller span length and a greater width-to-span ratio rendered unnecessary the Tancarville Bridge's remedies of truss continuity and midspan cable clamps; wind tunnel tests failed to generate flutter in 600-km/hr winds. Still, aerodynamics influenced the design in other ways. The main span truss is structurally connected to one of the 143-m side spans to relieve the towers of large horizontal forces, which instead are transmitted directly—via the truss—to the anchorages in extreme, but unlikely, cases of asymmetrical loading or longitudinal wind.

The 105-m tall towers—like those of the Tancarville Bridge—were built of concrete and, initially, were to take on a similar form. As built, however, they are simpler in outline, with no below-deck solid walls or angled upper legs. One portal brace is hidden within the truss, as in Ammann's Throgs Neck Bridge. Interestingly, the most unusual features of the Aquitaine Bridge are found underground. Poor soil conditions made deep foundations necessary. That for the west tower sits 34-m deep and was built with great difficulty. The base of its 50-m tall anchorage is equivalent in area to the footprint of the Arc de Triomphe in Paris, yet the 37,000-tonne anchorage is supported only at its corners via large circular caissons up to 32-m deep; one can actually peer beneath the anchorage at ground level! To avoid disturbing sensitive soils, prestressing tendons temporarily created horizontal forces later taken up by the main cables and truss; joints and jacks allow the anchorage to be raised up to 4 cm in case of future settlement. The smaller right bank anchorage also offered its share of challenges. Two large 54-m long shafts extend to 15-m below the level of the river, flaring out just behind the main Bordeaux-Paris rail line tunnel. These shafts carry almost all the forces transmitted by the 37 locked-coil strands in each main cable, assembled—as at the Tancarville Bridge—from hauling ropes without catwalks, but ungalvanized, as at the Williamsburg and Tancarville bridges. As it happens, this latter characteristic would prove costly for all 3 spans.

The $24 million Aquitaine Bridge opened in April 1967 after 4 years of construction. In many respects it is a scaled-down version of the Tancarville Bridge, whose influence extends even to the architectural detailing of the prominent south bank anchorage, which, unusually, contains elevators for the span's nonmotorized users. Here, the aesthetic challenge of tall anchorages is handled successfully.

As with New York's Williamsburg Bridge, the ungalvanized cables on both spans would generate major problems. Wire corrosion at the Tancarville Bridge had occurred as early as 1965, and in 1995, one strand actually broke. The Freyssinet Company, a world leader in cable stays, proposed an innovative solution. Previously, it had transferred the suspended structure of the Lorois Bridge from the existing main cables to newly installed replacements before removing the old cables. In 1996, it proposed a variation of the Dischinger hybrid to restore the Tancarville Bridge: each existing cable would be replaced by four smaller cables arranged in a box-like pattern; existing suspenders between the towers and the quarter-points of the main span as well as over the side spans would also be replaced with stays radiating from extended tower legs. To counteract compressive forces pulling the truss into the towers, the stays would be linked to a steel cable attached to the upper chord of the truss. The stays would stiffen the bridge, improving its aerodynamic stability, and each of the four smaller cables would be readily replaceable. Freyssinet claimed the system would be cheaper than a new but traditional suspension system. In the event, two new, smaller, cables of 90 helical wire ropes are being added on either side of and above each original main cable, both of which will remain in service. The midspan cable clamp—which had abetted corrosion—will be removed, aerodynamic studies suggesting it being safe to do so. Tuned mass dampers, two at midspan and another near the north tower, will also be added in this $32 million makeover. Yet even more ambitious is the complete replacement of the ungalvanized main cables of the Aquitaine Bridge, which have succumbed to unrecoverable corrosion. Work began in 2000 to spin new cables just outside of the existing cables; suspenders too are being replaced, and the deck is being modified to accommodate a widening to six lanes.

The Hungarian capital of Budapest was the site of two major post-war reconstructions. W. Tierney Clark's famous eyebar span across the Danube was rebuilt in line with its original visage and reopened in 1949. The new Elizabeth Bridge—completed in 1965—too, had an illustrious predecessor; its towers stand on the piers of the famous 1903 chain bridge that was irreversibly damaged during a 1945 German retreat that left only one tower and some chains intact. It was the last of the city's ravaged spans to be restored, and much debate preceded the decision to build either a replica or a new, modern span. A replica was found to require twice as much steel, and the use of the remaining tower would prevent the building of a wider bridge. The lavish ornamentation of the original was also seen as outdated, and ultimately a modern design prevailed.

The six-lane, 290-m span is notable on several counts. A 27.5-m wide orthotropic deck acts with three well-hidden 2-m deep longitudinal girders that taper slightly under the side spans; suspenders are attached to cross-frames every 10 m. The 50-m unsuspended side spans are simply supported but structurally continuous with the main span girders. Unusually, large buried counterweights between the towers and the anchorages are connected to the side span girders by large springs that reduce stresses on the rocker towers from any extreme loading of the main span. Each of the span's two main cables comprise 61 locked-coil strands.

The old span was a much-beloved Budapest landmark, and great attention was accorded its replacement's appearance. This concern is reflected in the simple portal-braced steel towers, which resemble those of the Rodenkirchen. Their exteriors are corrugated to eliminate the need for vertical stiffeners—thus saving material—but moreso to add texture. The slender but wide suspended structure is also graceful.

Unusually, its assembly involved loading to its final profile with 2,700 tonnes of gravel, whereupon suspenders were adjusted to their final lengths. Girder sections, which to that point had been temporary hinged, were then permanently joined. With greater traffic capacity, the new Elizabeth Bridge used 7,260 tonnes of steel, more than 30% fewer tonnes than its predecessor.

8.2 GREAT BRITAIN STIRS

The re-emergence of improved truss designs in America affected attitudes in Europe, where, after World War II, large suspension bridges began to receive serious attention. Europe would use different methods and materials, such as the concrete towers of the Tancarville Bridge. But the most dramatic developments would occur in Great Britain, where, in 1946, it was decided to proceed with two large bridges stalled by the war—a crossing of the River Severn between southern England and the coal and steel centers of Wales, and a bridge across Scotland's Firth of Forth alongside the triple cantilevers of Benjamin Baker and John Fowler's epic 1890 rail crossing. The design history of both spans would be deeply intertwined.

There was an interesting historical parallel. The Firth of Forth cantilevers had emerged in the wake of the terrible 1879 collapse of Sir Thomas Bouch's spindly wrought iron truss across the Firth of Tay just 50 km to the north; 79 lives were lost in that disaster. As a result, the Forth rail bridge featured well-considered margins against static wind pressure. The designers of the Forth Road and Severn bridges would operate in the shadow of the Tacoma Narrows Bridge failure, but the challenge of aerodynamics was more complex and subtle. The fragile economic conditions plaguing post-war Great Britain would also influence designers greatly in their quest for rigidity.

Ferries had plied both the Severn and Forth estuaries since about 1130. A tunnel across the Firth of Forth was mooted as early as 1740 and again in 1805, when twin parallel tunnels were proposed. In 1818, Edinburgh civil engineer John Anderson proposed a well-elucidated but technically improbable scheme of three 610-m cable-stayed or suspension spans. Influenced by Thomas Telford's 305-m span proposal of 1814 at Runcorn, Anderson incorporated its undesirable characteristics of an extremely low cable sag—imposing great axial stress in the wrought iron bar cables—and partial deck stiffening. Wilhelm Westhofen (1890) would later comment wryly "that on a dull day it would hardly have been visible and after a heavy gale probably no longer to be seen on a clear day either."

Great Britain's expanding rail system spurred an 1860 proposal for a bridge across the Firth of Forth, which, along with an 1864 plan for three 610-m span end-on-end suspension bridges, foundered. Yet plans by Bouch for a suspension bridge of two spans between 488 and 518 m were approved in 1873. Wrought iron chains would be hung from three wrought iron or stone towers centered by Inchgarvie Rock, where the central cantilever now stands. The $10 million scheme was remarkable given the prevailing skepticism over rail suspension bridges; John Roebling's Niagara Bridge, although successful, was one-quarter the size. The collapse of Bouch's Tay Bridge brought work on the foundations to a halt, and the scheme was abandoned a year later. Bouch's trusses were not as stiff as Roebling might have made them, but the design featured stays of the type used by Steinman more than 6 decades later. The present bridge—begun in 1882 and completed in 1890—cost $16 million and 57 lives. Only a single pier from Bouch's central tower remains.

The Severn Bridge, too, had a long and convoluted history. As early as 1825, Telford—with the Menai Bridge nearing completion—suggested a suspension bridge at Beachley. Many schemes were proposed; of interest is an 1845 proposal by Thomas Fulljames for seven end-on-end suspension spans carrying road and rail traffic. The British Admiralty objected to it as a navigational obstruction, but Fulljames also proposed a bridge of three larger spans, with cable stays extending to the towers from above and below the deck. This scheme was better received by the Admiralty, but it never advanced. Then in 1864, Fowler and Baker designed a rail bridge whose 305-m span and extensive use of steel were remarkably ambitious for the time. The River Severn was finally bridged in 1879 with a multispan rail truss 16 km upstream of Beachley, followed in 1888 by the Severn rail tunnel at English Stones about 5 km downstream of Beachley. Although the tunnel remains in use, the bridge was demolished during 1967 and 1968.

The growth in automobile traffic during the 1920s brought demands for road crossings of both the Firth of Forth and the River Severn. In January 1926, Great Britain's Ministry of Transport retained the firm of Mott, Hay & Anderson to investigate a road bridge at the Forth. Borings suggested sites about 1.6 km on either side of the rail bridge and a third paralleling it. The upstream site was rejected because of deep rock beneath the south tower and navigation concerns. Opposition from the railroad to a parallel bridge was anticipated, and in spite of good foundation conditions, this scheme was dropped, a fortunate decision from the standpoint of aesthetics. Mott, Hay & Anderson's 1929 report recommended the downstream site for a 731-m suspension span carrying a 12-m wide roadway and cantilevered footways. The following year, the firm examined an upper road deck on the Forth rail bridge, but high costs and certain rejection by the railroad caused it to be scrapped.

In 1931, the firm of Mott, Hay & Anderson was again retained to examine another site, this time 800 m upstream of the Forth rail bridge. Here, the south tower could be footed at a depth of 33 m. Preliminary plans called for a suspension bridge of 914-m span flanked by 411-m side spans. It would carry a 9-m wide roadway. Simultaneously, Sir Alexander Gibb & Partners proposed a downstream span. The Depression curtailed further planning, but in 1934, the Ministry of Transport concluded both 1931 proposals to be economical, with similar costs of about $17 million each. The upstream scheme of Mott, Hay & Anderson was eventually adopted in 1939, but by this time war loomed. The 1936 spanning of the Firth of Forth at Kincardine 25 km to the west also removed some impetus for a new crossing, although during World War II a bridge comprising 2 spans of 731-m each was proposed immediately east of the rail bridge.

Progress at the River Severn was slower, and not until 1935 were the firms of Mott, Hay & Anderson and Sir Alexander Gibb & Partners asked to independently study the question. Each firm recommended a longer but more direct crossing at English Stones instead of the narrower Aust-Beachley line; the latter route also required a smaller span over the River Wye. Mott, Hay & Anderson called for a 427-m suspension span—Sir Alexander Gibb & Partners, a slightly longer span of 450 m—both with two lanes. The British Parliament rejected a bill authorizing a bridge in 1936, but in 1943, the seeds were inadvertently sewn for the Severn Bridge's eventual construction when the Ministry of War Transport reviewed a long-discussed plan to build a tidal barrage across the estuary. Although a logical choice for a transport link, the barrage was considered a long-term proposition, and it was decided to pursue a bridge separately (a wise decision, for a barrage remains unbuilt).

The Ministry of Transport revived both the Severn and the Forth Road bridges in 1945. The Forth Road Bridge Joint Board was established in 1947, but the government's intention was to build the Severn Bridge first. Mott, Hay & Anderson and Freeman Fox & Partners—successor to Sir Douglas Fox & Partners, designer of the Sydney Harbour Bridge—were retained to design the bridge over the River Severn. The alignment was moved from English Stones to Aust-Beachley, where firm limestone existed and the Aust Cliff provided a logical east abutment. A 915-m span suspension bridge was suggested. Public hearings in 1946 brought a favorable response, and by July, construction was expected to begin by late that same year.

In 1948, Mott, Hay & Anderson was engaged to develop a design for the Forth Road Bridge. As both bridges were to have equal spans, the span over the Firth of Forth would build on the experience acquired in the design and construction of the Severn Bridge. Yet the failure of the Tacoma Narrows Bridge left many unresolved questions. Ambitions to begin construction were also tempered by the fact that the Severn Bridge's two main cables would consume the entire output of Great Britain's wire industry for 30 months. As a result, an exhaustive research program in suspension bridge aerodynamics was launched in early 1946. Ending in 1951, the program was also intended to generate a truss design for the Severn and Forth Road bridges. Initial research assessed the design, construction, performance, and aerodynamic qualities of the great American suspension spans, through which a close liaison with American researchers such as F. B. Farquharson (who had spent the latter part of 1946 in Great Britain) would prove influential. Christopher (Kit) Scruton would play a key role, under lead investigator R.A. Frazer. Scruton had studied aircraft flutter before World War II; in 1946, bridge aerodynamics was an interesting diversion, but it would come to dominate his career.

It was intended to test sectional models, as well as 15.2-m long, 1/100 scale full aerolastic models. The latter required a large wind tunnel, which was built with the aid of Italian prisoners of war in 1946 and 1947 in an aircraft hangar at Thurleigh, Bedfordshire, for the National Physical Laboratory (NPL). Initial testing involved 1/100 scale sectional models in a small wind tunnel at Teddington, while the large model—simulating a 915-m span with 311-m side spans and a 32-m wide, 7.6-m deep half-through truss—was built. This, the second full aerolastic suspension bridge model ever built, was ready for testing in 1948. The suspended structure was a half-through truss, with the upper of two planes of lateral bracing supporting a four-lane roadway with a median air gap. Three variants of this initial configuration were progressively devised on 1/100 scale sectional models. Air gaps placed between the roadway and flanking footpaths improved aerodynamic performance but increased the truss width to 36 m. The 7.6-m wide footways were then cantilevered from a narrowed 24-m wide truss with a slightly greater depth of 8.4 m, the footways serving to break up winds before they reached the truss. Finally, the roadway was shifted to the top of the truss. This last arrangement proved successful and roughly paralleled American developments, although its 1:120 depth-to-span ratio was less conservative. Test results with sectional and full models were remarkably consistent, and it was then decided to focus only on a larger 1/32 scale sectional model representing 183 m of span to fine-tune the final truss design, such details as handrails imposing surprisingly influential effects on aerodynamic stability.

Onsite studies at the River Severn wrought other changes. Borings and soundings in the winter of 1945-46 forced a shift of the Aust tower 91 m to the east, lengthening the main span to 1,006 m, which was now flanked by 305-m side spans. Initial plans called for 146-m tall diagonally braced towers, but a variant using three portal struts was then adopted, which would strongly resemble (although heavier and more ornate) those eventually built. The entire scheme—including a viaduct over the River Wye—was expected to cost $30 million. In August 1947, the government announced that construction would begin the following spring, with completion set for 1952 or 1953. Yet economic recession soon forced the government to slash expenditures, and in 1949, both the Severn Bridge and the span over the Firth of Forth were shelved indefinitely. Design work did continue, during which an 18-m eastward shift of the Severn Bridge's west tower for economic reasons fixed the span at 988 m.

By 1950, Gilbert Roberts (1899-1977) of Freeman Fox & Partners had assumed overall responsibility for the Severn and Firth of Forth designs. Roberts had graduated from London University after service in World War I, during which an injury had left him with a slight limp. In 1924, he joined Sir Douglas Fox & Partners and worked under the legendary Sir Ralph Freeman on the Sydney Harbour Bridge. During his tenure with the span's contractor, Dorman, Long & Co., Roberts helped develop high-tensile steels and electric arc welding; both would have profound significance for his future bridges. His first involvement in a suspension bridge came with the 320-m span Chirundi Bridge, completed in 1936. That year he moved to Sir William Arrol & Co., where the hollow box columns he introduced in power stations became standard in Great Britain. Roberts also improved dock gate design and, during World War II, helped develop all-welded marine landing craft. After returning to Freeman Fox & Partners as a partner in 1949, Roberts headed the design of the Dome of Discovery for London's 1951 Festival of Great Britain. Now, in abeyance of bridge construction, Roberts refined the Severn Bridge design, injecting improvements with technologies he had helped develop. In 1954, the design was essentially complete.

Preliminary design work on the Forth Road Bridge, although slowed, also continued. Borings were taken in 1949 and 1950 along the 1939 Mott, Hay & Anderson alignment. Several unsolicited schemes also appeared. An upper road deck was again suggested in 1950 for the rail bridge but was rejected by a Ministry of Transport committee comprising Roberts, H. Shirley-Smith, and J.F. Baker; it's 5.5-m wide roadway was too narrow, among other problems. An unusual proposal made in 1955 by G.A. Maunsell for a four-lane concrete tunnel supported by piles above the river bed was also rejected by another panel of Roberts, Shirley-Smith, and Sir William Halcrow.

In 1952, Scruton and Frazer's Severn aerodynamic studies were published. The findings, not surprisingly, paralleled American conclusions. Plate girders were found susceptible to both vertical and torsional oscillation. By simply covering the sides of the scaled 7.6-m deep truss of the full Severn model with paper, they had simulated vertical and torsional oscillations "identical in character" to those at the Tacoma Narrows Bridge (Frazer, 1951; Walshe, 1967).[2] Trusses were found to be largely immune to vertical oscillation, and a boxier truss with fairly conservative depth- and width-to span ratios could also quell tendencies toward torsional motion. Wind slots in the deck had been found as early as 1948 to greatly enhance aerodynamic stability; cantilevered footpaths also helped. Small details had a surprisingly positive effect on stability, including trussed—instead of solid plate—roadway stringers, wind bracing situated near the deck, and open handrailings. These helped to not only break up airflow but reduced the size of eddies generated in the lee of the truss and their consequent effect. Cable stays radiating from the tower tops as in the Bronx-Whitestone Bridge were also recommended, but they would not appear in either the Forth Road or Severn bridges as built.

An improving economy saw both the Severn and Forth Road bridges resurrected in 1956, and design work for the span over the Firth of Forth started in earnest the following year. The Severn Bridge was still slated for construction first, but in February 1958, the government—capitulating to intense political pressure—assigned priority to the Firth of Forth. Almost identical spans—988 m for the Severn Bridge and 1,006 m for the Forth Road Bridge—facilitated modifications enabling a similar bridge to be erected at the Forth, one major difference being an additional 9 m of vertical navigational clearance. Thus, had it not been for Great Britain's economic woes, the bridge that now spans the Forth would in all probability have been built across the River Severn.

The Forth Road Bridge would not be Great Britain's first large post-Tacoma suspension bridge. In fact, peculiar aerodynamics would envelope plans for a road crossing of the River Mersey at Runcorn-Widnes, where Arnodin's 1905 transporter suspension bridge was in dire need of replacement. A two-lane, truss-stiffened suspension bridge of 314 m span with unsuspended side spans was planned to closely parallel an existing rail bridge. The roadway sat on the bottom laterals of a 4.6-m deep through truss, with no upper lateral bracing. To save time, tower foundation work began in 1955 while Scruton tested 1/32 scale sectional models at the NPL. The stability of the design was confirmed. Yet in the lee of a sectional model of the rail bridge, the story was dramatically different. Eddies and vortices severely buffeted the suspension span, especially in scaled winds of 65 km/hr. Vertical motion also occurred in gentler winds, but higher winds often incited torsional motion in the model. Moreover, the largest waves doubled to an equivalent of 12 cm in the full-scale bridge when trains were present. The truss was subsequently deepened to 7.6 m and an upper plane of lateral bracing was added, but the writing was on the wall; the suspension bridge was abandoned in favor of a 330-m span fixed steel arch that, when completed in 1961, was the world's longest of its type.

Similar concerns characterized the planning of a suspension bridge over the River Tamar between Plymouth and Saltash in southwest England. The Tamar would closely parallel I.K. Brunel's 1859 rail suspension/arch Royal Albert Bridge. Plans for a high-level road bridge had been prepared as early as 1931 but had been rejected by the British Admiralty. A low-level bridge was promoted in 1953, and a 1954 study affirmed the feasibility of a crossing. Yet the central government expressed no interest in paying for it. Mott, Hay & Anderson was retained locally in 1955 and, in 1956, recommended a suspension bridge of 335 m span with 114-m side spans situated just 65 m away from Brunel's span. The firm had prepared the Runcorn-Widnes design, and wind tunnel tests with sectional models in 1956 bore out suspicions when buffeting emerged. Tests revealed vertical oscillations of up to 61 cm with a stationary train on Brunel's span in a small range of wind speeds. This situation was considered rare, but it required a shift in the height of the truss relative to the rail bridge to correct. A generous 4.9-m deep Warren truss was also crafted—its depth-to-span ratio well in excess of the 1:100 maxim of the period. The government approved the scheme in 1957 and eventually decided to fund it, allowing construction to begin in June 1959, 100 years after the Royal Albert Bridge opened.

Cleveland Bridge & Engineering's offer to build the Tamar Bridge in just 2 years—10 months less than planned—won it the contract. Pneumatic caissons were soon sunk to 14-m depths and anchorages tunneled 15 m into rock. The Tamar Bridge was the third large span, after the Tancarville Bridge and Norway's Varodd Bridge, to use concrete towers. Each tower is 79-m tall, with the walls of their legs up to 61-cm thick; concrete was also used for the cable bents. By early 1961, draping of the locked-coil main cables had begun, and by autumn, the truss—the first in a suspension bridge to be built of welded steel box members—was in place. Computers helped determine the center span-to-tower erection sequence to allow bolting as sections were added. The Tamar Bridge was opened in October 1961, 3 months later than Cleveland Bridge & Engineering's target date, largely because of major weather interruptions during excavation and truss raising. Still, its final cost of just $5 million was 20% less than the estimate. Remarkably, it was the first suspension bridge on British soil to exceed the 214-m span of Brunel's venerable Clifton Bridge, by then almost a century old. Wind-induced motion has been known to occur on the Tamar Bridge when trains pass over the adjacent rail bridge, but it has been considered inconsequential.

The Tamar Bridge has recently undergone a major metamorphosis to allow the bridge to support the 40-tonne trucks permitted under European Community directives; more capacity was also built into the span to support a 10-fold increase in traffic since the span's opening and thus avoid the need for a second bridge. Hyder Consulting (formerly Freeman Fox & Partners) was brought in to design an upgrade. New 6-m wide lanes would be cantilevered on either side of the truss, almost doubling its 15-m width. These outer lanes would channel traffic while the original deck was replaced by a stronger lightweight steel battledeck; the outer lanes would then be retained—one for local traffic and buses, the other for cyclists and pedestrians.

Unfortunately, the existing main cables could not, on their own, support the combined weight of the old concrete deck before its replacement and the new deck extensions. To provide the necessary strength, eight stay cables are being added from each tower top to the bottom chord of the truss over the main and side spans. In a variation of the Freyssinet Company's unrealized proposal for the Tancarville Bridge, these stays are linked over the bottom truss chord of the main span to form a suspension cable that lifts the truss and strengthens it considerably. In fact, the new stays limited the amount of new steel required to stiffen the truss to only 300 tonnes, compared to an almost doubling of its 7,900-tonne weight with traditional strengthening. As a final benefit, the stays also eliminated a slight sag in the truss that had developed over the years. This ingenious strategy reduced costs to $39 million, and thereby permitted the project—begun by Cleveland Bridge & Engineering in early 1999 and due for completion in early 2002—to proceed.

8.3 THE FORTH ROAD BRIDGE

By the time the Tamar Bridge opened in 1961, construction was well underway on the bridge over the Firth of Forth (*see* Fig. 8-1). Many of the fabrication and welding advances made under Roberts' tenure after re-initiation of the Forth Road Bridge design in the mid 1950s are found in its towers (*see* Fig. 13-3). Riveted multicellular towers typical of American practice had originally been envisaged, with a top horizontal strut and a curved below-truss brace favored by the architect, Sir Giles Gilbert Scott. This approach was subsequently considered to be structurally and economically unsatisfactory, and diagonal braces were adopted. A vestige of the original design motif can, however, be found in the arched concrete approach piers. Welded box construction—to which Roberts contributed greatly—now permitted larger cells, reducing the tower cross-section to only 11 cells. The contractor further reduced this number to five cells, aided by what has been suggested to be the first use of longitudinal stiffeners on steel plates. All steel members save the tower brackets that support the foot/cycle tracks were welded in the shop and then bolted in the field, a first for a span of this size. Each 156-m tall tower weighs only 2,450 tonnes—extraordinarily light by American standards—and was built from high-tensile steel. The legs are also extremely slender—only 2.9 by 7.3 m at their bases—and, unlike many, are parallel to one another. Their clean lines also benefit from slender welded closed-box diagonal bracing, bolted joint lines between adjacent boxes, and an absence of cruciform shaping common to American spans. In total, the towers were estimated to weigh 30% less than those of traditional design. The shaping of the tower piers is also unique and effective. The result is a fine example of diagonally braced towers, although the deep truss and cantilevered walkways obscure somewhat their clarity of form.

The 8.4-m deep, 23.7-m wide stiffening truss system also comprised welded steel box members with bolted connections (*see* Fig. 7-2). Its main span orthotropic deck, although not acting compositely with the truss, was the largest yet; it weighed half that of a conventional reinforced concrete deck and was estimated to save more than $2 million in tower, cable, and anchorage costs. Over the side spans, a standard concrete deck balanced loads on either side of the towers, avoiding the need to tilt the towers or move the tower saddles during spinning. One little-known contribution that reduced truss weight significantly was an advance in static analysis that allowed engineers to account for the torsional stiffness of the truss and deflections from shear. C.D. Crosthwaite and Frazer (the latter had directed Scruton's aerodynamic research) drove this change; Frazer performed almost 2 years of laborious hand calculations ending in 1951 just to verify the torsional aspects of the bridge. Oleg Kerensky (1967) would consider this the Forth Road Bridge's most significant achievement. D.M. Brotton also pioneered the use of computers for his analysis of the truss erection

Fig. 8-1. Forth Road Bridge
PHOTOGRAPH COURTESY OF LESLIE IRONSIDE.

sequence.

One subtle difference between British and American practice embodied in the Forth Road Bridge is the gentle camber of its truss, which, although confined to the main span in American bridges, extends over the side spans as well. The transition to the upward camber of the approach spans over the cable bent is the cleanest—perhaps save Denmark's Lillebælt Bridge—of any large suspension bridge. This effect is emphasized by the use of unobtrusive tunnel anchorages beneath the approach spans.

Gravity anchorages had in fact been originally planned, but great rock depths, especially under the south anchorage, made their cost prohibitive. Instead, four tunnels up to 77-m long were angled below ground, widening up to 13.7 m^2 at their bases.[3] The anchoring system also differed. Eyebars typical of American spans could not be easily produced in Great Britain. Instead, prestressed wire ropes embedded in the base rock extend to the tunnel portals through 456 concrete-encased steel pipes; here, the cable strands are looped around semicircular steel shoes. The 38,100 tonnes of anchorage are actually less than the combined total of the span's dead and live loads; rock does the rest.

Foundation work at the Firth of Forth began in late 1958. To foot the north tower, rock was simply blasted to enable the casting of a 17- by 51-m support slab. The south tower was, however, considerably more complex. Twin rectangular pneumatic concrete caissons enclosed in cofferdams were to have been sunk by workers excavating under compressed air. In the end, open excavation was adopted. A dewatered cofferdam was first excavated to a depth of 11 m. Two 18-m diameter circular reinforced concrete caissons were then sunk to a depth of 30 m, followed by the pouring of a 6-m thick connecting slab as support for the 9-m tall pedestals.

The superstructure consortium involved Cleveland Bridge & Engineering, Dorman Long & Co., and Sir William Arrol & Co. (the latter contractor had built the Forth Rail Bridge 8 decades before). Erection of the north tower began in June 1960, followed 4 months later by a start on the south tower. The Firth of Forth's strong winds soon made their presence felt in February 1961 when the incomplete north tower, now 122-m tall, began to oscillate. Winds of 30 km/hr induced motion with a double amplitude of up to 2.3 m at the tower tops with a period of just 4.5 seconds. Motion persisted for hours, and it not only seriously delayed work but made workmen ill. Concerns over structural health were also voiced. Vortex shedding was revealed as the culprit, but an ingenious arrangement of restraining ropes connected to a large sliding concrete block reduced motion to about 25 cm. By June, several months behind schedule but without further incident, both towers stood complete. Only after the bridge was completed did the contractors learn to their amazement that Scruton had advised the design engineers of the potential for tower oscillation as early as 1957 (Shirley-Smith, 1967); the fact that no one had ever advised the contractors evoked some rather pointed criticism!

The towers' wind encounters would soon pale in comparison to the experience of spinning the main cables. British cable spinning experience was virtually nonexistent, and in fact no one outside of the United States had ever undertaken spinning in its modern guise. Its complex and specialized equipment was expensive, and skilled labor was required. The approach adopted in the Chirundi Bridge was examined but was found to be unsuitable for use at the Firth of Forth. It appeared then that the only realistic approach was to import American expertise and use cable spinning. The planned use of the same equipment for the Severn Bridge—then beginning construction—would introduce economies, and the experience gained at the Firth of Forth would also benefit the Severn Bridge. Spinning was thus adopted to assemble the 11,618 wires into each 61-cm diameter main cable.

The venerable firm of J.A. Roebling's Sons, under Blair Birdsall,[4] would tackle the Firth of Forth's fickle and capricious weather. The firm had perfected the drawing of four wires over one wheel, thus doubling to eight the number laid per trip, and it had also developed cable saddles with stepped grooves and vertical stakes to maintain strand position during spinning. Yet the firm soon encountered unprecedented difficulties. Spinning did not start until late November 1961, on the cusp of winter. Attempts to spin in winds of up to 65 km/hr forced bridgeworkers to cling to catwalks that swayed up to 9.1 m out of alignment. Four-wire spinning was soon abandoned in favor of two-wire spinning, and a 50-km/hr wind limit was imposed. Yet with 40- to 50-km/hr winds the norm, sometimes only two or three wires were spun per day. The worst, however, was yet to come.

In early February 1962, after only a few strands had been spun, a fierce storm slammed into the Firth of Forth. Winds peaking at 160 km/hr battered the strands for 20 hours. The strands in the west cable over the north side span broke loose from their lashings, causing increasingly tangled wires to smash into catwalk railings, electrical equipment, and spinning gear. Over the main span, the catwalks and strands were pushed up to 15 m out of line. After 2 days the storm abated, but steady gale-force winds made only emergency repairs possible; much equipment lay in tatters. Damage was especially severe over the north side span, where equipment was actually lodged in the entangled cable strands; here, some 305 m of catwalk handrail from the east main span catwalk now lay in the Forth itself. Then only 3 days later a second storm of even greater intensity roared into the Forth. Winds exceeded 160 km/hr for 5-minute intervals; the main span cables swayed so violently that the east cable swung west of the normal position of its west counterpart. This storm left surprisingly little damage, largely on account of the interim repairs. But the havoc of the first storm had still to be rectified. The west cable was so badly tangled that engineers considered scrapping it. In the end, the 1,500 wires in all six strands were combed, layer by layer, in an operation lasting a month, with the result that spinning only resumed at the end of March. Strong winds ultimately forced spinning wheels to lie idle one-third of the time, extending a 3.5-month operation to one of 8.5 months ending—finally—in August 1962.

Concerns over the Forth's weather had ruled out the floating of truss units, and piece-by-piece cantilevering of the truss from the towers began in early 1963. This operation had been the source of some concern, as the lower chords would not be bolted until the entire truss was in place, thus reducing torsional stability. Wind tunnel tests prompted the decision to only partially install the steel battledecks (and the footways not at all) over the main span until the truss was in place, thus leaving a temporary 7.6-m wide central air gap; in the side spans, only the footways were placed. Still, there were problems. Minor damage from the sway of the unbolted truss plagued early operations, and a gantry later fell from the bridge, killing two. This was the second fatal accident—the first occurred when a tower-mounted winch moving a painting gantry failed. Both plummeted 137 m—the winch killing one bridgeworker and the 4-tonne gantry knocking a crane arm into the storm guys stabilizing the side span catwalk from which, miraculously, no one was thrown. Weather, as always, slowed work, but the truss was in place by year's end. The $32 million Forth Road Bridge opened in September 1964 as Europe's longest suspension span and the world's fourth longest; including approach viaducts, it stretches 2,515 m.

8.4 REVOLUTION ACROSS THE RIVER SEVERN

A lack of experience in the design and construction of large suspension bridges had in some ways freed the designers of the Forth Road Bridge from the traditions ingrained in the great American spans—witness, for example, the extensive use of welding and large box columns. Although an important advance, the Forth Road Bridge provided little hint of the remarkable developments that would occur at the Severn Bridge (*see* Fig. 8-2). Here, design work had continued with an intended 1961 construction start. Yet the designers—Freeman Fox & Partners for the superstructure and Mott, Hay & Anderson for the foundations and anchorages—were under great pressure to further economize. Roberts' team had managed to reduce the truss depth from 8.4 to 4.3 m—almost one-half that found in the Forth Road Bridge—by making the orthotropic deck act compositely with the truss to augment vertical, horizontal, and torsional rigidity. Over this period, a revolutionary new idea was also suggested—a full-width box girder. A single, large box with sloped edges (unlike the spine boxes used in the Duisburg-Homberg Bridge) would not only save considerable amounts of steel but offer a new means to achieve aerodynamic stability. In place of a truss that would break up the flow of wind, the box would divert wind around it in the manner of an airfoil (without the lift), thus minimizing the generation of vortices. Roberts and Dr. William Brown, a box girder expert with Freeman Fox & Partners since 1951, had been toying with the idea for some time. Although intriguing, it was radical—too radical for some, who felt it would be risky as a prototype for such a large structure; testing to verify and refine the concept could also threaten the schedule. An accident then occurred that would have profound consequences for not only the Severn Bridge but the future of suspension bridge design. During wind tunnel testing of the modified truss at the NPL at Thurleigh in 1959, a sectional model broke free and was devoured by the tunnel fan. A replacement would take time to build, so to fill this void, Roberts endeavored to test the box girder.

Consisting of solid steel plates joined into a rigid closed square, rectangle, or trapezoid, the box girder was not a new concept; it was introduced a century before in Robert Stephenson's Britannia Tubular Bridge (which was originally to have been suspended from wrought iron chains). Post-war box girder developments were led by West Germany and were closely related to advances in orthotropic decks. The box girder's economy derived from lightness, simplified erection and maintenance, and superior torsional strength. This latter attribute made them attractive for long-span suspension bridges, and their clean lines and thin profile offered improved aesthetics. Brown devised a shallow box profile and various edge shapings; the NPL, under the direction of Scruton and associate Dennis Walshe, built a simple plywood model to test the various options.[5] One arrangement, a six-sided trapezoidal box with tapered edges resembling an airfoil, possessed superior stability. Strips of wood simulating sidewalks were then cantilevered from various positions along the model's side, from which an intermediate location was found that added marginally to overall stability.

The form's superior resistance to vortex shedding and flutter convinced the Ministry of Transport to fund further testing and development. The design combined a high degree of torsional stiffness with the flutter characteristics approaching that of a flat plate because of its streamlined profile. Still, the shallow box was soon found to

Fig. 8-2. Severn Bridge
PHOTOGRAPH COURTESY OF HYDER CONSULTING.

be extremely sensitive to even minor design alterations. For example, closed railings impeded airflow and reduced stability. Minor oscillation would also occur in a small range of wind speeds around 25 km/hr. The riveted or bolted connections of a truss would slip harmlessly to structurally damp out such energy, but it was intended to weld the entire box structure together. The damping ability of a rigid all-welded structure, itself a relatively recent development and now envisaged on an unprecedented scale, was unknown. Additional damping was thought necessary, but it would have to emanate from elsewhere than the box girder. Under a conservative assumption of no structural damping at all, Roberts proposed to incline the suspenders longitudinally. This triangular connection between main cable and deck would damp out motion through a hysteresis (the absorption of energy through elasticity) effect. Inclined suspenders were thought to increase bending stiffness in the order of 5% to 10% and, in the first antisymmetric torsional mode (which is what brought down the Tacoma Narrows Bridge), dissipate three-quarters of the 7% of oscillation energy they would take up. The concept on which Modjeski & Masters' engineers had speculated in 1941 was now acquiring tangible form.

The design devised by Roberts and Brown was nothing short of revolutionary. The welded box girder, just 3-m deep over a clear span of 988 m, appeared to violate the principles so assiduously developed out of the Tacoma Narrows Bridge failure (*see* Fig. 8-3). Its 1:324 depth-to-span ratio approached the proportions of the original Tacoma Narrows Bridge, yet the principles underlying the Severn Bridge's design were so different as to render this traditional measure almost irrelevant. It cleverly integrated three means to enhance aerodynamic stability: superior torsional stiffness, aerodynamic shaping, and damping. The streamlined box profile would almost completely eliminate the conditions under which vortices and eddies are created; no midspan cable clamp was required. Shaped as an inverted wing, with the longer surface below, lift would not be generated. Moreover, the box girder's economy was impressive. It was estimated to cost 20% less than the shallow truss because of equivalent material and weight reductions—contributed in part by the composite orthotropic deck—and because of the ease and automation of welding.

With a depth less than a third of the Forth Road Bridge truss and less than two-thirds the weight, the Severn Bridge's box girder would be torsionally more rigid than its predecessor. Drag would be just one-third of the competing truss and one-fifth that of the Forth Road Bridge, relieving the towers of considerable lateral stress. Lighter towers, lighter cables, and smaller anchorages would add further economy, and the box surface would simplify maintenance and painting enormously—the amount of exposed surface was just 70% that of the truss. Being airtight, the interior was also sheltered from weather and water, further reducing maintenance and enabling the boxes to be floated to the bridge without the use of barges. All these extraordinary advantages did not lead to the immediate adoption of the box girder, and both it and the truss design were fully developed for tendering. Yet when superstructure bids arrived in early 1962, the welded box girder design produced an overall savings of 10% in comparison to the truss. On this basis, the box girder was selected.

Roberts' major accomplishment tended to overshadow his less visible but important refinements in the Severn Bridge's 136-m tall towers (*see* Fig. 13-3). The lower lateral static wind stresses imposed on the tower legs made it possible to

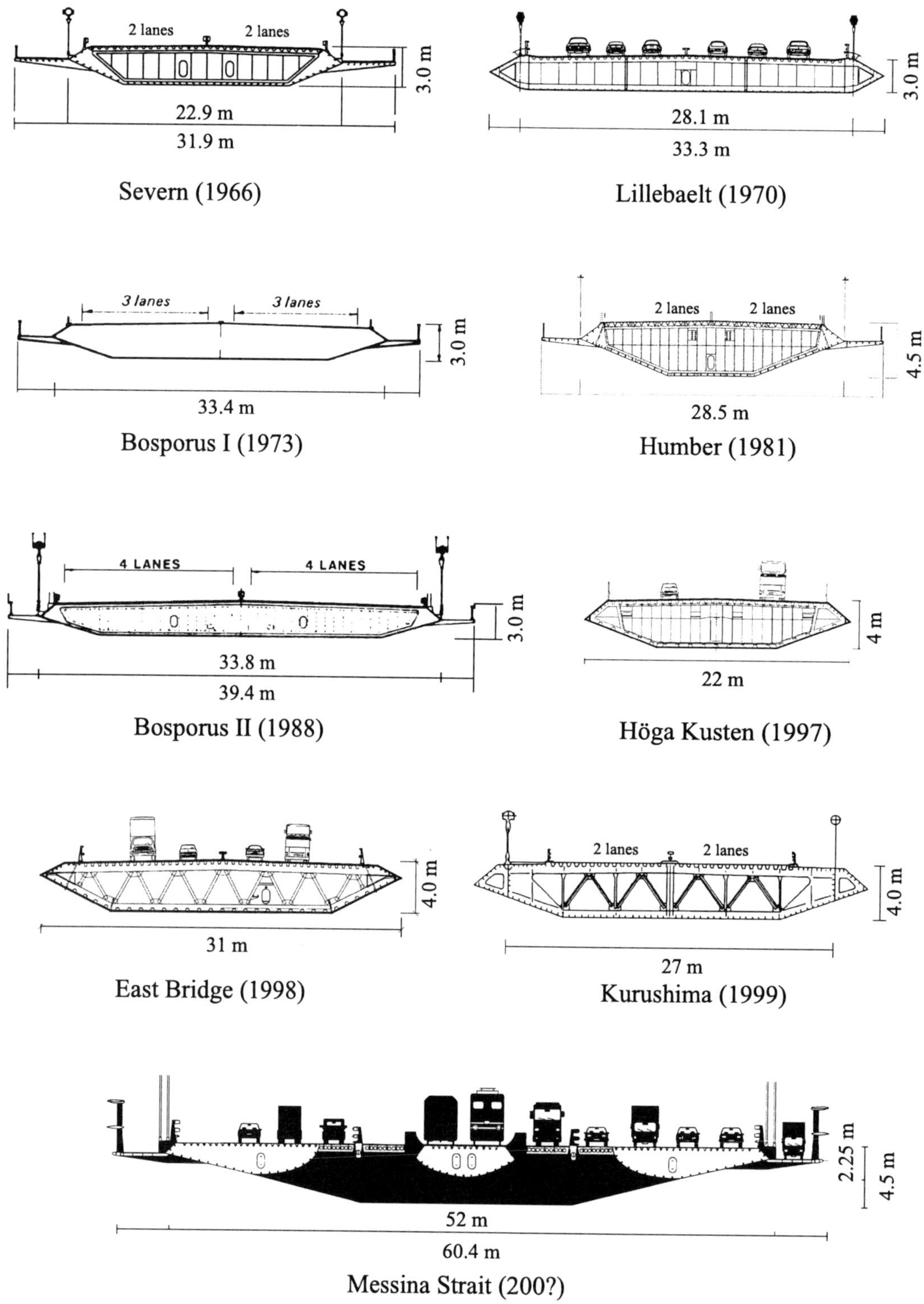

Fig. 8-3. Evolution of box girder suspended structures
MODIFIED FROM ORIGINAL DRAWINGS COURTESY OF THOMAS TELFORD LTD., COWI ENGINEERSAND PLANNERS A/S, HYDER CONSULTING, HUMBER BRIDGE BOARD, AND STRETTO DI MESSINA S.P.A.

fashion each one from a single rectangular steel hollow box formed from four welded stiffened plates. The use of portal struts also now became economical. Leonhardt's towers for the 1951 Cologne-Mulheim Bridge had featured single boxes, but the scale of the Severn Bridge was unprecedented and constituted a complete departure from American multicellular practice. Each 1,090-tonne tower was less than one-half the weight of a Forth Road Bridge tower. Moreover, with the exception of a few bolted connections, the towers would be all-welded. One final but subtle change was that the use of single cells also greatly reduced the extent of cruciform leg shaping.

Construction of the Severn Bridge began in 1961 with work on the tower and anchorage foundations. Unusually complex geology is matched by enormous tidal variations that reach 14 m—among the world's largest. Water levels change by 3 m in as little as 1 hour, and tidal currents approach 16 km/hr. Work on the 43- by 13-m Aust (east) tower pier could only be carried out for the few hours per day the site was exposed at low tide. Here, massive concrete blocks were stressed via steel bars to rock leveled to a depth of 5.5 m; these blocks support piers extending 19 m above rock. The more difficult Beachley tower pier, although outwardly identical to the Aust footing, is radically different. Two 18-m diameter cofferdams were first excavated to a depth of 20 m and then lined with concrete segments bearing directly upon rock; their interiors were then concreted. All told, this pier would devour more than twice the concrete of its Aust twin. The use of gravity anchorages also contrasted sharply with those at the Firth of Forth. Long prestressed steel bars restrain large steel distribution slabs that in turn anchor the threaded steel rods and semicircular strand shoes around which the cable strands are wrapped. Exposure of the Aust anchorage site only at low tide—as at the tower—restricted access during the early stages of its construction.

In March 1963, the same consortium then wrapping up work at the Firth of Forth began to erect the Severn Bridge's portal-braced towers. The towers' simplified design permitted lifts of 17-m tall sections, and work proceeded rapidly. Tower oscillation, however, had been predicted in winds ranging from 32 to 130 km/hr, and mindful of the travails at the Firth of Forth, engineers installed restraining ropes during the towers' ascent. As with the Forth Road Bridge, the towers' lightness also required they be stressed by steel cables to the pier bases (albeit temporarily), a strategy emanating from wind tunnel tests. Spinning of the 51-cm diameter cables took advantage of the equipment and personnel from the Forth Road Bridge. Although operations were subjected to oft-severe winds, to everyone's relief nothing approached the conditions encountered at the Firth of Forth. The cables were actually completed a few weeks ahead of schedule, whereupon the catwalk support ropes were removed for later use as the roadway guiderails. The 8,322 wires in each cable—almost 30% fewer than in the Forth Road Bridge—illustrate the magnitude of weight savings achieved with the box girder.

In late 1964, lifting of the epoch-making box girder began, from midspan back to the towers. Unusually, the entire main span was lifted into place before any lifts in the side spans, a measure that halved the number of required lifting struts from four to two. Temporary bulkheads made the 18-m long, 118-tonne box sections airtight, and units bobbed gently down the River Severn from nearby assembly yards before being nudged into position beneath the span. The tides and swift currents continued to prevail, and sections could only be launched during neap tides of high and placid water and be towed in currents of less than 10 km/hr. Such conditions occurred only six to 10 times per month; it was estimated that the entire suspended structure could otherwise have been lifted into place in half the 14 months it did take. For final effect, poor weather then set in to hamper welding of the lifted box sections.

The Severn Bridge's September 1966 opening concluded 5 years of construction, a noteworthy achievement given the challenging site conditions. The span stood at the leading edge of suspension bridge design, the most significant development since the George Washington Bridge more than 3 decades before. Danish engineer Niels Gimsing (1998) would later describe the Severn Bridge's box girder as the "most important innovation within suspension bridges in the 20th century."[6] Its total steel weight of 18,000 tonnes—11,160 tonnes of this comprising the box girder—constituted less than three-quarters of one Verrazano-Narrows Bridge tower and less than two-thirds of the steel in the Forth Road Bridge. The ancillary 235-m span over the River Wye was also the first cable-stayed bridge outside of West Germany to use a central plane of stays. The entire bridge—including the Wye—cost just $28 million, some 15% less than the Forth Road Bridge, a span already noted for its economy.

Visually, the Severn Bridge stood in utter refutation of the stiffening truss that had dominated post-Tacoma design. Its grace paralleled that of the ill-fated Gertie, but the span was flutter-resistant in winds of up to 240 km/hr. In smaller details, the Severn Bridge also reflected aesthetic refinement. Its single diagonal suspenders were simpler than the clusters of suspenders common to American spans. The lack of a midspan cable/deck connection also contributed to the grace of the bridge, and in profile its cantilevered walkways were, unlike those of the Forth Road Bridge, almost invisible.

The Severn Bridge generated tremendous interest—and controversy. Roberts, who had already been knighted in 1965 for the Forth Road Bridge, would earn patents for the box girder design and inclined suspension system. It was said that he and Brown devised new box arrangements at a rate of one per week. History has muddied the contributions of Roberts, Brown, and colleague M.F. Parsons to the conception of the Severn Bridge's box girder; more than 20 years later, Parsons (1988) would state that he suggested the idea to Roberts and that Roberts had instructed Brown to develop concepts (Heywood and Birchall, 1987).[7] Leonhardt also implied that his testing of streamlined box girders in the Teddington wind tunnel during the Severn truss investigations inspired Roberts' thinking. Yet Brown earned an Order of the British Empire for the Severn, confirmation that he was a key force behind its development.

Regardless, the Severn Bridge catapulted the careers of Brown and others associated with the span. Brown would assume a leading bridge design role in Freeman Fox & Partners, taking on the Bosporus and West Gate crossings in Turkey and Australia, respectively. Scruton would go on to perform wind tunnel tests for the bridge designs of Freeman Fox & Partners over the Bosporus and Humber. Although retiring from the NPL in 1970, Scruton would even contribute to Hong Kong's Tsing Ma Bridge before his death in 1990 at the age of 79. His name would live on through the Scruton number, a formula derived in the 1970s related to mass damping that figured prominently in the Severn Bridge investigations.

The advances embodied in the bridges over the Firth of Forth and the River Severn reverberated throughout the American engineering community as well.

American spans suddenly appeared to be outdated. The debate was sparked by two May 1964 editorials in the American weekly *Engineering News-Record (ENR)* questioning the economy of American suspension bridge design. The first editorial cited the slenderness and economy of the Severn Bridge and Freeman Fox & Partners' newly proposed and record-breaking Humber Bridge in northeast England. The second editorial lauded the Forth Road Bridge. *ENR* noted that the Severn Bridge's engineers had created a span that was 35% lighter and was then expected to cost $9 million less than the bridge over the Firth of Forth (it would end up costing about $4 million less); estimates for the proposed 1,396-m span Humber Bridge were just $35 million. The cost of the Verrazano-Narrows Bridge, it mused, reflected not so much the ability of American engineers to tackle enormous projects as it did a less rigorous approach to design innovation and cost control:

> The great discrepancies in cost between Britain's bridges and our U.S. world-beater need more explanation than the [low] Scottish wage rates . . . Actually, the costs per ton in place do not differ so much between the U.S. and Britain. The big difference is in the number of tons. The British figure the Narrows Bridge uses eight times as much steel as the longer Humber design calls for. This works out to 4.5 times as much per square foot of roadway . . . (Anonymous, 1964b).

A parallel span for the Delaware Memorial Bridge, on which construction had just begun, was noted to cost two and a half times the $40 million of the first bridge just 13 years earlier. The editorial queried, "Are the British becoming too daring in suspension bridge design? Did the Narrows Bridge cost 4.5 times what it might have? Is some value engineering in order in the U.S. on suspension bridges?" (Anonymous, 1964b).

The editorial provoked a swift response from American engineers. Ammann (1964) wrote in July, asserting the analysis "must be radically wrong." He argued that *ENR's* inclusion of bicycle and pedestrian paths in deck area calculations—amounting to 40% in the Severn and Humber bridge decks and almost 50% in the Forth Road Bridge, yet only 10% in the Verrazano-Narrows Bridge[8]—heavily favored the British bridges. He also cited the cost of the Verrazano-Narrows Bridge between anchorages as $184 million, not the $240 million quoted by *ENR*. Ammann wrote, "I am confident . . . that, all factors considered, recently built long-span suspension bridges in the U.S. are not wastefully designed." C.H. Gronquist (1964) of Steinman, Boynton, Gronquist & London concluded the per-weight steel costs of the Forth Road and Mackinac bridges to be roughly comparable. He intimated that, "a streamlined box section has been considered for suspension bridge design in the U.S." but that the Severn Bridge will "be observed with great interest." The debate also elicited a response from Roberts (1964), who confirmed that the box girder had reduced the Severn Bridge's cost by 10%.

Thirteen months after Roberts' letter came a lengthy response penned by Milton Brumer of the firm of Ammann & Whitney. It was published in *ENR* on September 23, 1965, coincidentally the day after Ammann's death. As a partner in the firm, Brumer, along with Leopold Just, had translated Ammann's concept for the

Verrazano-Narrows Bridge into a workable design. Brumer, of course, had earlier contributed to many other Ammann spans extending as far back as the Triborough Bridge.

Brumer cited the Verrazano-Narrows Bridge's greater height, which by necessity required more steel. He also wrote that the proportionately longer side spans of the British bridges translated into less steel per unit of deck area, whereas site constraints forced the use of shorter side spans in the Verrazano-Narrows Bridge. With equal span and height proportions, Brumer suggested, the cost premium per unit of deck area for the Verrazano-Narrows Bridge over the Forth Road Bridge was reduced to between 25% and 3%, depending on how one defined deck area. A lighter orthotropic deck had been rejected for the Verrazano-Narrows Bridge because of fabrication issues and the superior durability of a heavier concrete deck. Finally, a lower safety factor of 1.7 for allowable yield stresses also favored the British spans when compared to the 1.8 factor specified for American practice. Brumer stated that the use of the British safety factor could shave 10% off the weight of an American bridge.

Brumer then compared steel superstructure costs—$143 million for the Verrazano-Narrows Bridge versus the Forth Road Bridge's $25.2 million. Accounting for the Verrazano-Narrows Bridge's tripling of traffic capacity and the difficulties of construction in an urban area and congested harbor, Brumer suggested the cost differential per unit of roadway area was reduced to a factor of 1.3. That American wages were triple those in Great Britain, Brumer argued, was alone sufficient to render *ENR*'s analysis erroneous. He also believed government financing of the Forth Road Bridge made construction efficiency less imperative; the $40 million in interest alone that had accumulated during the construction of the Verrazano-Narrows Bridge exceeded the entire cost of the Forth Road Bridge. That this figure was so low, Brumer asserted, was because of a speedy 38 months of superstructure construction; it took 35% longer to erect the superstructure of the Forth Road Bridge, a bridge one-quarter the size. Cable spinning rates were six times those at the Forth Road Bridge (although without the Firth of Forth's malevolent weather), and deck concrete was placed at a rate 15 times higher than in the Forth Road Bridge's side spans.

Insofar as the box girder was concerned, Brumer stated that at the time of the design of the Verrazano-Narrows Bridge, welding procedures were neither sufficiently advanced nor skilled labor available in enough quantities to warrant its consideration. Truss fabrication and delivery, he claimed, were simpler. The greater aerodynamic sensitivity of a box girder to minor alterations (such as railing configuration) made it less flexible and more time consuming to design. Furthermore, Brumer expressed what would prove a prescient concern: that the Severn Bridge's inclined suspenders would require constant replacement. He felt these suspenders would be exposed to fluctuating stresses of up to 80% of their design limit, many times more than in vertical suspenders. Finally, Brumer argued that in spite of their external differences, the Severn and the Verrazano-Narrows bridges used the same structural concept—a torsionally stiff hollow tube. The fundamental difference between the two, he claimed, was in the Severn Bridge's orthotropic deck. This view was stretching the truth in some respects; the differences separating the two spans were not just of magnitude, but of kind.

A comparison of such contrasting spans built under different circumstances was in some ways fruitless; Brumer himself cited the differing conditions between the two countries and between the two spans themselves as rendering comparisons largely irrelevant. He (1965) concluded that, "It must be assumed that each bridge is the best solution for its particular location . . . a welded orthotropic structure, notwithstanding its lighter weight, would not have been economical in New York in 1960." This statement revealed a deeper condition extending beyond bridge engineering to national social and economic priorities, the steel industry and labor environments, and the circumstances demanding the experimentation that spawned the Severn Bridge. Great Britain, because of the war, had been forced to economize in its massive reconstruction program. This climate, in addition to a smaller steel industry in comparison to the United States, allowed Great Britain to more easily incorporate new techniques—such as those Roberts had developed under wartime incentives—to assist in its rebuilding. The massive U.S. steel industry retained a huge investment in plant, machinery, skills, and well-tried technology. A rapid transition to new methods would have been economically and logistically impossible. Moreover, higher American wages discouraged European methods that reduced material requirements but required more labor and specialized skills in, for example, the fabrication and welding of the large steel plates used in the Severn Bridge. This is not to say that the United States lagged in steel fabrication technology. Indeed, the Verrazano-Narrows Bridge introduced many developments, and welding was increasingly used in such bridges as the Vincent Thomas in California, completed that same year (*see* Chapter 10). Yet these developments did not approach the radical change embodied in the Severn Bridge.

In the end, the debate proved inconclusive. But the competition to span the Bosporus Strait in 1969 at Istanbul, Turkey would see the competing visions of Ammann & Whitney and Freeman Fox & Partners clash head-to-head in tangible form. The debate would continue even a decade later in 1979 when Ammann & Whitney engineers F-K. Chang and Edward Cohen felt compelled to defend the Verrazano-Narrows Bridge against Freeman Fox & Partners' Bosporus span; they attributed much of the Verrazano-Narrows Bridge's almost 50% heavier suspended weight (including cables) to its use of a concrete instead of an orthotropic deck.

Such vigorous defenses of American practice were in one respect an admission of a threat. Although some of Brumer's concerns over the Severn Bridge's design would eventually see the landmark bridge's luminescence fade, a century-long American domination of the suspension bridge was about to end. Moreover, other imaginative concepts would soon emerge, the originality of some even challenging the newly established orthodoxy established by the Severn Bridge.

ENDNOTES

1. Orthotropic decks had been used in Germany as early as 1934. Following tests at Lehigh University in Pennsylvania, these lightweight decks were used in several American spans built after 1936. They were further developed in World War II as ship battledecks.
2. The American Advisory Board on the Investigation of Suspension Bridges also cited this result as demonstrating vertical stiffness alone would not prevent oscillation.
3. This work provided a somber note when a murder victim was discovered in an abandoned access tunnel during construction.
4. J.A. Roebling's Sons was acquired by Colorado Fuel & Iron in the early 1960s and eventually closed in 1973; its Trenton, New Jersey, factory produced wire for almost 125 years. Bethelehem Steel would acquire the trademarks of J.A. Roebling's Sons.
5. Brown notes that Scruton and Walshe were quite distraught when informing Roberts and Brown at the NPL facilities of the model's breaking from the tunnel mounts. Brown and Roberts initially did not realize the model was destroyed, but in response to the news Roberts calmly suggested that the box girder cross-section Brown had developed for the Wye cable-stayed bridge (an ancillary part of the Severn crossing), or a variation thereof, be tested. Brown sketched out some possibilities that afternoon, and within a week, Walshe was able to report that one shape in particular–the six-sided box–offered promise. Scruton and Walshe would also become regarded for their 1957 development of strakes to reduce vortex-induced motion in smokestacks and other tall structures. These strakes have found use worldwide.
6. In 1998, the Severn Bridge earned a designation of "special architectural or historical interest," an accolade bestowed upon only 2% of British bridges (Anonymous, 1998).
7. Crosthwaite (1969) would write that an early attempt was made to test a streamlined box structure in 1945 with Freeman in a crude wind tunnel. Although its torsional stiffness and aerodynamic shape were recognized as advantages, the box was discarded as its aerodynamic damping was found to be minimal in comparison to the truss structures then under study. Crosthwaite ascribed the disappointing results to "not enough wind tunnel testing . . . for such an advanced conception."
8. In the Verrazano-Narrows Bridge, pedestrian use is limited to maintenance purposes only.

Chapter 9

A RENAISSANCE OF IDEAS

A wide range of means to ensure aerodynamic stability had now been devised both in America and in Europe, including torsionally stiff trusses to separate natural frequencies for vertical and torsional motion, longitudinal wind slots, steel grating, and truss/deck gaps, and the bolting of the main cables to the truss at midspan. The Severn Bridge's box girder was the most innovative of solutions yet devised to increase torsional stiffness by encouraging the attachment of a smooth wind flow around the suspended structure to reduce aerodynamic forcing effects. With the box girder's success, it appeared that trusses and depth-to-span ratios—including the 1:100 maxim guiding many truss-stiffened post-Tacoma spans—were just two of several means to resist vertical bending and torsional motion. This opened up a host of intriguing possibilities, many obtaining stiffness not from the suspended structure but from the suspension system itself. Interestingly, in some ways these explorations represented a return to approaches of the 19th century (Gasparini et al, 1999).

9.1 FRITZ LEONHARDT AND THE MONOCABLE SUSPENSION BRIDGE

Several radical suspension bridge concepts would provoke debate in the early 1960s. Most notable was the monocable suspension bridge advocated by outspoken German engineer Fritz Leonhardt, designer of the Rodenkirchen Bridge before World War II and a pioneer in West German cable-stayed and box girder spans after the war. Leonhardt's first—and largest—monocable design was submitted in response to a 1959 Portuguese competition to span the River Tagus at Lisbon. For the Krupp steel conglomerate, Leonhardt—with long-time architectural associate Gerd Lohmer—proposed a 1,104-m span hung from a single large cable along the line of the deck median. Large A-framed towers supported this "mono" cable, from which was hung a complex web of longitudinally and transversely inclined suspenders attached to the edges of a very thin steel deck plate.

The idea of median suspension was not entirely new. Norman Bel Geddes' Futurama exhibit for General Motors at the 1939 and 1940 New York World's Fair featured a fantastic four-level 30-lane monocable span that admittedly was more fantasy than engineering reality. In 1948, a West German contractor—in promoting ways to minimize the use of scarce steel—proposed a defacto monocable suspension bridge involving twin main cables suspended over the median of a hypothetical six-lane road and rail transit span. Torsional resistance was secured from a wide and deep box girder and from support at the tower pylons. A monocable suspension bridge was also proposed for Stockholm, Sweden in 1952. The latter two schemes featured single-pylon towers with up to two main cables; vertical suspenders were also restricted to the deck median, unlike Leonhardt's use of suspension from the deck edges.

Under Leonhardt, the concept matured. The aerodynamic research undertaken following the collapse of the Tacoma Narrows Bridge had influenced his unsuccessful 1952 proposal for an unorthodox triangular truss for France's Tancarville Bridge. Leonhardt subsequently expanded the idea of triangular stiffening to embrace an integrated deck and suspender system, and he then eschewed a stiffened suspended structure altogether by hanging a thin plate from a single cable. Stiffening was provided by the triangulated suspension system—inclined both along and across the deck—that took advantage of the shape's inherent rigidity. In transferring torsional resistance from the suspended structure to the suspension system, Leonhardt claimed steel savings of 30% in comparison to a conventional truss. He applied for a patent on this design in 1953.

The 22-m wide four-lane deck of his Tagus proposal comprised a simple orthotropic plate just 1.35-m deep with slightly thickened but rounded outer edges. This almost impossibly thin plate featured a depth-to-span ratio of 1:817—less than half that of the Severn Bridge—and was continuous between anchorages. Triangulated suspenders were superimposed over one another across the main span, and the monocable swung far above the deck at midspan to maintain traffic clearance. Over the 448-m side spans, the cable branched into two smaller cables on each side of the deck and was anchored in the traditional manner.

Not until the Tagus competition was Leonhardt able to subject a monocable to aerodynamic testing. Static model tests had earlier been undertaken in West Germany, but wind tunnel facilities there had been prohibited in the early 1950s by the occupying Allied forces. Leonhardt engaged the NPL in England, where Kit Scruton, at about the time of the Severn Bridge wind tunnel studies, tested the proposed Lisbon design. The results were not entirely satisfactory. Although the streamlined deck significantly reduced wind loads, its rounded deck edges created stability problems that were never entirely resolved. Leonhardt still claimed major advantages: the horizontal component of wind loads was just one-tenth—and the torsional component just 35%—that of a traditional truss. The Krupp bid failed for other reasons, most notably in not meeting demanding financing conditions imposed by the Portuguese. The small model of Leonhardt's monocable design now rests in the small museum dedicated to the history of the Tagus Bridge, in whose shadow it sits.

Undaunted, in 1961 Leonhardt and Lohmer proposed a graceful, low-level monocable span of 500 m over the Rhine River at Emmerich, again for Krupp. A single cable of 61 locked coil ropes supported an even more complex suspender network. Its 23-m wide, 1.2-m deep orthotropic plate deck was thinner, too, but featured a more conservative—at least in relative terms—1:420 depth-to-span ratio. Leonhardt also modified the edge shaping to smooth airflow across the deck. The cables over the 151.5-m side spans entered the anchor blocks near their rear faces, enabling all suspenders to be inclined. Prestressed steel cables would also stress the anchor block to the ground, a method pioneered by Leonhardt in his 1949 bridge over the Mosel River. Each leg of the 115-m tall triangular towers would, like those of the Severn, be built from single-celled steel columns. Static model tests were carried out at the University of Stuttgart, where Leonhardt had taught since 1958. In spite of the refinements, the Emmerich monocable was rejected as "too innovative," despite Leonhardt's claims (1982) that construction bids suggested it to cost 15% less than the trussed span eventually built. Twenty years later, he still considered the Emmerich his "most beautiful design."

The Emmerich would be Leonhardt's last major monocable proposal; a later scheme for a 110-m monocable pedestrian span at Mannheim, Germany was never built, nor was a monocable with stayed side spans proposed with Krupp later in the 1960s for Norway's Skjomen River. As late as 1986, Leonhardt was still promoting monocable suspension bridges. He would even live to see a large monocable built, albeit in modified form, when Osaka's Konohana Bridge was completed by Japanese engineers in 1990.

Pedestrian bridges would use various guises of monocable, including German engineer Jörg Schlaich's 110-m span Max-Eyth-See Bridge at Stuttgart, completed in 1989. A larger, more recent monocable pedestrian bridge resembling Leonhardt's concept (but without longitudinally inclined suspenders) was completed over the River Cher in 2001 in Tours France. This 235 m span is also notable for the launching of its deck into the span from the abutments, in place of direct lifting of deck segments.

The bridge that was built at Emmerich did in many ways diverge from traditional post-Tacoma truss design and is exceptionally graceful (*see* Fig. 9-1). W. Kingenberg and Hellmut Homberg used a shallow 4-m deep truss over a 500-m main span—which remains Germany's longest of any type. Lower transverse bracing is absent over the side spans, and the Emmerich Bridge was the first major suspension span to use the simplest form of Warren truss, one devoid of vertical members. The 17-m wide truss is neatly recessed behind a 3-m wide cantilevered walkway and floats freely through simple and elegant portal towers. The main cables sweep free of the deck at midspan, vindicating, at least here, Leonhardt's views on the aesthetic superiority of this approach. All these factors have imbued the Emmerich Bridge with perhaps the most uncluttered profile of any truss-stiffened suspension span. The orthotropic deck also acts compositely with the upper chord of the truss, marking a first for a truss-stiffened suspension bridge. The bridge opened in 1965.

In 1962, Leonhardt proposed a handsome twin-cable span for the Älvsborg Bridge across the Göta River in Göteburg, Sweden. Its longitudinally inclined suspenders overlapped one another, allowing the use of a wafer-thin steel plate deck. Leonhardt's aesthetic doctrines all found expression here—the simple but elegant portal-braced towers and pleasing main-to-side span proportions produced one of the more elegant suspension designs ever conceived. Yet this design was not selected—nor was a 1963 cable-stayed scheme—and in 1964, construction began on a conventional truss-stiffened span designed by Professor S.O. Asplund of the Chalmers Technical University of Göteburg. Asplund had earlier devised simpler adaptations of the Deflection Theory, and he also contributed to aerodynamic research following the Tacoma Narrows Bridge collapse. His 418-m span at Göteburg is notable for its concrete towers, which were the first in a suspension bridge to be slipformed (a

Fig. 9-1. Emmerich Bridge

process in which concrete is continually poured as the structure builds upon itself). A 4-m deep truss stiffens the six-lane bridge, which is flanked by unsuspended side spans of 167 and 163 m. It opened in 1966 at a cost of $8 million.

In the end, none of Leonhardt's post-1951 suspension designs were ever built, although his influence in other genres such as cable-stayed design was considerable. His wind tunnel tests at Teddington and at the National Research Council in Ottawa, Canada would later suggest satisfactory aerodynamic stability for box-shaped girders in which only the inclined faces were closed (Leonhardt, 1966). Yet Leonhardt's outspokenness and claims to have influenced the Severn Bridge through Scruton's tests of the Lisbon design—as well as Denmark's later Lillebælt Bridge—would prove controversial and remain unproven. He would also claim to have proposed prefabricated parallel wire strands in 1939, almost 3 decades before their first use. In spite of these stances—which were generally unpopular within the engineering community—Leonhardt remains notable for having pushed the limits of slender deck forms beyond the then-revolutionary box girder. Perhaps this was the time to do so, before suspended box girders would prove themselves in service and propel the form into the mainstream of suspension bridge design.

9.2 PRESTRESSED CONCRETE SUSPENSION BRIDGES AND CABLE TRUSSES

One other form that emerged in this period was the prestressed concrete suspension bridge. Such spans are rare, but Belgian engineer Daniel Vandepitte would build three diminutive examples in the 1950s. In these spans, the main suspension cables act as self-anchored, external prestressing cables in a parabolic shape. Vandepitte's first bridge of this type was completed in 1955 at Merelbeke, Belgium, with a 56-m main span and 18-m unsuspended side spans. Its suspended structure was formed from concrete I-beam girders with a depth of 1.6 m. In 1960, Vandepitte's larger Mariakerke Bridge was completed with three continuous spans of 40, 100, and 40 m. Unusually, its two 25-cm diameter cables were formed from one continuous cable of 2,070 wires wrapped around the ends of the 18.3-m wide, almost 2.4-m deep concrete suspended structure, again formed from I-beam girders. Moreover, its suspenders were made not of wire but from eyebars. The bridge cost about half that estimated for a continuous steel box girder span, and it was built deck first; the towers were then jacked up to stretch the suspension system and stress the deck together. Vandepitte's third span, again near Merelbeke, demonstrated further improvements in its 46-, 100-, and 46-m spans, with concrete box spine girders only 1.9-m deep and a more refined tower design.

Vandepitte considered concrete self-anchored suspension bridges suitable for spans of 350 to 500 m. He even tested the concept's veracity with American labor and material costs. In 1963, a consortium vying to build a crossing of the River Schelde at Antwerp, Belgium adopted his most audacious design—a high-level span of 350 m. Two sets of three lanes were stacked on two levels, the lower set buried in an almost 6-m deep concrete box whose sides were perforated for ventilation. The main cables would be formed from one continuous cable wrapped around the ends of the suspended structure, which cable stays would temporarily support until the two bridge halves were joined at midspan. Its $11.3 million cost was considered 10% cheaper than its closest rival, and the design was tendered for construction. An eleventh-hour decision to add two rail lines then pushed the entire project into a tunnel, and so ended Vandepitte's aspirations. Box girder and cable-stayed spans would emerge to dominate medium-length spans, and not until the 1990s was anything of a similar span built as a concrete suspension bridge. The three Belgian spans thus stand as rare examples of a novel concept.

In America, development of the cable truss continued after World War II. Acting as a suspension bridge under dead load and as a truss under live loads, the concept was somewhat similar to the monocable. Its multispan main cables helped to stiffen the structure, and inclined suspenders formed an integral part of deck stiffening—unlike in the Severn Bridge, where they were secondary contributors.

The firm of J.A. Roebling's Sons was the first beneficiary of its own research when in 1947 it completed a 210-m long, five-span cable truss across the Delaware River at Lumberville, Pennsylvania. Designed by Norman Sollenberger and Charles Sunderland, this configuration would characterize the firm's subsequent efforts; the prestressed concrete deck of the Lumberville span was also the first in any suspension bridge.

The reduced material requirements of the cable truss made it well-suited for remote areas, and the first large span—and the first for vehicular traffic—was built under these conditions in El Salvador across the Lempa River. Designed by Blair Birdsall and Sunderland of J.A. Roebling's Sons in collaboration with Sollenberger, the San Marcos Bridge used the standard configuration of five continuous spans—at 76, 159, 204, 159, and 76 m—extending a total length of 674 m.

A high water table and heavy seasonal rains made footing the piers for the three towers and two anchorages challenging. A severe May 1951 earthquake also tested the incomplete bridge after all but one of its towers had been built. The epicenter lay just 40-km away and literally flattened several nearby villages, but the bridge emerged unscathed. Delays in building the west anchorage then forced erection of the open-strand helical cables in the wet season. A placid 60- to 90-m wide river was now a raging torrent up to 365-m wide that rose and fell more than 1 m in 24 hours. Cable strands had to be hand-cranked by platoons of local laborers across the river, with a boatload of workers risking 20-km/hr currents to hack away mats of vegetation that often fouled the hauling rope. Strands were carried to the anchorages by long queues of locals, who also pushed cable bands to their final positions on the cables with their feet, the open strands serving as a de facto catwalk. A total of 64 trucks, each loaded to more than 14 tonnes, test loaded the span.

The San Marcos Bridge attracted considerable attention in American engineering circles upon its opening in 1952. Its other features also merited attention. Cables stretched between the anchorages along the outsides of the 91-cm deep girders also acted as the lower chord of the truss. Small wind gaps separated the roadway and sidewalks, and span lengths were chosen to avoid harmonic overtones. Sadly, the bridge was leveled by a bomb in 1981 during El Salvador's prolonged civil war; this fine example of a rare concept is now lost. J.A. Roebling's Sons did go on to design what would have been the largest cable truss ever—a five-span pipeline suspension bridge across the Mississippi River comprising three central spans of 1,080 m flanked by 540-m side spans. Model studies of the full structure were followed by the preparation of contract documents, but it was then decided to place the pipeline on a nearby road bridge.

America has not monopolized the cable truss form. Two large examples designed by Portuguese engineer Edgar Cardoso were built in the early 1970s over the Save and Zambesi rivers in the African country of Mozambique. The Save Bridge features three 180-m spans flanked by 90-m anchor spans.[1] The cable pattern is simpler and, unlike the crossed diagonals of the San Marcos Bridge, resembles that of the Severn Bridge. Cardoso built a 15-m long, 1:10 scale model to verify the design, and built the Save Bridge's slender deck not from steel but from precast concrete beams. The anchorages were also unique. Extending above the roadway in the form of large triangular portals, these raised anchorages maintained the truss action over the side spans.

The Tete Bridge—across the Zambesi River—was Cardoso's second span. This five-span bridge with three main spans of 180 m was threatened in 1978 when flooding undermined a tower caisson, tilting the tower by almost 1 m. The incident exacerbated poor construction and the abuse of the span by 200-tonne trucks—far exceeding its 50-ton vehicle capacity—during construction of the Cabora Bassa Dam.

The bridge remained open, but weight restrictions wreaked havoc on traffic between Zimbabwe and Malawi. Only 6 years later—in October 1984—was the caisson repaired and the tower propped back up to the vertical.

The former Soviet Union would produce some remarkable single-span cable trusses, including the world's longest. One early large example is the gas pipeline bridge across the Amu-Darya River in Russia. Completed in 1964, this single 390-m span supports two pipelines and a one-lane road. A more ambitious 660-m single-span cable truss was completed across the same river in 1974. Triangular A-framed towers extending to a height of 85-m support the twin main cables spaced just 2.8 m apart; the deep truss cradling the pipe is also braced with horizontal wind cables. A 720-m span completed in 1979 across the Dnieper River retains the world's span record for a cable truss *and* pipeline suspension bridge. As with the Amu-Darya span, the Dnieper crossing is suspended from triangular towers. Its longitudinally triangulated suspenders—as opposed to the San Marco Bridge's overlapping network—are typical of Soviet cable trusses. All superstructure steel was assembled in just 11 months.

The Dneiper River span, however, was not the longest cable truss ever built. This accolade goes to a temporary span of 874 m erected in 1956 during construction of the Volgograd Power Station. Four parallel cable trussed main cables supported eight 30-m wide steel cross-frames that in turn supported four parallel ropeways used to transfer more than 16 million tonnes of fill for the dam. The two towers—132-m tall and 42-m wide at their bases—rivaled those of large road spans. One issue: safety nets were required over more than one half of the main span to protect a shipping channel from falling debris. This in turn generated large asymmetrical loads during winter from snow and ice accumulating in the safety nets, but the cable truss proved an effective antidote against the threat of large deformations. The bridge was built in one year but was later dismantled and its steel re-used in a highway bridge across the same river. One other notable cable truss was built across the Volga River in the 1950s. Its main span of 470 m and its193-m side spans are impressive, but its four lanes also make this span the world's widest extant cable truss.

In 1966, Vandepitte suggested that, "trusses formed by the suspension cables, the deck and the inclined suspenders [would eliminate] the necessity for the deck to possess any rigidity of its own . . . One is thus lead to visualize a suspension bridge with a superstructure reduced to the bare essentials." Nowhere is this concept more vividly exploited than in pipeline suspension bridges. These unusual spans were built in profusion after World War II—particularly in the United States—but brought their own foibles of aerodynamic instability.

9.3 PIPELINE SUSPENSION BRIDGES AND OTHER IMAGINATIVE CONCEPTS

With little or no stiffness against static or dynamic wind pressure, pipelines—as suspended structures—have peculiarly sensitive aerodynamics, and cable trusses and other imaginative forms have been used to overcome their weaknesses. Single-cable suspension systems generally feature vertical and/or longitudinally inclined suspenders and complex horizontal tiedown systems, whereas two-cable systems with

Y- or V-shaped towers often use transversely triangulated suspender systems.

Pipeline suspension bridges are a relatively recent phenomenon. The first large bridge of this type was a 1926 effort extending 688 m across the Red River in Texas. The bridge, comprising several continuous central spans, was designed by the duo of Matthews & Kenan. By 1931, six other spans in the south central states had been built to their design, and they would come to dominate the form for more than 3 decades. The Little Blue River Bridge in Kansas, with a 152-m main span, was typical. Its main cables hung canted toward the pipe with inward sloping suspenders. Wide steel towers supported large brackets at their bases to which were attached horizontal wind cables. Connected at frequent intervals to the pipe over the main span, these wind cables helped to stabilize the bridge.

Matthews & Kenan soon eclipsed their own embryonic efforts. In 1931, their bridge of eight 97-m spans was completed across the South Canadian River in Texas. In 1932, their design for a single 390-m span was completed over the Missouri River in Nebraska. Then in 1940, they beat this record with a 402-m span across the same river near Sioux City, Iowa. This bridge was typical in that its unsuspended side spans terminated in "rest towers," beyond which the main cables extended to anchorages. Anchorage-to-anchorage distances could be impressive, and at Sioux City extended 1,079 m. This bridge—remarkably—was completed in just 135 working days.

Following World War II, pipeline bridge construction surged. Some spans took advantage of serendipity; a coal bin anchored one otherwise conventional coal conveyor span of 221 m completed in 1949 across the Monongahela River in Pennsylvania. However, only after the Tacoma Narrows Bridge collapse did a pipeline span reveal evidence of aerodynamic instability. The first victim was a 311-m span over the Colorado River near Blythe, California, one of five spans of up to 366 m built in 1947 as part of a Texas-California natural gas pipeline. Its 76-cm diameter pipe soon undulated in gentle winds with the "flexibility of a rubber hose" (Cain, 1975). Waves of up to 91-cm double amplitude were observed. Wind tunnel tests identified the culprit as eddies shed in winds of less than 10 km/hr. Triangular wind vanes were attached to the pipe to quell the motion, and were subsequently installed on another similarly sized span over the Gila River.

Similar low-speed oscillations of up to 91 cm occurred in 1951 on a new 213-m span natural gas span across the Coosa River in Alabama. Engineers had slung a second pipe beneath the original to add weight as a means to stabilize the bridge, but this was done to no avail. Rock-filled oil drums suspended from the pipes and anchored in the river also proved ineffective. David Steinman was then called in, and he engaged post-Tacoma colleague F.J. Maher of the Virginia Polytechnic Institute to subject sectional models to wind tunnel testing. The tests revealed an uncomfortably low critical velocity of between 8 and 24 km/hr. Wind vanes and fairings were considered, but Steinman superimposed a more effective triangular form of cable truss over the original vertical suspension system at up to a third of the cost. This was in fact the first practical application of his aerodynamic theories and rigidity coefficients, and his remedies here reduced vibrations to less than 1 cm.

Larger spans soon emerged with the completion of a 610-m span Matthews & Kenan bridge in 1951 across Louisiana's Atchafalaya River. The distance between rest towers was 1,112 m and between anchorages was a prodigious 1,426 m; even the

91-m wide brackets supporting the horizontal wind cables exceeded the 82-m tower height. This record was then eclipsed in 1955 by a 655-m span natural gas suspension bridge over the Mississippi River between Missouri and Illinois. This twin-pipe span is of classic Matthews & Kenan design. Its twin cables converge toward midspan, and wind cables are tied to the pipes and the main cables. The 100-m wind brackets easily exceeded the 75-m tower height. Another Matthews & Kenan twin-pipe span of 457 m was built across the Missouri River near Plattsmouth, Nebraska 2 years later; its suspended length exceeds 730 m, and its anchorages stand 1,006 m apart. This span secured the AISC's long-span award for 1957, outcompeting a 329-m span over the Colorado River between Arizona and California.

J.A. Roebling's Sons was very active in this field in its twilight years up to the early 1960s and promoted its activities heavily. The firm built a temporary 283-m span suspension conveyor during dam construction in Washington, a 110-m pipeline span across the Snake River in Idaho, and a 274-m eight-span bridge in Maine with tie cables. It also used temporary cable stays to support the cantilevering of one span in a manner now common to cable-stayed bridges. Equally imaginative erection methods were used elsewhere by other contractors. In 1962, a 635-m span near Portsmouth, Ohio was built over the barge-clogged Ohio River. River traffic could be interrupted for just 48 hours, a problem for which preassembly provided the solution. The 82-m tall inverted Y-shaped towers were assembled on the ground on hinged bases. All 701 m of pipe were assembled, tested, painted, and attached to the main cables and suspenders on shore. All were then lifted as a single unit by 17 mobile cranes to the shoreline and then swiveled 90° over a line of six barges strewn across the river. The towers, with single main cable attached, were then raised sequentially to create a self-supporting span with the pipe clearing the Ohio River by 32 m. The operation took less than the allotted 48 hours; a second pipe was later hung beneath the first. This span is one of a handful stabilized by a system of horizontal stays splaying out from anchorages situated some distance from either side of the towers.

Pipeline suspension bridges also proved popular in the Canadian provinces of Alberta and British Columbia (B.C.). One B.C. natural gas pipeline includes five such bridges, comprising two spans of 183 m, spans of 393 and 448 m across the Fraser River, and a 510-m span over the Peace River at Taylor Flats. A Matthews & Kenan natural gas span of 320 m crosses the South Saskatchewan River near Empress, Alberta. The Bow River near Cochrane, Alberta features two natural gas spans of 183 and 201 m, and an oil pipeline span across the Fraser River in B.C. is 329-m long. Engineering firm T. Lamb, McManus & Associates (designers of the Dunvegan Bridge) crafted the second Bow River span and also contributed to the Taylor Flats span. One interesting aspect of these northern pipeline bridges involved the curious aerodynamic benefits of ice build-up. This phenomenon was most pronounced on the 448-m span crossing the Fraser River, where it was found that a doubling of the span's weight from ice accumulation could improve its wind resistance by 160%. Elsewhere in B.C. near the town of Princeton stands a novel 290-m span over the 183-m deep gorge of the Similkameen River. This 1972 span, originally planned as an arch bridge, provides access to a copper ore body and supports three pipes stacked vertically (plus a walkway) rising at a 10% grade. Its main cables are discontinuous over the towers—anchorage being provided via separate backstays—and a nominal triangular cable truss is superimposed over the vertical suspender system. Asymmetric horizontal wind cables also span this spectacular canyon. In 1980, a nearby span of 404 m was built to the designs of the firm of Buckland & Taylor, in association with Robert McClellan. This bridge is also characterized by a sharp rise and is stiffened by wind cables. It is also the world's longest conveyor span.

Further north, construction of the Trans-Alaska pipeline in the 1970s produced a 366-m span over the Tanana River near Fairbanks. Its 122-cm diameter pipe is likely the largest carried by a dedicated suspension span. Crews labored through temperatures as low as -55° C to build the bridge. A similar 235-m span was also built north of the Arctic Circle. One other large American effort is a 585-m span carrying a natural gas pipeline over the Flaming Gorge reservoir in Wyoming. Its 86-cm diameter pipe rises at a dramatic angle, and is stabilized by wind cables extending from the base of its 67-m tall towers. Other American spans of note include the 307-m span Avalon Extension Aerial Crossing, the 259-m span Patterson Loop Aerial Crossing, and the 245-m span Calumet Loop Aerial Crossing; horizontal stays rather than parabolic wind cables restrain all three spans. A crossing of the Skagit River in Washington, as with the Flaming Gorge span, rises sharply over its 213-m span.

Europe has also produced major pipeline bridges. Italy's early 1950s Pont de Piacenze over the River Po featured two continuous 320-m spans flanked by 140-m end spans; here, large brackets extending from tall lattice steel towers restrained horizontal wind cables. An unusual design by the firm of Waagner-Biro was applied in spans of 260 and 320 m completed in 1957 and 1961, respectively. The two bridges stand about 5-km apart, south of Vienna across the Danube River. In both, the main cables are draped over wide latticed Y-shaped steel towers and converge at center span. This arrangement produces a sloping suspender system that enhances rigidity. The second span carries 13 petroleum pipelines and was built in just 11 weeks. Multiple spans remain rare, but in 1971, Italy completed a large methane pipeline span—again across the River Po—with three central spans of 255-m each. Tie cables in the manner of the early French suspension bridges connect its four towers. Italy also features other pipeline suspension bridges of up to 320 m in span.

Suspension bridges of unusual design are not confined to vehicular or pipeline spans. Buildings have also used the concept. Most notable is a 1964 self-anchored suspension bridge by the famous Italian architect-engineer Pier Luigi Nervi that suspends an enormous 250- by 30.5-m paper mill roof at Mantua, Italy. Two 50-m tall backward leaning inverted Y-shaped concrete towers support four cables over spans of 50, 150, and 50 m. Each "cable" is actually made from steel bars from which the 2.4-m deep roof trusses are suspended. With no load-bearing walls, the design offered flexibility for future expansion. Among other examples, U.S. engineer T.Y. Lin has used a suspension bridge to hang the roof of an elevated subway station in Taipei, Taiwan. The station—opened in 1993—is designed to mimic a traditional Chinese dragon boat. Its roof is suspended between two back-leaning concrete pylons over a span of 178 m; the cables are comprised of six strands of 222 wires each. Another more recent example is the speed skating oval at Salt Lake City, Utah for the 2002 Olympics. The roof of this structure is hung over a 100-m distance by a series of suspension cables using inclined suspenders; it, too, sports backward-raked towers. The U.S. office of the British engineering firm Ove Arup crafted the design.

One of the most unusual concepts ever conceived came from Lev Zetlin & Associates in 1966 and 1967 for a 14-lane crossing of Baltimore, Maryland's Inner Harbor. For this prominent location, the firm produced a 198-m span "tension cable roof" comprising backward-leaning Y-shaped pylons, multiple planes of main cables and suspenders, and three separate orthotropic decks spread over a width of 76 m. The unusual cable system, which relied on tension—not conventional anchorages—for support, permitted suspended structures just 1.2-m deep, much less than required for a truss-stiffened suspension span. Lev Zetlin & Associates also proposed prefabricated parallel wire cables, then under development by Bethlehem Steel (*see* Chapter 10). Only a model of the span was ever built—a less intrusive tunnel was opened in 1985.

A more recent 1990 scheme for a "sky shuttle" across the River Thames in London, England was also unorthodox. A suspension bridge of about 400-m span would carry trains suspended beneath tracks that served as the span's suspended structure. A consortium led by Acer (successor to Freeman Fox & Partners) made this unsolicited (and ultimately unbuilt) proposal for access to London's Docklands.

The pedestrian suspension bridge most reflects German engineer Schlaich's view (1990) that, "it is no exaggeration to say that modern structural design can pride itself to having cleared the way towards absolute freedom in the design of suspension bridges, including any such geometries which only a few years ago nobody would have dared to think of." His 1987 pedestrian bridge at Kehlheim over the Rhine-Main-Danube Canal proves the point—its main and side spans describe a horizontal arc of almost 180° over the channel. A slender concrete box girder suspended from only its inner edge is hung from mast-like guyed towers that lean dramatically into the curve of the span; its single main cable curves in both vertical and horizontal planes. The span's radical configuration has given it the unusual distinction of being simultaneously gravity-anchored and self-anchored.

Other pedestrian suspension spans have also pushed the envelope, including a Schlaich-designed Stuttgart span of three branched decks and another Schlaich span that even morphs into a lift bridge. Czech engineer Jirí Stráský has designed several innovative spans, including a record 252-m span pedestrian bridge completed in 1993 across Vranov Lake. The concrete suspended structure of this dramatic span is just 40-cm deep. Its deck widens from midspan to A-framed, backwards-leaning towers to enhance transverse stiffness. The bridge is, in fact, a hybrid form of stress ribbon (*see* Chapter 11) with external prestressing. Other unusual spans include a 1996 Austrian span of 138 m that is suspended from wide triangular towers. One of the largest pedestrian spans anywhere is a 375-m span Japanese suspension bridge completed in 1993. This truss-stiffened span features unusual serrated-edge towers whose effect is accentuated by the breathtaking mountain setting. London's Millennium Bridge, with a total suspended length of 370 m and a main span of 144 m, is another example of unconventional and dramatic design—too dramatic perhaps, for the bridge's June 2000 opening was immediately embroiled in controversy when vibrations forced the span's immediate closure. Finally, countries such as Nepal are home to spectacular, if more conventional, pedestrian suspension bridges in extremely rugged terrain.[2]

9.4 BRIDGE ACROSS THE TAGUS—INNOVATION IN CONVENTION

With the 1964 completion of the Forth Road Bridge, Europe's longest span had in a single leap increased by almost two-thirds. But as the towers of the Forth Bridge rose in 1962, work began on a span just 7-m longer at Lisbon, Portugal. It would be a largely American enterprise, but the period leading to its construction spawned many imaginative designs from European engineers that again challenged American orthodoxy.

Demand for access to the undeveloped areas south of the Portuguese capital of Lisbon across the River Tagus was rapidly increasing—by the early 1960s, 3-hour ferry waits were common. The closest road bridge was 25-km away, a rail crossing twice that distance. It is no surprise then that the legacy of proposals to bridge the Tagus at Lisbon was a long one. The first serious proposal had emerged as early as 1876 from Portuguese engineer Miguel Pais. By 1879, his design featured 76 truss spans—all but two of 60-m span—upstream of the present bridge. Road traffic would use the upper level of the truss, whereas rail traffic was tucked below on its lower level. In 1888, an American engineer proposed a rail bridge, and a year later French engineers proposed a 2,500-m road and rail bridge with seven alternating steel deck arch spans of 300 and 160 m. An 1890 German proposal then called for multiple 200-m span steel cantilevers situated near the Pais alignment. Benefiting from newly obtained geological information, this design was the most realistic yet and was under discussion even 20 years later.

A bridge resurfaced in 1913, and in 1919, a 4,500-m tunnel was proposed. Two years later, Spanish engineer Alfonso Peña Boeuf proposed a two-level road/rail bridge comprising 14 arch spans of 150-m each. The plan was seriously considered but after a lengthy debate was rejected because of its obstruction to shipping. During this debate, however, the first suspension bridge proposal emerged. This 1926 plan for a record 1,240-m span hung from ponderous gothic stone towers overcame navigational concerns; even its 750-m side spans would have established a span record. Yet for all the activity, only in 1930 did the national government sanction a Tagus crossing. Borings were taken along the 1876 Pais alignment, and 3 years later guidelines were developed as a basis for soliciting proposals for a bridge between Lisbon and Montijo to the southeast. However, none of the 1934 proposals—including a scheme put forth by contractor U.S. Steel—met the financial requirements. U.S. Steel re-emerged in 1938 with a multispan truss proposal, and serious discussions ensued to settle its financing. Agreement was apparently close at hand when negotiations fell apart; war clouds also loomed, and the proposal died. The same fate awaited a 1938 suspension bridge proposal and another scheme proposed in 1942 during World War II.

The first post-war proposal came from Peña Boeuf, who in 1951—3 decades after his first effort—proposed an unconventional suspension bridge of three 600-m spans flanked by end spans of 300 m. Perhaps anticipating the reaction of shipping interests, he then suggested an ambitious single span of 2,000 m. This, and an allied bridge over the Gibraltar Straits, never left his drawing boards. That same year, Portugal's then-longest bridge was completed across the Tagus 20-km upstream of Lisbon, but the need for a link in the city itself was becoming urgent. In June 1953, the government appointed a commission to investigate the matter. It reported 4 years later in April 1957 that various combinations of road and rail bridges and tunnels were economically and technically feasible. The government endorsed a road and rail crossing at a point where the river is 2 km wide. Further studies pointed to a bridge, and in May 1958, the Gabinet da Ponte sobre a Tejo (Tagus River Bridge Office) was established to administer the project. Six months later, the crossing was included in the national development plan for 1959 through 1964. The Bridge Office developed specifications on which alternative designs would be developed and, in April 1959, launched a design-and-build competition. It cast the net widely—specifications were issued in Portuguese, French, English, and German. Bidders were given almost unlimited scope in devising solutions within the parameters of a main span of at least 1,000 m, side spans no less than 350 m, and a 70-m vertical clearance at midspan, the highest of any large suspension span yet. Contestants were required to develop road-only and road/rail transit designs and would also have to finance the estimated $28 million project.

Under Cardoso (whose subsequent design of the Mozambique cable trusses would be greatly influenced by his work on the Tagus), the Bridge Office initiated parallel studies to assist in its evaluation of the entries. This work was very imaginative; of 10 concepts, only one used a classic suspension bridge. Four concepts combined inclined and vertical suspenders in various arrangements in spans of up to 1,200 m. Two variants avoided deep rock off the south shore with a central tower up to 305-m tall splitting two 1,000-m spans. Although the trusses and diagonally braced towers of these twin-span configurations appeared dated, their anchorages were almost identical to those of the Tancarville Bridge. Cardoso had been exposed to Leonhardt's work, and three monocable schemes with slender steel deck plates supported by A-framed towers were also developed, one with a fantastic shore-to-shore span of 2,000 m, the others with 1,000- and 800-m spans. The suspension system, however, acted more like a cable truss, with prestressed cables along the deck edges to help stiffen the structure. Two cable-stayed schemes were also devised, one with a 1,100-m span, and a second unorthodox scheme with a central Y-shaped tower splaying out over two 1,000-m spans. Highly theoretical by intent, these concepts were ahead of their time; even today, some appear futuristic.

Four bids were submitted by the competition's late February 1960 deadline. A Portuguese-led team proposed an ambitious truss-stiffened, cable-stayed span of 1,050 m that would have trebled West Germany's then-record 301-m Severin Bridge in Cologne. This remarkable road-only span was priced at $45.5 million. Leonhardt's 1,104-m monocable span was one of three schemes submitted by a West German/British/American consortium spearheaded by steelmaker Krupp. Most conventional of the three was a truss-stiffened road/rail span of 1,125 m with 450-m side spans. A third truss-stiffened offering was more unusual than even Leonhardt's monocable. Trellised Y-shaped towers separated the main cables over a distance several times the deck width, producing transversely inclined vertical suspenders whose inclination diminished as the cables narrowed to deck width at midspan; tower stays similar to those in Steinman's Deer Isle and Thousand Island bridges were also envisaged. A largely American offering led by U.S. Steel featured a traditional truss-stiffened span of 1,011 m flanked by 469-m side spans. This design placed two rail transit lines on the truss bottom chord. Backstay spans—like the Krupp bid—were also used, those on the north shore extending half the length of the side span. Steinman's firm, now Steinman, Boynton, Gronquist & London, developed the preliminary design. The fourth and largely German bid featured a 1,032-m suspension span with 516-m side spans devised by Kurt Klöppel. Its lattice—instead of the usual Warren—truss was stiffened with two cable stays radiating over main and side spans from diagonally braced towers.

Ironically, much of the Bridge Office's analysis had little to do with engineering. The Portuguese consortium was disqualified by its lack of a road/rail variant of its cable-stayed scheme. The Krupp-led and German bids failed to meet financing terms. U.S. Steel proposed the only eligible financial package, even though its design was one of the more expensive submitted. This, undoubtedly aided by the American steelmaker's and Steinman's experience with large suspension bridges, saw the bid accepted and, in May 1960, the government approved a provisional contract to design and build the bridge worth almost $62 million. Cardoso would claim almost 2 decades later that his own concepts were subsumed by a political desire to foster good relations with the United States. Leonhardt would complain, too. Regardless, the bridge as built was not as conventional as its appearance might suggest (*see* Fig. 9-2). One reason for this was the built-in flexibility of the U.S. Steel design. Cost considerations had led to a postponement of the lower-deck dual rail transit line, but the ability of the design to allow for its later addition with only minor alterations to the bridge had also helped U.S. Steel snare the contract. The Tagus Bridge would thus be built initially with four lanes; U.S. Steel was given 300 days to devise a final proposal.

Over the next few months, Steinman, Boynton, Gronquist & London, with Bridge Office staff, developed the final design. The bridge's main and side spans were fixed at 1,013 and 483 m, respectively. It would rank as the world's fifth longest span, the longest outside of North America, and the longest span designed to accommodate rail traffic. This latter feature, and its staged construction, would make the Tagus Bridge unique among long-span suspension bridges. Sadly, Steinman would not have a hand in crafting the details. He died in August 1960 after having participated in preliminary designs, and detailed design unfolded under associate R.M. Boynton. Yet Steinman's influence remained. Strength and rigidity requirements imposed by rail loads were reflected in the 21-m wide and 10.7-m deep truss. Its depth-to-span ratio of 1:95 is one of the deepest of any large post-Tacoma span. The inner two traffic lanes and external service walks were built of open steel grating, and the trusses extended beyond the roadway edges. The main cables were also clamped to the truss at midspan to restrain longitudinal and torsional motion. In wind tunnel tests on a 1:50 scale sectional model at the Langley Research Station wind tunnel in Virginia, the truss was found stable in winds of up to 233 km/hr; the model builder was none other than F.B. Farquharson. Flexibility also permeated the road deck design. Future removal of the median barrier could add a fifth lane, and extension of the roadway edges would allow for the eventual addition of a sixth lane.

The bridge in its initial state would appear thoroughly conventional. Yet the addition of the lower deck rail lines would create an unprecedented transformation. The superimposition of cable stays and additional suspension cables would see the

Fig. 9-2. The Tagus (25th of April) Bridge during the late 1990s truss and cable strengthening to support rail traffic. Note the (planned for) sag of the truss.
PHOTOGRAPH COURTESY OF R.A. DORTON.

Tagus Bridge supported by three independent and dissimilar structural systems. The use of cable stays was very much influenced by Cardoso's Bridge Office studies. Four sets of large stays would radiate from the towers to the bottom chords of the truss over the first quarter of the main and first half of the side spans. A third set of stays acting as de facto auxiliary suspension cables would be attached to the bottom chord of the truss at five locations over the central portion of the main span and at three locations over the shoreside halves of the side spans. A longitudinal strut—doubling as support for the rail line—would ensure that horizontal forces from the new cable systems would not be transmitted to the original truss, which remained suspended by the original main cables. Only the towers had to be initially built to their ultimate strength; the stiffening truss system would require more internal members and brackets. Such a complex and finely tuned convergence of structural systems presented an intricate analytical problem that a few years earlier would have been rendered virtually unsolvable without the use of computers.

Steinman's preliminary design had placed expansion joints at the towers, but rail loads made the truss bend too much. The joints were then moved to the cable bents and finally over the anchorages, providing a continuous truss over a then-record distance of 2,277 m to help resist deformations from the passage of trains. Each expansion joint must provide for up to 1.4 m of movement. The continuity of the truss, which placed the tower legs beyond the truss edges, also made expansion of the upper road deck possible. The 190.5-m tall diagonally braced towers would for a short time be Europe's tallest free-standing structures, rivaling Lisbon's Christ the King statue overlooking the estuary from the south shore hills just east of the bridge. Their great height resulted from the navigation requirements, the deep truss, and the need to reduce horizontal tension emanating from the tower stays and secondary cables. The legs were slightly canted, and because of the span's structural continuity, no brackets support the truss as it passes through the towers. The final design included one other major change; at the insistence of the Bridge Office, the 945-m long high-level northern approach bridge soaring up to 64 m over the city would be built of prestressed concrete instead of the steel trusses proposed by U.S. Steel. Its spans would use Portuguese concrete and familiarize workers with this structural form. The lower-level rail spans, however, would be built at a later date using steel box girders. In May 1961, a $60 million contract was signed that specified the design and construction schedule. The final design was submitted in November 1961 and approved in May 1962, 2 years after the provisional award. Construction began that November.

Notwithstanding its innovative structural system, the Tagus Bridge is probably best known for its tower foundations. The span's south tower pier 500-m offshore had to penetrate 27 m of water and 52 m of mud and sand before resting on basalt at a depth of 79 m—deeper than any bridge pier sunk to that time. The siting of the north tower pier on shallower but steeply sloping rock also greatly complicated its design and construction. Added challenges included strong tidal currents, tidal variations of up to 4.5 m, and earthquake resistance. A 1755 earthquake had virtually leveled the city, and minor tremors had struck as recently as 1955. Earthquake-induced tsunamis were also a threat; most of the destruction and many of the 60,000 deaths in the 1755 event resulted from a massive water displacement that almost emptied the estuary and then returned to inundate the city. Steinman, Boynton, Gronquist & London retained Tudor Engineering, a world leader in earthquake engineering, to design the two tower foundations. Founder Ralph A. Tudor had designed the Transbay's piers to similar depths and earthquake exposure. Like Steinman, Tudor died before the design was finished, but his influence was profound. New methods of seismic analysis used computers to simulate structural response from an actual earthquake—a 1940 event in El Centro, California—in place of dynamic models.

For the south tower pier, Tudor refined the air dome concept so brilliantly used 30 years before in footing the central anchorage of the Transbay Bridge. The 24- by 41-m steel caisson cradled 28 large open-bottomed dredging wells. Adjustments in air pressure within various combinations of wells maintained buoyancy and an even keel during descent, even righting the caisson in the improbable event of its overturning. The first 9 m of caisson was floated to the pier site and then slowly sunk by injecting concrete into gaps between the wells as layers of steel were added on top. Twice, sudden listing of the caisson required frantic adjustments of well pressures to avoid exceeding a critical 3-m variance from the horizontal. These and other obstacles were overcome and in August 1964, after 18 months of work, the pier was ready for tower steel. Throughout the entire operation, not one person descended to the riverbed.

The north tower pier had been completed 9 months earlier. Here, Tudor contoured the caisson base to undulating rock to deal with a 7-m elevation difference between its opposing corners. The caisson first descended with a deliberate list through 21 m of water to permit its angled bottom to contact the seabed. It was then canted upright during the excavation of 21 air-domed wells through 7.5 m of mud. Six cylinders excavated at each pier end through another 6 m of rock were then concreted. Undulating bedrock also affected the footing of the north cable bent, which is even deeper than the north tower. Steel H-piles had been planned but the Portuguese, fearing corrosion from static electricity shed by the span's future electric rail lines, rejected them. Instead, 58 concrete piles were cast into casings driven up to 57-m deep, a difficult operation that saw pier work extend for 15 months. The north anchorage foundation was even more vexing. It sits astride an ancient but heavily fractured fault. Initially, a concrete-filled cofferdam was envisaged. But soil studies could not be undertaken until a factory on the site could be acquired and demolished. Subsequent borings revealed extensive fissuring and severely undulating bedrock. In the end, 12 large concrete caissons were driven to depths of up to 34 m around the foundation perimeter to form a cofferdam. Piles were then driven in the rear half of the cofferdam to help temporarily support a huge 37- by 52-m concrete base slab. The horizontal force of the main cables would eventually render these piles superfluous, although they would remain. Should the fault ever become active, the dozen perimeter caissons will adjust independently to ground movement while leaving the 162,000-tonne anchorage intact. Of the major piers, only the south shore anchorage was built in the dry, high up on the slopes. Virtually all of its 100,000-tonne mass—66-m long and up to 38-m wide—is buried beneath the roadway.

The north tower began to rise in December 1963, followed in August 1964 by the south tower; the longer task of footing the south tower pier allowed the same creeper crane to assemble the towers sequentially. The diagonally braced towers had been substantially modified since preliminary designs, when a rigid-frame tower not unlike that used at the Verrazano-Narrows Bridge had been considered—until the additional expense signaled its demise. Rapid developments in the shop welding of large steel plates alone had seen the tower legs reduced from a 20-celled base cross-section to one of only five cells. The logistics of getting tower steel to the bridge site were complex. Cells weighing up to 60 tonnes were fabricated in Ambridge, Pennsylvania; barged down the Ohio and Mississippi rivers to New Orleans, Louisiana; and then shipped by freighter across the Atlantic Ocean to Lisbon. By January 1965, both 6,030-tonne towers stood complete, each held together by 150,000 high-strength bolts.

Spinning of the 58-cm diameter, 11,248-wire main cables began that same month—ahead of schedule—and was completed after just 76 working days in May 1965. Truss erection began that July. Welded connections are found throughout the truss, as they are in the cable bents, with a total of a half-million bolts used for field connections. A new high-strength structural alloy steel was also used in more than 50% of the truss. Lifting involved very long units comprising 46-m long, four-panel sections weighing up to 330 tonnes each. This strategy, a first for a large suspension bridge, almost doubled the lift lengths used at Mackinac and permitted all 1,980 m of suspended span to be placed in just 41 lifts ending in January 1966. A 600-m tunnel was also placed under the southern approaches to avoid disruption when the rail lines were eventually built.

The $75 million Tagus Bridge (first officially named the Salazar Bridge and since 1975 known as the 25th of April Bridge) was opened in August 1966, 6 months ahead of schedule. Given the immense foundation challenges, this was a major achievement. Up to 3,000 Portuguese workers were on the bridge at the peak of activity, with no more than 70 American nationals onsite. Most workers were recruited from rural areas, and high turnover occurred during planting and harvesting periods. Although four workers were killed during construction, labor strife was absent as Portugal's military regime—for better or worse—had rendered strikes illegal. The Bridge Office's research program was also continued during and after construction; just before opening, 350 m of main span was crammed with 1,600 tonnes of heavy trucks to cause a discernible deflection of more than 3 m. The span was heavily instrumented to ascertain its behavior under wind, traffic, and earthquakes and to optimize—or even change—the rail design, when its time arrived.

The Tagus Bridge was the world's longest continuous truss and the longest single span designed for rail use for more than 30 years; it also remained the world's second longest span between anchorages (behind the Mackinac Bridge) until 1998. Its towers are remarkably slender for their size and the loads they carry, although the latticed diagonal bracing appears somewhat dated in comparison with the simple closed boxes of the towers of the Forth Road Bridge. The span's somewhat stiff appearance is emphasized by a minimal truss camber—required to accommodate rail traffic. Its red-orange patina—designed to complement the blues of Lisbon's sun-drenched skies and water and the pinkish hues of its buildings—emerged from more than a dozen color schemes examined by writers, artists, and art critics. The sculpting of the north anchorage was also carefully considered. Dramatic views of the bridge are revealed from the span's circuitous north approach; however, the span and its approaches uprooted 900 families—a symptom of what superhighways had already wrought in

America.

Perhaps the most memorable event associated with the Tagus Bridge opening was the November 1966 International Symposium on Suspension Bridges held in Lisbon. This conference remains the largest exclusively devoted to the subject. It attracted 177 engineers, including such influential figures as Gilbert Roberts, William Brown, Alan Davenport, Christian Ostenfeld, Manabu Ito, Arne Selberg, Asplund, Leonhardt, and Niels Gimsing. It was here that Leonhardt produced a stunned silence when he publicly lambasted Portuguese officials for rejecting his Tagus monocable proposal.

The story of the Tagus Bridge would not end upon its opening. A reversible fifth lane was placed over the open steel median in the mid-1970s. In 1983, Steinman, Boynton, Gronquist & Birdsall (the latter of San Marcos fame) was retained to design modifications to accommodate a sixth lane, but this project did not move ahead. Only in 1990 would plans for the sixth lane and a lower rail deck be resurrected, but a requirement that the bridge now carry heavy freight and double-deck passenger cars would increase rail loads by two and a half times those contemplated in the original design. The Steinman firm was again retained in 1992 to design a retrofit. The towers and foundations could accommodate the additional 16,000 tonnes of dead and live loads, but the 1960s hybrid cable-stayed/suspension system could not; its sliding bearings—allowing the two systems to move independently—were also considered a maintenance nightmare. Four concepts were then considered: the 1960s plan, an altered cable-stayed system, self-anchored suspension cables, and two traditionally anchored suspension cables. All options cost roughly the same, but aesthetics, constructability, and maintenance favored the last option. A total of 200 new, stronger lower truss laterals would support an orthotropic rail deck, and the sixth lane would extend the upper deck over the former outer air gaps. The increase in truss weight would be supported by the new cables; rail expansion joints at the extremities of the continuous truss would move up to 1.5 m. Tests on taut strip (*see* Chapter 10) and sectional models validated the aerodynamic stability of the reconfigured truss in winds of up to 259 km/hr, well beyond the 216-km/hr requirement.

The job started off with a controversial September 1995 award of a $238 million contract to a German-led consortium; its original bid was the highest of five submitted. In spite of threats of legal action from the low bidder, the British contractor Trafalgar House, the contract stood unchallenged, and work began in January 1996. American Bridge[3] would spin the cables and even resurrect some hoisting equipment used in building the original span. The bridge would remain open at all times to the 140,000 vehicles crossing it daily, but maintaining precise tolerances in the face of major steel additions and constant motion under traffic and wind would pose a major challenge. The two new cables would be draped over new saddles housed in extensions of almost 4 m to the tower legs and more than 5 m for the cable bents. The divergence of the new cables at the cable bents to new anchorages beside the originals necessitated an above-deck horizontal tie between the bent extensions to avoid their being pulled apart.

The two 4,104-wire, 35.5-cm diameter cables—about half the diameter of the originals—were spun almost 4 m above the existing cables between April and July 1997. American Bridge had planned to use traditional spinning methods, but the Steinman firm's involvement with "controlled-tension" spinning on Denmark's East

Bridge (*see* Chapter 13) convinced it otherwise. Nonetheless, the firm did not want to make a large investment in sophisticated equipment for such a modest spinning effort and instead created a hybrid approach that still eliminated one entire work shift. The simultaneous—rather than sequential—activities of cable spinning and adding 9,100 tonnes of lower deck steel met a demanding schedule but created complications, as all steel would initially be suspended by the original cables, deflecting the truss temporarily at midspan by up to 4 m until the new cables were in place. Media reports in August 1997 even suggested the now 3-m sag—well within planned tolerances—threatened collapse, causing a furor that died down only somewhat when officials explained the plan (*see* Fig. 9-2). A second engineering challenge involved suspender installation. Weight distribution over the span had to be maintained as the new suspenders—braced to pass around either side of the original cable—were attached to the bottom of the truss at points between the original suspenders. The railroad—including 900 m of steel box girder spans below the north approach viaduct—was opened in early 1999.

Traffic capacity across the River Tagus has now been greatly enhanced, not only because of the expanded suspension bridge, but also with the opening of a six-lane, 420-m span cable-stayed bridge 15-km upstream the previous year.

ENDNOTES

1. An early proposal presented at the 1966 Lisbon International Symposium on Suspension Bridges featured three 210-m spans flanked by 93-m anchor spans.
2. One of the largest and more entertaining of pedestrian suspension bridges is the Bandar Sunway Bridge, situated in a theme park outside the Malaysian capital of Kuala Lumpur. The bridge comprises an impressive 352 m main span with a single 82 m anchor span. Its timber deck follows the line of the main cables, and the absence of any lateral restraining cables enhances the thrill of crossing this deliberately unstable bridge. The span has conspired to both delight and alarm visitors since 1993.
3. American Bridge was spun off from parent U.S. Steel as an independent entity in the 1980s.

Chapter 10

THE TWILIGHT OF AMERICAN DOMINANCE

With the Verrazano-Narrows Bridge, the challenges of enormous size, span, and weight had been surmounted on a scale not likely to be repeated on U.S. soil for some time. Yet British engineers now challenged the American hegemony in long span design, most particularly in the Severn Bridge's radical box girder. Notwithstanding the Verrazano-Narrows Bridge's domination of American engineering interest in the early 1960s, that nation's—and Canadian—engineers would also produce a number of notable midsized suspension bridges in efforts spanning an entire decade.

10.1 RETURN TO THE WEST COAST

In 1961, construction began on a suspension bridge across the entrance to Los Angeles' far-removed harbor at Long Beach. The 457-m span Vincent Thomas Bridge would be California's first large suspension bridge since the Golden Gate Bridge. The state's Division of Highways teamed up with Glenn B. Woodruff, who 30 years earlier was engineer of design for the Transbay and later served on the Carmody Board following the Tacoma Narrows Bridge collapse, to design the span. The Vincent Thomas Bridge brought innovations to American practice that in some respects make it more technically advanced than the Verrazano-Narrows Bridge—particularly in its extensive use of welding. Yet its September 1963 opening after just 26 months of construction occurred in the shadow of the larger bridge; in consequence, these achievements are not widely known.

Financial and legal obstacles delayed construction twice before work finally began. Buyers of bonds to cover half the span's $14 million cost couldn't be found in the autumn of 1960, after construction tenders were issued. A second bond sale in March 1961—after the re-advertisement of foundation contracts—foundered when the sole bidder was the state's financial adviser on the sale itself. The bonds were finally sold in April, allowing yet a third round of bidding, and foundation work began that July. Neither cofferdams nor caissons support the towers, cable bents, or anchorages because of great bedrock depths. The 102-m tall towers stand on welded steel H-piles driven to depths of 40 m; in total, some 990 piles support the piers of the main suspension span. The slipforming of the concrete cable bents at the end of 154-m side spans was also significant, the twin 37-m tall columns of one bent being poured in one continuous 89-hour push. Perhaps the most notable construction development associated with the bridge was the extensive use of shop welding. Kaiser Steel, holder of the $11.4 million superstructure contract, used large welded steel boxes in the span's towers, a first for American practice. Each tower leg cross-section was comprised of just four boxes in cross-section that, when bolted together, formed a cruciform shape; each tower—braced by five open-trussed portal struts—weighs only 558 tonnes. Welding was also used extensively elsewhere, including over the approach spans. This signaled, at least in America, the coming of age of a technique offering many advantages over riveted or bolted construction—fabrication, clean appearance, and simplified painting.

Further economies stemmed from the sequential assembly of each tower using the same equipment. More unorthodox economies emerged during erection of the stiffening truss. Pre-assembled truss units were first placed on land beneath their final position in the side spans and attached to suspenders by temporary pendants. The main span was then hoisted into place without any counterbalancing side span lifts; the pendants transmitted the weight of the unerected side spans to the main cables. Once the main span was in place, the same equipment was used to erect each side span, again sequentially. One other minor design anomaly is the slight inward cant of the main cables, which are 2.1-m closer together at center span than at the towers. This feature was intended to reduce lateral sway of the truss.

The 4.6-m deep Warren truss reflected the post-Tacoma practice of a 1:100 depth-to-span ratio. Yet one surprising revelation emerged during its design. Early configurations placed the deck just below the level of the top chord in order to lower the grade of the approach spans. Wind tunnel tests on sectional models then revealed that the open steel slots between the road lanes could actually make the span more aerodynamically sensitive and induce torsional oscillation in winds as low as 48 km/hr. It was a completely unexpected finding but one of sufficient import to warrant the omission of the slots and prompt engineers to place the roadway atop the truss' upper chord. Another unusual feature of the suspended structure is the connection of the structurally discontinuous main and side spans at the towers to permit longitudinal movement. Restraining devices at the towers guard only against vertical and transverse motion.

One other stillborn feature would help lay the seeds for future cable spinning developments. In planning the spinning operations, J.A. Roebling’s Sons engineer John Nixon had sought to simplify wire adjustments. He suggested that tension adjustments be used as a way to adjust the sag of a wire over the main and side spans simultaneously. The idea showed sufficient promise to allow experimentation in the early stages of spinning, but a lack of time and money saw most of the cables spun traditionally. Nixon’s concept, however, would prove an insight of great prescience.

The fine lines of the Vincent Thomas Bridge matched its technical success. The span is one of the more graceful medium-span high-level suspension bridges yet built, its appearance only marred by the recent addition of tall safety fences over the suspended spans. The heavily instrumented bridge has also furnished valuable data emanating from the 1987 Whittier and 1994 Northridge earthquakes for seismic response and the structural damping of vibrations; information in this latter area is rare, but is of crucial importance to bridge aerodynamicists.

In remote northeast British Columbia stands a small bridge across the Peace River whose 207-m span belies its great ingenuity. In the Hudson Hope Bridge, Canadian engineers designed what would remain the world's longest prestressed concrete box girder suspension bridge for more than 30 years. A bridge here was required to reduce construction-hauling distances from Vancouver to the massive Peace River power project by some 320 km. The bridge site had no rail access, and roads were primitive, rendering the transport of bridge steel difficult. Initial options from a 1961 study by Phillips, Barratt & Partners suggested four options: a 172-m span steel deck arch, two variants of a continuous steel deck truss on two different sites with 104-m spans, and a 206-m span concrete box girder suspension bridge. River ice and the proximity of the future dam saw both truss schemes rejected. Arch footings—although on land—would be difficult to build. In the final analysis, the suspension design proved the most feasible.

The suspension bridge was in fact a relatively late arrival. H.H. Minshall, who specialized in remote construction in the province, had independently conceived the use of precast concrete box girder segments stressed together to form a suspension bridge back in 1959. The consultants investigated, and this concept won out; in addition to a projected cost of just $1.2 million, it avoided piers in the 24-m deep chasm. Moreover, raw material for the 32-m tall concrete towers and 34 deck segments could be obtained onsite. Work began in September 1963 on a 207-m span—there being no side spans—and was completed only 1 year later, despite a 2-month hiatus during the area's severe winter.

Each of the Hudson Hope Bridge's main cables are comprised of 20 exposed prefabricated helical strands—wrapping was unnecessary because of the region's dry climate. The strands are anchored 61-m behind the towers to 80 steel rods, which in turn are anchored at the base of 18-m shafts drilled into rock. An ingenious anchor with serrated edges expanded into the rock as loads increased on the cables during erection of the suspended structure. This latter element—the Hudson Hope Bridge's major innovation—was formed from 34 concrete segments cast on both sides of the river, lifted into position, and then stressed together to form the main span. Although the segments were only 6.1-m long and about 1.2-m deep, each weighed 82 tonnes. This weight required larger cables and anchorages but contributed to the span's aerodynamic stability, predicted using David Steinman's rigidity coefficients. Erecting the box segments was another challenge. Models helped predict changing cable stresses, as there was concern—unfounded, as it happens—that the weight of the first few segments could bend or even permanently kink the main cables. Each segment was swung out from a casting bed, suspended next to the tower, and then swung progressively via adjacent suspenders to its intended position in the main span. Incredibly, all 34 segments were placed in just 10 days. The segments were then stressed together by tightening 70 steel cables passing through ducts along the length of the span.

The Hudson Hope Bridge was completed in September 1964 at a cost of less than $2 million. Its builders predicted similar spans could be built in as few as 6 months, and they considered the concept suitable for spans of up to 244 m. Minshall would earn a patent for his efforts. Visually, the Hudson Hope Bridge proved the equal of its technical prowess. Its roughly finished towers complement the rugged setting, as does the single leap of the deep chasm, the dominance of the main span being reinforced by the lack of side spans. In spite of this success, prestressed concrete suspension bridges remain rare; their weight is a drawback, especially for longer spans, and only certain circumstances favor their use.[1] Not until the 1990s would the span of this ingenious bridge be eclipsed.

Two other small California spans in similarly remote settings were more traditional in design, but they were innovative in their own right. The first—in the Sierra foothills near the gold-rush town of Oroville—was opened in 1965 near the historic Bidwell Bar Bridge, thought to be the first American suspension span west of the Mississippi River. This pioneering 75-m span served until the late 1940s, but a new dam would inundate both it and a subsequent replacement span.[2] A 338-m span suspension bridge was designed to soar 191 m above the Middle Fork of the Feather River—until filling of the reservoir, when its clearance would be shorn to just 14 m. The span, designed by Arthur Burnes and Ernest James of the California Department of Water Resources, out-competed alternative designs for concrete arch and steel cantilever bridges.

The site's steep canyon walls forced the unsuspended plate girder side spans to curve out from under the main cables. They also influenced methods used to build the bridge when work began in early 1963. Bethlehem Steel, which held the $4.4 million construction contract, transported material via a logging railroad at the bottom of the gorge, where concrete for the span's seven piers was also mixed. Concrete and steel were lifted using a 580-m long cableway strung across the gorge that assembled virtually everything—towers, cables, and the 4.2-m deep stiffening truss. One tower pier was shoehorned into the canyon walls. Both piers rose 55 m to grasp the 47-m tall steel towers, whose legs were formed from single welded steel boxes connected by four open trussed portal struts. Assembly of the 37 helical strands in each 28-cm diameter cable was performed without a catwalk; workers in a small carriage suspended from the cableway unreeled two strands simultaneously from cage-enclosed drums that accompanied them across the gorge. To avoid slippage of the anchorage blocks as loads increased with the addition of the truss, prestress was induced by connecting the cable strands to prestressed steel rods housed in pipes extending through the anchorage concrete. Bethlehem Steel would later use a variation of this unique approach in the Newport Bridge later that decade.[3]

Another innovation protected the main cables. Since 1952, Bethlehem Steel had sought a more weather-resistant and easily maintained system than steel wire wrapping. The steel maker teamed up with du Pont in 1960 to develop a synthetic wrapping, and it then convinced the state to use this wrapping on the Bidwell Bar Bridge. The new fiberglass cloth acrylic resin system was topped with a sand-sprinkled syrup for secure footing, and it was estimated to last 50 years and require no painting. This system constituted the first major change in cable protection since John Roebling introduced wire wrapping in 1845.

In Oregon, the building of the Round Butte Dam on the Crooked River generated plans for a smaller 179-m span completed in 1963. The bridge is largely conventional, save diagonally braced steel towers perched atop slender hammerhead piers. This rather precarious visual effect is diminished by the low vertical clearance of the half-through truss. A similar span crosses the nearby Deschutes River.

More innovative was a replacement for the 110-m plate girder suspension span of 1940—destroyed by flood in 1965—across California's Klamath River. It would be the second American suspension bridge to use a box girder and—unusual for

America—featured towers of reinforced concrete. A.P. Bezzone, Jr. and Ostap Bender of the state highway department hurriedly produced the design to allow work to begin in September 1965. Their choice of a suspension bridge was not just out of sentiment for the former span—the Klamath River's turbulent, debris-laden waters and steep slopes rendered falsework unsuitable for other span types. A shallow but stiff suspended structure was also required to clear high water and funnel heavy logging trucks; this pointed to a box girder for the 131-m span. The strength of a box girder almost 50% deeper than the 1.75-m depth used was achieved through various means—a concrete deck acting compositely with the box girder, 4 internal longitudinal plate girders supported on the stiffened bottom flange of the box girder, and transverse floor trusses. The unsuspended side spans of the former bridge were also replaced by 52-m suspended side spans. The choice of concrete towers was unusual because complex formwork was required to mimic the arched gothic motif of the original steel towers. Yet even with this factor and the use of a solid wall to connect the tower legs below deck, concrete proved the more economical material.

Cable and box girder erection also reflected ingenuity. Helical strands were towed across the river, attached to the anchorages, and only then hoisted atop the towers. Although only 6 strands were required in each main cable, work was delayed for more than 3 months because of an inexplicable loosening of wires in some strands that required their remanufacture. Unusually for an American span, the strands were not wrapped, with the two layers of exposed strands in each cable held in place by suspender clamps. Main span box girder segments were then hoisted into place. To permit the welding of the side span box segments into a single unit, the barges used to haul the cable strands across the river were placed on the main span and filled with water. Later, by adding and removing water, these water-filled barges stabilized the center span during paving of the side spans. The bridge opened in 1967.

The Klamath Bridge won a national award from the AISC, as had its 1940 predecessor. The bridge is of even more refined appearance than its progenitor, particularly its concrete towers and clean steel box girder. Both features would not appear in another American suspension bridge until after the turn of the 20th century. One other noteworthy California suspension bridge completed in 1968 is an unusually long pedestrian span of 183-m in the state capital of Sacramento, the Guy A. West Bridge. Aesthetics drove the choice of a suspension bridge over a plate girder span; the bridge is also notable for the welded pipes that comprise its stiffening truss.

American engineers were also engaged in other overseas work aside from the bridge over Lisbon's River Tagus. Venezuelan authorities had planned a major crossing of the Orinoco River to help open up the isolated but mineral-rich southern reaches of the country. Construction of the 712-m span Angostura Bridge began in 1964. Conditions were primitive—and dangerous—as the Orinoco River harbors crocodiles and piranhas. Huge tidal variations are compounded by pronounced wet and dry seasons to vary water levels by as much as 15 m, a mixed blessing for the span's engineers—Sverdrup & Parcel—and the foundation contractors. As such, the concrete pedestals supporting the tower legs rise as much as 21 m above low water and are braced to one another for stability. The dry season made possible excavation of the south tower pier without any need for caissons or cofferdams. The twin cofferdams of its northern counterpart were founded on rock 18 m below low water.

Here, rapidly rising waters spurred contractors to establish a 28-hour record for tremie concreting. The span's large 69-m long by up to 32-m wide anchorages anchor two 47-cm diameter, 7,190 wire main cables, which were spun in just 2 months. U.S. Steel supplied and erected all superstructure steel under subcontract to a Venezuelan consortium; all steel was shipped directly from the United States up the Orinoco River to the bridge site, with transfer to barges during periods of low water. All 44 Warren truss units were assembled onsite.

The Angostura Bridge's 7.6-m thick welded trusses are relatively deep in proportion to the main span, and they strongly resemble those used by Steinman in the Mackinac Bridge. Open steel grating was used over the two inner lanes, and gaps separate the truss edges from the four-lane roadway. This similarity is perhaps not surprising, as lead designer Carl Ulstrup had worked for Steinman on the Mackinac Bridge. One unusual aspect of the span's 134-m tall portal-braced towers is the placement of the lower strut within the truss, a feature adopted by Othmar Ammann in the Throgs Neck Bridge; a restrained curved upper portal brace provides the only visible connection between the tower legs. The towers thus appear somewhat precariously perched during periods of low water when the pedestals are fully exposed. Relatively long 280-m side spans enhance the Angostura Bridge's proportions, and Venezuelan-designed prestressed box girder approach spans diverge from the deep trusses that prevailed in American practice at the time. In spite of often difficult conditions, the Angostura Bridge was completed in January 1967—6 months ahead of schedule—at a cost of $35 million. It was then the world's ninth longest span, but this relatively little-known span remains South America's longest.

10.2 THE RISE OF PREFABRICATED PARALLEL WIRE CABLES

The success of the 655-m Delaware Memorial Bridge following its 1952 opening was dramatic—even almost alarming. By 1955, traffic was almost twice that expected and, by 1956, another crossing was thought necessary to meet the traffic needs of 1960. In 1959, a parallel twin bridge acquired momentum but it was predicted to cost more than twice the $45.7 million cost of the original span. Completion was hoped for in 1965, but funding disagreements between Delaware and New Jersey stymied progress until 1962, when agreement was finally reached to build a twin bridge.

It was obvious that a second span would mimic the appearance of the first. In 1962, Howard, Needles, Tammen & Bergendoff—designers (with Ammann & Whitney) of the 1952 span—and E. Lionel Pavlo, were retained to prepare detailed plans. Their January 1964 report placed the second bridge just 76 m north of the first. Although outwardly identical, the estimated $70 million bridge (*see* Fig. 10-1) embodied technical advances made since the end of World War II—stronger low-alloy steels, larger (and thus fewer) steel plates, and improved construction methods. The foundations would also differ. With bedrock up to 168-m deep, the towers and Delaware anchorage would be supported on steel H-piles instead of the caissons of the first bridge; only the New Jersey anchorage would use a cofferdam. A thicker, more durable, but heavier concrete roadway also required 9,196 wires in each cable, versus the first bridge's 8,284 wires.

Fig. 10-1. Delaware Memorial twin bridges. The first bridge was completed in 1951, and the second span in 1968.
PHOTOGRAPH COURTESY OF THE DELAWARE RIVER AND BAY AUTHORITY.

In early 1964, Bethlehem Steel shocked all comers when it made an unsolicited bid to build the bridge for $72.5 million. This surprise offer reverberated all the way to the governors of both states, who declined it. Bethlehem Steel had been motivated by a new method of assembling parallel wire cables that—potentially—offered significant cost and time savings over in situ spinning, which had dominated for more than a century since Roebling's 1855 bridge at Niagara. Spinning, for all its hegemony, had several disadvantages. The process was both complex and laborious. Sophisticated tensioning systems at the anchorages were required to ensure a proper running out of the wires, and the method demanded the individual adjustment of thousands of wires to the proper sag. All this occurred high in the air in often-turbulent weather, where, as experience at the Forth Road Bridge had demonstrated, havoc could result.

Land spinning, used by Herman Laub around 1900, later across the Ohio River in the U.S. General Grant and Fort Steuben bridges of the late 1920s, and in 1929 by Roebling engineers in a 244-m Arizona span suspension bridge, was not suitable for long spans. Nor was the method used by British engineers in the late 1930s to assemble the cables of Africa's Chirundi Bridge. Helical cables were no panacea either—they stretched more easily under load than did parallel wire cables and required more than 10% more wire, even with prestressing. This and their high costs of manufacture eliminated any erection economies for long-span applications. The method Bethlehem Steel now proposed for the Delaware River crossing combined the faster installation methods of helical strand cables and the superior strength of parallel wire cables. The firm had first considered the concept in 1959. It now proposed to prefabricate entire parallel wire strands in the factory. Precut strands—complete with anchor sockets—would be rolled around drums, transported to the bridge, draped into place on catwalk-mounted rollers, anchored, and then lifted into the tower saddles. The method promised to revolutionize cable assembly. Firstly, labor requirements could be reduced by as much as three-quarters. Moreover, strand assembly would not be tied to progress on the bridge, and erection time on the bridge itself would be reduced in an operation involving less equipment and less weather sensitivity than spinning. Factory assembly also promised cables of higher quality after compaction. The method was thought to make parallel wire cables competitive with helical strands in spans of up to 488 m.

To simplify manufacturing and reeling, strand sizes were reduced to less than 100 wires. A painted guidewire ensured strands did not twist during unreeling on the catwalk. But great care was required to avoid the slippage and kinking (or "birdcaging") of wires during shop reeling and onsite unreeling. This hurdle was overcome in June 1962 with tests on a 152-m long, 37-wire strand at Bethlehem Steel's Williamsport, Pennsylvania plant. Strands with 61, 91, and then 127 wires were then successfully tested. By mid-1964, Bethlehem Steel considered the tiger tamed.

When Bethlehem Steel bid competitively to assemble the main cables of the second Delaware Memorial Bridge in the summer of 1964, its price for prefabricated parallel wire strands was a stunning 40% less than the engineers' estimate for spun cables. In place of 19 spun strands of 484 wires in each cable, Bethlehem Steel proposed 152 strands of 61 wires in a slightly enlarged cable of 9,272 wires. The number of individual sag adjustments would drop precipitously from more than 55,000 to just more than 900. Bethlehem Steel claimed 3 months of spinning could be pared down to 1 month, with just 25 workers per shift instead of the 100 required for spinning. Yet for all the promise of the new method, the Bridge Authority was not convinced that the new Delaware River span should act as a prototype. That December, its consulting engineers rejected Bethlehem's cable bid, citing insufficient evidence of accurate manufacture and assembly. Nevertheless, Bethlehem Steel won the cable contract and spun the cables at the same low price. In early 1965, it also snared contracts for the towers and stiffening truss, again at rock bottom prices.

Meanwhile, foundation work had begun in the summer of 1964. The record for continuous tremie concreting established in the first span was shattered in a 14-day pour in March 1965 of a 9-m thick base for the New Jersey anchorage cofferdam, although a formwork collapse in the same anchorage later that spring killed two and injured nine. In October 1965, Bethlehem Steel began to erect the 127-m tall towers, 3 months behind schedule because of substructure problems. An uneventful spinning of the main cables followed the towers' April 1966 completion, but in the summer of 1969, an oil tanker rammed into one of the almost-completed fenders protecting the tower piers of both spans. The accident—curiously reminiscent of the 1951 ramming of the New Jersey tower of the first span—left the towers untouched but caused $1 million in damage. The twin span was opened in September 1968—more than a year behind schedule—and brought traffic relief for just 3 days when the first bridge was closed for a year-long, $11 million deck rehabilitation. By December 1969, each span was carrying four lanes of one-way traffic. The Delaware Memorial Bridge remains the largest parallel twin suspension bridge in the world, and offers its users the unusual experience of viewing from afar a carbon-copy of the bridge they are crossing. The powerful imagery of the twin spans suits the surrounding industrial landscape, notwithstanding the diminished prominence of the first span, whose fine lines have suffered somewhat as a result.

The rejection of Bethlehem Steel's prefabricated parallel wire strand cables at the Delaware Memorial Bridge appeared to only strengthen the company's resolve. In early 1965, it engaged the firm of Steinman, Boynton, Gronquist & London to evaluate the method. A model bridge and cable erection system was built at Bethlehem Steel's Steelton, Pennsylvania plant, where three 91-wire strands were assembled. The Steinman firm's November 1965 report endorsed the method but cited a need for extreme care in its use. Moreover, its recommendation for a modest prototype, with 91-wire strands in a 91-strand cable, might still not have enhanced Bethlehem Steel's chances at the Delaware Memorial Bridge. In fact, the firm was not alone in devising the method. One little-known parallel development was the test manufacture and draping of a 122-m long prefabricated strand of 91 wires over a mock tower by British contractor Dorman Long in the spring of 1965. The firm successfully overcame birdcaging in reeling and unreeling, and it considered the method feasible. Japanese contractors would conduct similar tests, leading to advances that no one could then predict.

Notwithstanding the Steinman firm's cautious endorsement, Bethlehem Steel soon received a second chance to use the method. In early 1966, superstructure contracts were advertised for a suspension bridge across the East Passage of Rhode Island's Narragansett Bay. Bethlehem Steel's $18.9 million bid to build the steel superstructure—including the main cables—was accepted that April; the Newport Bridge would become the first to use the revolutionary new method of cable erection. A main span of 488 m centers over 3,400 m of bridge in one of the smaller options devised by the firm of Parsons, Brinckerhoff, Quade & Douglas. It had been among the first to suggest a bridge, whose absence during the notorious hurricane of 1938 had prevented the evacuation of Rhode Island's lower east coast. World War II killed any momentum for a crossing, but in 1954, the state created the Rhode Island Turnpike and Bridge Authority. Apart from the usual challenges of financing, the largest obstacle was the U.S. Navy, whose concern over its upstream installations kept the project in check. It is said that the U.S. Navy underwent a remarkable turnaround the day after a fog-enshrouded ferry carrying Mamie Eisenhower was nearly run down by a U.S. Navy destroyer. By 1960, agreement was reached on the parameters of a bridge—a minimum span of 305 m and a liberal vertical clearance of 65 m, just 2.5 m shy of the Golden Gate Bridge.

Tunnels, bridges, and combinations of both were initially studied. Great depths soon ruled out a tunnel. A 914-m span suspension bridge that would have placed the east tower on land was rejected; apart from cost, its high-level approaches would have razed a good portion of Newport, the wealthy summer resort city the span was intended to serve. A more northerly alignment was chosen, and alternatives were pared down to cantilever and suspension spans. Although a 305-m span cantilever was thought to be less expensive, borings indicated prohibitive rock depths. The final choice came down to the most economical span length for a suspension bridge. An 853-m span was considered, but the benefit of shallower rock beneath its towers was eliminated by the need for larger anchorages. In the end, a 488-m span was adopted, flanked by 210-m side spans and no less than five different bridge types spread over 2,520 m of approach span. Financing difficulties and then a state-wide referendum stalled a planned autumn 1962 start on the estimated $33 million bridge. By the time approval was secured, the bridge had become a $45 million undertaking. An April 1965 start was then further postponed because of additional legal complications.

Construction finally began that September on a $17.3 million foundation contract. Footing the piers would not be easy. Water up to 43-m deep at the towers masked bedrock depths of up to 132 m, rendering caissons too costly and time-consuming to sink. Parsons, Brinckerhoff, Quade & Douglas instead elected to drive piles to depths of up to 50 m, shear them off above overburden, and plunk down a cofferdam on top of the cut piles to support the tower bases. More than 500 friction piles support the east tower alone. Divers torching off all 1,286 pile tops over a total of five piers were submerged for periods of up to a week in a Westinghouse diving bell. A steep slope beneath the 34- by 41-m west anchorage added to the challenges. Here, in a dewatered cofferdam, 72 cylindrical caissons were drilled 6 m into rock; many were inclined to resist the future cable forces. Steel piles protruded from the caissons to grasp anchorage concrete.

By late 1967, the Newport's gothic steel towers had attained their 122-m height. Each leg was built from four box sections, with a fifth cell formed when assembled onsite. All tower steel—as well as that in the suspended and approach spans—was shop welded and bolted in the field. Bethlehem Steel's cable work then began. Parsons, Brinckerhoff, Quade & Douglas had envisaged in situ spinning of the 4,940 wire cables, but both Bethlehem Steel and rival American Bridge had proposed prefabricated parallel wire strand cables when contracts were tendered in early 1966. Bethlehem Steel's 61 wire strands increased the number of strands in each cable from 19 to 76, but the method would avoid thousands of spinning wheel passes and sag adjustments. Bethlehem Steel also proposed to use a larger diameter wire, reducing the number of wires in each 39-cm diameter cable by more than 300 to 4,636. The contractor intended to shave more than $500,000 and 4 weeks off the $3.1 million and 9 weeks it had estimated for in situ spinning. On the recommendation of Parsons, Brinckerhoff, Quade & Douglas, the Bridge Authority consented to the change. Bethlehem Steel would now see its activities above Narragansett Bay become the focus of engineering interest worldwide.

Strand fabrication began in November 1966 and was completed in 8 months. Installation began in February 1968 but was immediately stopped in its tracks when some wires in the very first strand slipped, badly deforming it. Bethlehem Steel engineer Jackson Durkee, the driving force behind the system (and son of engineer L.R. Durkee, who had witnessed the Tacoma Narrows Bridge collapse), specified tighter binding during unreeling. This solved that particular problem, but technical glitches persisted. High winds, snow, ice, and temperatures that plunged to -17° C also plagued operations. After 6 weeks of intermittent work, fewer than half the strands were in place at a time when the cables should have been completed. Conditions then improved, and in early May, the cables were finally finished. Given the weather, spinning may not have fared much better. For all the difficulties, the Newport Bridge's cables constitute a major achievement, even moreso in view of their great length—1,375 m. As a final touch, Bethlehem Steel also used the cable wrapping system it had pioneered on the Bidwell Bar Bridge.

The anchorages extend the Newport Bridge's list of innovations. Bethlehem Steel had modified the plans of Parsons, Brinckerhoff, Quade & Douglas by introducing the pipe anchorages it had developed in 1964 in conjunction with its prefabricated parallel wire strand cables. Each strand was threaded through one of 76 tightly clustered tubes in two large prefabricated anchorage assemblies embedded in concrete at the rear of the anchor block. This arrangement capitalized on the compressive properties of concrete, simplified erection, and reduced reinforcing steel and the mass of the anchorages themselves.

The Newport Bridge's truss reflects the conservatism of post-Tacoma design. Its 4.9-m depth is 1/100 of the main span. The 20-m wide truss also features 1.8-m gaps beyond the roadway edges and sports a midspan cable tie. Its perforated curbs—with 15-cm high, 1-m long slots spaced every 1.4 m—are unusual; they emerged from wind tunnel tests on 1:50 scale sectional models at the U.S. Bureau of Public Roads in Maclean, Virginia. One counterintuitive result of subsequent tests was the discovery that a 94-cm high solid median barrier actually enhanced aerodynamic stability (although such a barrier was never used on this or later spans).

The Newport Bridge was opened in June 1969 at a cost of $61 million. Visually, it is one of the more successful high-level midspan suspension bridges. The clean Gothic arches of the towers resurrected a style made famous by the Brooklyn Bridge and not seen since Steinman's ornate St. Johns Bridge of 1931. The anchorages are tucked inconspicuously beneath thc deep approach trusses, and the disparate bridge types that make up the crossing are visually well coordinated. But one feature did prove contentious. Bethlehem Steel had shop-painted all steelwork with an epoxy coating, specified by Parsons, Brinckerhoff, Quade & Douglas, that was designed to eliminate field painting for 10 years. By October 1970, paint was being shed in sufficient quantities to warrant its complete removal and replacement. Allegations flew between the Bridge Authority, Bethlehem Steel, and the bridge designers over the cause, but by 1971, Bethlehem Steel faced a $2 million lawsuit and nonpayment of the final $800,000 of its contract. Years of legal wrangling finally ended at the U.S. Supreme Court, which in 1984 ruled in favor of the Bridge Authority in a judgment costing Bethlehem Steel several million dollars. The replacement paint—the same as originally specified by Parsons, Brinckerhoff, Quade & Douglas—has since performed well. Another incident in February 1981 saw a tower pier hit by an oil tanker in a thick fog; the pier emerged undamaged but the ship, minus almost 3 m of bow, did not.

In spite of this contentious epilogue, Bethlehem Steel's pioneering cable work at the Newport Bridge, although not without problems, proved the viability of prefabricated parallel wire strand cables and set the stage for later developments. The bridge was the crowning achievement of Alfred Hedefine, an engineer with the firm since 1948. Hedefine had designed Niagara's Rainbow Bridge (1940) when with the firm Hardesy & Hanover and would go on to head the design of the 383-m span Fremont tied arch (1972) in Portland, Oregon.

Curiously enough, for all its efforts, Bethlehem Steel never again used its prefabricated parallel wire strand technology. This can be attributed largely to a dearth of American suspension bridges following the completion of the Newport Bridge. In fact, Bethlehem Steel even pulled out of bridge building in 1984, although still acting as a supplier of bridge steel. Ultimately, the Japanese would expand the scope of prefabricated parallel wire cables to lengths and sizes that even Bethlehem Steel engineers might not have imagined.

10.3 CANADIAN DEVELOPMENTS

In Canada, two suspension bridges of note were opened in 1970. One is Halifax's A. Murray MacKay Bridge (*see* Fig. 10-2), an innovative span of 427-m designed by Roger Dorton of the Montreal firm Pratley & Dorton.

The two lanes of P.L. Pratley's Angus L. Macdonald Bridge had wrestled unsuccessfully with burgeoning traffic across Halifax's harbor since its opening in 1954. Studies initiated under son Hugh in 1962 led to a 1963 recommendation for a three-lane suspension bridge over the Narrows in the city's west end, near the line of an unbuilt 1929 scheme. Although this general locale was adopted in 1964, controversy erupted in 1965; other proposals, including a south-end tunnel and a twinning of the existing bridge, were mooted. The provincial government stepped in, and by year's end, a Narrows alignment was confirmed for a four-lane suspension bridge forming part of a planned ring road around the city. Dorton went on to design the bridge throughout 1966, and groundbreaking took place in June 1967.

Although not overly large, the bridge established several precedents. Its composite steel orthotropic deck was the first for a new North American suspension bridge, and only the second in a trussed suspension span (Germany's 1966 Emmerich Bridge preceded it).[4] Wind tunnel studies also added significantly to the understanding of testing methods and aerodynamic stability during erection.

While the choice of a suspension bridge was obvious, the type of suspended structure was another matter. The Severn Bridge had proven the case for a box girder, but even with Denmark's 600-m span Lillebælt Bridge entering construction, it was not clear how far its advantages extended to smaller spans. Dorton examined three options—a standard truss with a concrete deck, an orthotropic truss, and a flat-bottomed orthotropic box girder similar to that used in the Lillebælt Bridge. A composite steel orthotropic deck reduced truss depth dramatically—from 4.25 to 2.9 m—which translated into a weight savings of 15%. Surprisingly, the box girder was dropped before studies were even completed when it was found to require 16% more steel than a truss. Dorton's studies suggested its economies only appeared in spans longer than 549 m. Dorton also toyed with the idea of a twin parallel box system as a way to reduce steel requirements but the concept had unknown aerodynamic qualities—interestingly, this configuration has since re-emerged in connection with ultra-long spans. The preliminary analysis was aided by the use of Steinman and

Ammann's stiffness indices. In the end, the slender orthotropic truss was adopted with a depth-to-span ratio of just 1:147 (*see* Fig. 7-2).

The weight savings of the orthotropic deck truss extended to the towers, main cables, and anchorages. For example, the 61 helical strand, 37-cm diameter main cables are just slightly larger than those used in the Angus L. Macdonald Bridge. Although slightly less efficient than parallel wire cables, helical strands could be easily obtained from Canadian manufacturers, whereas prefabricated parallel wire strands were available only from Bethlehem Steel or American Bridge. The cables were strung between 88-m tall towers, for which portal and diagonal braced options were studied. The former commanded a 20% premium in steel, and although more struts would have reduced this margin, they were thought less pleasing. Diagonal braces were adopted in clean towers with legs of remarkably slender proportions.

Arne Selberg's methods of aerodynamic analysis were used to proportion the 17.4-m wide four-lane truss. The truss has a high degree of torsional stiffness because of the orthotropic deck, which also produced one of the lightest suspension trusses ever, as measured per unit of deck area. This characteristic and the uncertain aerodynamic effects of placing the solid orthotropic deck on truss sections before lifting into the bridge generated some concern over aerodynamic stability during erection. The uniqueness of the design led to testing at the University of Western

Fig. 10-2. A. Murray MacKay Bridge

Ontario's Boundary Layer Wind Tunnel in London, Ontario under Alan Davenport, an authority in bridge aerodynamics who had founded the facility in 1965.

Davenport was retained after the design had been completed, but his studies added knowledge in three areas—the effect of turbulent versus smooth wind flows on aerodynamic stability, the relationship between full aerolastic models and sectional models, and the aerodynamic behavior of partially erected suspended structures using full aerolastic models, a field that to this point had been largely ignored but where the risk was potentially greater than in completed bridges.[5] With the first full aerolastic model in North America since the studies for the second Tacoma Narrows Bridge, Davenport tested the stability of the truss at different stages of erection with varying coverages of orthotropic deck. The tests revealed that full decking did lower critical velocities but not enough to prevent its full installation before lifting into the bridge.

For the bridge in incomplete and finished states, Davenport modeled turbulent flows in the near-ground "boundary layer" atmosphere. All full aerolastic modeling to this point had involved smooth wind flow, and Davenport's studies in this regard are considered a major step in the development of bridge aerodynamics. To create the turbulence to which bridges are exposed, he crammed the 24-m long wind tunnel with blocks to create a rough surface texture. The results were intriguing, if perplexing. Whereas the model in turbulent flow revealed only random buffeting at lower wind speeds and no torsional instability in strong winds, in smooth flow there was no buffeting at all but there was a susceptibility to a violent and catastrophic torsional motion in very high winds. The results suggested the span's critical velocity to be higher than predicted under smooth flow (more discussion of this phenomenon is found in Chapter 18). Davenport found that sectional models exposed to smooth flow also significantly underestimated a bridge's aerodynamic stability in comparison to full-length models in turbulent flow. This was a major concern, as tests on sectional models in smooth flow by F.B. Farquharson and George Vincent, as well as those by Kit Scruton and R.A. Frazer, suggested a correlation with full model behavior.

The inconsistent results of the testing for the A. Murray MacKay Bridge lead Davenport to conceive and—with colleagues Hiroshi Tanaka and Guy Larose—test "taut strip" models, comprising many deck segments supported by taut wires or tubes that simulate the mass and inertia of the entire main span. These hybrid models—which really came into their own during studies of the Bronx-Whitestone and Golden Gate bridges in the early- to mid-1970s—would in essence bridge the gap between sectional and full aerolastic models. They are of a smaller scale than full aerolastic models, and the demands for accurate modeling of wind turbulence and deck cross-section details are high. Nonetheless, taut strip models have proven effective, simple, and inexpensive proxies for full bridge models—and provide more accurate results in turbulent wind than do sectional models. Moreover, they remove the unwanted influence of towers and cables that "muddy" the aerodynamic characteristics of the suspended structure alone, and thus permit a better comparison with sectional models. Finally, taut strip models eliminate the need to consider Froude's Law in determining scale relationships (*see* Chapter 5). Davenport and others have since used taut strip models in fine-tuning the design of suspension and cable-stayed bridges around the world. The laboratory—and Davenport—have long been recognized internationally.

Davenport's studies validated Dorton's design, and they also helped predict wind conditions and their effects on the construction schedule—a first. A year of foundation construction was uneventful. Each tower was then to be erected sequentially using the same creeper crane beginning in the summer of 1967. Yet the Halifax tower was plagued with poor fits between steel sections and was not finished until December. Fortunately, unusually benign weather allowed a planned winter shutdown to be almost completely averted during erection of the delayed Dartmouth tower. Nonetheless, the catwalk cables sat at the bottom of the Narrows for 2 weeks awaiting the readying of the tower. All 61 helical strands in each cable were placed in little more than a month, with each strand draped and anchored in just an hour. The truss was also lifted efficiently; two 54-tonne units were lifted into the main span each day, although piece-by-piece erection was required in the 157-m side spans over land. A temporarily cantilevered approach span did provide some anxious moments when it was gripped by vortex-induced oscillation (Scanlan and Vellozzi, 1980).[6] Epoxy-asphalt paving followed, and the $13 million bridge was opened in the summer of 1970. Interestingly, the bolting of its orthotropic sections, unlike the welded connections in the Severn Bridge, was intended to add to the span's structural damping capacity. The slender truss—almost as thin as the girder approach spans—well-proportioned steel towers, and unobtrusive anchorages tucked beneath the approach spans make the A. Murray MacKay Bridge one of the more visually appealing of the midspan suspension bridges.

A second, larger Canadian suspension bridge was completed that same year across the St. Lawrence River beside the historic 1918 Quebec cantilever bridge at Quebec City. The 668-m span Pierre Laporte Bridge was of more conventional design than the A. Murray MacKay Bridge, but it remains Canada's longest single span.

Road traffic first crossed the river on a narrow roadway that had been squeezed onto the 549-m span rail cantilever span in 1929. In 1945, P.L. Pratley devised preliminary plans for a downstream span where the river was substantially wider. By 1947, Pratley's plans called for an $18 million bridge whose 975-m span would have ranked as the third longest in the world. Although this scheme died, traffic growth did not, and in 1952, one rail line on the cantilever gave way to an expanded 9-m roadway as a stopgap measure. A number of bridge options were studied during the 1950s, as was a tunnel and ere still more lanes on the existing cantilever bridge. Yet a new bridge became increasingly imperative, with the depth and width of the river, combined with its steep banks, suggesting a suspension span. Potential sites ranged from the Citadel near the heart of old Quebec (whose visual effects—let alone its physical disruption—might have been even more ruinous than Steinman's stillborn 1958 span across Upper New York Bay) to a point midway between the Citadel and the 1918 cantilever bridge.[7] By the fall of 1961, a site just 198 m west of the cantilever was adopted. A joint venture of Demers-Vandry-Gronquist (the latter involving Steinman, Boynton, Gronquist & London) designed the six-lane bridge. Not surprisingly, the deep stiffening truss over the main and flanking 187-m side spans reflect American tradition.

Construction began in June 1966 in an environment of strong winds, fast tidal currents, and a 6-m tidal range. The foundations soon presented a large headache when bedrock was found to slope steeply beneath the north tower. To avoid bedrock, it was hoped that a shallow overlying layer of glacial till would support a thick concrete slab. It did not, and to further frustrate efforts, this same overburden resisted efforts to blast through to bedrock. With winter shutdown looming, it was finally decided to drill 225 piles 3 m into bedrock to support the base slab. The south tower pier provided its own, if different, challenges. Dense till and large boulders forced workers to dig all 23 m to bedrock, blast a level surface, and dump almost half a million tonnes of sand to a height of 5.5 m before the caisson could even be installed. The caisson was then sunk by excavating 30 dredging wells similar to those used in the caissons of the Tagus Bridge in Lisbon.

The tower foundations were not completed until early 1968, although by contrast the 48-m long and 44-m wide anchorages were built without major difficulty. The 128-m tall towers then began their ascent. They are of conventional appearance, yet their simple arched portal braces departed from Steinman tradition and are more suggestive of an Ammann lineage. The cruciform legs were built from only four outer cells to form a five-celled welded cross-section. Each tower weighs only 2,880 tonnes, with more than 225 tonnes in each avoided through the use of new high-strength steels. The European technique of prestressing the tower bases to their pedestals to secure them more firmly in their free-standing position was also used.

The Pierre Laporte Bridge's 12,580 wire, 62-cm diameter cables were the first in Canada to be spun in situ—all previous large spans had used helical strands. Out of respect for the gusty St. Lawrence River and with memories of the Firth of Forth still fresh, crews tied completed strands to the catwalk. Spinning began in early spring 1969 but almost immediately became ensnared in a 2-month province-wide strike that pushed the schedule perilously close to a year's delay if spinning could not be finished before winter. Yet within weeks of the workers' return, spinning was back on schedule, and the cables were completed in another 7 weeks; efforts were aided by technical improvements from British contractor Cleveland Bridge & Engineering that permitted unprecedented speeds of wire unreeling. Crews even managed to hoist several truss sections into place before cold weather halted further work.

The 6.7-m deep truss reflected the 1/100 ratio of depth-to-span orthodoxy of many post-Tacoma spans. Yet Steinman's trademark open steel grating and air gaps were absent, likely on account of a generous 28-m truss width; even then, wind tunnel tests on sectional models predicted a critical velocity of 257 km/hr, 45% greater than required. Closed-box truss pieces were fabricated by Cleveland Bridge & Engineering in Great Britain and protected against salt exposure on the trans-Atlantic crossing—upon the pieces' arrival, any salt was immediately washed off. Steel was then welded and bolted into 196-tonne units on barges at the foot of the 1918 cantilever. At high tide, the 24-m long sections were floated out to the bridge and lifted into place.

One topic of some discussion involved the road surface. A conventional concrete deck was adopted on the basis of studies for the Newport Bridge that had concluded orthotropic decks to be uneconomical; the threat of corrosion from road salt that could penetrate the orthotropic deck's thin asphalt overlay was another concern. Yet at about the same time, Pratley & Dorton had come to the opposite conclusion for the A. Murray MacKay Bridge. Their studies suggested an orthotropic deck to be 15% cheaper, and they believed that epoxy asphalt would provide sufficient protection. At the time, Dorton would speculate that a lack of American familiarity with orthotropic decks may, too, have biased deck selection for the Quebec span.

The $40 million Pierre Laporte Bridge—named in honor of a politician slain during the political crisis that then gripped the province of Quebec—was the world's 12th longest suspension bridge when it opened in November 1970. Its proximity to the rail cantilever is somewhat detrimental to the appearance of both spans, for they are difficult to comprehend as individual structures, as one can, for example, at the Firth of Forth. Nonetheless, the two bridges provide a visually dramatic contrast of two different engineering eras, transport modes, and bridge types.

10.4 AMERICAN TWILIGHT

The great era of post-war American suspension bridge building ended in the early 1970s. Its twilight began with the 1971 completion of the Dent Bridge, a 320-m span in Idaho over the future Dworshak Reservoir. This bridge would be only one of a handful of American post-Tacoma spans to use plate girders. Its engineer, the firm of Howard, Needles, Tammen & Bergendoff, had designed one other, the 1954 self-anchored Paseo Bridge in Kansas City.

The Dent Bridge's 44.5-m tall steel towers stand on concrete piers up to 58-m high that—like those of the Bidwell Bar Bridge—would be later submerged. The 76-m side span girders were erected first as simple beams to the tower piers, whereupon they served as platforms to assemble the multiple-braced steel towers. Faced with a precipitous 183-m drop to the gorge bottom, contractors placed the 19 helical strands in each cable from an endless hauling line draped over the towers. Each strand was anchored in tunnels blasted up to 49 m into rock that were later plugged—save a small access shaft—with concrete. In raising the 3-m deep, 12-m wide plate girders, engineers adapted a simple "slack-line" system used in logging. All 21 main span units were hoisted into position from the tower bases by means of lines attached to movable collars on the main cables; the units were "slid" into place, working from midspan to the towers. The side span beams were then transformed into suspended spans with the addition of suspenders. The $7.8 million Dent Bridge opened in November 1972 after just 22 months of work, its innovative construction methods helping it to earn the AISC's long-span bridge award for 1972.

The last large American suspension span of this era was a twin for the 1952 Chesapeake Bay Bridge, which in its first year funneled 50% more traffic than predicted. A second span would have a brief but controversial gestation. In 1964, the state government recommended a second crossing of Chesapeake Bay. It rejected a site near Baltimore, Maryland as too expensive and a southerly site as too far removed, but it then sought to build an $80 million tunnel across Baltimore's harbor. Southern legislators pushed for the rejected southerly crossing. A March 1966 decision to proceed with a parallel span for the 1952 bridge sparked more debate, only resolved with a decision to build yet another Baltimore crossing, the 366-m span Francis Scott Key Bridge, a cantilever truss opened in 1977.

The firm of J.E. Greiner, engineers of the first bridge, designed the second (officially, both comprise the William Preston Lane, Jr. Memorial Bridge). But unlike the Delaware Memorial Bridge, duplication was not an obvious choice (*see* Fig. 10-3). For many reasons, the 6.5-km long bridge differs in almost every respect save the span configuration of the suspension bridge. Concern for navigation safety made it desirable to minimize the distance between the two bridges, but the ship protection islands surrounding the low-slung anchorages of the first span pushed the new bridge 137 m away. The expense of such islands was avoided in the second bridge altogether by the use of 20-story concrete monolith anchorages extending to deck level. Backstay spans were also eschewed, but to maintain symmetry, the 200-m side spans of the first bridge were adopted.

The new bridge carries three lanes versus the original span's two, one lane being reversible. The roadway is also perched atop a truss about 30% deeper than the first bridge, improving views, which in the first span are obscured by the through truss. The towers of the new bridge are drastically different—the diagonal braces of the first span were replaced by portal struts for improved appearance and simplified maintenance. The desired effect was achieved in spite of an unusually large number of struts—two above the roadway and four below (one of which is buried in the truss). Unfortunately, the visual problem of tall anchorages in midsize spans was not entirely solved here. Differences extended to the approach spans, while still maintaining a smorgasbord of archetypal American spans. The 238-m span subsidiary cantilever mirrors the unusual haunched form of Baltimore's larger Francis Scott Key

Fig. 10-3. William Preston Lane, Jr. Memorial Bridge. The first bridge (*left*) was completed in 1952, and the second span in 1973.
PHOTOGRAPH COURTESY OF THE MARYLAND TRANSPORTATION AUTHORITY.

Bridge. Plate girder spans of up to 92 m in the curved high-level approaches are of simpler appearance than the trusses of the first span; they capitalized on stronger steels and fabrication advances. Simpler solid steel bents supporting the approach spans also complement the towers of the newer suspension bridge. Finally, all steel was shop- welded and then bolted in the field.

By the time the design was ready for tendering in 1969, costs had escalated to an estimated $97 million. A low bid of $42 million for the foundations then came in almost 20% more than the estimate. Moreover, a May 1969 contract award almost immediately became mired in problems that would dog foundation work to the bitter end. First was the discovery of hundreds of thousands of cubic meters of unexpected silt that had to be mucked out of the perimeter cylinders of the 35- by 57-m anchorage cofferdams. Strong tidal currents then made it necessary to temporarily brace the unfilled cofferdams to the seabed. Tropical storms and winter weather stopped work several times, compounding an already uncooperative climate of wind, rain, and fog. To regain lost time, concrete was even ferried via giant Sikorsky helicopters to remote approach span pier frames and in the upper reaches of the anchorages.

American Bridge won the $38 million superstructure contract in November 1969. With the foundation delays, a late 1972 completion date was now fiction. Tower assembly from barge-mounted cranes was followed by the firm's first use of prefabricated parallel wire strand cables, another change from the helical strands of the first span. What American Bridge had estimated as a 4-month spinning operation was completed in just 2 months in December 1972. The 61 wire strand size pioneered by Bethlehem Steel in the Newport Bridge was used in 55 strands within each of the 3,577-wire, 33-cm diameter main cables, along with six smaller strands of 37 wires. All strands were anchored in an assembly dubbed the rocket launcher, similar in concept to that devised by Bethlehem Steel, which was then immersed in helicopter-delivered concrete.

Extensive use was made of pre-assembly. Approach deck trusses up to 183-m long were barged to the bridge, and more than half of the 238-m main span of the cantilever truss was winched into place in a single 1,000-tonne unit. American Bridge also used a new neoprene cable wrapping, which it had first applied in 1969 in a trial replacement on Steinman's Mount Hope Bridge. With a weight one-fifth that of traditional wire wrapping, the new material was claimed to have eliminated 100 wires from each of the Verrazano-Narrows Bridge's four main cables had it then been available. An unusual top dressing of crushed walnut shells embedded in the topcoat of paint provided firm footing for maintenance workers.[8] The William Preston Lane, Jr. Memorial Bridge twin span was finally opened in June 1973 at a cost of $118 million, far more than the initial estimates. Yet additional costs loomed. In December 1972, the foundation contractors had sought almost $19 million in damages, alleging that inaccurate design information precipitated delays. American Bridge, too, claimed compensation for the costs it incurred during the 12 months it could not get onsite because of the foundation problems. Almost $5 million was eventually paid to the foundation consortium, although J.E. Greiner, the firm that designed the bridge, was not found culpable.

It is somewhat ironic that for all the efforts of Bethlehem Steel and American Bridge to develop prefabricated parallel wire cables, neither firm has yet to use them again. The maturing of America's road network meant that few large bridges remained to be built; the William Preston Lane, Jr. Memorial Bridge would be the last large suspension bridge opened in the United States for the next quarter century. Yet America would soon carve out an important niche in the rehabilitation of aging suspension bridges. The most acute problems have been found in New York City, where the three oldest East River bridges—the Brooklyn, Manhattan, and Williamsburg bridges—have vividly demonstrated the costs of neglect. The Williamsburg Bridge was actually closed in 1988 for emergency repairs to its corroding steelwork. That same year, the bridge was even a candidate for complete replacement in an international design competition. This exercise generated many innovative cable-stayed and suspension solutions, including those that could be superimposed over the existing bridge or built next to it and then slide into place. In the end, a rehabilitation of the existing bridge was adopted. The stiffening truss of the nearby Manhattan Bridge has been punished by asymmetrical loads from a design fault—the placement of the heavy subway lines near the truss edges—in addition to major neglect. The Brooklyn Bridge—the oldest of the three spans—requires the fewest, if still extensive, repairs, largely because of more diligent and earlier care. Extensive cable (including suspender replacement), truss, and tower rehabilitations on the Williamsburg and Manhattan bridges, in particular, will cost well more than $1 billion and some 15 years of work before completed in 2005.

New York City has also hosted one of the more radical of suspension bridge upgrades when from 1960 to 1962 the long-planned lower deck of the George Washington Bridge was added. A total of 76 pre-assembled truss units were lifted—19 into the side spans and 57 in the main span—by Bethlehem Steel without ever disrupting upper deck traffic. No small feat, considering each unit weighed 200-tonnes and measured 33-m wide, more than 27-m long, and 9-m deep. Side span units were winched up 60 m and then rolled beneath the deck along temporary beams that were later reused as stringers to support the lower deck roadway. More than 12,500 tonnes of structural steel were lifted, and all main span units were placed in just 3 months. Almost 50,000 new bolt holes were drilled or existing ones reamed out in the original upper level deck to attach the truss; the towers had another 9,000 holes drilled into them. Much of the $145 million cost was devoted to a complex network of approach ramps. In aerodynamic terms, the second deck increased the overall torsional stability of the George Washington Bridge by a factor of three. The 14-lane span stands today as the world's most capacious suspension bridge.

On the other side of the continent, transit and road authorities have long coveted the unused lower lateral bracing of San Francisco's Golden Gate Bridge. This additional capacity is available thanks to the huge main cables (whose salt-ravaged suspenders were replaced in the 1970s). This has never come to pass, but the 1989 Loma Prieta earthquake, although leaving the bridge undamaged, revealed its vulnerability. The bridge is currently undergoing a seismic upgrade expected to cost an incredible $300 million when completed in 2007. The approach spans alone will consume two-thirds of this cost. The west spans of the Transbay are also undergoing more than $200 million of seismic retrofitting; a half million rivets will be replaced by high-strength bolts, truss steel will be strengthened, and dampers will be added between the towers and truss. Overall, some 7,700 tonnes of steel will be added. The

Vincent Thomas Bridge also underwent a $25 million seismic strengthening in the late 1990s involving the installation of 48 motion dampers in various parts of the span, in addition to foundation strengthening and numerous other improvements.

Other major work has involved the replacement of deteriorated concrete decks with lightweight orthotropic deck units. These lighter steel decks reduce stresses in the main cables and trusses, and they also increase live load capacity. To minimize traffic disruptions, concrete deck pieces are cut out at night and immediately replaced with modular orthotropic units before the morning rush hour. This expensive and exacting surgery was first performed in the mid 1970s on the upper deck of the George Washington Bridge. It has since been applied on the Golden Gate Bridge, which underwent a $55 million deck replacement in the mid 1980s; on Philadelphia's Benjamin Franklin Bridge in a $56 million facelift; and on New York's Throgs Neck Bridge. The largest job of this type so far is reserved for New York's Triborough Bridge, which is in the midst of a mammoth $250 million redecking project designed by the firm of Ammann & Whitney; included is the replacement of the first freeway-to-freeway interchange in the United States and many approach spans. New concrete decks are also being installed on suspension bridges; that planned for the Verrazano-Narrows Bridge could cost $45 million. Major deck work on other spans is contemplated; rehabilitation of the Walt Whitman Bridge, which began in the late 1990s, could ultimately cost up to $150 million.

Perhaps the most drastic deck makeover was reserved for Vancouver's Lions Gate Bridge when in 2000 work began on the first-ever total truss replacement. In this $59 million job, the bridge’s single-lateral through truss is to be replaced with an orthotropic deck truss. Only 40 night closures are planned, during which contractor American Bridge is to cut out entire truss sections and lift in new truss sections. Maintaining stability between the old and the new truss has proven especially challenging, and this innovative project will take more time than expected to complete. On Canada's east coast, Halifax's Angus L. Macdonald Bridge also received a two-year, $38 million facelift that cantilevered bicycle and pedestrian paths off both sides of the span and widened the existing two-lane deck to three lanes. This work was completed in 1999.

Smaller spans have also come under the surgeon’s knife. In Central America, the corroded main cables and suspenders of Santo Domingo's 175-m span Puente Duarte are being replaced with four new cables carried in tower leg extensions. Work began in 1999 on this 2-year, $32 million program that also includes the replacement of many truss members. One particularly bad area of tower corrosion was found below the deck from the decades-old habit of people urinating off the bridge—a difficult social engineering problem!

The cost of maintaining these aging spans is huge; painting and steel repairs alone on the George Washington Bridge were estimated in the late 1990s to cost between $35 and $50 million. Yet to replace such spans would involve almost unfathomable costs and traffic disruption, and the time required to obtain stringent environmental approvals, especially in congested urban areas, can take years. In any event, with most of America’s road network in place and the domination of midsized American spans by cable-stayed structures, the need for new suspension bridges waned following the completion of the William Preston Lane, Jr. Memorial Bridge. The focus on large suspension bridges would soon shift to other continents—first Europe, then Asia—where great strides in both suspension bridge engineering and cable fabrication would be made.

ENDNOTES

1. A similar self-anchored concrete box girder span of 152 m with 49-m side spans was under design in 1966 for the Stikine River in British Columbia (Minshall, 1966). It was never built.
2. In 1966, the 1856 Bidwell Bar Bridge was dismantled and, in the following year, it became only the second bridge recognized as a National Historic Civil Engineering landmark. The bridge was reassembled on a nearby ridge in 1974.
3. German engineer Fritz Leonhardt (1966) claimed to have pioneered such a system in his 1949 to 1951 replacement of the Cologne-Mulheim Bridge.
4. The 1987 to 1989 truss replacement of Quebec's 177-m span Beauharnois Bridge featured an orthotropic deck.
5. Studies of the erection sequences for the suspended structures of the Forth Road and Severn bridges had used sectional models.
6. Robert Scanlan found out about this episode through a quirk of luck. His daughter happened to be at the Bedford Institute of Oceanography, which sits next to the bridge; seeing the oscillation and knowing of her father's interest, she grabbed a movie camera and filmed the motion. Scanlan later sent the film to Dorton.
7. A 1968 proposal for a $130 million second suspension span into the old historic port of Quebec City—now a World Heritage Site—also never succeeded.
8. This system would not endure well; its use in the second complete cable replacement for the General U.S. Grant Bridge across the Ohio River at Portsmouth in 1978 would not help to save the span (dismantling of the bridge began in 2001 as construction began on a nearby cable-stayed span that is to open in 2004). An early 1980s application of the system on Charles Ellet, Jr.'s famous Wheeling Bridge also failed and was replaced in 1999.

Chapter 11

THE BOX GIRDER COMES OF AGE

The Severn Bridge had proven the case for suspended box girders. In Europe, Freeman Fox & Partners would build on its initial success in larger spans, one of which would rank as the world's longest when completed. However, Danish engineers would build the second bridge to use them and, with this effort, set the stage for a long-term ascendancy in the design of long-span, box girder suspension bridges.

11.1 THE LILLEBÆLT BRIDGE, DENMARK

No fewer than 483 islands (of which 100 are inhabited) augment Denmark's large peninsula of Jutland. In consequence, this small nation of 5 million has built many fine bridges since its first major span was completed across the Lillebælt (or Little Belt) in 1935. The 600-m span successor to this first bridge was opened in 1970, just 4 years after Gilbert Roberts' triumph at the Severn Bridge. The significant features of this suspension bridge—after its box girder—include reinforced concrete towers, buried anchorages, and the longest locked-coil strand cables ever erected. Yet having evolved in the Severn Bridge's shadows, this smaller but noteworthy Danish span has not received the acclaim it deserves for its engineering and aesthetic prowess (*see* Fig. 11-1).

The Lillebælt is the smallest and most westerly of three channels that funnel waters of the Baltic Sea to the Kattegat—a salient of the North Sea. It separates the Jutland peninsula from the island of Funen. The 1935 span, an 825-m long overhead continuous truss, carries road and rail traffic and crosses the channel where it narrows to less than 1 km. Its central 220-m span remains one of the longest of its type. The bridge is also notable for the footing of its foundations in open-sea conditions and its clean appearance, reflecting the Danes' long-standing tradition of excellence in public architecture. By the 1950s, however, its narrow two-lane roadway had reduced the span's status to that of a major bottleneck. In 1959, the Department of Public Works commissioned the Danish engineering firm of Ostenfeld & Jønson to examine a second crossing. The firm's founders, Christian Ostenfeld and W. Jønson, had played a key role in Denmark's first large bridge-building drive during the 1930s and would soon design two (albeit small) Finnish suspension bridges, a 220-m span truss-stiffened span across Kirjala Sound and the Uittamp-Sääksmäki, a 155-m plate girder span—both completed in 1964.

Alternative routes and bridge types were studied between 1959 and 1962. Two alignments were considered—one immediately west of the existing bridge, and a

Fig. 11-1. Lillebælt Bridge

second 1,200-m crossing 5 km to the east. A 450-m span truss-stiffened suspension bridge was considered for the shorter western alignment, but its long approaches and severance of forests on Funen made it unappealing. The 1,200-m Lyngs-Odde line to the east thus became the favorite. Of eight schemes devised for this location, suspension bridges of 600 and 670 m span were the front-runners. Five less likely alternatives would have nonetheless established span records: self-anchored suspension, cable-stayed, and overhead truss bridges with 320-m spans, and concrete box and steel haunched plate girder bridges of 220 and 280 m span, respectively. A 300-m span steel overhead arch rounded out the field. All three 320-m spans were thought to be up to 15% less expensive than the 600-m suspension span, assuming the trouble-free footing of seven channel piers to depths of up to 30 m. Yet the suspension spans required only two shallow piers. A lack of suitable bedrock and exposed working conditions convinced Ostenfeld & Jønson to further examine the 600- and 670-m span suspension bridges.

The 600-m span featured 220-m side spans, and the 670-m span—with 250-m side spans—offered improved navigation safety and placed the Jutland anchorage on land. Although more expensive, this larger scheme was also considered aesthetically superior, a factor given strong weight in the evaluation. Ship collision was an important consideration given the curved navigation channels and strong currents, and this led to what has been claimed to be the first use of such analysis in long-span bridge design. This work eventually confirmed the suitability of the shorter 600-m span. Danish authorities endorsed this scheme in November 1961, opting for a six-lane crossing costing between \$22 and \$28 million. So strong was the desire to build the bridge that funding had been appropriated in the summer of 1960, before any design and cost had been fixed. Legislation was passed in March 1962, and Ostenfeld & Jønson then began a 2 year program of detailed design.

Over this period, several innovations significantly altered the span's character. First, the side spans were lengthened to 240 m. More importantly, a dearth of bedrock had meant that the large concrete gravity anchorages would require support from an extensive system of piles. An unorthodox solution to this problem was devised that not only eliminated above ground anchorages altogether, but enhanced the span's overall appearance and economy. Each anchorage is buried, tucked beneath the approach viaducts about 170- and 190-m back of the Jutland and Funen cable bents, respectively. The anchorage itself is merely a 1.5-m thick concrete slab, which in shape (and partially in function) resembles a massive ship's anchor that splays out to a 90-m width along its rear edge. It slopes up at a 10° angle over its 53-m length to match the inclination of the incoming cables, but each 13,500-m^3 slab is virtually unseen. Nor is it sufficient to restrain the 17,000-tonne pull of the cables; the frictional resistance of a 50,000-m^3 compacted gravel pad and the weight of the approach spans—transmitted via an approach pier through the anchorage—help to resist sliding forces.

Caissons resting on a somewhat plastic Lillebælt clay had been initially planned to support the 118-m tall main towers. Yet foundation conditions were unfavorable. In fact, one caisson on the 1935 span had dropped almost 28 cm in 20 years; worse still, some piers had tilted as they descended. To avoid such pratfalls, engineers used the concept rejected for the anchorages—caissons resting on piled foundations. The Jutland pier sits atop 211 concrete friction piles—its Funen counterpart sits atop 220—rammed to depths of up to 49 m. More than one-third are inclined to distribute tower loads. Parts of the work would, however, prove to be unexpectedly difficult. Piling was badgered by an unusually frigid January of 1966, and ice was soon lodged in the 200-m trestle connecting the piling frame for the Jutland tower to shore. The subsequent removal of the piling frame to the safety of shore effectively stopped further work. Then, when the ice did break apart, parts of the trestle were damaged. Pile driving did not resume until early March, but by early December 1966, all piles had been driven into place.

Meanwhile, a partially complete concrete Jutland tower caisson had been built in a dry-dock excavated near the bridge. It was launched in September 1966 and was then placed on a temporary platform in 12 m of water, where its 21-m high walls were completed. The Funen tower caisson was then cast. The Jutland tower caisson was placed in April 1967, but the August 1967 journey of the Funen caisson required two towing attempts to overcome strong winds. The piles were then concreted to the 56- by 22-m caisson bases—which quickly taper to 46 by 10 m—in 19 m of water. This plan required the now-unusual activity of underwater work under compressed air. No Danish regulations governed the practice, and in the end, a West German firm undertook the meticulous 2-month job. Two prestressed concrete blocks and a connecting slab were then poured to serve as bases for the tower legs.

The Funen tower began its ascent in October 1967, its Jutland twin the following March. Spurred by Norwegian successes in smaller spans and by the French in the Tancarville Bridge, engineers had added concrete as an option to the steel towers originally envisaged. Each 14,000-tonne tower in concrete would overwhelm the only 1,500- to 2,000-tonne weight estimated for each tower in steel. Yet with little steel manufacturing capability of its own, Denmark would have had to import steel at relatively high cost, a factor that overcame concrete's longer construction period and its need for stronger foundations. Contract bidding verified this thinking when concrete proved the cheaper of the two by almost 30%. The towers' simple appearance masks a maze of steel reinforcement within their inclined legs and large amounts of prestressing wire in the two portal braces.

Both towers were essentially complete by November 1968. Contractors had been offered in situ spinning and locked-coil strands on which to bid; Ostenfeld & Jønson were aware of Bethlehem Steel's work with prefabricated parallel wire cables but considered them untested. European familiarity with spinning was minimal and primarily the preserve of the British. It therefore came as a surprise when, in late 1965, most French and West German firms specified in situ spinning in bids for the steel superstructure. The leading bids appeared to confer little economic advantage of one method over the other. Although locked-coil cables required 10% more steel, they were adopted to help minimize the chances for weather-imposed disasters of the type that had plagued spinning across the Firth of Forth. Some 4,000 tonnes of British wire were supplied to Monberg and Thorsen, the Danish firm awarded the $14 million superstructure contract in November 1965. Fifty-five locked-coil strands were bound with six smaller strands to form a 58-cm diameter cable; between five and seven strands were draped across the Lillebælt daily. Steel rods prestressed into the anchor slabs restrain the main cables, a European practice that varied from American eyebar connections. Aided by long backstay spans, the Lillebælt's 1,500-m long cables exceed the Tancarville and Chesapeake Bay bridges' helical strand cables by a large margin. They will likely retain their distinction as the world's longest locked-coil cables for many years, although one disadvantage—a susceptibility to stretching—had dropped the girder at midspan by 70 cm by the early 1990s.

The evolution of the Lillebælt's streamlined steel box girder in some ways paralleled that of the Severn Bridge's. A 6-m deep Warren truss that adhered to the 1:100 post-Tacoma depth-to-span maxim had been contemplated. Influenced by Kit Scruton's early aerodynamic studies, Ostenfeld & Jønson placed 1.5-m wide swaths of open steel grating along the truss edges and in the median and, like the Forth Road Bridge, adopted a steel battledeck to reduce weight. Surprisingly, wind tunnel testing on sectional models at the Norwegian Institute of Technology (NTH) under the direction of Arne Selberg revealed the longitudinal slots to confer no advantage; in fact, they lowered the truss's critical velocity and presented a concern for snow and ice blockage. This result in some ways duplicated the test results on America's Vincent Thomas Bridge, notwithstanding the different truss configurations. A modified truss design was subsequently prepared for tender—without slots.[1]

Spurred on by the promising results of Roberts' box girder tests on the Severn Bridge, Ostenfeld & Jønson then decided to explore its feasibility at the Lillebælt. Selberg was again engaged and, with the assistance of fellow Norwegian Erik Hjorth-

Hansen, subjected some 22 cross-sections to sectional model wind tunnel testing. Theoretical work in bridge aerodynamics by A.G. Frandsen also influenced the design. The result was a different shape from the Severn Bridge's trapezoidal box—a flat, six-lane orthotropic box girder 3-m deep and 33.3-m wide (*see* Fig. 8-3). Wind tunnel investigations at the NPL in England on the Severn box girder had just been completed, and the two groups agreed to exchange information, which, as it turns out, independently corroborated the stability characteristics of suspended box girders. As a result of static model tests, Ostenfeld would tout a 45% reduction in bending and a 30% higher torsional rigidity in comparison to the truss design; this further separated vertical and torsional frequencies, thereby minimizing flutter potential.

The factors influencing the Lillebælt design were different from those at the Severn Bridge, and these influenced its shape. The girder would support six—not four—lanes. There were no sidewalks, whose cantilevering at the Severn Bridge marginally improved aerodynamic stability. The Lillebælt Bridge's larger width, smaller span, and high degree of stability also prompted the use of vertical instead of inclined suspenders. Cable stays were found to offer only marginal value, although lugs built into the tower saddles permitted their later use should it be deemed necessary. There were more subtle differences, including a rounded transition between the deck and the angled edges. Small wind vanes about 5-cm above the deck curve over this transition and deflect winds downward to reduce low-speed vortex shedding oscillation; consultations with aircraft designers helped to formulate this measure. A tall midspan triangular brace also connects the main cables to the box girder. One other feature pioneered in the Lillebælt Bridge was the dehumidification of the box girder and anchor chambers to reduce steel and cable strand corrosion (this latter malady would cost millions of dollars to remedy in the Brooklyn and Williamsburg bridges). This feature proved so successful that subsequent spans of the firm's design eschewed even the precautionary interior painting applied at the Lillebælt Bridge.

One controversial aspect of the design was Fritz Leonhardt's claim at the 1966 Lisbon Symposium that the Lillebælt was changed "after his intervention." Leonhardt had met with Ostenfeld in Copenhagen, Denmark in November 1964, but this followed an August recommendation to offer both truss and box girder designs for construction tenders. Leonhardt's association with German contractors—potential bidders on the span—had prevented Ostenfeld from revealing the strategy. He and Jønson (1966) stated emphatically at the Lisbon Symposium that Leonhardt "had neither direct nor indirect relation to the design of any part of the new Lillebælt Bridge." Frandsen, who had also attended the meeting, would echo their remarks.

Construction bids reflected those at the Severn Bridge. Eight percent lighter and with simpler fabrication, the box girder was bid at a price some 20% less than that of the truss. The box girder was selected, but Monberg and Thorsen would use different methods to transport the box segments. Fabrication occurred more than 180-km from the bridge site, and so to avoid prolonged salt-water immersion of the boxes, the 270-tonne, 24-m long units were shipped on a specially built catamaran barge. Lifting began in the autumn of 1969 and, except for a 6-week hiatus forced by severe winter weather that clogged a normally free-flowing channel with ice, was completed the following summer. Overland side span boxes were hoisted from the barge and progressively winched into place via succeeding suspenders. After paving and clean-up, the $38 million bridge was opened in October 1970.

A strong tradition of aesthetics in Danish public works produced an elegant structure. In fact, Danish law required the design's blessing by an aesthetics board. As a result, close cooperation between engineers and architects characterized the design's development. The results are impressive; the buried anchorages combine with an almost seamless transition between suspended and approach spans to produce an unbroken line between abutments. Textured approach piers enhance an element normally considered of secondary aesthetic importance. The slight inclination of the main tower legs and their clean integration with the portal beams were obtained at no additional cost. With the use of concrete came simplicity—the cruciform leg shaping commonly associated with steel towers was absent. Finally, the sweep of the cables above the deck at midspan—even with the large cable clamp—adds to the airiness of the bridge.

The Lillebælt Bridge provided critical knowledge by which to turn sights to Denmark's much larger Store (or Great) bælt. The firm of Ostenfeld & Jønson, renamed COWIconsult in 1972—the year Ostenfeld retired—would become intimately involved in this scheme as it acquired momentum in the second half of the 1970s. Yet neither founder would live to see the dream realized. Ostenfeld (born in 1900) would die in 1976, and Jønson (born in 1907), 4 years later. Even then, greater political and economic barriers would frustrate overcoming the largest physical obstacle to Denmark's unification. Other bridges would emerge before this great channel was finally spanned.

11.2 SPANNING THE BOSPORUS AT ISTANBUL, TURKEY

The first of these spans would link Europe and Asia at Istanbul, Turkey, where the strait known as the Bosporus narrows to as little as 1 km. One of the world's busiest waterways with immense historic, economic, and strategic importance, this 22-km long channel links the Black and Marmara seas and provides the sole outlet to the Mediterranean for the southern republics of the former Soviet Union. Yet as much as it facilitated waterborne commerce, the Bosporus severed Turkey, greatly inhibiting efforts to both modernize the larger Asian part of this nation and to balance the growth of Istanbul.

The Bosporus had been spanned once before. Persian ruler Darius the Great—through the Samian architect Mandrocles—is said to have lashed 200 to 300 boats together in 512 B.C. at Kalkedon to invade Greece in his war with the Scythians; his forces later retreated across the same bridge in the face of a Greek onslaught. Not until the latter part of the 19th century did several crude schemes emerge, but all proved beyond the scope of contemporary engineering. One German proposal for a multispan, chain suspension/arch bridge would give inspiration to the modern cantilever. The first half of the 20th century brought periodic spates of interest. A fantastic 1,000-m span reinforced concrete arch inspired by Eugene Freyssinet, the great French pioneer in prestressed concrete, was promoted in 1930. That decade, the city of Constantinople (now Istanbul) solicited proposals and favored an outlandish American-Anglo scheme for a 29-m wide three-level suspension bridge of about 800 m span. One level was reserved for rail traffic, a second for a road and tramway, and the third for a pedestrian promenade lined with shops, cafés, restaurants—even cinemas—all offered at an impossibly low cost of $10 million.

World War II reconfirmed the Bosporus as a major obstacle to land transport, and international interest in a crossing was revived once peace was achieved. In 1953, a suspension bridge described as second in span only to the Golden Gate Bridge was proposed by West German steel giant Krupp; by late 1953, the signing of a $68 million contract appeared imminent for a span with an anchorage-to-anchorage distance of 1,500 m and a 65-m vertical clearance. Hoped-for international financing to cover half its cost never emerged, and the project died.

Between 1950 and 1954, traffic on the ferries crossing the Bosporus almost tripled. In 1956, the Turkish government retained the American firm De Leuw Cather to examine national transport needs, including a Bosporus crossing. Two locations were studied, and a line just north of Istanbul was selected. Tunnels and bridges were examined—two bridges in detail—with the aid of the Baltimore, Maryland firm J.E. Greiner. One, a 292-m span overhead cantilever truss, required deep water approach spans. A four-lane suspension bridge spanning 675 m with 262-m side spans halved the number of deep water piers from four to just the two supporting the towers, which would stand more than 100-m offshore in water more than 30-m deep. The diagonally braced towers funneled a generously deep 8.5-m truss with slots of open steel grating fashioned over its 19.5-m width as in the second Tacoma Narrows Bridge. Tunnel anchorages would be burrowed into the steep slopes on either shore—that in Asia behind a 61-m backstay span. Although the suspension and cantilever bridges were found to be roughly equivalent in cost, the suspension bridge won out, largely on the basis of improved appearance and fewer deep-water piers. A May 1956 report laid out an ambitious 36-month construction schedule from early 1957 to 1960. With ever-increasing traffic, the span's $51 million cost would be recouped by 1977.

The government accepted the recommendation and sought an engineer to prepare detailed designs. This prize fell to David Steinman, who, supposedly without any overtures on his part, was retained in October 1957. A preliminary design released the following month revealed his stamp, but with a major surprise—the 132-m tall portal braced towers, which very much resembled those of the recently completed Delaware Memorial Bridge. It is curious that in the twilight of his career, Steinman would embrace a motif synonymous with Othmar Ammann's bridges and that Steinman himself had never before used. Steinman also altered the truss, realigning secondary members and reducing its depth to 7.5 m (and widening it marginally to 20 m). Air gaps would also separate the four lanes and two sidewalks from the truss edges. Finally, Steinman added five wire stays radiating from the tower/deck juncture to points along each main cable over each of the main and side spans for additional aerodynamic and earthquake resistance. These stays, of course, reflected Steinman's approach to wind stability, and he had also included them in a span proposed in 1950 for Italy's Messina Strait (*see* Chapter 16). Construction was now estimated to begin in the summer of 1958, with opening as early as autumn 1961.

This configuration would not last. Borings revealed the rock gorge underlying the Bosporus to be wider than originally thought, requiring the tower piers to be footed below the already considerable 51-m depths envisaged. Steinman surmounted this obstacle in final plans submitted in 1959 by lengthening the main span by almost 40% to 942 m. The tower piers now stood about 61 m offshore, which also placed them further from the busy shipping lanes. The side spans shrunk commensurably—by 55 m to a length of 207 m—and the towers grew by 26 m to a height of 158 m to account for the longer span. The truss, too, was altered—again—its depth now 9.4 m, although the 50-m navigational clearance and 1:100 depth-to-span proportions of Steinman's initial design were retained. The arrangement of secondary members was also revised to a cross-section almost identical to Steinman's recently completed bridge across the Straits of Mackinac. Studies revealed that the tower stays and roadway slots could be dispensed with; Turkish authorities had not favored the latter treatment. Tunnel anchorages were also eschewed in favor of above-ground gravity anchorages, and the Asian backstay span was eliminated.

Detailed plans were completed to the point that construction contracts could be advertised, but the project became embroiled in the 1960 national elections. Its $50 million cost, to be financed through American banks, became an issue, and in May, the new administration shelved the project. Steinman died that October; Mackinac would be the last large suspension span he would see built. Somewhat ironically, an image of his Bosporus span graced the cover of his reverential biography, published the year before his death.

Others had also attempted to span the Bosporus during this period. In 1957, French concerns sought to build a $35 million suspension bridge with a span in the 610-m range. In 1958, the West German firm Dyckerhoff & Widmann achieved more inroads with an unusual scheme downstream of the American alignment. Ulrich Finsterwalder, a pioneer of balanced cantilever construction in prestressed concrete, devised the design. Finsterwalder proposed a unique marriage of cantilever and primitive suspension concepts in a three-span bridge with spans of 450, 600, and 450 m. Two large prestressed concrete cantilevers—40-m deep at the piers—extended outward 150 m and supported a four-lane roadway atop its upper "chord." In the 300 m distance between the cantilevers of the large central span, the deck hung of its own accord in the form of an incredibly thin, 26-cm deep prestressed concrete slab containing suspension cum prestressing cables—a "stress ribbon." In profile, the deck undulated—dropping by more than 11 m between the massive piers—although it still maintained a 50-m midspan navigational clearance.

The Turkish government was intrigued by the scheme and established a commission to study its merits. Two areas of concern were wind and foundations. Finsterwalder's extraordinarily thin deck had included road shoulders of open steel grillage, and although no wind tunnel studies were performed, calculations suggested its aerodynamic stability to exceed that of the Cologne-Mulheim Bridge. Nonetheless, the commission recommended changes, and Finsterwalder subsequently increased the ribbon's depth to a still-impossibly thin 31 cm. A second commission reiterated concerns about foundation depths and urged an increase in the main span to 800 m. This increased the sag of the stress ribbon excessively, and although Finsterwalder convinced the commission to abandon this option, the scheme ran afoul of the Turkish Council of Architects and Town Planners over its appearance. Finsterwalder then revised the design yet a third time. He reduced the depth of the cantilevers at the piers by 50%—to 20 m—by reducing span lengths to 396, 408, and 396 m. He also reduced the sag of the central 190-m span stress ribbon to just 1.5 m.

Finsterwalder's efforts, however, were to no avail, for abandonment of the Steinman plan also signaled an end for its competitors.[2] Dyckerhoff & Widmann had also offered another daring Finsterwalder design—a 1,000-m span monocable suspension bridge with unsuspended side spans. The slender box girder was formed from prestressed concrete and was hung from a single plane of vertical suspenders along the span's median. A model of the proposal was built but the scheme went nowhere. Finsterwalder died in 1988 at the age of 90, having submitted one of his last proposals in the 1988 competition to rebuild New York's Williamsburg Bridge.

Although a crossing of the Bosporus remained officially dormant, it was not entirely so. In the mid-1960s, German engineer Max Herzog developed preliminary designs for a unique 1,000-m span suspension bridge with 250-m side spans for the Turkish Ministry of Public Works. Four lanes and two footpaths sat atop a 24.5-m wide lozenge-shaped orthotropic box girder, unlike the trapezoidal cross-section used in the Severn Bridge. For a girder depth of just 2.9 m, Herzog calculated a critical velocity of up to 400 km/hr, more than twice the 162-km/hr maximum recorded in the Bosporus. Steel tonnage was impressively low too—the 170-m tall towers required only 3,000 tonnes of steel each, and the box just 12,000 tonnes—allowing two main cables just 60 cm in diameter. Herzog claimed live load capacity per meter of span to be three times that of the Verrazano-Narrows Bridge at a cost of just $41 million—impressive figures if true. His scheme went nowhere.

The hiatus did nothing to quell rampant traffic growth or alleviate the congested ferries, and in 1967, the government launched new studies. Technology had now advanced to the point that Steinman's 1960 design was shelved; Freeman Fox & Partners, Ammann & Whitney, and Steinman, Boynton, Gronquist & London were approached for new proposals. Perhaps disillusioned by its prior experience, Steinman, Boynton, Gronquist & London declined the invitation. In any event, all plans submitted in October 1967 followed the 1960 Steinman alignment. Ammann & Whitney proposed a truss-stiffened suspension span patterned along the Verrazano-Narrows Bridge with a six-lane upper deck and a lower level designed to support six future lanes. Roberts of Freeman Fox & Partners adapted the epoch-making Severn Bridge design in a $25 million, six-lane box girder suspension bridge of 1,100-m span. Knowledge of subsequent (but unforeseen) traffic growth might have swung the pendulum the other way, but Ammann & Whitney's proposal was rejected as too costly and, in January 1968, Freeman Fox & Partners was selected to design the span. Ammann & Whitney engineers later exonerated themselves when a second span appeared necessary—a lower deck, they said, would cost a fraction of the $90 million required for another box girder suspension bridge (Cohen *et al*, 1980).

One economy of the Freeman Fox & Partners' design derived from footing the towers on land. Studies suggested major cost savings and a reduced construction period of 3.5 instead of 5 years. The side spans would be unsuspended and structurally independent of the main span. This configuration, not used in a long-span bridge since the Ambassador Bridge, allowed the side spans to be built quickly and provided construction access to the main span. It also reduced bending stresses in the towers induced by traffic loads, making possible further weight reductions there. In all, the use of unsuspended side spans was estimated to save $1.2 million.

Reinforced concrete had been considered for the towers, but inferior earthquake resistance—an important consideration—and a longer construction period favored the adoption of hollow steel box towers similar to those at the Severn Bridge. The Bosporus towers are wider and taller on account of the six-lane deck, longer span, and higher navigational clearances.[3] Tunnel anchorages were also considered, but rock depths were too great. The steep slopes did, however, allow gravity anchorages to be buried; they incorporate the type of prestressed anchoring system used in the Forth Road Bridge. Freeman Fox & Partners had also designed a 6-m deep Warren truss with a steel battledeck and—unusually—inclined suspenders. The heavier truss required stronger towers, thicker cables, larger anchorages, and steeper grades to achieve a 64-m navigational clearance. It was hoped, however, that the truss would attract American contractors whose experience might generate time and cost savings that would defray increased material costs. As it happens, no one bid on this design.

By the early 1970s, delays of 48 hours for trucks and up to 6 hours for automobiles trying to cross the Bosporus were not unusual; fog could stop the ferries altogether. As a result, Turkish authorities dictated an extremely tight 42 month schedule to build the bridge. Freeman Fox & Partners completed detailed designs in a frenetic 9 months, aided by experience with the Severn Bridge, the extensive use of computers, and foundation data from the 1959 Steinman proposal. Roberts had retired in 1969 (he would die in 1977), and William Brown assumed responsibility for the project, which in its final form became a 1,074-m main span flanked by unsuspended side spans of 231 and 255 m. Its 3-m deep steel box girder incorporated welding improvements and was even slightly thinner in proportion to span length than the Severn Bridge's (*see* Fig. 8-3). In fact, the resulting depth-to-span ratio of 1:358 gave the Bosporus the thinnest proportions of any long-span suspension bridge built to that time Yet its greater six-lane, 33.4-m width offered improved aerodynamic qualities over its predecessor. Testing of 1:50 scale sectional models of a 100-m length of box girder at the NPL wind tunnel in Teddington, England, revealed a critical flutter velocity of 216 km/hr, far greater than the 162 km/hr required.

Difficulties in completing the complex financing arrangements stalled the advertisement of construction contracts until June 1969. Further delays were encountered when the bids, received in September from European and Japanese consortia, included conditions the government could not accept. Four months of protracted negotiations with an Anglo-German joint-venture comprising the Forth Road and Severn bridges veteran Cleveland Bridge & Engineering (for the towers, cables, and box girder) and Hochtief AG (for concrete work) finally produced a $33.4 million contract in January 1970.

Completion was still planned for March 1973, now allowing only 36 months for construction. Efforts were set back almost immediately when excavations for the Asian tower indicated a need for deeper footings; tower work was also delayed by the late delivery of erection equipment. Even the delivery of tower steel was complex—panels fabricated in Sicily, Italy required special trains and the temporary removal of overhead electricity lines for the 15-km journey to port. Onsite, separate cranes crawling up each 165-m tower leg replaced the traditionally-used creeper cranes; each unit weighed half those used at the Severn Bridge but possessed twice the lifting capacity.

The spinning of the two 58-cm diameter main cables introduced major changes. Operations were completed in a remarkably short period of 3.5 months ending in October 1972. Freeman Fox & Partners had sought to use prefabricated parallel wire strands; the British firm Dorman Long had, of course, developed its own system in 1965 at the same time as Bethlehem Steel. Confidence in the method had developed rapidly, in spite of having been used only in the Newport Bridge and planned for Japan's 712-m span Kanmon Bridge (*see* Chapter 14). However, Cleveland Bridge & Engineering and Hochtief AG's bid for spun cables was $1.7 million cheaper, largely because much of the spinning equipment (including the catwalks) used at Quebec's Pierre Laporte could be adapted for use at the Bosporus. Otherwise, prefabricated parallel wire strand cables might have been used; British contractors would not be introduced to the new method for another quarter century.

Each main cable contains 10,412 wires, with the 548-wire strand size being the largest used since the Transbay Bridge; four additional 192-wire strands were spun into each side span cable to take on higher tensile forces. Several innovations sped operations. Spinning wheels moved almost continually because of the development of an integrated reeling/unreeling system that took the cable wire from the bobbins and passed it directly to an unreeling machine that then fed the wire on to the spinning wheels. This new system reduced the splicing of cable wire to a one-minute operation from one of up to 15 minutes under the formerly separate operations, which had also required spinning stoppages. The system also saved a huge amount of space and reduced machinery costs to about one-fifth those at the Firth of Forth, not even accounting for inflation. Spinning wheels often covered 700 m per minute and averaged 560 m per minute, half again as fast as on the Severn and Pierre Laporte bridges. A sophisticated signal system quickly identified foul-ups and helped to bridge the language gap between British contractors and the local Turkish workforce. The spinning performance was exceptional, and it beat an already tight deadline.

Erection of the main span's 60 box girder sections, each weighing 136 tonnes, began in late 1972. The discovery that the salt-water immersion of the Severn Bridge's boxes had severely deteriorated their antirust paint prompted the use of barges for transport. Although assembly of the main span progressed without incident, operations over the side spans were not so fortunate. In February 1973, a fierce storm swept six of the unerected viaduct spine box girders into the Bosporus from a quay where they had sat in storage. The boxes, weighing up to 40-tonnes each, sank in about 10 m of water. They were subsequently fished out of the water by floating cranes but were found to have suffered a severe beating, with considerable denting and tearing. Yet after just 3 weeks of repairs, the boxes were ready for hoisting into the bridge, avoiding an otherwise 3-month delay for remanufacture.

By this time, the planned March 1973 opening was now unattainable, but in mid-August, the bridge was sufficiently complete to allow traffic to cross. Turkish authorities delayed formal opening ceremonies until October 30 to coincide with the 50th anniversary of the founding of the Turkish Republic by Kemal Atatürk (for whom the bridge is officially named). Nonetheless, completion of the span in just 42 months by a previously unskilled Turkish workforce numbering 450 at the peak of construction constituted an impressive achievement. Added to this was a complex division of steel fabrication proportioned to international loan support—tower steel was fabricated in Italy, deck units in Italy and Great Britain, and the main cable wire in West Germany. The bridge would cost $36.1 million—more than budgeted for—but was still remarkably inexpensive.

Ironically, the outpouring of enthusiasm during the span's opening celebrations almost quite literally proved its downfall. Pedestrians were given full title to the bridge a day before its opening to vehicular traffic. The crush of spectators proved beyond expectation, and tens of thousands poured onto the bridge at a rate of up to 2,000 people per minute. Soon, up to 100,000 people were crammed onto the span with 400,000 more waiting their turn (*see* Fig. 11-2). Swarms of people taking in the newly-created views began churning from one side of the span to the other, causing the box girder to sway and twist so that the tilting of light standards was clearly discernible to those on shore. Aghast engineers from Freeman Fox & Partners persuaded local police to stem the onrush, and the bridge, subjected to a density of people later estimated to exceed by three times its design capacity (30,000 pedestrians in addition to a full vehicular traffic load), emerged unscathed.

The Bosporus Bridge opened as Europe's longest span and the world's fourth longest. Its simplicity of line and slenderness contrasts starkly with Istanbul's fine-grained qualities, which might otherwise have been overwhelmed by a deep and more visually-complex truss-stiffened bridge. The Bosporus Bridge leaves a more powerful visual impression than the Severn Bridge, its dramatic setting enhanced by the single bold shore-to-shore leap of the main span. The towers, although similar to the Severn Bridge's, are in profile thicker, and they sturdily frame the main span despite lighter, thinner portal struts; the higher elevation of the box girder enhances this impression of strength. The tautly stretched side span cables communicate the enormous tensile forces supporting the main span, and the thin unsuspended side spans—supported by concrete columns of almost unbelievable slenderness—appear to almost float in midair. Standing astride a historic waterway littered with temples and mosques, the Bosporus Bridge provides a striking but fitting contrast between the traditional and the modern in Turkey, a contrast that emphasizes the span's modernity in engineering terms.

11.3 THE HUMBER BRIDGE, ENGLAND

As the Bosporus Bridge neared completion, excavations for what would be the world's longest suspension bridge—the Humber—were beginning. A long and frustrating history of false starts extended back almost a century before construction finally began in 1973. Yet the frustrations would not end here, for the acrimonious experience of building this bridge would leave scars that remain today.

Where the River Humber flows past the industrial center of Hull on England's northeast coast, fully one-fifth of the nation's river water is disgorged into the North Sea. Its endless deposition of huge volumes of silt has created ever-changing patterns of shoals and channels that have always challenged those navigating its shallow waters. Local folklore refers to the fording of the river in Roman times; this even occurred in the early 1900s prior to the dredging of navigation channels, which today funnel an abundance of waterborne commerce.

With the onset of the Industrial Revolution, Hull grew at a frenetic pace and

Fig 11-2. Bosporus Bridge: opening-day crowds in October 1973
PHOTOGRAPH COURTESY OF HYDER CONSULTING.

worldwide oceanic trade developed. The river, although facilitating shipping, also blocked land commerce. By mid-century, 70,000 people crossed the river annually on a ferry system that had plied the muddy waters between Hull and the south bank town of Barton since the 1300s. The ferries, however, proved unable to cope with a growing goods movement between London and Hull. A bridge here, where the river is about 2-km wide, was discussed as early as 1865 and again in 1867. Yet the first serious scheme—in 1872—involved a rail tunnel. An untested plan to sink tunnel sections from pontoons (a method widely used today) aroused opposition from local shipping interests, which no doubt feared losing a lucrative monopoly on cross-river traffic. Still, the scheme had broad local support, and armed with a 10,000-signature petition, advocates almost succeeded in obtaining London's blessing; a vote by the British Parliament the following year left the tunnel only one vote short of approval.

An 1,800-m long bridge emerged again in 1883, the year work began on the monumental Firth of Forth cantilevers to the north. A total of 32 truss spans of 46 m each centered two larger trusses each 76 m in span, all carrying rail traffic 27 m above the river. A rail tunnel also surfaced, but within a year, both projects had succumbed to the political influence of upstream towns. In 1912, another tunnel was proposed, but not until 1927 was serious consideration again given to a crossing when the city of Hull retained Sir Douglas Fox & Partners (predecessors of Freeman Fox & Partners) to investigate bridges and tunnels. Here, Sir Ralph Freeman proposed an 11-m wide, road-only bridge comprising a 275-m cantilever span flanked by a multitude of deck truss spans; rail was rejected, as it would have doubled an estimated 1930 cost of $10 million. Construction of the bridge was estimated to take 5 years. Notwithstanding the now-predictable criticism of shipping interests—which alleged that the multiplicity of piers would silt up navigation channels—legislation received the blessing of a British House of Commons committee that did, however, stipulate stringent requirements for shipping protection. It also appeared that the government would fund three-quarters of the span's cost. Construction was seemingly imminent in 1931 when economic crisis brought on by the Depression brought down the government. With its demise went the bridge. Local political squabbling also killed a 1932 tunnel proposal—the last, as it turns out.

The great American suspension bridges of the 1930s brought a single-span crossing of the River Humber within the realm of the possible. Freeman examined the design of the Golden Gate Bridge and, in 1935, devised an ambitious suspension bridge with a 1,372-m span that would not only dissolve arguments against his cantilever but stand as the world's longest. At that time, I.K. Brunel's 214-m span at Clifton remained Great Britain's longest suspension bridge, but Freeman now proposed a span almost seven times as long. Economic and political factors conspired against its construction, and the idea languished during World War II. Freeman then refined the design, which by 1946 had evolved to a half-through truss hung from diagonally braced towers and grounded by ponderous, classically detailed anchorages. Yet pressing post-war reconstruction demands and commitments for bridges at the River Severn and the Firth of Forth saw no action taken on this $26 million proposal.

In 1955, Gilbert Roberts of Freeman Fox & Partners proposed a new $39 million suspension bridge of identical span to Freeman's schemes of 1935 and 1946. The design, which drew on the aerodynamic research undertaken in connection with the Forth Road and Severn bridges, featured a truss similar to that later adopted at the Firth of Forth. Roberts' design served as a catalyst, for pressure soon mounted on the British Parliament to support the proposal. Parliament did create the Humber Bridge Board in 1959, but this victory was more apparent than real as the Bridge Board could not finance the bridge, estimated to cost $45 million. Moreover, Parliament viewed the bridge as providing only local benefit and refused to provide any funds.

At this time, construction had just begun on the Forth Road and Tamar bridges;

the Severn Bridge was in the final stages of design with construction virtually assured. Money, as always, was in short supply, but as the Severn Bridge's economy became apparent, Roberts performed radical surgery on his Humber Bridge design.

Revised plans released in April 1964 by the Bridge Board revealed a streamlined 3-m deep welded steel box girder hung from inclined suspenders. Two 161-m tall single-cell steel hollow box towers similar to those astride the Severn Bridge framed a slightly lengthened main span of 1,396 m. If the Severn Bridge's slenderness had been surprising, the Humber Bridge's 1:458 depth-to-span ratio—more than 40% thinner—was sure to startle. Steel tonnage of the suspended structure was reduced by 20% in comparison to the truss, and commensurate reductions in cable, tower, and anchorage weight saw overall costs plunge by almost 25% to an estimated $35 million. One unusual feature of the design was its marked asymmetry. The Hessle (north) tower dictated a short side span of 280 m, but the open water Barton tower required a 411-m side span to site its anchorage at a more favorable location just offshore.

These substantial cost reductions did little to stir the British Parliament's enthusiasm; prospects appeared even dimmer with a 1965 report recommending a moratorium on construction for at least a decade. In the interim, the Bridge Board continued planning and conducted soil borings throughout 1967. Two years later, in 1969, another study was released, and this time the news was good. Completion of the bridge by 1976 was urged to promote regional growth; its $55 million cost was to be recouped by tolls from the increased growth the bridge would spur. A century of frustration finally came to an end in May 1971 when a newly elected national government agreed to lend the Bridge Board 75% of the now-estimated $57 million cost; the Bridge Board would finance the remaining 25% on the open market. Completion was expected in 1976, a date that in light of subsequent events would prove grossly optimistic.

A planned 1972 construction start accelerated the preparation of detailed designs. Significant modifications to the 1964 design by Roberts were made under Bernard Wex, although the streamlined box girder and inclined suspenders remained. Wex lengthened the main span to a record 1,410 m and stretched the south (Barton) side span to 530 m, increasing an already pronounced asymmetry (*see* Fig. 17-1); this strategy placed the south anchorage on dry land, simplifying its construction. Given the long legacy of problems later encountered in sinking the Barton tower caissons, the decision proved a wise one whose consequences extended far beyond those contemplated at the time it was made.

A second significant change was the adoption of reinforced concrete towers in place of steel. Concrete suspension bridge towers were, of course, not new, but the Humber Bridge was on another scale altogether. Freeman Fox & Partners sought to use slipforming, a method used to build tall silos, chimneys, and building cores; only in Sweden's Älvsborg Bridge and America's Vincent Thomas Bridge had it been previously used in a suspension bridge. The method had been ruled out at the Bosporus Bridge in the late 1960s but was now estimated to offer substantial savings. A marginally lower tower height of 159 m was also achieved—in spite of the longer suspended span—because of a relatively low 30-m navigation clearance at midspan. The towers are, in consequence, unusually low for a span of such size (*see* Fig. 13-3). With the change in tower material came a fourth cross-brace. The last major departure from the 1964 design was the adoption of a deeper box girder. To achieve the required level of aerodynamic stiffness in torsion for such a long span (i.e., the separation of vertical and torsional frequencies), the box girder had to be almost 50% deeper than contemplated in the original scheme. It's 4.5-m depth was a compromise between increased torsional stiffness and additional weight and drag, yet it still provided an extremely economical alternative to a truss (*see* Fig. 8-3).

In spite of these measures, inflation had elevated the 1971 estimate of $57 million to more than $62 million in 1972, when detailed designs were completed. The early 1973 signing of the steel superstructure contract with British Bridge Builders (BBB—a consortium of the same firms that built the Forth Road and Severn bridge superstructures) jacked the cost up again, to $65 million. Yet the problems, as it turns out, were just beginning. Work on the bridge began—behind schedule—in March 1973 when John Howard & Company began its $18.5 million contract to build the foundations, towers, and anchorages. A well-jointed and shallow bed of chalk on the north shore provided an ideal footing for the Hessle tower and anchorage, where an 11.6-m thick concrete tower support slab was footed at a depth of 8 m. Trouble-free progress saw the two tower shafts topped out in May 1974, awaiting only the pouring of their cross-braces.

Yet the firm's experience in sinking foundations at the Firth of Forth and River Severn was no match for the obstacles it would soon face on the south side of the estuary. Two 24-m diameter reinforced concrete caissons were to support the south tower at a depth of 36 m on a bed of stiff but heavily fissured Kimmeridge clay, overlain by soft alluvium, boulder clay, sand, and gravel. In order to bridge these fissures, cellular bases were adopted, as they were for the Barton anchorage foundation. An ominous harbinger of future events emerged in the form of a nationwide labor dispute that delayed steel deliveries for the south tower access trestle. Once that issue was resolved, steel sheet cofferdams were rammed into the mud around each caisson site and filled with sand. Turbulent currents then scoured away much of the protection placed around the cofferdams. Three months were lost attempting to rectify the problem; only the dumping of 12,000 tonnes of chalk and sandbags reduced scour to manageable levels.

The caissons were built inside the cofferdams, and sinking proceeded on schedule, aided by a bentonite slurry that lubricated the steel cutting edge to reduce friction. Hard Kimmeridge clay was then reached sooner than expected and, this barrier slowed penetration to about 2 cm per day. Then, at a depth of 26 m, the west caisson struck an artesian well that striped away the bentonite skin. With the cutting edge wedged firmly in muck, the bentonite was impossible to replace. High-pressure water jets eventually did the trick, but to facilitate sinking, the caisson walls were extended, thereby adding weight. Some 6,000 tonnes of steel billets were also temporarily stacked atop the recalcitrant west caisson. This latter measure produced spectacular results, for within only hours the caisson had sunk more than 1 m.

By June 1975, the south tower pier was now a year behind schedule; the north tower had stood ready for cable spinning for an even longer period. In September 1975, it was decided to halt the west caisson 1-m short of its intended goal. Concrete work was finally wrapped up in March 1976, 3 arduous years after construction had begun. The exercise almost ruined John Howard & Company, which had to be supported by the Bridge Board to stave off bankruptcy. Moreover, these difficulties occurred during a period of rampant inflation, which by June 1976 had swelled the span's cost to more than $83 million. Critics began making unenviable comparisons with another period cash cow—the Concorde supersonic airliner.

Two 12-hour shifts laboring 7 days a week on the south tower benefited from the experience gained in slipforming the north tower. The builder—Tileman & Company—had built the Lillebælt's towers and completed both legs of the Barton tower in less than 11 weeks, this in spite of denser steel reinforcement to enable cable spinning to begin sooner with only the upper two cross-beams in place. Even so, all 124 box girder sections had been fabricated before a single loop of cable wire was draped across the estuary. A revised schedule now contemplated hoisting of the box girder in late 1977 and early 1978, with the bridge opening possibly late that year.

Prefabricated parallel wire strands for the two 14,946 wire[4] main cables had been considered, but in situ spinning was adopted. Spinning was planned for a 7-month period beginning in September 1977, but in fact 21 months would be required to complete the task. Rain, snow, and high winds that winter were persistent annoyances but nowhere near the magnitude experienced at the Firth of Forth. More problematic was low productivity. Daily quotas, although ambitious, were considered realistic, especially in light of the performance at the Bosporus Bridge by a largely unskilled labor force. Yet production invariably amounted to one-third the targeted amount. Worker morale deteriorated. Public impatience over the lack of progress did little to ease tension between the Bridge Board, contractors, and bridge workers. The controversy came to a head in the summer of 1978 when Freeman Fox & Partners refused to authorize a $1.5 million payment from the Bridge Board to BBB, whose retaliatory court challenge threatened to bring work to a complete standstill. Pay incentives did generate productivity improvements, but not until June 1979 were the cables completed. In the midst of this controversy emerged suspicions that the 1969 study that had opened the door for the bridge had exaggerated traffic projections to secure the span's approval; the subsequent construction delays and inflation that had bloated the span's cost fueled more criticism of its ability to pay for itself.

Erection of the box girder began in October 1979, its early stages notable for the ingenious use of water containers to help avert oscillation. Yet not even this final phase of construction was devoid of major incident. Box segments were raised in traditional fashion via hoisting ropes extending from reeling engines at the tower bases, up over the tower tops, along the main cables, over lifting struts supported by the main cables, and finally down to a waiting box. Once a box was lifted, the struts would be shifted to the next lift. Work progressed relatively smoothly, and by March 1980, almost two-thirds of the boxes were in place.

Then, during a routine late-night shifting of the lifting struts up the steeply raked cables of the Hessle side span, disaster struck. Without warning, one of the four 30-tonne lifting struts slipped off the cables and headed for the ground. As the unit fell, it broke in two. One piece bounced off two suspended deck sections and sheared the temporary connections holding them together, leaving them dangling precariously at grotesque angles from their suspenders. The other piece plummeted 76 m to the ground and smashed into an unlifted segment, causing considerable damage. Miraculously, no one was killed, although three bridgeworkers were injured. A gaping hole was punched into one of the two suspended sections, but within a week both had been winched back into place and their damaged .deck panels replaced. Further lifting was halted until the cause of the accident was determined, and recriminations among the various parties wrought additional delay. When work did resume, only nine more boxes were lifted before an April 1980 dispute over bonus payments imposed yet another shutdown.

In June, the last box was lifted, and welding began. Poor weather in late 1980 then prevented the completion of deck paving before winter. A mid-April completion date was then threatened by work restrictions around the towers during the dismantling and lowering of the four tower cranes in early 1981. Not until late June 1981 were the two central lanes of the bridge finally opened. Even this event perpetuated a history of discontent and delay when a bomb threat—a hoax, as it turns out—postponed the opening for an hour while the bridge was scoured. By the July 1981 opening ceremonies, the bridge was more than 4 years behind schedule, and costs had escalated to $180 million, more than three times the estimate of a decade earlier. Ironically, much of the publicity surrounding the Humber Bridge's construction had stemmed less from its record-breaking span than from its litany of technical problems, labor disputes, weather delays, cost overruns, and concerns over financial viability. This latter preoccupation would prove devastatingly accurate. By July 1983, the bridge's debt had risen to almost $290 million and, by 1993—even though traffic volumes had grown steadily—had reached a horrifying $650 million. A $90 million bailout was mooted while the debt remained capped, but today the span's finances remain perilous.[5] Ironically, by 1997, tolls had poured some $175 million back into government coffers, almost the amount it actually cost to build the bridge.

However acrimonious its construction, the Humber Bridge is largely an aesthetic success, in part because of idiosyncrasies—some with intended visual effect, others a consequence of unrelated factors (*see* Fig. 11-3). The marked asymmetry of its side spans appears logical given the tower locations. Unusually, the deck profile flattens over the main span, and the deck camber is more pronounced over the long Barton side span (although somewhat marred by a slight kink at the south tower). The box girder, although more than 60% deeper than at the Severn and Bosporus bridges, still presents a slender profile because of the lengthy main span, but also because two thirds of its cross-section is sloped, a byproduct of aerodynamic requirements; its flat bottom appears only as a narrow spine. The backward-raked and staggered panels of the south anchorage emphasize resistance to the cable's pull; its flanking curved southern concrete box girder approach spans are of similar depth to the suspended spans and afford motorists excellent views of the suspended spans before passing onto them. One other feature is the distinctive extension of the tower legs above the top portal strut, accentuating height in a span characterized by relatively low navigational clearances. The large circular caissons supporting the south tower are, however, less appealing. Preconstruction renderings featured the low rectangular profile of the north tower footings, but the extension of the caissons to aid their sinking altered the design to one whose bulk is in stark contrast to the small pedestals of, for example, the Mackinac Bridge.

The Humber Bridge remains a technical triumph—if not a financial one. It has also helped highlight a problem affecting many cable-stayed bridges—the vibration of

Fig. 11-3. Humber Bridge
PHOTOGRAPH COURTESY OF DAVID LEE PHOTOGRAPHY, BARTON-UPON-HUMBER, U.K.

its inclined suspenders from vortex shedding induced by wind and adhering raindrops. Concerns over long-term fatigue were sufficient to warrant the installation of special restraining clamps in the early 1990s.

ENDNOTES

1. Most of those who commented on the subject at the 1966 Lisbon Symposium viewed slots as beneficial, even in winter climates.
2. Finsterwalder tried again in 1961 with a 294-m span, 24-cm deep stress ribbon for the Zoo Bridge in Cologne, West Germany, which he anchored with huge 152-m long below-ground concrete blocks. He also proposed a 25-cm deep stress ribbon across Lake Geneva in Switzerland with three 460-m spans alternating with two 200-m spans. Neither span was built. The sole vehicular stress ribbon is a three-span, 150-m bridge across the Maldonado River in Uruguay.

   Since the 1970s, stress ribbons have become increasingly popular for pedestrian bridges in Europe, Japan, and, more recently, America. German engineer Jörg Schlaich and Czech Jirí Stráský have become noted practitioners of the art. Spans have reached 170 m. Perhaps the most bizarre stress ribbon of all is a water pipeline built high on a glacier in the Swiss Alps in 1982. The tower centering the bridge's 611- and 247-m spans had to be allowed to creep 50-m downslope with the glacier ice in a 3-year period, during which time the longer span would grow to 661 m and the smaller span would decrease by a commensurate amount to 197 m. A 10-m lateral migration had also to be planned for. At this time, the tower would then be lifted back upslope to repeat its 3-year journey.

   A variation in which the stress ribbon supports a number of piers of varying height (that in turn support the deck) is termed an underspanned suspension bridge. In form, it suggests an inverted arch bridge. Of a limited number of underspanned suspension bridges—in China, Norway, Japan, and Germany, the first—a 1972 Costa Rican vehicular span of 144 m across the Rio Colorado—is the largest.
3. One idea that did not see the light of day was an elevator that would treat visitors to panoramas of the city.
4. An additional 800 wires in two strands were spun into each of the steeply raked north side span cables to deal with higher tensile stresses.
5. The Humber Bridge was not alone. By 1990, the Severn Bridge was almost $200 million in arrears; the Forth Road Bridge, by contrast, paid off its construction costs in 1995.

Chapter 12

THE BOX GIRDER—PROBLEMS, POLITICS, AND PSEUDOBOXES

12.1 SEVERN UNDER SIEGE

Even before the Humber Bridge entered construction, the wave of optimism surrounding the Severn Bridge had begun to evaporate. Fatigue and corrosion, as Ammann & Whitney engineer Milton Brumer had predicted in 1965, would soon lead to the premature replacement of its inclined suspenders. Brumer's concern over less rigorous British loading standards, although incidental to the Severn Bridge's design concept, would reflect a much more profound issue. In fact, the question of design loads would place the landmark bridge in the center of an engineering debate that would ultimately see it undergo a fundamental but controversial strengthening.

The first harbinger of future events emerged in 1970 with a tragic string of construction collapses in three steel box girder spans in Milford Haven, Wales, in Vienna, Austria, and in Melbourne, Australia. A fourth bridge collapsed during construction at Koblenz, West Germany the following year. The worst failure involved Melbourne's West Gate Bridge over the Yarra River, a cable-stayed span of 336 m designed by Freeman Fox & Partners. In August 1970, a high-level approach span buckled without warning and collapsed, killing 35. The accident touched off a furor in Great Britain; it followed the collapse at Milford Haven—which had killed four—by only 4 months. Interim findings of Great Britain's Merrison Enquiry, called to investigate box girder design and construction, placed all large single-celled box girder spans under suspicion—including the Severn Bridge. These findings precipitated a June 1971 closure of two of the Severn Bridge's four lanes pending structural evaluations. The Merrison Enquiry's final report in 1973 fostered a simmering debate within Freeman Fox & Partners that would influence suspension bridge design more than a decade later.

A routine inspection in 1971 first uncovered fine cracks in the Severn Bridge's orthotropic deck stiffeners. Temporary bulkheads that provided buoyancy during the floating of box girder sections had not been removed in an effort to save time and money; their removal was now imperative, and any savings were more than lost. The box girder then underwent a $1 million strengthening in 1975 to meet new standards, but by 1977, two of the Severn Bridge's four lanes were again temporarily closed. This time, it appeared that the deck supports at the towers and anchorages could fail if traffic was concentrated on one side of the span. Expansion joints were then replaced in 1978. As if this were not enough, an inspection of the suspenders that same year revealed some wire breakage; bending stresses—as predicted by Brumer—were thought responsible. This discovery prompted the replacement of 49 suspenders. Further inspections in early 1982 then revealed wire corrosion of such a nature that two of the span's four lanes were for a third time slated for closure, this time during the Severn Bridge's notorious morning traffic jams, when traffic loads peaked. It now appeared that all 340 suspenders might require replacement at a cost of $9 million.

As public frustration mounted, Freeman Fox & Partners felt compelled to defend its design. The firm stated that the large increases in truck weights implicated as one of the problems could not have been foreseen at the time of the Severn Bridge's design. Yet some asserted that the suspender/deck connections, although permitting longitudinal movement, made no provision for lateral sway induced by wind. Freeman Fox & Partners had altered the connection design in the Humber Bridge, but Hellmut Homberg, the outspoken German designer of the rebuilt Rodenkirchen (1951) and Emmerich (1965) bridges and a pioneer of single-plane cable-stayed bridges, suggested that inclined suspenders acted more like truss members and therefore coped with loads for which they were not designed. Yet Homberg's prediction that the Bosporus Bridge might be similarly afflicted was not borne out; Turkish authorities stated that no such problems had emerged. Bernard Wex, a partner at Freeman Fox & Partners who had spearheaded the Humber Bridge design, stated that Homberg offered "no new knowledge" (O'Neill, 1982a). Yet in its design of a 608-m span steel box girder suspension bridge across the Panama Canal in the late 1970s (a span that was never built), Freeman Fox & Partners had reverted to vertical suspenders. Inclined suspenders, once described as "vital" for the Severn Bridge, were now considered "important" by colleague William Brown (O'Neill, 1982a).

This controversy paled in comparison to the next crisis. In June 1982, an evaluation by engineering consultant Flint & Neill raised concerns over the entire bridge; its towers, suspenders, splay saddles, and rocker boxes urgently required work. The main cables, piers, anchorages, box girder, and foundations would require strengthening—(although analysis of the towers during the Merrison Enquiry had found them to possess adequate strength). Huge increases in traffic loads were singled out as the culprit. The number of trucks plying British roads had increased 14-fold between 1962 and 1976, and up to 70% of the 32,000 vehicles crossing the Severn Bridge daily in 1982 were heavy trucks. Yet numbers represented only half the story. The Severn Bridge had been designed to handle trucks of up to 24-tonnes in line with the 1954 standard BS153. But 2 years before the bridge opened, in 1964, the standard was raised to 32 tonnes, and in 1982, it was increased to 38 tonnes (in 1999, it went up again to 40 tonnes and will soon increase to 44 tonnes). As a result, traffic using the bridge was thought to impose live loads more than double those for which the span had been originally designed.[1] Dr. Anthony Flint, whose firm prepared the analysis, had been involved in the Merrison Enquiry and had contributed to the strengthened 1978 code.

In one sense, the findings acquitted the Severn Bridge's revolutionary design—and, by implication, Freeman Fox & Partners—for many other spans were similarly afflicted. Yet the Severn Bridge's lightness made it particularly susceptible. Live loads constituted up to 30% of the total carried by its main cables, unusually high for any bridge. The span was described by bridge failure expert Bernard Clark as a "too perfectly designed structure with no margin for unknowns" (O'Neill, 1982b).

Then, in November 1982, another Flint & Neill study concluded that the Severn's towers could be threatened under (admittedly) extreme traffic loads. One option was to jack four small columns inside each single-celled leg to the cable saddles, as well as a change to vertical suspenders to relieve stresses and to improve restraint against longitudinal movement of the main span. Vertical suspenders would, however, reduce the span's critical velocity by 20% and require compensatory measures. Regardless, all suspenders would have to be replaced and the box girder extensively rehabilitated. In February 1983, the government announced its intention to repair the bridge at a cost of up to $45 million, pending a final diagnosis. Flint & Neill's final report in May 1983 confirmed past conclusions but brought more bad news. Most shocking was its conclusion that the towers had a 10% chance of collapsing in strong winds over a 120-year period. The adjacent cable-stayed Wye Bridge was also singled out for rehabilitation; stay replacement and additional stays could push the total cost of the refurbishment to more than $54 million. The firm reiterated its belief in vertical suspenders, and suggested counter weighting and anchorage reinforcement to improve safety factors against an almost impossible specter of anchorage overturning.

Freeman Fox & Partners was incensed by the findings. The firm had not been consulted in the work to-date and claimed to be rebuffed in its efforts to contribute to the analysis. It alleged that Flint & Neill's standard of safety—three times that for which the span was designed—was "plucked out of the sky" and "ludicrously high" (Anonymous, 1983a); it argued that every long-span bridge in Great Britain would be rendered deficient by this standard. The firm contended that Flint's scenario was tantamount to having 130 fully loaded 38-tonne trucks crammed on to the span, whereas it believed the bridge rarely carried more than a sixth of the loads it was designed to accommodate. Freeman Fox & Partners felt that although such "a freak extreme loading could occur once or twice in a lifetime," traffic control measures would be cheaper and rarely—if ever—required. The firm also felt that suspender corrosion resulted from water seepage; only a small number suffered minor fatigue and it felt wind—not traffic vibration—was responsible. Brown (1983) dismissed the threat of any toppling of the Severn Bridge's towers as "even more ridiculous."

Flint & Neill had retained Mott, Hay & Anderson to review its proposals. Yet Mott, Hay & Anderson's October 1983 assessment raised more concerns than even Flint & Neill's. It warned that heavy traffic had so weakened the bridge that high winds could knock the towers over "with little or no warning" and that in an unloaded condition the span might not survive a wind of 160 km/hr (Anonymous, 1983c). It warned that, "enough heavy vehicles to risk failure could pass onto the bridge in about three minutes or less." Flint & Neill viewed such possibilities as infinitesimal and downplayed the results, but morning lane closures became 24-hour fixtures until a strengthening was completed. The number of trucks on the bridge at any given time was also strictly regulated, and the span would be closed in high winds, offering motorists the alternative of a 50-km detour. Most urgent was the immediate need to replace 36 suspenders and to replace the remainder within 2 years.

Adverse publicity surrounding the Severn Bridge was reducing investor confidence in an already economically depressed Wales, and the pressure to act mounted. In February 1984, the government announced it would follow virtually every one of Flint & Neill's recommendations, save one. Six years and $55 million were to be invested to shed the span of its liabilities; 2 years would be required to firm up a rather tricky design, and up to 4 years were needed for the rehabilitation itself. The work would be meticulous—especially the jacking of four cylindrical columns inside each tower leg. The portal struts immediately above and below the deck would be buttressed with steel plates to alleviate the towers' wind load problem. Saddles at the towers and anchorages would be strengthened, webs added in the box girder near the anchorages, and welding repairs made in the remaining boxes. Only Flint & Neill's recommendation to replace all suspenders would await further study.

Freeman Fox & Partners reacted with incredulity. It labeled the program "a complete waste of money" that responded to "doomsday loadings" (Middelboe and Ferguson, 1984). Sir Alan Harris, a noted British civil engineer, commented sarcastically on the prospect of trucks lined up "nose to tail with a typhoon blowing" (Anonymous, 1984). Flint & Neill downplayed the loading issue, pointing out that only one-third of the budget was related to strengthening, with much of this destined for the cable-stayed Wye Bridge.

While the decision appeared to put some finality to the controversy, clamor for a second crossing intensified. The Severn Bridge's incessant traffic tie-ups brought mounting frustration. Its finely tuned aerodynamics also prevented the use of windscreens that would have allowed truck crossings in high winds.[2] In July 1984, the government launched a study of options, including a twinning of the original span. In the end, a six-lane, 456-m cable-stayed span was adopted. This new bridge was opened in 1996 on the southerly 5-km alignment at English Stones—the site of the first serious proposal to bridge the River Severn, in 1935.

After almost 3 years of design, a $54 million contract was awarded in April 1987 for the Severn Bridge's strengthening; including the Wye Bridge and other work, the total cost was now $65 million. The Severn Bridge's inclined suspenders would remain, largely because of the complex changes vertical suspenders posed for the deck and cable band connections. Nonetheless, all 340 suspenders—including the 100 replaced during the past decade—would be superseded by ones of larger diameter. More than 22 km of deteriorated welds in the roof of the box girder would be ground out and rewelded—all at night when traffic was light. An additional steel wall would also strengthen each box girder segment. Most meticulous was the jacking of four 41-cm diameter steel cylindrical columns into each 120-m tall tower leg. Space was horribly confined—6-m lengths of column were inserted through a small door at the tower base and welded to the bottom of previously jacked sections that ultimately extended to the cable saddles. Then, without stopping traffic, the new columns and cable saddles were jacked up 15 mm to relieve the original tower of 2,000 tonnes, or 22% of the dead load they formerly carried, and one-quarter of the span's live load.[3] The tower saddles were also braced. In the Wye Bridge, each pylon was extended to acquire a second set of cable stays.

By the time work ended in 1991, the cost of strengthening the Severn and Wye bridges had exceeded $120 million. There were human costs as well. As work drew to a close in the late summer of 1990, a deck scaffold sheared loose near the Aust tower, carrying two workers 50 m to their deaths; a third miraculously survived the plunge.

In spite of its huge cost, the rehabilitation added a mere 2%—or 750 tonnes—to the Severn Bridge's weight, while tripling its load-carrying capacity. Even more remarkable is that most work occurred sight-unseen, with only minimal disruption to the 40,000-plus vehicles crossing the span daily. Virtually no alterations were made to the appearance of this landmark bridge, save windscreens near the towers and a coat of white paint that is designed to reduce thermal stresses. Most responsibility for the Severn Bridge's resurrection lay, however, with its main cables. They had not only weathered the crisis unscathed; Flint & Neill found them able to withstand almost three times the maximum loading conditions envisaged at the time of the span's design.

The approach that had guided Flint & Neill's analysis of the Severn Bridge and the remedial action taken to shore up the bridge had emanated from Anthony Flint's work with British bridge design codes following the Merrison Enquiry. It departed radically from traditional bridge analysis. The commonly used permissible-stress method applied a single safety factor to the overall load the bridge would sustain. This single factor accounted for the unchanging weight of the bridge—its dead load—and the more variable live loads imposed by traffic and wind.

Flint felt the approach failed on two counts. First, the higher degree of safety required for less easily predicted live loads was imposed over the entire bridge, when in fact dead loads—which form the bulk of the weight a bridge must carry—are easily calculated and don't fluctuate as much. As a result, a bridge design could be overly conservative. Second, because live loads do fluctuate so widely, certain parts of a structure could be stressed beyond capacity if designed to an "averaged" loading standard. Flint sought to separate dead and live load safety factors, reducing that for the former to a factor (usually up to 1.2) greater than the weight of the bridge, but using higher live load factors for wind and traffic loads that more accurately reflect their individual variability. The approach—called limit-state design—was felt to reflect the actual stresses imposed on a bridge and allow engineers to design bridges to a more consistent level of safety.

The approach was not universally supported. Freeman Fox & Partners' Michael Parsons, who had worked on the Forth Road and Severn bridges (and would later stake a claim for pioneering the latter span's box girder) described this procedure as "taking metal away from the towers and cables and putting it in the deck" (Heywood and Birchall, 1987). Describing the suspended structure as not critical to the survival of the bridge itself, Parsons argued that it made sense to provide additional safety in those elements that were critical—the towers and cables (although his consideration of problems in suspended structures as being "minor" was not shared by all [Boyd, 1987]). Brown (1983) was incredulous that British authorities would use a method he described as "incorporating arbitrary factors as yet unproven or calibrated for bridges of this [the Severn's] type." Some engineers believed Flint's methods would actually add to costs. Parsons claimed that had the first Bosporus Bridge been designed according to Flint's criteria it would have been in greater peril of collapse during its unforgettable opening day crush. Parsons stated that event to have exceeded the design load by a factor of three, still within the bounds of safety for which the bridge had been designed under the permissible-stress method.

The Severn Bridge's woes are pause for reflection. Its innovative weight-saving design rendered the span particularly sensitive to unexpected increases in traffic loads and drew it into the larger debate over steel box girders. Flint himself acknowledged that Freeman Fox & Partners bore no responsibility for the problems that besieged the span, which was still resilient enough to endure a tripling of the loads for which it was originally designed. Moreover, at no time were its revolutionary box girder or hollow box towers—as design concepts—called into question. These features have successfully endured in spans designed by Freeman Fox & Partners and others, attesting to the foresight of Gilbert Roberts, Brown, and Parsons in developing the concept and battling tradition to secure its emergence in the Severn Bridge.

12.2 INTRIGUE ACROSS THE BOSPORUS

The Severn Bridge was not the only box girder suspension bridge designed by Freeman Fox & Partners to come under the skeptical gaze of American engineers. The firm's 1973 span across the Bosporus at Istanbul would also be criticized—not for any engineering deficiencies, but for an inability to cope with its phenomenal popularity. An almost Byzantine intrigue would surround efforts to build a second span, and the outcome would ultimately drive a wedge through the unanimity displayed by Freeman Fox & Partners in its defense of the Severn Bridge.

The 1973 opening of the Bosporus Bridge had inundated Istanbul with traffic. The effects of this onslaught on the fine-grained fabric of this historic city had been predicted by critics; the city's Chamber of Architects had labeled the span part of a "bridge trap" even before it opened and predicted a second bridge would be needed within the decade. Yet even this dire prognosis would prove conservative. Within 28 months of the span's opening, and 15 years before expected, the cost of the bridge—including interest—had been recouped by tolls. By 1977, traffic volumes were 10 years ahead of projections, and automobile ownership in Istanbul had doubled in just 4 years. Ironically, a mere 10% of Istanbul's (then) 4 million inhabitants used the bridge. Intercontinental truck traffic destined to and from the Middle East was the big problem. The rapid rise of petroleum prices in late 1973 produced a windfall for European contractors as the Middle East underwent a massive building boom; the Bosporus Bridge became a lynch pin in the transport network connecting the two regions. With intolerable traffic congestion, it was certain that a second crossing would also prove a financial windfall. Its allure is reflected in the frenetic maneuvering of authorities, designers, and contractors over the relatively short but turbulent period leading to its construction. Building the bridge, too, would generate controversy.

The first tangible move toward a second bridge came in October 1976 when the Turkish government retained Freeman Fox & Partners to investigate two sites for a crossing—one near the existing span and a second downstream near the Sea of Marmara. In the autumn of 1977, it was announced that a second span would be built—although its location remained unresolved. Istanbul's growth had now made a span near the original bridge improbable, but opponents predicted the desecration of the most scenic and historic parts of the Bosporus for a favored site 5 km to the north. If uncertainty clouded the span's location, optimism surrounded its financial prospects; tolls were expected to pay for the estimated $52 million bridge in 1 year.

Freeman Fox & Partners' experience with the original span no doubt influenced its January 1978 appointment by Turkish authorities to design and supervise construction of the second. The expected similarity between the two bridges would accelerate design, although the first span's designer, Gilbert Roberts, would not contribute; he died on New Year's Day 1978—about the time of the design award—at the age of 78.

Some urgency was attached to the now $85 million, 1,166-m span. An ambitious schedule called for ground breaking on May 29, 1979, a date with decidedly political overtones—the 526th anniversary of Constantinople's conquest by the Ottoman Empire; construction was to take just 3 years. Freeman Fox & Partners took a more pragmatic view, commenting (Anonymous, 1978) that, "it would be breaking all records if people move as fast as that." The firm would be perversely rewarded for its prescience, for political infighting saw its design contract squelched not long after its award. Not until 1983—a year after the original completion date—was the scheme resurrected. That November, Freeman Fox & Partners again received the prestigious design assignment, but under controversial circumstances that for a second time snatched what seemed certain victory from the firm of Steinman, Boynton, Gronquist & Birdsall. The competition to secure the design contract—which included 14 km of approach roads—was intense. All five contestants were leading designers of suspension bridges: apart from Freeman Fox & Partners was the American firm of Ammann & Whitney, which, along with the Steinman firm, was seeking its first new suspension bridge commission in almost 20 years. Rounding out the field were Mott, Hay & Anderson, which had recently completed designs for a now-aborted crossing in Hong Kong, and the Danish firm COWIconsult (formerly Ostenfeld & Jønson), designer of the Lillebælt Bridge.

Ammann & Whitney's fee came in the highest—more than three times that of Steinman, Boynton, Gronquist & Birdsall—which was the lowest. Not surprisingly, Freeman Fox & Partners' selection generated some invective from Steinman, Boynton, Gronquist & Birdsall. The firm had undercut the Freeman Fox's bid by a large margin, yet Turkish authorities had disqualified it on a technicality. It then emerged that Freeman Fox & Partners would have been awarded the job the previous August without any competitive bidding whatsoever; only at the eleventh hour were bids requested, with the government said to hold "an extremely strong preference" for the British firm (Anonymous, 1983d). Although Steinman, Boynton, Gronquist & Birdsall charged that Freeman Fox & Partners had been allowed to lower its price, Freeman Fox suggested the government to be "exceedingly happy" with the first span. Amid (unrealized) threats of a formal protest by the American government and legal action by Steinman, Boynton, Gronquist & Birdsall, Freeman Fox & Partners began to sculpt its fifth steel box girder suspension bridge.[4]

At the time, Freeman Fox & Partners was vigorously defending its Severn Bridge design. Loading standards in the first Bosporus Bridge were only 10% higher than in the Severn Bridge, yet it had suffered none of the Severn Bridge's maladies.[5] Freeman Fox & Partners intended to adopt the same standard in the second span. In the event, any issue regarding design loads was fortuitously avoided by a new Turkish requirement that the deck support closely spaced 145-tonne tanks. This would now make the new box girder 50% heavier than the first Bosporus Bridge and exceptionally strong. Still, Freeman Fox & Partners intended to adopt many features of the original Bosporus Bridge—a six-lane cross-section, portal braced towers, and unsuspended side spans. Improvements would be made to the streamlined steel box girder and inclined suspenders. Yet resemblances began to fade once design began. First, the deck was widened to eight lanes. To ease construction, the towers were planted atop the steep slopes of the Bosporus to create a span of 1,090 m, even though the strait itself is only 800-m wide. This restyling produced a peculiar configuration in which the tower legs do not extend below deck level at all.

William Brown—a partner at Freeman Fox & Partners since 1969—was slated to lead the design. His performance at the first Bosporus Bridge in the early 1970s had much to do with the Turkish government's preference for Freeman Fox & Partners in designing the second span. Yet although the first bridge was a technical success, Brown viewed the second as a platform for radical innovations in streamlined suspended structures. His work, beginning in 1975 with J. Roy Richardson of Great Britain's National Maritime Institute (NMI—formerly the NPL), had led to the concept of two smaller parallel boxes in place of a single large box (it was Richardson who had suggested altering the center of gravity of the Humber Bridge's incomplete box girder in changing winds through the alternate draining and filling of water bags along its edges to counter oscillation). Each box—supported by separate tower and cable systems but braced to one another—would be two-thirds the depth of a single box. Brown stated that weight reductions and simpler fabrication, transport, and installation would generate significant cost savings. But the most significant benefit of the split box was derived from its aerodynamic stability.

The concept of a split box was not new—Swiss engineer J. Ackeret had suggested the idea in the early 1940s following the Tacoma Narrows Bridge collapse. Now, however, Brown and Richardson pursued the idea more rigorously. The concept emerged from the observation that streamlined box sections are somewhat susceptible to uplift, behaving much in the manner of an airfoil. Cross-winds maximize lift forces about one-quarter across a box girder's cross-section (as Theodore von Kármán had surmised with the plate girder of the Tacoma Narrows Bridge immediately following its collapse). Counteracting the resulting torsional forces became increasingly expensive as spans increased—and prohibitively so for spans greater than 3,000 m, such as that contemplated by Brown for Italy's Messina Strait.

The breakthrough was the result of inspired thinking.[6] Aerodynamicists at the NMI (site of the ground-breaking work on the Severn Bridge) were subjecting Freeman Fox & Partners' design for the second Bosporus Bridge to wind tunnel testing. It was observed that slots in the deck reduced air pressure concentrations and, hence, increased the flutter critical velocity. Wider slots produced even higher critical velocities. It was a logical step to widen the slots to the point of having two separate—but rigidly connected—boxes. The transparency of the cross-section at its quarter-points would render air pressure concentrations harmless, meaning the parallel boxes together would require a significantly reduced level of torsional stiffness. Additionally, the lift forces on a parallel, downstream leading edge would balance those at the leading edge, thus reducing torsional forces. Numerous cross-struts would ensure the twin boxes acted aerodynamically as a single unit.

Brown now characterized the slotted decks on many truss-stiffened suspension spans as "crude arrangements" (Anonymous, 1983b). In these earlier spans, slots had been placed where permitted by traffic, rather than where pressure concentrations occurred. Wind tunnel tests suggested the new approach was both technically and economically attractive for extremely long spans, such as the 3,300-m span Messina Strait proposal and schemes to span the Gibraltar Straits and English Channel with multiple 2,000-m spans. Its advantages just brushed the span length envisaged for the second Bosporus crossing, but Brown predicted that for this bridge the concept could slice up to 20% off the cost of a traditional box girder design. Moreover, he predicted refinements would make the concept competitive for spans as small as 305 m. A split box would in fact be proposed—unsuccessfully—for Greece's Corinth Strait crossing in the late 1980s in a 1,220-m span with boxes just 2-m deep. At one time it was even suggested that the trouble-plagued Severn Bridge be transformed into a twin box structure. Richardson would take the concept further, envisaging catamaran-like structures with boxes separated by a distance of four to five times their width; he even stated (1981) that the need for torsional resistance in the parallel boxes was "entirely eliminated." Brown considered this concept to be of economic advantage in spans greater than 3,200 m.

Senior colleagues at Freeman Fox & Partners were not so predisposed to the radical concept; nor were, apparently, Turkish authorities. Others predicted twin decks would be overly flexible, suffer from fatigue cracks, and be uncomfortable to cross; the connection of the boxes was also seen to be problematic by some. Brown lost that battle, and then his tenure as design leader was truncated; he fell ill soon after the basic design of the single box structure was fixed and was replaced by Parsons, who said the firm had "stuck to what we'd done on all the other bridges" (Heywood and Birchall, 1987). The convention established in the first was thoroughly embedded in the second, although Parsons did check critical parts of the design using Flint's limit-state design approach. One other noteworthy change was a retrenchment from inclined suspenders. Freeman Fox & Partners stated that the increased box size and stiffness made this switch possible. It also stated that vertical suspenders were "less prone to fatigue," a tacit admission that the Severn Bridge's problems had influenced its design philosophy (Anonymous, 1986). Upon returning, Brown was presented with a fait accompli. His public objections to Flint's Severn Bridge findings, speculation of disagreement within Freeman Fox & Partners over the its preliminary design for the giant 3,300-m suspension span across the Messina Strait—which Brown headed—and rumored differences of opinion over the firm's direction all suggested an unhappy marriage. Yet Brown's skills and experience were considerable, and the Turks valued his presence. He was appointed to head the span's construction.

The design was virtually complete by September 1984. Building the bridge and some 200 km of approach road generated fierce competition among the 14 consortia bidding for the job, particularly in the area of financing. One offered its own design that would have relieved Freeman Fox & Partners of any further involvement. Yet the April, 1985 tender opening generated disbelief when a consortium led by Japanese firm Ishikawajima-Harima Heavy Industries (IHI)—with Italian and Turkish firms—bid $551 million for a project widely regarded to cost at least $800 million. The consortium had undercut the next lowest bid—from Bosporus Bridge veteran Cleveland Bridge & Engineering—by $120 million, even though the British price for the bridge alone was 10% less than IHI's $121 million figure. The selection of the IHI consortium that May touched off a war of words between Japan and Great Britain that embroiled even its prime ministers. British officials described the Japanese financing package as "foolish" (Barrett, 1985). Cleveland Bridge & Engineering remained hopeful that its experience with the first span would land it a subcontract, and British Prime Minister Margaret Thatcher even wrote to Japanese counterpart Yasuhiro Nakasone on behalf of the firm. But with Great Britain's attacks on its aid package, Japan intimated the firm's chances were remote. Thatcher did little to improve prospects when she said in reference to the bridge that Japanese trade tactics "stick in the gullet." The entire episode would prove ironic, for more British steel would ultimately find its way into the second bridge than into the first. Later, in October 1986, the Turkish government intimated that if Cleveland Bridge & Engineering developed an acceptable proposal to finance, build, and operate yet a third bridge, the firm would win an exclusive franchise without any competition. This offer was viewed in Great Britain more as a gesture of reconciliation than a serious proposal.

The span's cornerstone was laid on May 29, 1985, again on an anniversary of the Ottoman conquest of Constantinople. Brown was on site and in charge of the project. That November, Brown was still in Istanbul and still in command, but no longer employed by Freeman Fox & Partners; Brown had resigned from the firm. Turkish authorities had given him total control over the bridge's construction, and Freeman Fox & Partners was to provide staff—under Brown. By the fall of 1986, Freeman Fox & Partners' two senior site engineers were gone after "several warnings" over what Brown was quoted as describing their "negative approach to the project" (Heywood and Birchall, 1987). The final link with the firm Brown had served for almost a quarter century was now severed. Freeman Fox & Partners expressed concern over the ability of Brown's small team to cope with the daily travails of building the world's sixth longest suspension span and almost 30 km of approach road. Yet Brown sought to change what he considered a design orthodoxy that did not sufficiently account for constructability. He wasted little time in applying his ideas. Within a year he stated that "very little other than the road alignment is totally in accordance with the tender documents" and that a full year had been shaved off the span's construction schedule (Byrd, 1986).

Brown's changes did little to alter the appearance of the span, but they were substantial. Most visible was the replacement of the 25-m backstay spans with huge walled embankments extending to the towers. This change alone was estimated to save $1 million. Brown also made all tower steel of equal thickness, enabling each 105-m tall tower to be shipped from Japan in only eight prewelded sections. Once on site, the sections were simply stacked into place on top of one another in just 2 months by cranes planted on the embankments. Whereas 4 to 6 months had been previously scheduled using tower-mounted creeper cranes, Brown claimed each tower could have been assembled in as little as 8 days. Each tower leg was prestressed into a separate concrete pad extending to bedrock, the base elevations of the pads differing by up to 7.5 m. By intent, the towers mimic those of the original span, with two portal struts connecting slender single-celled steel tower legs. Their greater width does, however, give them a squatter appearance. Low Turkish wages had actually made more labor-intensive (but simpler) concrete towers less expensive to build, but in hoping for favorable Japanese financing terms, Turkish authorities wished to attract

Japanese contractors and included as much steel in the design as possible.

Brown met an early test. The anchorages are buried, hidden 210-m behind each tower. Yet hard rock on the European side was found to be 9-m deeper than originally thought, adding more excavation and 3 months to the schedule. Brown altered the design to reduce the amount of extra excavation and changed the anchoring arrangement of more than 1,900 prestressing bars to accelerate the onset of cable spinning. The extent of Brown's changes concerned Freeman Fox & Partners, for he had not consulted the designer of record at all. Brown harbored no such anxieties; he dismissed any role for Freeman Fox & Partners and undoubtedly benefited from what was regarded as the exceptionally high quality of the Japanese work.

Almost 600,000 m^3 of material would be removed for the tower and anchorage foundations. By November 1986—after just 18 months of work—the towers were completed and anchorages sufficiently advanced to allow the stringing of catwalks. Here, Brown used large winches to haul all 22 catwalk cables across the Bosporus in only 3 days instead of the previously planned 2 weeks; the busy channel, funneling 400 vessels daily, was closed for just 19 hours. The 76-cm diameter cables were substantially larger than those of the first span and the 504-wire strand size—using wires of British manufacture—was surpassed by only the first Bosporus Bridge and Japan's Shimotsui Seto Bridge, then also under construction (*see* Chapter 14). As in the first Bosporus span, four smaller strands were added between the towers and the anchorages to handle higher tensile forces. Cleveland Bridge & Engineering had proposed prefabricated parallel wire strand cables in its failed bid, so it is at first glance surprising that IHI shunned what was now Japanese leadership in this field and instead adopted cable spinning. Yet Japanese improvements in spinning technology promised large efficiency gains. Proper wire placement in the strand was assured by controlling tension during spinning, thereby eliminating individual wire adjustments. At the second Bosporus Bridge, tension was set at 50% that of a free-hanging wire, a large increase over earlier Japanese efforts. Japanese-inspired increases in wire diameter also reduced the 18,648 wires in each Freeman Fox & Partners-designed cable to 16,128. Spinning wheels traveled in opposite directions over each catwalk, laying wire into two strands simultaneously. The improvements produced spectacular results; crews placed more than 9,500 tonnes of cable wire in 4 months from March 1987, a rate more than 40% higher than the impressive 3-month showing on the first bridge.

The eight-lane roadway sits atop a streamlined steel box girder 33.5-m wide—still the largest used in a suspension bridge (*see* Fig. 8-3). Cantilevered footpaths and a 3-m depth—identical to that of the first bridge—imbue the box with a distinctive Freeman Fox & Partners' cross-section. All 62 boxes—each weighing almost 220 tonnes—were placed in just 1 month in the summer of 1987, with up to three boxes lifted daily. Boxes near the towers were temporarily suspended, then, in pendulum-like motions, swung into place above the steep slopes. Brown estimated his techniques shaved 3 months off the box erection schedule.

Celebrations for the late May 1988 opening of what is officially the Fatih Sultan Mehmet Bridge occurred without the firm that designed it—Freeman Fox & Partners. A remarkable 30-month construction period beat the schedule by almost 6 months and stayed within the $122 million budget (although huge fluctuations in exchange rates during construction fueled speculation that Japanese contractors had actually incurred costs of up to twice that amount). Brown had managed a project some thought beyond the ability of a single person to manage. Most who worked with him held him in awe, the head of the IHI consortium describing "a sort of magic" that Brown generated on-site (Heywood, 1988). The fast-paced effort benefited from Brown's successful efforts to overlap operations that normally occur sequentially in suspension bridge building, such as the construction of the anchorages simultaneously with that of the catwalks and then the main cables. The new bridge was now open, or almost; it stood unused for more than a month while Brown rushed the approach roads to completion. The bridge served its first customers in early July on four of its eight lanes, and it was only fully opened at year's end.

The second Bosporus Bridge stands as one of the more unorthodox of the large suspension bridges (*see* Fig. 12-1). The slopes drop away precipitously at the truncated towers, catapulting bridge users immediately over one of the world's busiest shipping lanes. From beneath, the span is more subdued; the olive patina of the towers and box girder—similar to that of the first bridge—serves to reduce the span's visual domination of the densely wooded hillsides.

In the frenzy to complete the bridge lay the seeds of yet a third Bosporus span. Projections a year before the opening of the second Bosporus Bridge suggested even it could be deluged as early as 1995, when 94 million vehicles were predicted to cross

Fig. 12-1. Fatih Sultan Mehmet Bridge (the second Bosporus Bridge)
PHOTOGRAPH COURTESY OF HYDER CONSULTING.

the Bosporus annually over both spans. Yet the estimated $440 million cost of a third bridge met a wary Turkish government. It felt a third crossing—if required at all—should await completion of the second. Furthermore, it favored private financing. This recalcitrance was not shared locally, and in late 1987, Istanbul's mayor decided to go it alone. Some labeled the move a gesture to influence pending national elections. Brown, although viewing the opportunity with some interest, was quoted as saying the scheme was "all dreams as far as I can see" (Anonymous, 1987). Yet the project soon acquired a surprising amount of momentum. The winning bidder would not just build the bridge, but design, finance, and operate it; once tolls had paid for its construction, ownership would revert to the city. Bidders were also given virtually free rein to develop their own designs; only the span's location, just 450 m south of the original bridge, was fixed. The February 1988 opening of the four bids received by the city generated shock waves when the Turkish contractor Sezai Turkes Feyzi Akkaya (STFA) working with IHI on the second span priced the bridge and 8 km of approach roads at just $171 million, more than 30% lower than a bid from Cleveland Bridge & Engineering (now a part of British construction giant Trafalgar House). Rumors abounded that STFA was able to secure IHI's equipment on the second span at bargain prices, but Trafalgar House insisted it had "an exclusive agreement" with IHI to collaborate on the third bridge (Goltz and Reina, 1988). Istanbul's mayor announced a decision would be forthcoming by the end of the month. In spite of its higher bid, Trafalgar House was touted as the favorite, largely because of expected financial aid from the British government.

All bidders took full advantage of the engineering flexibility offered them. Most striking was the revolutionary nature of the Turkish design, for which Brown, then wrapping up the second span, proposed a split box suspension bridge. He had examined up to 20 parallel boxes, but the final concept resembled his aborted proposal for the second Bosporus Bridge. Twin boxes of three lanes, each cradling a light rail transit line, would just exceed 1,000 m in span. By paring the box depth down to 2 m, Brown produced an almost impossibly thin depth-to-span ratio of 1:500. He predicted a weight savings of up to 15% in comparison to a conventional box girder. A more conservative Trafalgar House design prepared by COWIconsult called for a 1,168-m span flanked by suspended side spans of 284 m; this scheme resembled a 1987 proposal from the Danish consultant for Denmark's Storebælt crossing. Concrete towers with a single above-deck portal strut would be slipformed; the slender box girder—about 10% wider than Brown's split box—would also carry light rail transit lines and be dehumidified to eliminate interior painting, a company trademark. Large gravity anchorages would be buried out-of-sight in the steep slopes.

Progress soon slowed, but that summer the Turkish government suggested a third span could be financed as part of another highway scheme. Although providing the project more credibility, the offer would delay construction until 1989. A 2-year austerity program then halted further progress until the fall of 1992, when design firms were invited to compete for supervising the span's design and construction. This initiative foundered, too. A third bridge remains stillborn; rail and road tunnels have since been since mooted, although in 1998, a $700 million bridge was again in the works. Debate continues, and both existing spans remain gridlocked. The often unpredictable, complex politics of bridging the Bosporus may yet produce another span.

The Bosporus was not the only venue to disappoint Freeman Fox & Partners during this period. In the late 1970s, the firm designed a box girder suspension bridge near the Pacific coast terminus of the Panama Canal. The new bridge would stand upstream of the Bridge of the Americas, a steel arch bridge of 344-m span completed in 1962 and the sole crossing of the canal. Construction—to be funded by the Venezuelan government—was to begin in 1980, with completion slated for 1983. Yet all was not well. Right-of-way acquisition was delayed for more than a year. In October 1980, a construction contract was actually signed, but uncertainty over the future widening of the Canal stymied further progress. Approval to proceed was once again obtained in November 1981, with 39 months of construction expected to lead to the bridge's opening in January 1985. The bridge had now acquired a 608-m center span as insurance against a second set of shipping locks. A slender four-lane steel box girder passed through portal braced towers; the side spans would be unsuspended. Freeman Fox & Partners had adopted vertical suspenders, unusual in light of its vigorous defense of the Severn Bridge at the time, and also proposed the use of weathering steel to minimize repainting. Unfortunately, Panama's insistence on an 11-m reduction in the 67-m midspan clearance ran afoul of American canal authorities. Agreement was eventually reached on a 54-m vertical clearance that would shave costs by some $30 million. Yet when work finally began in the summer of 1983, Venezuela devalued its currency in an effort to quell rampant inflation. The project was then abandoned, and only in 2000—17 years later—was a second Canal crossing resurrected; this time, it would be a cable-stayed bridge of 420-m span situated about 20 km north of the Bridge of the Americas. The span, designed by the American firms T.Y. Lin International and The Louis Berger Group, is slated for completion in 2004.

At this time a long-time Freeman Fox & Partners' contemporary, the firm of Mott, Hay & Anderson, had experienced a major disappointment of its own. The firm had been absent from long-span suspension bridge design for some time, but it had recently prepared plans for a bridge in Hong Kong that would have overtaken the Humber Bridge as the world's longest span—had its construction not been postponed. Nonetheless, the project would ultimately see Mott, Hay & Anderson capitalize on its experience at the Firth of Forth and the River Severn in the design of a unique—and record-setting—suspension bridge.

12.3 HONG KONG'S BOX-LIKE TRUSS—THE TSING MA BRIDGE

The rapid growth of Hong Kong in the half-century since World War II has seen it emerge as one of the world's great shipping and trading centers. Hong Kong also benefits from a spectacular setting, where skyscrapers inexorably crawl up the precipitous slopes that overlook the city's famed harbor. Yet its prosperity and rugged topography have always placed a premium on land to enable the continued expansion of a city literally bursting at its seams. To this end, in 1973, planners first cast their gaze toward the large, undeveloped, but isolated island of Lantau to the west. Studies affirmed its suitability for development, but a fixed connection to the island was required. Simultaneously, a study to replace Hong Kong's congested single-runway airport at Kai Tak was underway. These and subsequent studies recommended Chek Lap Kok, 17 km west of Kai Tak off the north shore of Lantau, for a new airport. Government approval came in 1977, but access from the mainland would involve island-hopping on an enormous scale. A British consortium headed by Mott, Hay & Anderson was selected in July 1978 to recommend a route and the form of this crossing.

Mott, Hay & Anderson had not seen a long-span suspension bridge of its own design built since the Severn Bridge, and even there it had focused on the foundations. In 1972, the firm had designed a 252-m span self-anchored concrete box girder suspension bridge over the River Foyle in Londonderry, Northern Ireland. Considered more sabotage-resistant than an earlier cable-stayed design and intended to create more jobs locally, the span also featured A-framed concrete towers. Concerns over sabotage led to the use of parallel three-lane box girders separated by a 1.5-m gap, each hung independently from two cables. The bridge would have displaced the Hudson Hope as the world's longest (and largest) prestressed concrete suspension span, but it was never built; instead a 234-m span haunched steel box girder bridge was opened in 1983.

The Lantau crossing was another scale altogether. A single crossing from the New Territories at Sham Tseng to the island of Ma Wan was rejected because of the steep flanking slopes. The only other option was to hop across Tsing Yi and Ma Wan islands via the 1,500-m wide Ma Wan channel—the main shipping route to China—and the smaller 500-m wide Kap Shui Mun channel between Ma Wan and Lantau. A deep submarine trench under the Kap Shui Mun ruled out a bored tunnel. A four-lane sunken tube tunnel that in places emerged above the undulating seabed slightly undercut the $239 million cost of two four-lane suspension bridges. Notwithstanding their higher cost, the bridges emerged as the victor because of maintenance considerations and their flexibility for expansion; four more lanes could be added at little extra cost. The government accepted this strategy in April 1979.

The rugged topography was destined to place the Lantau crossing among the world's most spectacular bridges. Its centerpiece, the 1,350-m span Tsing Ma, would be exceeded only by the Humber. Unsuspended side spans would lead to a viaduct soaring up to 45 m above Ma Wan's rugged north coast, where a 486-m suspension span would cross the Kap Shui Mun channel to Lantau Island. In the long-term, a 390-m span cable-stayed bridge would link Tsing Yi to the New Territories.

Typhoon gusts up to 306 km/hr exceeded by a third those for which most European suspension bridges were designed, as were average hourly winds of up to 223 km/hr. Although box girders featured low drag, their stability in gusts exceeding 200 km/hr had not been confirmed. Initial studies confirmed that critical velocities were indeed below those required. But Mott, Hay & Anderson engineers who had worked on the Forth Road and Severn bridges some 25 years earlier devised the novel idea of combining the advantages of the vented Firth of Forth truss with the streamlined Severn Bridge box girder. The resulting concept was without precedent, but it ensured both aerodynamic stability and traffic movement during severe weather. In appearance, the suspended structure suggested a trapezoidal box girder. Yet its fairings were simply fascias that enclosed—and were supported from—an interior vented Warren truss. This ingenious marrying of box girder form and truss structure produced a shallow and economical suspended structure just 6.3-m deep that simultaneously reduced drag and increased rigidity. Over its 38-m width, transverse Vierendeel trusses supported two orthotropic upper-deck roadways of two (expandable to three) lanes separated by a generous 9.4-m wide central air gap. An enclosed lower-deck two-lane roadway would funnel traffic when high winds closed the upper deck. Small air gaps straddled the lower road lanes, and slotted horizontal stabilizers extending 6 m from the stainless steel fairings enhanced stability.

Detailed designs were completed in a 2-year period, ending in mid-1982, at Mott, Hay & Anderson's offices in the London suburb of Croydon. During this time, the expansion of the upper-deck roadway to six lanes was abandoned in favor of lower-deck rapid transit lines straddling the central two-lane roadway. This in turn reduced the width of the truss. Yet the most visible alteration was a lengthening of the main span of the Tsing Ma Bridge to ensure against ship collision. The towers were first placed 50 m beyond the shipping channel; the 203-m tall Ma Wan tower would stand in shallow water and its Tsing Yi twin on land. Yet the short unsuspended Tsing Yi side span was found to be susceptible to the bending and twisting of the main span. The Tsing Yi tower was then moved further inland—in fact, its footing 50 m up the hill removed the side span altogether. The main cables would now dive into tunnel anchorages 290-m behind the towers. This strategy, by coincidence or otherwise, nudged the main span to 1,413 m, a mere 3 m longer than the Humber Bridge. To further reduce ship collision risk, the Ma Wan side span became a 333-m suspended span that would terminate in a huge gravity anchorage requiring the excavation of 800,000 m^3 of rock.

Slipformed reinforced concrete towers with diagonal bracing were initially proposed, but multiple portal braces were later adopted. An ovoid profile for the tower legs would smooth wind flow and reduce the potential for tower oscillation. The Ma Wan tower would lack a base slab; each leg would stand independently in the coastal shallows. The 90-cm diameter main cables would comprise 26,000 wires—28,000 over the side spans—in the largest application yet of prefabricated parallel wire strands, although the door was left open for in situ spinning. The smaller Kap Shui Mun Bridge would rise over its 486-m main span and suspended side spans of 153 and 135 m, producing towers of dissimilar height. At one point, multispan trestles, a causeway and a floating bridge were suggested for this smaller crossing, but the suspension bridge was ultimately recommended.

Wind tunnel tests at the NMI in England involved Kit Scruton and D.E. Walshe, veterans of the Severn Bridge studies. One concern was to prevent winds from blowing along the inside of the Tsing Ma Bridge's hollow suspended structure. Various horizontal stabilizers were found to offer only marginal improvements and were dropped from the final configuration, a 32.6-m wide, 6.4-m deep truss with a 1:225 depth-to-span ratio, significantly lower than other twin deck suspension bridges. The narrower truss also reduced the width of the central air gap, but to a still substantial 5.4 m.

The design was completed at the end of 1982, and the crossing in its initial arrangement of just four-lane upper deck lanes was estimated to cost $400 million. Construction was expected to begin in 1983 and last 4 years. Yet the crossing, designed down to the last detail, was never built. The economy was now mired in recession, and budgets were pinched. Political uncertainty over the 1997 handover of the British colony to Chinese rule also threw a cloud over the project. In 1983, the airport project was shelved, and with it died the Lantau crossing.

Hong Kong's expected growth could not be ignored for long. The 1984 Sino-British agreement on the future Chinese administration of the British colony boosted confidence, and in 1987, the airport issue was re-examined. These studies led to the October 1989 launch of a huge $16.3 billion program comprising an airport at Chek Lap Kok, a nearby 200,000 person new town, port facilities, and 34 km of high-speed rail and road links from central Hong Kong to the airport. All were required to open in 1997, in time for the British handover of Hong Kong to China.

Mott MacDonald (successor to Mott, Hay & Anderson) was given the task of updating the 1982 plans for the Lantau crossing. This time, however, the firm would only develop an outline design, as the government intended the winning contractor to prepare detailed plans and build and operate the span. Then in May 1990 came an unsolicited proposal for a bizarre suspension bridge from the ebullient German engineer Hellmut Homberg in a two-page advertisement carried in *New Civil Engineer*, a British weekly. Homberg proposed to thread a 71-m wide, 4-m thick steel deck supporting six road lanes and two rail lines between four 210-m tall columns over a span of 1,413 m; a ventilated membrane would ensure all-weather use. Homberg claimed the scheme would cost $615 million, 20% less than Mott MacDonald's plan. In a jibe at British practice, Homberg (1990) stated that, "naturally the hangars are vertical installed and not incline [*sic*]." He is said to have approached Cleveland Bridge & Engineering with the proposal, but he was apparently spurned. Homberg died that July at the age of 80.

Politics then clouded the project again. Chinese concern over the fate of the colony's deep financial reserves was heightened by Hong Kong's October 1990 decision to increase public funding of the airport program in order to meet the 1997 target; the British had not consulted China on this change of strategy. By early 1991, it was feared that if China could not be assuaged, the banks might shy away and scuttle the project. In July, China did affirm its support for the project, but at the cost of greater influence in future decisions. Even then, this issue was not entirely resolved until late 1994.

The change of funding strategy now saw Mott MacDonald charged with developing detailed designs. As political debate continued, the firm quietly introduced major changes. The Tsing Ma Bridge was shifted 700 m south of its 1982 line to mesh with access roads for a new bridge to the New Territories and to better fit a six-lane expressway into Tsing Yi's steep slopes (which nonetheless required the removal of more than 9 million m^3 of material that now sits under the new airport at Chek Lap Kok). The entire crossing—including the Tsing Ma Bridge, a 500-m viaduct across Ma Wan, and a 441-m suspension span over the Kap Shui Mun—would extend 3.5 km. Although the new Tsing Ma Bridge extracted many concepts from the 1982 design, particularly its unique suspended structure and the tower motif, there were major changes. The main span was reduced to 1,377 m, which moved the Tsing Yi tower down to the shoreline. This in turn created a Tsing Yi side span, unsuspended to accommodate ramps converging at the tower; the main cables would be funneled down huge 80-m long, 18-m wide tunnel anchorages. The Ma Wan tower stood 400-m offshore, protected by a large artificial island, and was flanked by a 355-m suspended side span and an 80-m backstay span extending to a largely buried gravity anchorage. Both 206-m tall towers would support two record 115-cm diameter main cables. Mott MacDonald intended to use 127-wire prefabricated parallel wire strands—329 in each main span cable—with an additional 19 strands over the side spans.

The suspended structure underwent subtle but significant change as well. It was widened to 41 m to support six upper-level lanes, was deepened to 7.4 m, and would clear the water by 62 m. The orthotropic upper and lower decks and non-structural stainless steel fairings remained fundamental to the design, but the size and placement of the air gaps required rethinking. Some 18 different configurations for the upper central air gap ranging in width from 2 to 9.4 m were tested in the NMI wind tunnel and the Boundary Layer Wind Tunnel at the University of Western Ontario in Canada. Smaller gaps maintained acceptable critical velocities and even reduced vortex shedding, but eliminating the gap altogether produced a precipitous drop in critical velocity. A width of 3.5 m emerged that was smaller than the 1982 design, and a 12-m central vent was also placed beneath the lower-deck transit tracks, which now sat between two separated emergency lanes (*see* Fig. 7-2). A "turning vane" was also placed at the roadway edge similar to that used on Denmark's Lillebælt Bridge to reduce instability from vertical gusts. Slotted windscreens extending 60 m on either side of the tower were added to minimize vehicle buffeting, features used only on the revamped Severn Bridge. All of these measures combined to provide an impressive flutter critical velocity of 300 km/hr.

The bridge had also been beefed up to permit 140-km/hr train speeds, with the truss made to be continuous between anchorages. It would be fixed at the Ma Wan anchorage—all expansion would be taken up at its Tsing Yi counterpart 2,135-m away. Road movements would reach 1.75 m, and for the rail line, 2.2 m; the latter required special expansion joints of unprecedented size. Avoiding fatigue from the pounding of 200 trains daily would require particular attention to welding details. Comparisons with the Humber Bridge were revealing: the 50,000 tonnes of steel in the Tsing Ma's suspended structure were more than triple—and the 550-tonne tower saddles more than 10 times—the weight found in equivalent features of the British span. The 33,400 wires in each Tsing Ma cable would more than double the number in each cable in the Humber Bridge. Resistance against typhoons, earthquakes, and heavy truck and rail loads wrought an extremely strong bridge. The Tsing Ma would in fact be the world's longest road and rail span and only second in span to the Humber Bridge on its planned opening in 1997.

Construction tenders for the Tsing Ma Bridge were called in July 1991, with work planned to begin in the spring of 1992. The government intended to place all risk—from inflation and typhoons to unforeseen foundation conditions—on the backs of contractors, an unpopular move that caused one to drop out before the December 1991 deadline. A British-Japanese consortium lead by Trafalgar House was then awarded a $933 million contract to build the bridge in May 1992. The award sparked severe criticism from Korean contractor Hyundai, which had bid just $600 million. Yet a last-minute government demand for $250 million in financial guarantees had precluded its bid, arousing suspicions that British interests were being favored. The government defended its decision, stating that a guarantee of on-time performance was crucial because the airport could not open without the bridge.

Trafalgar House had actually submitted two bids—one for the outline design, and a second bid some $105 million lower proposing a gravity anchorage on Tsing Yi in place of the tunnel anchorages and the use of spun—instead of prefabricated parallel wire strand—cables. This second bid took the prize. The bridge would have to be structurally complete in 4 years, allowing rail installation to begin by the summer of 1996. Work began almost immediately. Virtually all materials—including 230,000 m³ of concrete for towers, anchorages, and other piers—arrived by sea to congested work sites. Almost 1 million m³ of material was gouged from Tsing Yi to create a 55-m deep hole for its buried gravity anchorage. Its Ma Wan counterpart, up to 76-m long and 75-m tall, displaced 170,000 m³ of rock. Each Ma Wan tower leg rests on separate concrete caissons built on a barge off Tsing Yi and placed in April 1993 on a 12-m deep leveled rock platform. Over-excavation had required some rather tricky underwater concreting to level the platform, but once the caissons were in place, a 5-hectare artificial island was built around them. The pier was finished that August. The simpler Tsing Yi tower footings were taken to a 7-m depth on reclaimed land just offshore.

The 6-m wide tower legs feature rounded corners and, in thickness along the line of the span, taper sharply from 18 m at their bases to 10 m at deck level. Four portal struts decrease in depth with increasing height. Unusually large plinths support the cable saddles, which in turn sport large illuminated stainless steel covers for visual effect; most spans, by contrast, de-emphasize the saddles. The tower legs were slipformed with concrete designed to last 120 years in Hong Kong's marine climate. This demanding specification caused unending headaches, for it required months of testing on more than 100 concrete mixes at the contractor's expense. A mix thought to meet the specification was then discovered to set too slowly for slipforming and could only be used in the anchorages and piers. A June 1993 acceptance of a new mix finally saw the Tsing Yi tower begin to rise, almost 3 months behind schedule.[7] In spite of torrential rain and two near-misses with typhoons, the tower was completed in just 15 weeks at the end of September. Unfortunately, the early delays had scotched plans to use the same formwork to erect the Ma Wan tower. It was built with its own formwork in an operation ending in December 1993, leaving sufficient time to meet a July 1994 target for the start of cable spinning. Steel forms for the portal braces measuring up to 12-m deep and 34-m wide and weighing up to 420 tonnes were lifted into place as the slipformer passed; these forms were then filled with concrete. The tower legs alone would ultimately consume more than 40,000 m³ of concrete, and each complete tower weighs 65,000 tonnes.

Trafalgar House had predicted a $29 million savings with spinning in lieu of prefabricated parallel wire strand cables. There would be typhoon risk, but planned efficiencies would help meet a tight 40-week schedule in spinning the 91 strands of up to 368 wires into each 33,400-wire cable (6 additional strands of 304 wires each would also be spun into each side span cable). Plans called for the cable strands to be anchored via stressed cables to the back face of the anchorages up to 45 m from the dehumidified splay chambers. But the tight schedule now required that spinning begin before the Tsing Yi anchorage was finished; this was ingeniously accomplished by building up the anchorage from its bottom and midpoint elevations simultaneously. Catwalk assembly was then launched with a heart-stopping moment in March 1994 when a catamaran bolted over the pilot wire just as it was being pulled from the water; fortunately, no harm came to either. Spinning began that July. Trafalgar House had eschewed the increasingly common controlled-tension method to capitalize on low labor costs, using a core of 120 disciplined Ghurkas accustomed to great heights. Eight wires were laid per run from four loops over the spinning wheels, doubling the per-run output of the Severn, the original Bosporus, and the Humber bridges. Completed strands were also lashed to the catwalk to guard against high winds. Although only one typhoon warning was issued—sparing the cables and the several days required to secure and unsecure them—record rains in July and August slowed progress to the point that the cables were soon 6 weeks behind schedule. The training of the hundreds of workers involved in the spinning operations also proved time consuming. Crews eventually overcame these teething problems and would spin 160,000 km of wire in 9 months. In March 1995, the largest cables built to that time were finished, a remarkable 2 weeks ahead of schedule.

Suspended structure steel was fabricated in Great Britain, Japan, and the United Arab Emirates, and the stainless steel fairings were shaped in Australia. Truss steel was assembled into 36-m long units in China along the Pearl River about 80 km north of the bridge site. Some units weighed as much as 1,000 tonnes—the largest lifted into a bridge to that time—although initially lifts of 72-m long sections had been considered. Strand jacks on the lifting struts eliminated the traditional tower-based hoisting engines; the first unit at center span was lifted in August 1995. Crews closed the main span in March of the following year, the month Trafalgar House was absorbed by Norwegian industrial giant Kværner. By April 1996, all 51 units were in place, on schedule, as planned 4 years earlier (*see* Fig. 12-2).

Meanwhile, a $207 million, 441-m span cable-stayed bridge over the Kap Shui Mun had superseded Mott MacDonald's outline design for a suspension bridge of identical span. This bridge, built between 1993 and 1997, remains the longest road and rail span of its genre and one of the strongest ever built. The entire Lantau Link was completed—on schedule—in April 1997 at a cost of $1.25 billion, and it was opened to traffic the following month, just before the transfer of Hong Kong's administration to China. Some 1,800 people worked on the Tsing Ma Bridge at the peak of activity, but sadly, three lives were lost during construction. The spectacular opening ceremonies, presided over by former British Prime Minister Margaret Thatcher, were replete with lasers and a waterfall of fireworks cascading from more than 1,000 m of main span. No fewer than 8,000 permanent lights over the Lantau Link are orchestrated nightly to create dramatic patterns; those on the outer faces of the towers change color with the seasons, and the curved tops of the saddle covers are marked with a band of brilliant white light.[8] A nearby visitor center overlooks the Tsing Ma Bridge, whose enormous strength and capacity are masked by the exceedingly slender profile of its suspended structure; the sturdy towers and prominent saddle covers communicate the bridge's strength. On a more prosaic note, a wide array of sensors monitors wind and earthquake response. One unwanted feature: an unsightly fungus that completely covered the towers of both the Tsing Ma and Kap Shui Mun bridges within 2 years of their opening; all concrete had to be

Fig. 12-2. Tsing Ma Bridge – lifting of suspended structure
PHOTOGRAPH COURTESY OF MOTT MACDONALD.

thoroughly scrubbed and coated with an antifungal agent.

Hong Kong's bridge building program, rather than abating, has since accelerated. The spectacular $225 million Ting Kau Bridge, built between 1994 and 1998 to link the Tsing Yi terminus of the Tsing Ma Bridge to the New Territories, comprises dramatic end-on-end cable-stayed spans of 448 and 475 m joined by a large central pylon. More suspension bridges are planned for greater Hong Kong, which by 2007 will also host what is to be the world's longest cable-stayed span. The most ambitious of these schemes—a 27-km bridge across the Pearl River delta—is solely a Chinese initiative. This project in particular demonstrates the huge strides made by China's engineers in a matter of a little more than a decade. These advances are described in Chapter 16.

ENDNOTES

1. Brown questioned the need to increase the design live load, as he did not think the actual loads on the bridge had even reached the level of the original design load.
2. The problems of high wind on the Severn Bridge were underscored by the devastating storm of January 25, 1989, considered Great Britain's worst in 300 years. The Severn Bridge was closed, and a truck overturned on the span.
3. The towers of the Forth Road Bridge endured a similar $14.5 million surgery in the mid-1990s to relieve each tower of about 60% of the loads it formerly carried.
4. Ironically, Steinman, Boynton, Gronquist & Birdsall had just reviewed Freeman Fox & Partners' design for a 608-m Panamanian box girder suspension bridge (also discussed in Chapter 12); another of the defeated—COWIconsult—would be retained in 1984 by Turkish authorities to check Freeman Fox & Partners' design of the second Bosporus Bridge.
5. The relative health of the Bosporus I span has been cited by Brown as key evidence that the Severn's strengthening was unnecessary. Comparisons were made during the late stages of the Bosporus I design with (then) interim and more stringent British box girder design standards. As a result of this analysis, Brown concluded their application to create an overly conservative structure. No changes were made to the original box girder design, and the completed bridge—soon carrying 50% more traffic than predicted—has served well since.
6. Brown recounts that in 1974 his concern over the required depth of the Humber box girder lead him to seek alternatives. He devised and built a simple wind tunnel in the basement of Freeman Fox & Partners' offices, where he subjected a strip of balsa wood approximating a flat plate to various tests. Brown was able to confirm a relationship between flat plate theory and the Severn box girder cross-section, but at this time, his work on the ultralong Messina span suggested a need for more stability. One strategy–to split the plate into narrower parallel strips–had the effect of increasing flutter speeds significantly. Brown eventually obtained patents for the split deck concept, including twin- and multiple-box variants.
7. In 1997, the concrete contractor sought more than $70 million in compensation for the difficulties encountered with the tower concrete specifications; the case remained under arbitration as late as 1999.
8. Several suspension bridges have been spectacularly illuminated since David Steinman first established a trend with the cable lighting of the Mount Hope Bridge in 1929. Since 1987, subway trains crossing Philadelphia's Benjamin Franklin Bridge have been "chased" by a system of almost 600 floodlights. The lights, situated at the base of the suspenders and along the main cables, flash in sync with the trains' passage across the bridge. Other Japanese spans have received interesting lighting treatments (*see* Chapter 14).

Chapter 13

THE SCANDINAVIAN EXPERIENCE

If the 1960s and 1970s had seen British engineers emerge as a dominant force in suspension bridge design, the following 2 decades would see Scandinavian engineers come into their own. Norwegian engineers in particular have carved out a unique niche for small, narrow suspension bridges stiffened by both trusses and box girders. The 20th century would also end with enormous spans in both Denmark and Sweden that would push box girder suspension bridge design to new frontiers.

13.1 A UNIQUE NORWEGIAN LEGACY

One nation whose contributions to the art have been largely overlooked is the sparsely populated country of Norway. Suspension bridges have proven well-suited for bridging the deep fjords that punctuate a rugged coastline exceeding 21,000 km in length, and since 1900, the nation has built some 200 of them. Although no bridges to-date have rivaled American, European, or Japanese behemoths, Norwegian engineers have mastered the aerodynamic challenges associated with long, narrow suspension bridges that rarely exceed two lanes in width.

The nation's first recorded suspension bridge dates back to 1842; one, the Bakke Bridge, remained standing 110 years after its 1844 opening. The few spans of the 19th century featured trusses of increasing rigidity to avoid the afflictions of the flexible English and French spans. With improved cable wire after 1900, suspension bridges became more common. In 1904, the 97-m span Gulsvik Bridge was opened, and between 1906 and 1918, five spans of up to 95 m were built. Three of these spans, beginning in 1913, signaled a trend toward flexibility with solid cross-girders in place of trusses and little longitudinal stiffness. In the late 1920s and 1930s, suspension bridges became a popular and inexpensive solution that not only avoided deep water piers and scaffolding, but log booms and shipping. Their size increased, too; the Atna, a truss-stiffened bridge opened over the Glomma River in 1923, spanned 150 m.

O. Stang, chief bridge engineer of the Norwegian Public Roads Administration from 1912 to 1939, sought to avoid the higher costs of rigid spans by using flexible rolled I-beams. Stang first experimented with the idea around 1920 and applied it in several spans, beginning with the 183-m span Rånåsfoss and the 130-m span Bingsfoss bridges, both completed near the capital city of Oslo in 1927. The 150-m span Framnes (1931) and the 160-m span Elverum (1932) bridges were of a similar design. The latter span featured a depth-to-span ratio of just 1:458, and its immunity to aerodynamic oscillation would later perplex American investigators (Balog, 1943).

Further economies emerged in the Vrengen Bridge—a 168-m span completed in 1932. Its continuous stringers acted compositely with the deck to enhance stiffness, although some oscillation did occur (*see* Chapter 4). These early efforts spawned an inexpensive, standardized design for small suspension bridges. Over the 12 years to the start of World War II, 40 flexible I-beam suspension bridges were built, ranging in span from 50 to 230 m.[1]

The Norwegian Bridge Office published Stang's approach—which relied on fairly rudimentary mathematics—in 1934. Younger engineers demanded a more theoretically based approach, and it was here that Arne Selberg would fill a void. Selberg had graduated from the Norwegian Institute of Technology (NTH) in Trondheim in 1934 with a civil engineering degree. The following year he began a career with the Norwegian Public Roads Administration and, at age 27, designed Norway's longest suspension bridge—the 230-m span Fyksesund Bridge (1937). Its unruly aerodynamic behavior profoundly affected Selberg, for he began to develop the theoretical basis for suspension bridge design previously lacking from Norwegian practice. Selberg's doctoral thesis *Design of Suspension Bridges* was submitted to the NTH in 1941, the year after his return to the institute. Selberg kept one foot in the Bridge Office while World War II delayed the granting of his doctorate until May 1946; in 1949, he formally severed his government ties to accept the post of Professor of Structural Mechanics at NTH. With the 1949 completion of a wind tunnel at NTH, Selberg began to investigate the aerodynamic behavior of suspension bridges. His research would earn him a reputation as one of the world's leading experts in the field. During a distinguished career at NTH, ending in 1979 at age 69, Selberg would also become associated with such notable bridges as the Lillebælt. He died in 1988.

One span was built during World War II—the Tynset, a 130-m span completed in 1943. Both it and the 160-m span Elverum Bridge[2] featured a more modern suspended structure incorporating a reinforced concrete deck slab. Yet initial post-war efforts reflected more pre-war tradition than aerodynamically sound design. The Tana (194 m, 1948), Brandvalbru (185 m, 1949), and Glomma River (108 m, 1951) spans were typical in their use of slender rolled cross-beams; wind tunnel tests after the Tana Bridge's completion revealed a sensitivity to high winds. In 1948, the Kongsvinger Bridge, a 108-m span with wide arched portal towers, was completed. The Alta Bridge (104 m, 1948) was a rare Norwegian example of a through truss. Then, in 1956, the first of a new generation of Norwegian spans reflecting Selberg's aerodynamic research was completed. The 337-m span Varodd Bridge, a striking two-lane truss-stiffened bridge designed by Reidar Ingebrigtsen (who died before the span was completed), was Europe's longest until France's Tancarville Bridge opened in 1959. Its 2.7-m deep truss instituted a trend in Norway that would last a quarter century. The design evolved from wind tunnel tests at NTH under Selberg's direction. A median slot of open steel grating was found to significantly increase the critical velocity of the truss, but it was considered too expensive; in any event, the use of top and bottom lateral bracing proved sufficient. The Varodd Bridge's marked asymmetry also typifies Norwegian practice. Its concrete portal-braced towers stand 57- and 71-m tall and funnel the steeply rising truss; they are also solid below the truss and in this respect resemble the pylons of the Tancarville Bridge. There is only one suspended side span, of 141 m; a steep slope forces a sharp curve in its counterpart before anchorage. The Varodd's fine lines were obscured in 1994 by a parallel 260-m span concrete box girder—among the world's longest, and a second similar bridge will soon replace the suspension span.

The Varodd Bridge's concrete deck was cast in 20-m long slabs and contributed nothing other than its own weight to the rigidity of the suspension system. Wear of its slab joints—by the early 1980s, the deck required annual repaving—led to the use of continuously reinforced concrete decks in later spans that also enhanced torsional stiffness. Less secondary stiffening was required, and the upper wind truss found in the Varodd Bridge was absent in subsequent spans. Truss construction also evolved; closed box members well-suited to welding offered a cleaner appearance and simpler maintenance.

Flexible suspension bridges were not completely abandoned. Between 1960 and 1968, seven spans ranging from 125 to 180 m were built. One, a 150-m span hung from A-framed steel pipe towers with twin cables, resembles a monocable. Yet the nadir of the Norwegian type of flexible deck was approaching; tests at the NTH suggested it to be aerodynamically unsuitable for longer spans. Concrete box girder cantilever bridges also began to make inroads into the bottom end of the span range occupied by flexible spans. The era of small suspension bridges was coming to an end.

The next major truss-stiffened span following the Varodd was the Tromøy Bridge, a 240-m span completed in 1961. The Brevik Bridge, an exceptionally handsome high-level span of 272 m, was completed in 1962. The aesthetic challenge of small, high-level suspension bridges was handled with great success here; reinforced concrete towers are solid below the deck as at the Varodd, and the steep slopes enabled the inconspicuous anchorage of the main cables in chambers tunneled into rock. Other notable truss-stiffened spans of this period include the 228-m Erfjord (1963), the 325-m Rombak (1964), the 290-m Tjeldsund (1967), and the 225-m Folda (1969) bridges. One unusual span is the Kjerringstraumen Bridge. This 1969 truss-stiffened bridge features 180- and 200-m spans flanking a single concrete tower footed on a mid-fjord outcrop. The finely shaped central pylon and a spectacular mountain backdrop accentuate the simplicity and elegance of this bridge.

Most of the 19 truss-stiffened suspension spans of more than 150 m built from 1956 to 1982 are found along Norway's coast. A depth-to-span ratio of about 1:100 was common, and all spans greater than 200 m used reinforced concrete towers, initially built with fixed forms. In 1970, climbing forms that are jacked up the tower legs were introduced in the 103-m tall pylons of the Sotra Bridge, a two-lane 468-m span completed in 1971 west of Bergen. This bridge typifies another Norwegian trademark—the use of locked-coil strands, unbound and exposed. Their speed of installation is an important consideration given Norway's truncated construction seasons. Surprisingly, not until the Sotra Bridge was a catwalk first used.

The Sotra Bridge's tenure as Norway's longest span lasted just 1 year when, in 1972, the 525-m span Skjomen Bridge was completed. In this bridge, continuously moving slipforms were used to erect the 100-m high towers and have since dominated. It was here that Fritz Leonhardt with German contractor Krupp unsuccessfully proposed yet another monocable. In 1975, the 260-m span Kjellingstraumen Bridge was completed. Slipforming was then adopted for the towers of the Kvalsund Bridge, a 525-m span completed in 1977 and one of the world's most northerly suspension bridges. It is markedly asymmetrical, with a two-lane roadway rising 16 m between 81- and 97-m tall towers. Both unsuspended side spans curve away from under the main cables, which are anchored in rock up to 138 m behind the towers. Another peculiarity of the Kvalsund Bridge is the cantilevered support of the 5.5-m deep truss over a 54-m distance into both side spans, presaging its appearance in Japan's much larger Shimotsui Seto Bridge in the 1980s; this feature was also present in the Skjomen Bridge.

What could be Norway's last trussed suspension bridges were completed in the early 1980s—the 325-m span Nærøysundet (1981) and the 250-m span Haglesundet (1982) bridges; the latter span introduced precast reinforced concrete deck panels to form an easily built continuous roadway. A hiatus in construction then marked the duration of the 1980s, but this period would spawn a revolution in Norwegian design practice.

Norway entered a new era of suspension bridge building in the 1990s with the introduction of streamlined steel box girders. A box girder devised as early as 1965 for the Skjomen Bridge for research comparisons had suggested superior economy, torsional resistance, and ease of maintenance. By the early 1970s, orthotropic decks were also being mooted as a lightweight alternative to concrete slabs for spans from 450 to 550 m. Although the two concepts were well-suited to one other, one concern was the aerodynamic behavior of long, narrow box girders. Widths not exceeding 17 m would, for the spans contemplated, reduce width-to-span ratios to a range unseen since the original Tacoma Narrows Bridge. Norway's coastal winds and the effects of snow blockage of deck parapets were added concerns; subtle changes in deck profile could significantly alter aerodynamic behavior, as demonstrated in early tests of the Severn Bridge's box girder. Open railings, however much favored for aerodynamic stability, compromised vehicle safety in high winds, a dilemma underscored in March 1977 when a caravan of vehicles on the Sotra Bridge was overturned in gusts of 195 km/hr.

By the early 1980s, most design problems had been overcome. Revenue from North Sea oil fields also began to fuel a major expansion of Norway's bridge program. Engineers sought to develop a standardized box girder profile applicable to a wide range of spans. For 600- to 900-m spans, a 3-m deep, 12-m wide trapezoidal box was developed. For spans of 900 to 1,200 m, the prognosis was not yet clear. Spans less than 600 m were now considered to be the preserve of other bridge types—most notably cable-stayed spans. Norway quickly embraced this alternative bridge form and in 1991 completed the 530-m span Skarnsundet Bridge, a record-setting cable-stayed span for which a truss or box girder suspension bridge was once considered.

In 1992, Norway completed its first box girder suspension bridge. The 623-m span Gjemnessundet Bridge features a 13-m wide, 2.5-m deep box girder weighing just 2,300 tonnes; its main span is flanked by unsuspended side spans of up to 210 m. Construction began in 1989 as part of the $180 million Krifast link, which includes a 5-km undersea tunnel and a 933-m pontoon bridge that narrowly beat out a 775-m box girder suspension span. Each cable is arranged in the standard exposed configuration, here involving two horizontal rows totaling 14 locked-coil strands.

Construction proceeded smoothly for the most part, but soon after the first 36-m box section was lifted into the main span in January 1992, a vicious storm cut off power to the area for a week; stranded bridgeworkers spent one night *inside* the box girder to escape the storm's fury. Then, as more sections were added, the girder profile threatened to exceed desirable deflection limits. Further lifting had to be suspended until the 13 sections that had been installed could be temporarily welded together. All boxes were in place by the end of April, and the Gjemnessundet Bridge opened in August 1992 as Scandinavia's longest suspension bridge. Its slenderness is almost alarming—the narrow box girder floats ribbon-like at heights of up to 49 m through slender 108-m tall portal-braced concrete towers. The uninterrupted splaying of the main cables from the tower tops to the anchorages is especially dramatic, as is the sharp vertical curvature of the main span itself. This 3-year venture cost $31 million.

Just months later—in December 1992—a $61 million bridge to the island of Askøy near Bergen acquired the title of Scandinavia's longest span (*see* Fig. 13-1). Completed after only 30 months of construction, the 850-m span of the Askøy Bridge stands less than 5 km from the Sotra Bridge. Precipitous slopes tumbling to depths of 320 m limited the range of suitable span types. Deep rock tunnel, floating submerged tube tunnel, and cable-stayed crossings were considered, but a suspension bridge proved the most economical. The span's two (expandable to three) lanes are supported by a 15.5-m wide, 3-m deep orthotropic box girder. Upon expansion, pedestrian traffic will be transferred to new 2.5-m wide wings clipped to the girder edges, giving it a Severn-like cross-section. Selberg's formulae assisted in predicting a critical flutter velocity of 245 km/hr, verified in wind tunnel tests on a full aerolastic bridge model at the University of Western Ontario's Boundary Layer Wind Tunnel in London, Canada.

The Askøy Bridge is markedly asymmetrical, with unsuspended side spans of 173 and 34 m; the latter curves into a tunnel as the cables disappear into rock above and beside the roadway. Steel tendons carry the cables through individually drilled 25-m long conduits to a concrete slab 35-m below road level that pushes against the intervening rock. Virtually only the rock restrains the span, reflecting a simple, efficient, and inexpensive anchorage used in a number of Norwegian suspension bridges. The 152-m towers are slender portal-braced frames. The south tower is shoehorned into a steep slope, with one leg footed 19-m higher than the other, a distance greater than the tower width. The 21 locked-coil strands in each cable are distributed over three horizontal layers by exposed rectangular frames from which the suspenders hang. This arrangement simplified design and maintenance, despite increasing wind loads by up to twice that of a bound, circular cable. Two additional strands were draped over the side spans. The tower saddles are unusual; their cores are concrete but they are framed by steel plates below (to permit sliding) and on top (curved, to guide the cable strands). Dehumidified housings avoided the need for painting and sealing the strands over the saddles, and the saddles themselves were transformed into architectural embellishments several magnitudes higher in cost than required for purely functional reasons. The inside of the unpainted box girder is also dehumidified, as are the anchorage cable splay chambers.

The Askøy's towers were topped out in May 1991, with slipforming interrupted only to allow its portal struts to be placed. A helicopter then laid the first catwalk

Fig. 13-1. Askøy Bridge
PHOTOGRAPH COURTESY OF MONBERG AND THORSEN A/S.

wires to avoid shipping and the channel's deep waters. Six months later the main cables were in place, with each 76-tonne strand draped across the span by trolleys mounted on wooden tracks laid on the catwalk. Yet the most impressive performance was reserved for the box girder. Whereas conventional erection into the span had been estimated to take several months, all 24 box segments were instead lifted into place by floating crane in just 8 working days. Up to five 180-tonne, 36-m long boxes were placed in a single day; those near the mainland were swung into place like pendulums from adjacent suspenders.

The Askøy Bridge opened 5 months ahead of schedule, an achievement rewarded at the opening ceremonies by the winter's first sleet storm. Its slenderness, like that of the Gjemnessundet Bridge, is breathtaking, in the slender and narrow box girder, its pronounced vertical curvature, the tall, thin towers, and the clean sweep of the main cables unfettered by any rigid midspan cable clamp.

Norway's ambitious plan for an 800-km coastal highway linking Stavanger, Bergen, and Trondheim should firmly entrench the box girder in Norwegian practice. A box girder replaced a planned truss in the Osterøy Bridge, a 595-m span completed in 1997 northeast of Bergen near the Sotra and Askøy bridges. Unsuspended side spans of 151 and 171 m in prestressed concrete flank its 121-m tall concrete towers. Foundation work began in September 1994, but a bid of $17.4 million for the superstructure in early 1995 was double that budgeted for. A second attempt that summer brought a $15 million bid; sensing the inevitable, authorities awarded a contract that October. The unusual cable configuration—a single flat plane of 12 large locked-coil strands—not only simplifies maintenance but reduces wind loads, as well as snow and ice accumulation. All strands were erected over the summer of 1996. The larger reels of cable strands—and the crane used to lift them—squeezed through the many tunnels leading to the bridge with literally centimeters to spare. The 13.6-m wide box girder features a typical Norwegian arrangement of two lanes and a pedestrian track. It is just 2.5-m deep and was lifted into the main span in 36-m long sections in just 3 weeks in March 1997. The $39 million span was completed later that year, and it shares the Norwegian propensity for simplicity of outline and slenderness. Interestingly, this bridge has been prone to minor oscillation, and wind vanes were planned for installation soon after the turn of the century. A box girder span of similar visage to the Osterøy Bridge is the 446-m span Lysefjorden Bridge, also completed in 1997, near Stavanger.

Box girders also feature in the two suspension bridges of the $260 million Triangle Link project 70-km south of Bergen. The 677-m span Storda and the 577-m span Bømla bridges, completed in late 2000 and early 2001, respectively, both feature a 2.7-m box depth and an 11.7-m box width. One tower and anchorage on the latter span stand on different islands, and the deck curves away toward yet a third. Construction began in the late 1990s where, in a departure from tradition, spun cables of 32- and 30-cm diameter were selected by contractors in preference to locked-coil strands as a result of recent—and dramatic—efficiency improvements in Denmark's giant East Bridge; those of the Storda Bridge were spun in just 7 weeks in spite of a loss of 25% in spinning time to inclement weather. A trend of increased competitiveness with cable-stayed spans has continued in the Feda Bridge, a smaller 331-m suspension span currently under design; its box girder is just 2.6 m deep. The Dalsfjorden Bridge is another slender span of 500 m under design with a slender box of the same depth.

Norway's most audacious proposal involves a two-lane box girder suspension bridge of 1,325 m span across the rugged Hardanger Fjord that, if built, will take narrow box girder suspension bridge design to new frontiers. A 20-m wide, 4.5-m deep box girder is required for torsional stability. Wind tunnel testing in 1991 and again in 1995 had to account for a chaotic wind environment. Curved guide vanes as featured in the Lillebælt Bridge may be used; Erik Hjorth-Hansen, who worked on the Danish bridge 3 decades earlier, has been involved in this work. Erection of the span's narrow 181-m tall concrete towers will also be challenging. This $165 million span has been debated for many years, and its construction has yet to commence.

New structural forms have, in turn, wrested once-planned suspension bridges from other deep-water fjord crossings. Floating bridges outcompeted suspension spans in the Krifast and Salhus projects. In the latter, a 1,090-m span box girder suspension bridge was replaced by a floating/cable-stayed span opened in 1995. In the mid 1990s, floating bridges initially competed against four suspension spans of 1,100, 700, 900, and 850 m in southern Norway's Tysfjord crossing. All lost out to a tunnel.

Deep-sea oil-platform technology—in which Norway holds considerable expertise—has been mooted for a crossing of the deep Høgsfjord on the country's southwest coast. Here, a box girder suspension bridge with twin 620-m spans centered by a 135-m deep tower, or a single 900-m span with tower footings up to 125-m deep, has been considered. Noted Danish engineer Niels Gimsing has also suggested that suspension bridges with floating towers anchored to the sea bottom with tension cables might be suitable for Norway's deep fjords. He has calculated that a continuous floating suspension bridge of three 400-m spans would be significantly cheaper to build than a single 1,200-m span, with cable steel dropping from 6,000 to 1,400 tonnes and box girder steel from 13,000 to 9,500 tonnes.

It is fitting that Norway's longest span—the Hardangar Bridge—will clear the same waters as one of that nation's groundbreaking efforts, Selberg's Fyksesund Bridge. These two bridges represent the innovation of two periods of Norwegian suspension bridge building. Future spans promise to maintain Norway's considerable but often overlooked heritage of suspension bridges, and to entrench its unique niche as a prolific producer of long, narrow spans.

13.2 THE STOREBÆLT EAST BRIDGE, DENMARK

The completion of the Lillebælt Bridge in 1970 left one, if formidable, obstacle to Denmark's physical unification—the Storebælt, or Great Belt. This 18-km wide channel separates the nation's two main islands of Zealand (home to Copenhagen, its capital) and Funen. It is one of the widest bodies of water ever bridged, and comprises a shallow 8-km west passage and a 9-km east channel; the latter cradles a 55-m deep gorge that funnels copious volumes of shipping destined to and from Baltic ports. Separating the two channels is the diminutive island of Sprogø, a vital stepping-stone for all proposals. Although low-level, multiple-span bridges dominated plans for the West Channel, many innovative long-span concepts would be generated in the quest to span the daunting East Channel.

Efforts to span the Storebælt came tantalizingly close to fruition on several occasions. Ferries had plied the strait since the 16th century, and in 1855, a tunnel for military use was mooted. The completion of the Lillebælt and Storstrøm bridges during the first great spate of Danish bridge building in the 1930s rekindled aspirations for a fixed link. A 1934 proposal for a rail bridge generated much discussion and was followed 2 years later by a scheme from three Danish contractors for a two-lane road/rail truss and girder bridge as part of their promotion of a national freeway network. The centerpiece of the East Channel crossing was a 400-m span cantilever bridge. The Danish government then announced a 10-year, $250 million program to bridge the Storebælt and the 15-km Øresund—the latter to link Denmark and Sweden. The Danish State Railways prepared the first government report for the

Storebælt in 1937 and, for the East Channel, proposed an overhead through truss with 300-, 350-, and 300-m spans inspired by the 1935 Lillebælt Bridge. In 1939, funds were earmarked for initial studies, but World War II intervened before this work could begin. It has been speculated that had war been averted, a bridge might have been completed as early as 1950.

The bitter winter of 1947 rekindled post-war interest in a crossing, and in 1948, a commission was established to study the question. Its work continued throughout the 1950s and 1960s as Denmark entered a second period of large bridge building. A five-fold increase in cross-Bælt traffic between 1948 and 1959 helped buttress economic arguments in support of a bridge. The commission's 1960 report proposed a two-level truss of identical span to the 1937 scheme, with four lanes supported by an upper deck and two rail lines below. Borings taken in 1963 even revealed foundation conditions to be far superior to those at the Lillebælt.

In 1965, the project appeared to acquire serious momentum when the government launched an international design competition. Two minimum spans of 325 m, or a single minimum span of 450 m, were required to cross the East Channel. A total of six road lanes and twin rail lines would also be required to clear the channel by a generous 67 m. The competition spawned an incredible 155 entries from 28 nations—and unexpected results. Of the 18 winners announced in 1967, 11 involved cable-stayed bridges of up to 600-m span, a surprise given the only recent evolution of this bridge form in its modern guise. A 450-m cable-stayed span from the American firm Sverdrup & Parcel captured one of four first prizes. West German engineer Ulrich Finsterwalder proposed a triumvirate of 350-m span concrete cable-stayed bridges with a deck varying in cross-section from just 0.4- to 1-m deep. Record span box girders in concrete and in steel of up to 400 m were also proposed. The sole suspension bridge finalist was a bizarre 488-m self-anchored span involving an 18-m diameter tube enclosing stacked three-lane roadways. American engineer Paul Weidlinger concocted this all-weather crossing.

It was hoped that road and rail links across the Storebælt and the Øresund would be open by 1982, but several factors would slow progress. By 1968, the threat of wayward vessels became apparent; the Storebælt's navigation channels punctuated the line of the proposed crossing at oblique angles and changed course on either side. That same year, three variants of an 800-m suspension bridge were examined. One was of conventional configuration, another featured stays radiating to the main cables over the main and side spans from the tower-deck juncture in a manner reminiscent of Steinman's arrangement, and yet another involved a Dischinger suspension/stay hybrid. The truss featured inclined struts connecting a narrower lower rail deck to a 33-m wide upper roadway. A tunnel was also mooted at this time.

Work then intensified with the 1970 establishment of a technical committee headed by Niels Gimsing that was charged with developing design concepts and cost estimates. At this time, suspension bridges were not considered sufficiently rigid to support rail traffic. One concept featured a sleek 1,670-m overhead truss bridge with two 375-m navigation spans separated by a central 300-m span; the triangular truss enclosing the rail lines could support three-lane roadways clipped onto either side at a later date. Much effort was devoted to tandem cable-stayed bridges splitting the two navigation channels. Two 600-m spans supporting six road lanes and two rail lines were favored, a remarkable situation given that no cable-stayed bridge then exceeded 350 m in span. Several contractors were asked to prepare preliminary cost estimates.

In 1972, the committee changed course and opted for a rail-only crossing comprising a low-level West Channel bridge and a tunnel beneath the East Channel; car shuttle trains would provide the most immediate relief for the lowest cost and still permit a road crossing in the long term. But the idea proved unpopular, and in 1973, legislation supporting a combined road and rail bridge was passed. The Arab oil embargo that autumn postponed a launch of the project organization Statsbroen SB, and design issues were soon subsumed by concerns about cost. Prolonged political debate finally came to an end in May 1976 with the creation of Statsbroen SB. Studies began immediately, with the link intended for completion in 1985. COWIconsult—formerly Ostenfeld & Jønson—led a design effort peaking in 1978.[3] The one- or two-span debate was finally put to rest with a decision to use a single span of at least 750 m. The alignment was fixed in June 1977, but the bridge's capacity proved contentious. The 1973 legislation had called for six lanes, but Statsbroen SB was not convinced of the need and adopted a four-lane cross-section in November 1977. This decision did not sit well politically, and in March 1978, Statsbroen SB acquiesced; the bridge would carry six lanes and two rail lines.

Preliminary designs for a 780-m span cable-stayed bridge were developed. Shipping safety had formed a rallying point for bridge opponents, so it was decided to also investigate a 1,500-m span with superimposed suspension and cable-stayed systems; Japan's plans for a 1,780-m span over the Akashi Kaikyo had now convinced Statsbroen SB that long-span road/rail suspension bridges were feasible. After examining 1,200-, 1,500-, and even 1,800-m spans, Statsbroen SB settled on a 1,416-m span flanked by relatively short 360-m side spans to enhance stiffness. Two main cables—each containing 34,432 wires—would be spun in situ or use prefabricated parallel wire strands. Several truss variants and a slender box girder were devised, but a 31-m wide, 10.4-m deep trapezoidal truss was adopted for both the suspension and cable-stayed options; its optimum depth was even less but for the need to accommodate the railway's electrical system. Sloped box members connected the six-lane upper orthotropic road deck to the narrower twin-track lower rail deck. The truss was also continuous through the towers to strengthen it against localized deformations from train loads.

The suspension bridge imposed a 10% cost premium, but its improved navigation safety convinced Statsbroen SB to tender it with the 780-m cable-stayed span; either type would establish a span record. Much attention was directed to their appearance and, in particular, the towers and anchorages of the suspension bridge. The 220-m tall concrete towers would be modeled on those astride the Lillebælt. From dozens of anchorage configurations emerged a unique open triangular frame of remarkable transparency in which angled legs channeled the cable strands down to the rear of the anchorage. Unfortunately, these designs emerged under a darkening cloud of political uncertainty. A February 1978 report suggested the government finance the $1.1 billion Storebælt link or a $2 billion North Sea natural gas pipeline, but not both. A construction start planned for May 1978 was delayed, but Statsbroen SB continued design work and, by August, was negotiating contracts to launch 5.5 years of construction. National elections that month then produced a coalition government bent on reducing the national debt and, on August 31, the new government shelved the project. The decision stunned Statsbroen SB; the design was virtually complete, and four international consortia had been short-listed to build the bridge over the East Channel. Contract bids were to be called on October 15, just 6 weeks away. Now, the project was to be postponed for at least 5 years. Statsbroen SB completed the detailed design of the bridge, but by late 1979, the organization longer existed.

Four years later, in 1983, the government authorized new investigations, again headed by Gimsing. Three years of study by the same COWIconsult-led team produced major changes for the East Channel. By paring down capacity to four lanes and reducing train weights and speeds, engineers now considered a trapezoidal steel box girder to be feasible. A 35-m wide deck with road lanes cradling two rail lines evolved from the Farø Bridge box girder, a COWIconsult-designed 290-m cable-stayed span completed in 1985. Further study suggested a bored road tunnel to be prohibitively expensive but a bored rail tunnel feasible. This pointed to a road-only bridge over the East Channel, which without the railroad could be lighter, longer, and less expensive. An 1,800-m box girder span with taller versions of the 1978 towers was considered possible, but plans to replace approach spans with massive embankments all the way to the anchorages raised concerns among many over blocked currents and salinity disruptions in the Baltic Sea. This was no small concern, as almost 75% of its volume flows through the Storebælt. This issue alone would profoundly shape the crossing's design.

Political agreement still remained elusive as the government coalition's leading party supported a rail crossing only. Only in June 1986, when it was agreed to give the rail link a 3-year head start over the road link, was the necessary compromise reached. This decision separated the modes of transport, and it became almost certain that the East Channel road crossing would be a bridge. Legislation enacted in May 1987 finally paved the way for the estimated $2.3 billion project; work on the low-level West Channel road/rail span and East Channel rail tunnel would begin in 1988, and the East Channel road crossing would commence 3 years later and open in 1996, 3 years after rail link. A so-called "Zero Solution" also compensated for water blockage from the bridge piers and an almost quadrupling of the 40-hectare island of Sprogø by way of massive amounts of dredging; a 1,200-m embankment off Zealand was also abandoned and others shortened. Preparatory work began in May 1988, and that November, a $470 million contract for the 7.8-km East Tunnel was awarded. A controversial June 1989 award of a $410 million contract for the West Bridge then landed Storebælt A.S.—successor to Statsbroen SB—in the European Court of Justice over charges of favoritism toward Danish bidders. The new company was eventually compelled to compensate the losing bidders.

Meanwhile, design work on the East Channel road crossing intensified. By October 1988, the last tunnel option—an immersed tube—was discarded. COWIconsult reworked the 1978 suspension and cable-stayed schemes into four-lane spans. For both options, it replaced the double-deck truss with a 4.5-m deep, 30-m wide box girder. In the suspension bridge, COWIconsult lengthened the side spans from 360 to 480 m. Taller 233-m concrete towers and the open triangular anchor blocks were retained, although refinements the following year eliminated the tower's lower strut, and submerged anchorages were even considered.

Storebælt A.S. still intended to offer suspension and cable-stayed spans for bidding. Yet navigation safety remained an issue. This prompted further research, which by early 1989—and benefiting from studies for Japan's Akashi Kaikyo Bridge—indicated a 780-m span to be inadequate. It also now appeared that the most economical span lengths would be around 900 and 1,200 m for cable-stayed and suspension bridges, respectively. CBR, a joint-venture of COWIconsult and Rambøll (the latter an amalgamation of the firms that had previously worked with COWIconsult on the 1978 designs)—with Acer Freeman Fox (formerly Freeman Fox & Partners) and Leonhardt Andrä und Partner—devised outline designs for record-breaking cable-stayed spans of 916 and 1,204 m and for 1,448- and 1,688-m span suspension bridges.

The cable-stayed spans avoided the potential navigational risk presented by large anchorages. Although the 916-m cable-stayed span was considered feasible (COWIconsult had undertaken detailed design of the steel box girder for the 856-m span Pont de Normandie), uncertainty surrounded the 1,204-m span. More than half the wind load borne by the box girder would be transmitted from the cable stays, and to ensure stability during the 600 m of free cantilevering during erection, the box girder would have to be widened by almost 6 m. The 310-m tall towers introduced other construction imponderables. In the end, the unknowns, although considered manageable, made the larger cable-stayed span too risky (although by 1991, COWIconsult claimed 1,200-m cable-stayed spans to be feasible).

Meanwhile, navigation simulations suggested only the 1,688-m suspension span would maintain existing levels of shipping safety without protective measures or the straightening of navigation channels. Mathematical modeling also favored a span of at least 1,500 m. With shorter spans requiring more compensation dredging and deeper and well protected piers, the 1,688-m span became competitive in cost, and this was the bridge—flanked by 660-m side spans—recommended by Storebælt A.S. in February 1990. Storebælt A.S. also urged a partial straightening (aided by some dredging) of navigation channels for added insurance. Approval was obtained in March for a slightly reduced span of 1,624 m that would eclipse the Humber Bridge by 15% and be the first suspension bridge to surpass the magic mile (*see* Fig. 13-2).

An unusually large sag ratio of 1:9 reduced horizontal—or sliding—forces at the anchorages and thus costs; it also produced tall 254-m towers in either steel or concrete. Steel towers, with three struts above and a single strut below—but not supporting—the box girder could be built more quickly, but the concrete towers were simpler, with just two struts. Side spans of 535 m were flanked by approach spans extending 1.5 km from an embankment off Sprogø and 2.5 km from Zealand; concrete (at 124 m each) and steel (at 168 m each) spans were offered as options. By May 1990, an outline design was ready for tendering. The prequalification of eight consortia by Storebælt A.S. then touched off a second tempest when French contractor Bouygues, which had been at the center of the West Bridge controversy, protested its exclusion to the European Community. It claimed to possess the financial and technical muscle to carry out the job, including Briton William Brown and the Japanese long-span specialist firm Chodai. This time, the protest fell on deaf ears.

A summer 1991 contract award was planned, but obstacles elsewhere threatened

Fig. 13-2. East Bridge, Denmark
PHOTOGRAPH 4411-13 BY JAN KOFOD WINTHER, COURTESY OF STOREBÆLT BILLEDARKIV.

the schedule. Tunneling suffered delays, and technical and labor issues dogged progress on the West Bridge. It also appeared that the tight East Bridge schedule had pushed up the bids. It was decided to extend construction by 6 months to the spring of 1997, retaining the railroad's politically ordained 3-year lead intact. Then, in the summer of 1991, the government halted contract negotiations until the tunnel problems were resolved. Preliminary works sat idle, including a huge 25-m high embankment off Sprogø, protective islands near the anchor blocks, and excavations at the tower bases.

In June, another serious—and potentially ruinous—dispute boiled over. The East Channel provided Finland's only deep-water access out of the Baltic Sea for leviathan oil rigs destined for North Sea oil fields. Finland had expected a redesign as early as 1989, but navigation clearances had already been raised in response to Soviet concerns. Maritime nations had been notified as early as 1977, with no Finnish concerns. Yet in the spring of 1991, Finland sought an injunction before the

International Court of Justice at The Hague to prevent further design work. It also claimed the 65-m clearance would be inadequate to allow for the increasing size of Finnish-built luxury liners. Denmark rejected Finland's offer to help fund a bascule bridge—or a tunnel—and the Finns similarly rejected Denmark's offer to fund the dismantling and re-assembly of oil rigs. The International Court of Justice turned the injunction down that July, but it ordered further negotiations that produced a September 1992 agreement whereby Denmark would pay Finland $15 million in compensation.

The superstructure contract posed another dilemma. A consortium led by Italian contractor CMF Sud had bid almost 25% less than anyone else. Yet the firm had virtually no bridge-building experience; apart from some deck fabrication, it would subcontract work to other companies. CMF Sud had retained the Steinman firm—now part of the giant Parsons Transportation Group and vying for its first large new suspension span since the Tagus Bridge 25 years earlier—to bolster its bid. It was rumored that CMF Sud had bid to enhance its chances of building Italy's long-discussed Messina Strait Bridge. But only with performance guarantees issued by the Italian government did Storebælt A.S. accept its bid. Two contracts totaling $831 million were finally signed in October 1991 with a German/Dutch/Danish joint venture for foundations, concrete towers, and anchorages, and with the CMF Sud consortium for the main cables, suspenders, and box girder suspension and approach spans. This momentous event was clouded by the fact that it followed by just 1 week a devastating flood that brought work on the East Rail Tunnel to a halt for several months.[4] Completion of the East Bridge was now slated for late 1997.

CBR undertook detailed design from 1991 to 1993, with COWIconsult (now known as COWI) focusing on the steel superstructure under the direction of Klaus Ostenfeld, a nephew of the firm's co-founder. Tests on a 1:80 scale sectional model were used to refine 16 different box shapings into a 4-m deep, 31-m wide trapezoidal box girder. Taut strip modeling at the University of Western Ontario was undertaken to confirm the design in turbulent flow, and later a new boundary layer wind tunnel at the Danish Maritime Institute tested a 14-m long aerolastic model of the bridge; this model now sits in the exhibition center at the project's east end. The box girder was designed to withstand flutter in winds of more than 266 km/hr, twice those ever measured in the Storebælt. The main cables are attached to the deck at midspan by a 6-m long triangular clamp, and hydraulic buffers at the anchorages restrain the box girder. These buffers should reduce total longitudinal movement in the bridge's life to 10% or less than that without buffers. Railing design demanded care, as snow blockage was found to reduce the span's critical velocity by up to one-third. Nonetheless, a 3.5-m high perforated windscreen can be added to enhance traffic safety with only a 15% penalty in critical velocity. In total, these measures permitted a shallower box than at the Humber Bridge in spite of the longer span.

Other features include the dehumidification of all 7 km of suspension and approach span girders and of the anchorage cable splay chambers. No expansion joints are found over the 2,694 m of suspended span between anchorages—a record—nor over the approach spans. The structural continuity of the box girder reduced deformation from traffic loads by 25% but required joints at the front of each anchorage to take up record thermal expansion of up to 2 m. The box girder hangs free through the towers with no portal strut or vertical links; buffers serve only to limit lateral motion, although torsional restraint can be provided for if deemed necessary in future. The design avoided the need for stronger steels to counter otherwise large bending stresses in the suspended structure near the towers, a measure required in Japan's Bisan Seto spans. Transverse trusses to improve the fatigue resistance of the orthotropic deck also replaced initially planned solid bulkheads inside the box girder (*see* Fig. 8-3).

The triangular anchorages and the towers were also refined with the aid of the architectural firm Dissing & Weitling. Lighter anchorage braces were devised, the anchorages shortened, and the large ballast pier in the 1978 scheme replaced with a slender pylon. Most of the 325,000 tonnes of concrete, steel, and ballast (mostly sand) in each anchorage sits below water. The use of just two above-deck portals in the concrete towers is unusual (*see* Fig. 13-3). Wind loads from the suspended structure and cables normally taken up by a lower strut are absorbed by a flaring of the tower legs, which are seamlessly integrated into a pier that extends 21-m above water for ship protection. Here, as with the anchorages, simplicity masks enormous scale. Each tower contains some 95,000 tonnes of concrete; in contrast, the taller steel towers of Japan's Akashi Kaikyo Bridge each weigh just 23,000 tonnes. With ballast, each East Bridge tower from seabed to tip weighs 190,000 tonnes.

Two huge drydocks 70 km from the bridge were excavated to pour the 18 approach span piers and four main caissons. At the tower piers, overburden was first removed and 10-m thick stone pads then laid, compacted, and leveled. The 78- by 35 m east tower caisson was sunk to a depth of 25 m in April 1993, after which some 65,000 tonnes of sand ballast were then placed within its 32,000-tonne shell. Two weeks later, the west tower caisson was in place. The 36,000-tonne, 112- by 55-m anchorage caissons were placed in October. Their front and rear thirds bear on triangular gravel pads wedged 22 m into clay, but their middle portions lack any support from below.

As it happens, the west anchorage caisson encountered some rather unusual obstacles. Its arrival on site was first marked by the discovery that its gravel bed was encrusted with sand and mussels. Failed attempts to vacuum the mussels off forced the caisson back to the production site. Then as the mussels were scraped off, the weather deteriorated. When the caisson did return, strong currents prevented its sinking. Only six days following its initial arrival was the caisson finally in place, alleviating fears of a postponed sinking until the following spring. Waste concrete, iron ore, olivine, and copious amounts of sand were then dumped into the anchorage caissons, followed by almost 2 years of concreting. Meanwhile, the towers had begun to rise in mid-1993 and stood completed in August 1995.

Tests for the 4-km long cables of Japan's Akashi Kaikyo Bridge had led Storebælt A.S. to consider prefabricated parallel wire strands as well as in situ spinning for the two 83-cm diameter main cables. Yet in spite of an expected savings of 3 months, prefabricated parallel wire strand cables proved uncompetitive. Spinning was adopted in the winning bid; each 18,648-wire main cable would comprise 37 strands of 504 wires and, with a length of 2,965 m, be the longest ever spun. Spinning was to begin in the midwinter months of 1995 to 1996—an unfortunate consequence of the tunnel delays and political obligations concerning the rail link. But other problems soon

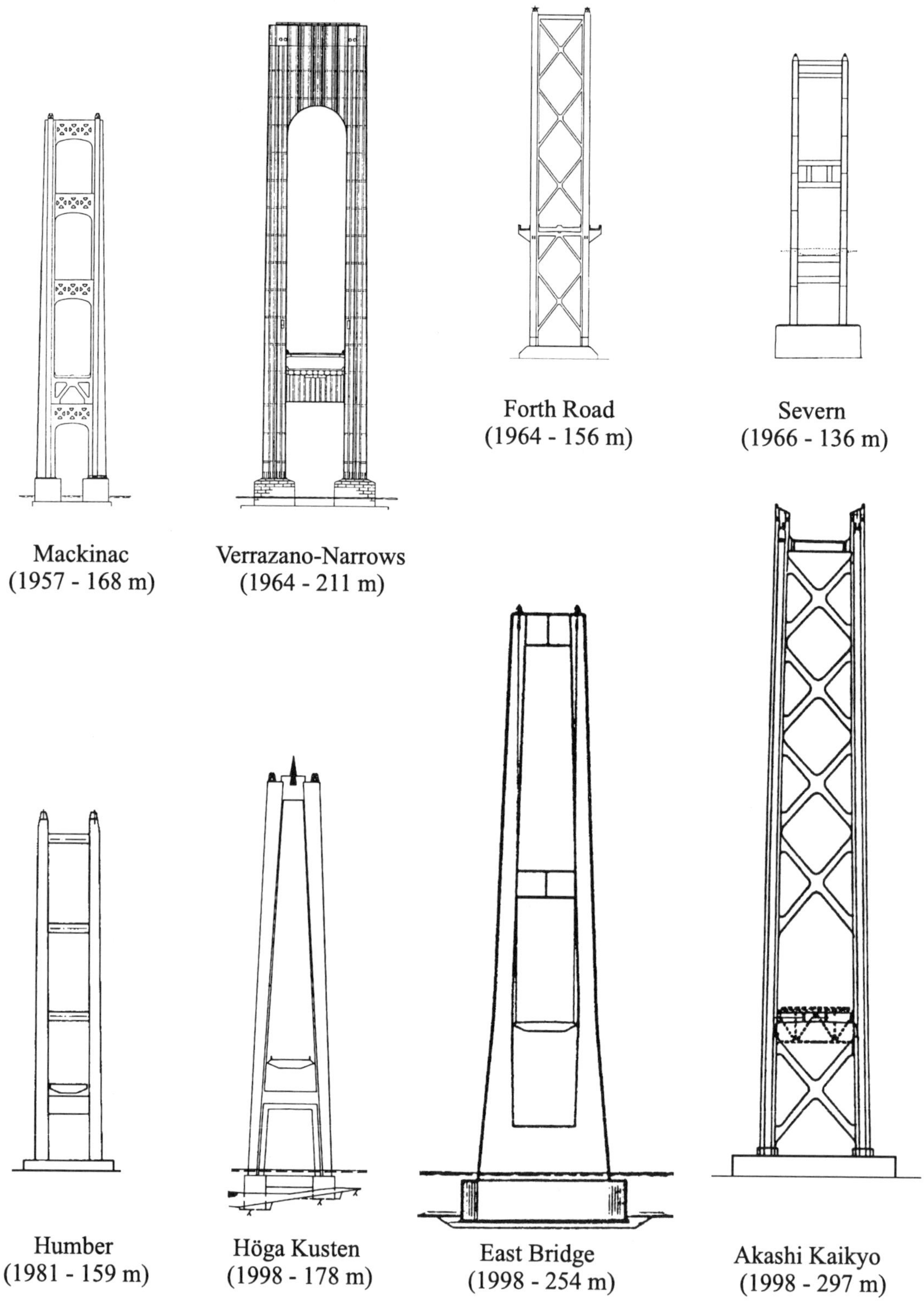

Fig. 13-3. Suspension bridge towers

MODIFIED FROM ORIGINAL DRAWINGS COURTESY OF PARSONS CORP. (STEINMAN), MTA BRIDGES AND TUNNELS SPECIAL ARCHIVE, THOMAS TELFORD LTD., HUMBER BRIDGE BOARD, COWI ENGINEERS AND PLANNERS A/S, SUND AND BAELT A/S, AND THE HONSHU-SHIKOKU BRIDGE AUTHORITY.

emerged to impose yet further delay. Cooling pipes built into one anchorage to prevent the overheating of freshly poured concrete failed and froze. The brutally cold winter of 1995 and 1996 then brought ice-clogged channels that restricted transport. Work—when it could be done at all—inched along in wind chills of -30° C. It was almost March 1996 before the first 100-tonne tower saddle was rolled up its tower leg. To cap things off was the discovery of an explosive containing more than 300 kg of dynamite between the west tower and anchorage, one of more than 180 dropped in the East Channel by the British during World War II; it was towed away and detonated (a sunken British World War II bomber had also been blasted away during construction of the West Bridge).

It was now clear that the bridge would not open in 1997. Yet 1996 also brought major achievements. In August, the last of 26 pre-assembled steel approach spans erected during a 19-month period was placed. These approach bridges had been a source of additional concern. Coinfra (successor to CMF Sud) had increased approach span lengths from 168 to 193 m while retaining the original 6.7-m box depth. Aerodynamic testing then predicted vertical oscillations in specific—if rare—circumstances that, although not structurally threatening, might disturb motorists. COWI devised moveable "tuned mass dampers" that hung from springs inside the girders and were restrained by hydraulic pistons to counteract this sway. The firm had already retrofitted a number of bridges in this way, including the Humber and the Severn bridges (Bolton, 1995). As part of Storebælt A.S.'s research effort, the Farø Bridge was even deliberately "excited" to observe its response. The fix, although effective, cost $5 million.

Cable spinning finally began in July 1996 in the relative calm of summer, with the distinctive arc described by hundreds of catwalk lights soon earning the span the nickname of "Big M." Spinning improved on the controlled-tension method developed by the Japanese (*see* Chapter 14) for the Hirado (1977), the Shimotsui Seto (1988), and the second Bosporus Bridge (1988); 504 adjustments of individual wires in each strand were now replaced by a single adjustment of a complete strand. To further simplify adjustments, wire tension was increased to 80% of its free-hanging tension, much higher than the 25% used in Japan and the 50% in the second Bosporus Bridge. Still, the method's complexity required 4 years of preparation before the first wire was strung, and each wire's precise placement within the cable had to be programmed into a computer. In spite of these demands—and in part because of them—the payoff would be tremendous.

The large open base of the east anchorage served as a wire handling area, from which spinning wheels drew four wires at a time. Two "dead" wires drawn from the bottom of the wheel were laid into strand formers that cradled and separated the strands as they were built up. The "live" wires drawn over the top of the spinning wheel were placed on rollers beside the strand former. On its return trip from the west anchorage, the empty wheel transferred the former live wires from the rollers into the strand formers—again at computer-controlled tension. During this time, another wheel traveling from the east anchorage laid down four new wires. Completed strands were adjusted and clamped firmly into the strand former in its final position within the cable cross-section, thus adding stiffness to the catwalk. French nationals that built and operated ski lifts in the Alps were recruited to spin the cables and proved well-suited to the task, laying up to 259 tonnes of wire in a 24-hour period. Spinning continued even in late autumn winds exceeding 55 km/hr, and only gusts of up to 150 km/hr toward the end halted work. The completion of the cables ahead of schedule in late November halved traditional spinning costs and shattered productivity records; spinning efficiency was a whopping 47% higher than achieved in the Verrazano-Narrows Bridge. William Brown, who advised the contractors, observed that this spinning rate could even erode the supremacy of cable-stayed bridges in smaller spans. Indeed these advances, aided by Norwegian equipment, would upset Norwegian suspension bridge cabling tradition, as described earlier in this chapter. The completed but uncompacted cables were then wrapped in red plastic to prevent ice forming between the strands. This tactic avoided the frustrations faced by the J.A. Roebling's Sons Company when spinning the cables of the second Tacoma Narrows Bridge; Blair Birdsall, who at age 89 helped the Steinman firm, with others, devise the spinning strategy over the East Channel, had personally experienced that debacle. Birdsall, who began his career with the Roebling company on the cables of the Golden Gate Bridge in the 1930s, would die in 1997.

By the spring of 1997, some 30,000 tonnes of box girder were ready for lifting into the suspended spans. Steel fabricated in Italy and Portugal was shipped by sea to northern Jutland for storage (although inauspiciously several approach span deck panels fell into a storm-tossed Mediterranean from the first shipload in December 1992). Suspended span boxes were typically 48-m long and weighed 550 tonnes. In April 1997, the month the rail tunnel opened, an initial 57-m section weighing more than 1,000 tonnes was lifted into midspan. An unusual erection sequence used previously in the Severn Bridge halved the normal number of lifting struts; all 33 main span boxes were placed first, working from midspan to the towers. Poor weather hampered early operations, and 2 weeks were lost to repairs of a lifting apparatus when the main span was almost fully placed. Not until December were all 24 side span boxes in place. The box segments passing through the towers were first suspended from the lower portal strut, but were only lowered into place after the entire suspended span was welded; they are not suspended, but are instead supported by adjacent deck units.

The suspended spans also managed to generate additional aerodynamic research. Wind tunnel tests in 1993 had revealed a minor vortex shedding problem; guide vanes bent around the bottom corners of the box girder were devised to quell the instability, which, although structurally innocuous, might disturb motorists. It was decided to observe the completed bridge before proceeding, as turbulence was thought to possibly inhibit vortex production and hence the motion. However, during welding of the box girder the phenomenon was incited in winds of up to 36 km/hr striking the bridge perpendicularly. Paving was then thought to increase structural damping as a defense, but it did not. So the vanes went on. Some 1,400 m of main span were graced with 2-m wide curved vanes set 60 cm from the girder sides. Installation began in April 1998, but a 2-week labor dispute saw the last steel placed just 2 days before the bridge's opening in June. Ironically, accurate measurements of their effects soon ended with the theft of the on-site monitoring computer, but visual observation suggests the vanes have successfully quelled the vortex shedding problem.

The East Bridge opened in June 1998, just 2 weeks shy of a full decade of construction on the entire $3.2 billion link. The bridge would suffer from cost overruns that brought its total cost to almost $1.1 billion, and it would also open 8 months behind schedule. This, of course, paled in comparison to the tunneling problems, where delays had the effect of whittling down the planned 3-year lead over the road bridge to just more than a year. These woes did little to quell enthusiasm for the link; some 275,000 people ran, walked, cycled, or otherwise crossed it sans auto during the opening celebrations.

The East Bridge is a technical and an aesthetic triumph. It is the longest box girder span yet built and the longest suspension bridge with a continuously suspended structure; it is also the first large suspension bridge since Mackinac to feature long over-water approaches. In total, some 37 million tonnes of material would be moved for the entire project—14 million tonnes for compensatory dredging alone to maintain the Zero Solution. Yet for its massive size, the grace of the East Bridge is unmatched The approach and main span girders describe an incredibly thin line over a distance of almost 7 km. The remarkably transparent triangular anchorages improve on a shape characteristic of Ammann's bridges. An unfettered passage of the box girder through the simple but elegant towers is exceptionally clean, and the towers' unusual flaring is a visually superior solution for ship protection than "dolphins" or stone islands. A dearth of roadway lighting is also unique for a long-span suspension bridge.

The 1986 agreement that brought the Storebælt project into being also required the study of a road/rail crossing of the Øresund, the 15-km channel separating Sweden and Denmark. Agreement to build this $2.1 billion crossing was reached in 1991; the link—which features a 3-km immersed tube tunnel, a 510-m cable-stayed span, and several kilometers of low-level bridge—was opened in 2000.

A road and rail crossing of the 19-km Fehmarnbælt between Lolland in southern Denmark and Germany also came under study as a result of a Swedish condition in the Øresund agreement; this link would greatly reduce distances from Sweden to continental Europe. One early concept devised in 1964 had envisaged twin Dischinger-type hybrid double-deck spans centered by an open-framed central anchorage. Initial studies following the 1991 Øresund agreement suggested the link to be less technically demanding than the Storebælt project. Water depths, for example, do not exceed 30 m. These studies generated concepts for multiple cable-stayed bridges, a suspension bridge, and a combined tunnel/low-level bridge. At the end of 1992, COWI—with the German firm Lahmeyer—examined several schemes, including a 1,300-m span box girder suspension bridge with two-lane roadways straddling a single rail line in the median. The scheme's 250-m tall multiply braced towers resembled the unbuilt steel pylons of the East Bridge. Almost 15 km of 150-m approach spans were envisaged for this $3 billion crossing, as were construction techniques developed for the Storebælt crossing. By 1995, the field had been whittled down to two options—a cable-stayed bridge with three 724-m spans and a 1,752-m span suspension span flanked by 612-m side spans. The configuration of the suspension bridge was in the spirit of the East Bridge save for a 15-m deep truss accommodating two lower-deck rail lines; it, too, was continuous between open triangular anchorages. Further study had by 2000 favored the cable-stayed option, but the economics of this crossing remain uncertain and a decision to proceed remains to be made.

13.3 HÖGA KUSTEN BRIDGE, SWEDEN

Six months before Denmark's East Bridge opened, neighboring Sweden completed a major suspension bridge on its east coast 500-km north of Stockholm. The nation's largest previous effort, the 418-m span Älvsborg Bridge, had been completed more than 30 years earlier, in 1966. The Höga Kusten—or High Coast—Bridge would be almost three times the size, but it was conceived, planned, and built within a single decade.

The need for a crossing stemmed from the inability of the Sando Bridge, a record 265-m span concrete arch bridge completed in 1943, to support the same increases in permitted truck weights that had affected the Severn Bridge. The issue prompted a study of solutions in 1987. A Sando Bridge upgrade would not resolve deficiencies in its approaches, but a crossing 10-km downstream near the mouth of the Ångermanälven River showed promise. Ferries had plied this 1.5-km wide gap since 1620, and a crossing here had been suggested in the 19th century. The opening of the Sando Bridge had put paid to further discussion, but now a 5.6-km tunnel and 1,170-m span suspension bridge came under scrutiny. With water depths reaching 90 m, the tunnel was found to be almost 70% more expensive than a 1,170-m span, $116 million suspension bridge. The bridge, adopted in 1990, would rank among the world's top 10 in span. Initial designs called for two lanes supported by a 13-m wide steel box girder suspended between 159-m tall concrete towers. Unsuspended concrete box girder side spans of about 200 and 100 m would extend to anchorages tunneled into the rugged slopes. Completion was planned for 1997.

The alignment was not without controversy. A mélange of forests plunging down steep hillsides to the ocean, myriad bays and inlets, and picturesque pockets of farmland had earned the area a designation of national scenic interest. The Swedish road administration's Council for Culture and Beauty strongly influenced the design of the 32-km project, and the appearance of the suspension bridge was a crucial concern. Studies proceeded under the Stockholm-based firm Kjessler & Mannerstråle, with COWI responsible for detailed design. Changes were soon injected into the preliminary scheme. A shift in alignment to enhance views from the bridge lengthened the main span to 1,210 m, whereas suspended side spans of 310 and 280 m were adopted to encourage a clean uninterrupted profile for views of the bridge. The concrete towers were also placed just offshore in the shallows, which also reduced ship collision risk. Only two portal struts are used over their 178-m height; the lower strut sits well below the box girder, which cleanly passes through the towers to a height of 40-m above water. Unusually, the battered tower legs thicken across their faces as they rise and slope inward; they plunge dramatically into the water uninterrupted by pier caps, but they are exceptionally strong to withstand ship collision (*see* Fig. 13-3). Adorning the top portals are 11-m tall steel-framed glass pyramids—a modern version of the flèches from Steinman's unbuilt Liberty Bridge of the 1920s in New York City. Although costlier, the revised bridge design received government blessing (*see* Fig. 13-4).

The orthotropic box girder received special aerodynamic treatment. One early decision was to widen the box from 13 to 22 m to allow expansion to four lanes. Its slender 4-m deep trapezoidal shape emerged from five configurations tested at the

Fig. 13-4. Höga Kusten Bridge
PHOTOGRAPH COURTESY OF MONBERG AND THORSEN A/S.

newly opened boundary layer wind tunnel of the Danish Maritime Institute. A key concern was ice and snow accumulation. A sharply sloped edge molding that shed snow and ice more easily than rounded or horizontal edges was adopted after wind tunnel testing on sectional models. Provision was also made for heating cables in the molding, and paint was even formulated to prevent snow sticking to the girder edges. Finally, open railings were smoothly shaped to minimize snow blockage. Other COWI trademarks included dehumidification of the box girder to eliminate the need for interior painting, as well as structural continuity over the 1,800-m between anchorages, where hydraulic buffers take up expansion of up to 1.7 m (versus 2.9 m had there been none). The main cables are fixed to the box girder at midspan with a 6-m long clamp, and bearings at the towers restrain lateral movement. So armored, the box met the target flutter critical velocity of 180 km/hr—even with snow-blocked railings—although provision has been made for tuned mass dampers at midspan should snow blockage affect aerodynamic stability. To date, it hasn't (*see* Fig. 8-3).

Final approval to build the bridge was obtained in June 1993. Work began that November, with completion slated for the summer of 1997. Swedish construction giant Skanska headed foundation and tower work, and the Danish contractor Monberg and Thorsen built the steel superstructure from material of diverse origins—Swedish steel for the box girder, Spanish wire in the main cables, French-made suspenders, and cable saddles forged in Denmark. The east tower required the excavation of just 5 m of rock, whereas its western twin stood on exposed rock 17-m below water. The 37 strands in each main cable are anchored in rock via prestressed steel tendons using Norwegian techniques found in such spans as the Askøy. Each tendon plunges down a 37-m long drilled shaft, emerging in a chamber to be stressed against the intervening rock, which forms a cost-free counterweight to the bridge's weight. All 4 anchorage cable splay chambers are dehumidified.

By February 1995, the towers had reached a height of 60 m in a complex operation that required changes to leg shape and thickness for all 42 formwork lifts. Work was on track for a September topping-out when in April a fire broke out during concreting operations at a height of 80 m in one leg of the south tower. During the incident, five workers were forced to seek refuge beneath the tower's lower strut to avoid being hit by burning debris. Six meters of damaged tower leg proved beyond repair and had to be lopped off and recast. Not until December 1995 were both towers completed. The 60-tonne welded—instead of the usual cast—steel cable saddles were then rolled up the outside of the tower legs, and in March 1996, the pilot rope for the catwalks was placed high above the river by helicopter.

Prefabricated parallel wire strands and in situ spinning were both offered to contractors bidding on the 65-cm diameter main cables. Spinning was adopted, with each 11,248-wire cable assembled from strands of 304 wires each; four additional strands totaling 480 wires were spun into each side span cable. Spinning operations, which began in July 1996, eschewed the controlled tension method utilized on Denmark's East Bridge, and the simultaneous but different spinning methods on both spans generated much rivalry. The previous harsh winter and tower delays had raised fears that spinning might not be finished before winter, but crews put in a memorable performance; by September, daily spinning output was 50% higher than the 150 tonnes expected. All 8,500 tonnes of wire were in place by late autumn.

The 14,000 tonnes of box girder were welded in Finland into 48 segments—each typically 40-m long and weighing 320 tonnes. Unusually, the side spans were placed in September 1995 before the towers were even completed. In this 10-day operation, boxes were placed onto temporary steel supports by floating crane and then simply slid into position. During cable spinning in the summer of 1996, the boxes were welded together, but only with the lifting of the 31 main span boxes were the side span boxes attached to the main cables, becoming suspended when the weight of the main span lifted them off the falsework. Concerns about aerodynamic instability during the early stages of main span box lifting had prompted wind tunnel testing of a

12-m full aerolastic model of the bridge at various stages of box erection at the Danish Maritime Institute; rapid installation of the first few sections was seen to overcome any problems. In the event, erection proceeded at incredible speed, with all 1,210 m of main span lifted into place by floating crane over just 11 working (and 18 calendar) days in April and May 1997.

Notwithstanding the speed of spinning and box erection, the span's December 1997 opening was 5 months behind schedule. Nonetheless, the construction period of just 4 years was remarkable given the span's northern location—just 150 km shy of the Arctic Circle—with short working seasons, bitter cold, and snow, ice, and wind that predominate for up to 5 months a year. The absence of any serious accident was unprecedented for a bridge of this size, and it was all the more remarkable given the international flavor of the workforce—Swedish, Finnish, Norwegian, Danish, and Dutch—that was largely familiar with only their native tongues. But perhaps most outstanding is the Höga Kusten's striking appearance. The $167 million span is perhaps the sleekest long-span suspension bridge of all. The unfettered towers and pencil-thin deck are incredibly transparent. The three-color ensemble of terra cotta deck, concrete towers, and light gray main cables is also unique. One other accolade—the Höga Kusten is the world's longest toll-free suspension bridge.

ENDNOTES

1. The Bingsfoss, Framnes, and Vrengen bridges have since been replaced by other span types.
2. Although F.H. Frankland (1934) describes the Elverum Bridge as a 1932 span of 180 m, Selberg (1950) cites a 160-m span that, with the Tynset, represents a class of "recent Norwegian suspension bridges." Selberg's span length is used.
3. The other firms were B. Højlund Rasmussen and Rambøll & Hannemann.
4. The tunnel would open more than 3 years late at a cost $860 million, 80% more than contracted for.

Chapter 14

JAPAN'S REMARKABLE EVOLUTION

One remarkable post-Tacoma Narrows Bridge phenomenon has been the emergence of Japan as a leading builder of suspension bridges. Only in 1955 did Japanese engineers surmount the 200-m span barrier with a 216-m steel arch—the Sakai Bridge—near Nagasaki. Japan's progress, however rapid, has been inevitable. Seas and wide straits separate its four major and countless subsidiary islands. Moreover, the nation's emergence as a major industrial power has enabled it to amass the enormous financial and technical resources required to overcome these immense physical barriers. The costs have been staggering, amounting to tens of billions of dollars, but in the process many technological and construction management frontiers have been breached by a nation that in 1960 had not one large suspension bridge. This remarkably rapid evolution has occurred, paradoxically, at a carefully calculated pace.

14.1 EARLY BEGINNINGS

Small suspension bridges of modern vintage first spanned Japan's mountain valleys. The first was the Katsuyama Bridge, a 1916 span of 93 m. Larger bridges followed—the 86-m span Niyodo (1921), the 110-m span Kawamoto (1922), and the 139-m span Shoki (1924). Two bridges stand out from this formative period. The Miyoshi Bridge, a 140-m trussed span completed in 1927, was then the Orient's longest suspension bridge. Its designer, Jun Masude, had studied with the great J.A.L. Waddell (who had taught at Tokyo's Imperial University from 1882 to 1886) and the Miyoshi Bridge's trussed towers resembled those of an American span, the Bear Mountain Bridge.[1] The second was the Kiyosu Bridge across the Sunida River in Tokyo. Completed in 1928, this self-anchored span of 91 m resembles Germany's 1915 Cologne-Deutz Bridge. To suspend the span, engineer Yutaka Tanaka used neither cables nor eyebars; each "cable" comprised six layers of flat rolled steel plates arranged vertically, eyebars being confined to the final link in the 46-m side spans. Twin 2.6-m deep steel box spine girders supported a 26-m wide deck and constituted the first use of higher-strength steels in Japan. The span became well-known throughout the country for its technical and aesthetic prowess.

World War II severely punished Japan's infrastructure, and the 1950s marked the nation's first real foray into suspension bridge building, mainly in connection with dam construction. One of the more unusual bridges of this period was the Tabisoko Bridge, a two-lane 114-m span completed in 1954, and only the fifth to use David Steinman's Florianopolis truss (and the second of the type to use wire rope cables).

Other trussed spans include the Shin Odomari Bridge, a handsome 150-m span with double-braced portal towers (1957), the 102-m span Daito Bridge (1959), and the Yakumo Bridge, a clean single portal-towered 114-m span with a slender deck truss (1961). The largest suspension bridge of this era was the Kawazu Bridge, a 178-m span notable for a triangular deck truss similar in concept to Fritz Leonhardt's Tancarville proposal of the early 1950s.

One other span of note is the Konaruto Bridge, an unusual four-span truss-stiffened structure of 440 m completed in 1960 on the island of Shikoku near the Naruto Straits. It is a rare modern example of a true tandem suspension bridge devoid of a central anchorage (the cable truss being a hybrid). An A-framed steel tower separates two 160-m spans, which in turn are flanked by smaller anchor spans. New high-strength steels were also used in portions of the bridge. In spite of its peculiar geometry, the Konaruto is a handsome structure. That same year, work began on a Tokyo span of similar visage to the venerated Kiyosu Bridge. Yet the 69-, 142-, and 69-m spans of the girder-stiffened Kasai Bridge are more of a Vierendeel truss[2] than a suspension bridge, for the flat steel plates forming the span's upper chord "cables" are discontinuous at midspan. In this way, the bridge resembles Germany's bridle-chord Duisburg-Homberg Bridge. The Kasai Bridge opened in 1963. Other small suspension bridges of this period include the 129-m span Nishihaga, the 108-m span Miye, and the Owatari bridges; all used a novel procedure to reduce erection stresses by permitting the re-adjustment of suspender lengths following deck placement.

In 1955, studies began on what would become Japan's first large suspension bridge. The Wakato Bridge, across Dokai Bay in northern Kyushu, connects the cities of Wakamatsu and Tobata. Cantilever, movable, and a number of suspension bridge concepts were initially considered, as was a tunnel and even a cable-stayed bridge, but ultimately a two-lane, 367-m suspension span was recommended. Studies on issues with which Japanese engineers had little exposure were undertaken—wind tunnel analyses at the University of Tokyo under the direction of Atsushi Hirai (who, spurred on by the collapse of the Tacoma Narrows Bridge, had initiated aerodynamic studies as early as 1942), seismic studies, and experiments with wire rope cables and suspenders by government engineers and Kobe Steel. Hirai's wind tunnel work was in fact the first to experimentally determine the aerodynamic forces on a bridge section, which would help lead to the concept of aerodynamic derivatives (*see* Chapter 18).

Post-Tacoma Narrows Bridge orthodoxy strongly influenced the truss design, which wind tunnel tests suggested to possess a generous critical velocity of 614 km/hr Open steel grating was placed in the median and in two flanking 1-m wide swaths that separated the roadway from 3-m wide sidewalks, and gaps separate the roadway and truss edges. High-strength steels used in portions of the truss also kept its depth down to 4.5 m. Construction began in early 1959, and the $14 million span was opened in September 1962. Other notable features include the towers, which were originally to have been riveted but ended up being assembled with the more modern techniques of welding. Between anchorages, the Wakato Bridge is 680-m long, including 89-m suspended side spans and 41-m backstay spans.

The Wakato Bridge was a benchmark in Japanese suspension bridge engineering and, upon opening, it was the longest suspension bridge in the Orient. Its two 51-cm diameter cables—each comprising 61 strands—were the largest in Japan to use locked-coil wire strands. Aesthetically, too, the Wakato Bridge was a formative experience. Its bright red color enlivens an industrial setting, but the 85-m tall portal-braced towers and the cable bents are rather stiff in appearance; the anchorages are also massive, an effect exacerbated by the perching of offices on their tops. The offices were eventually razed in a 1984 to 1990 widening that also saw the walkways make way for two additional lanes. Traffic could not be impeded during the conversion, and so overhead gantries were used to lift portions of the old roadway off the truss and install a lighter, wider steel battledeck. This intricate $120 million job took 6 years, but aesthetically the span has benefited.

With the success of the Wakato Bridge, larger suspension spans soon received serious attention. In 1963, plans were unveiled for a 305- to 500-m span over Tokyo's inner harbor. By 1967, the $200 to $300 million project had evolved to a 606-m span with 203-m side spans, stiffened by a 6-m deep truss. Massive concrete anchorages would be footed at depths of up to 59 m, and hybrid portal/diagonal towers soar 102-m skyward. The crossing was eventually built as a tunnel. By mid-decade, the first of the truly large proposals emerged with a 1,300-m span to link the islands of Shikoku and Honshu across the Seto Inland Sea; this scheme is described later in this chapter. In 1965, a record-breaking 1,521-m span was proposed for the 8-km wide entrance to Tokyo Bay. Side spans of 681 m brought the total suspended length to 2,883 m, exceeding the Mackinac Bridge by almost 750 m. Its 233-m tall steel towers would also be of record height, with one of two variants resembling those of the Golden Gate Bridge, which they would surpass. A six-lane, 11.5-m deep truss braced by a Byzantine array of secondary steel would hang from four main cables hung in pairs; flanking air gaps occupied a full 13 m of the 40-m truss width. The design resulted from aerodynamic studies conducted by Hirai on full aerolastic models at a large wind tunnel opened in 1964 at the University of Tokyo; this would be the first such model since the early post-war investigations for the Severn Bridge. In total, the bridge would consume some 100,000 tonnes of steel.

That same year, a United Nations-sponsored report proposed a second national transportation corridor involving a crossing of the 20-km wide Bay of Ise to bypass the city of Nagoya, and a tunnel to Shikoku beneath the 11-km wide Kitan Straits, bypassing Osaka to the south. The Ise Bay scheme acquired a more tangible—and fantastic—form in 1968, with suspension bridges of 600-, 1,100-, and 1,200-m span, and a record span of 1,500-m flanked by 570-m side spans. The 220-m tall towers of this largest span supported four 78-cm diameter paired main cables from which was suspended a four-lane truss similar to that proposed at Tokyo Bay. Five cantilever bridges from 300- to 400-m span would also form part of this ambitious scheme. The high level of Japanese interest in long spans at this time was manifested by a significant presence at the 1966 Lisbon International Symposium on Suspension Bridges; Japanese engineers formed the second largest national contingent and delivered many of the symposium's papers.

It was clear that long spans would require parallel wire cables in place of the locked-coil strands of Japan's recent efforts. Government and industry research—involving Nippon Steel and Kobe Steel—began in 1961. A major concern was the vulnerability of cable spinning to typhoons. Efficiency improvements could reduce on-site spinning time, but partially completed cables would still be required to weather major storms. By contrast, prefabricated parallel wire strands promised to reduce cable erection time by up to 50% in comparison with spinning, while at the same time reducing labor needs and increasing safety. Larger anchorage faces and exacting tolerances in strand manufacture were necessary, but most pressing was the need for practical experience.

Several small spans in the second half of the 1960s provided it. High-tensile cable wire manufacture was perfected in 1964, and Kobe Steel spun its first experimental parallel wire cable in 1966. In 1967, Nippon Steel spun Japan's first in situ parallel wire cables in the 150-m span Kanaya Bridge and then repeated this performance in the Hakogase Bridge, a 206-m span completed in 1968. Prefabricated parallel wire strands were introduced that year in the 175-m span Konpira Bridge, the first span after the Newport Bridge to use them. The stiffening truss of this bridge was also the first in Japan to be bolted, and it also featured tubular members. Prefabricated parallel wire strand cables were used again in the Hachiman Bridge, a 160-m span completed in 1969 north of Tokyo. Its deck was simply a flat concrete slab straddling shallow steel girders. In 1970, Nippon Steel used prefabricated parallel wire strands in the cables of the 175-m span Wakise Bridge on Kyushu. It then spun one cable and used prefabricated parallel wire strands in the other in the Kami-Yoshinogawa Bridge, a truss-stiffened span of 253 m completed in 1971. Each prefabricated strand of 204 parallel wires remains the largest ever used in a suspension bridge. More conventional 91-wire prefabricated parallel wire strands were then placed by Nippon Steel in the Ryugatake Bridge, a 174-m span completed in 1972. One other notable span of this period is the 209-m span Shin-tsu Bridge in Oita, completed in 1969.

14.2 CROSSING THE THRESHOLD—THE KANMON AND OTHERS

In 1968, construction began on the Kanmon Bridge, which links Honshu with the southerly island of Kyushu across the Kanmon Strait. Its 712-m span would almost double Japan's longest span, and its design would benefit from the lessons learned in the Wakato Bridge and from research for the spans planned between Honshu and Shikoku. Its use of prefabricated parallel wire strand cables would also reflect the remarkably rapid progress of Japanese engineering.

The first serious proposal to span the Kanmon Strait emerged in 1896 as a symbol to commemorate the end of the Sino-Japanese War. The idea was revived in 1916 with an ambitious rail cantilever bridge whose 560-m span would have been the world's longest—at that time, no Japanese bridge had anywhere near that span. Plans in 1920 and 1931 (the latter for a rail suspension bridge) foundered. Not so for a single-track rail tunnel, built from 1936 to 1942, that was followed by a parallel rail tube between 1940 and 1944. Over this period momentum for a road crossing grew. A 1937 proposal envisaged a suspension bridge of 720-m span with 288-m side spans. A four-lane roadway sat on the bottom (and only) lateral brace of a 22-m wide, 8-m deep truss soaring 65-m above the strait. Its towers—resembling complex versions of those at the Transbay Bridge—would rise 145-m. Had this bridge been built, its span would have ranked in the top five in the world. Yet the plan was opposed by the military, and boring soon began instead on a two-lane road tunnel. Work proceeded until 1946 but was then suspended until 1952 while Japan coped with post-war reconstruction. The tunnel was finally opened in 1958. It soon proved its worth, for by 1964, study of a parallel crossing began in the wake of worsening traffic congestion.

In December 1965, the project received the go-ahead, and by 1967, a six-lane, 712-m suspension span with 178-m side spans had emerged after a review of bridges, tunnels, and even causeways. A 3.2-m deep streamlined steel box girder was a serious contender, but its behavior in typhoon winds was uncertain; shipping volumes of up to 1,000 vessels per day also rendered the lifting of deck units to 61-m heights hazardous. A 29-m wide Warren truss with a 6-m—later increased to a 9-m—depth was adopted whose 1:70 depth-to-span ratio gave the Kanmon Bridge the deepest proportions of any post-Tacoma Narrows Bridge span to that time. Other aerodynamic measures—open steel grating over the 2.5-m wide median and outer 2-m roadway edges and air gaps to the truss edges—produced a critical velocity of 235 km/hr. Concern about shipping safety was such that its piece-by-piece cantilevering out from the towers was accompanied by no fewer than three layers of safety netting.

Construction began in 1968. Only the 23-m deep Kyushu tower foundation used a pneumatic caisson; all other piers were excavated on land, which avoided shipping and 15-km/hr tidal currents. Each 143-m tall tower weighs only 3,000 tonnes, and their legs are comprised of just 3 cells in cross-section built from welded boxes connected by high-strength bolts. In an unusual departure, the Kyushu tower was erected with a land-based crane up to 160-m tall.

The choice of prefabricated parallel wire strands for the main cables emerged after much debate and testing, but increased safety and reduced weather susceptibility swayed the decision in their favor. After the successful manufacture and unreeling of a test strand in 1970, Nippon Steel hauled the first of 308 strands across the strait in September 1971. Although the 1,160-m cables are 215-m shorter than those of the Newport Bridge, the Kanmon Bridge's suspended structure is longer, wider, and heavier. As a result, each main cable is 66 cm in diameter and contains 14,014 wires—more than three times the number in each Newport Bridge cable—and this even accounts for a slightly larger wire diameter. Yet with more than double the Newport's strands, the cables were erected in only 2 months. Unusually fine weather did prevail, but a 91 wire strand size—versus the Newport Bridge's 61—reduced strand numbers and increased efficiency; as many as four strands were draped per 8-hour shift. Pipe anchorages similar to those at the Newport Bridge were also used in the Kanmon Bridge's 130,000-tonne anchorages.

One other notable aspect of construction involved truss assembly. To avoid typhoon damage and spectacles of the type endured by Halifax's hurricane-battered Angus L. Macdonald Bridge 20 years before, engineers permanently bolted the truss as it grew, distributing just a few temporary hinges to take up cable deformations. The exaggerated truss distortions flattened out once erection was completed, whereupon the hinges were then replaced with bolts. The method proved successful and was adopted in subsequent Japanese spans.

The $79 million Kanmon Bridge opened in November 1973 after 5 years and 5 months of construction. The bridge was, with the Angostura, the world's 10th longest span, and it would propel Nippon Steel to a position of world leader in the prefabrication of parallel wire strand cables; the firm would secure work on almost every major Japanese suspension bridge to the end of the century. The Kanmon Bridge was also instructive in aesthetic terms, particularly in the low placement of the towers' diagonal bracing, which unfortunately blocks the views of bridge users. Later spans would avoid this fault, but it is a minor complaint in view of the Kanmon Bridge's significant and successful role as a test bed for many technologies that were later incorporated into the design and construction of the Honshu-Shikoku bridges.

Further advances emerged in the Namhae Bridge, a 404-m South Korean span built between 1968 and 1973. This Japanese designed- and built bridge was only the third suspension bridge to use a streamlined steel box girder. Deep water and strong currents necessitated a suspension span, but typhoon exposure demanded careful consideration of aerodynamic stability. The two-lane, 1.6-m deep lozenge-shaped box girder was the narrowest and thinnest yet used in a suspension bridge, but wind tunnel testing predicted a critical velocity of 381 km/hr, more than 15% higher than a truss. The box girder was also almost 50% lighter. Extensive pre-assembly reduced costs by up to one-third of traditional construction methods. Most spectacular was the assembly of the 60-m tall, 230-tonne multiply braced portal towers which were welded in Japan, barged across the Sea of Japan and erected, in a single lift and complete with cable saddles, by floating crane. Nippon Steel spun the span's two 26-cm, 2,128 wire main cables, and vertical suspenders support the box girder, pre-assembled in 40-m long units. The Namhae Bridge was the most visually pleasing Japanese suspension bridge to that time; it consumed just 3,200 tonnes of steel and cost just $3.2 million. Its opening ceremonies also attracted 100,000 people, remarkable for a span of this size. Ito (1999) reports, however, that the girder has since suffered some vibration-induced damage from typhoons and traffic.

Pre-assembly again figured prominently in the Higashi Ohi Bridge, a 264-m Japanese span opened in 1976. Its towers, as at the Namhae Bridge, were placed in one lift, and Nippon Steel used prefabricated parallel wire strands for the main cables. In the Hirado Bridge, a two-lane, truss-stiffened span of 465 m built between 1973 and 1977 off Kyushu, the 88-m tall towers were also lifted in large sections by floating crane; the 6-m deep truss was placed in just 2 months, again by floating crane. The towers and brilliant red patina of this bridge suggest the Wakato Bridge, but the Hirado Bridge is of much improved proportions. The anchorages, for example, that flank unsuspended side spans of 105 and 139 m are unobtrusively tucked beneath the roadway.

The Hirado Bridge's major advance, however, lay in the improvements wrought by Nippon Steel in its main cables. As part of the company's research, tests on the Kanmon Bridge during its construction pointed to the restraint of the cable strands, and Nippon Steel subsequently field-tested a single tower, 210-m span arrangement complete with catwalks and cabling machinery. Here, engineers strung four 127-wire prefabricated strands and spun a single 600 wire strand. Nippon Steel then spun the two 4,332-wire, 37-cm diameter Hirado Bridge cables in 1975. It expanded the role of strand formers in securing the 19 strands in each cable, and this allowed spinning to continue in winds of up to 54 km/hr. Most significant was the relaxation of tension in the wire as it was being spun. Traditional spinning placed free-hanging wire under high tension (corresponding to a wire hanging under its own weight), so that its sag roughly matched that of the main cable to which it would be later bound; each wire required individual adjustment. At the Hirado Bridge, Nippon Steel instead used computer-controlled tension-sensitive devices at the anchorages to maintain just a small amount of tension in each wire as it was spun.

With this method, each of the 228 wires in each strand could be placed directly into its final position, thus providing support for additional wires as the strand was spun. The adjustment of hundreds of individual wires was now replaced by a simple adjustment of an entire strand. Automation also removed almost all labor from the catwalk; workers were now required only at the cable saddles and anchorages. Wire coils more than double the size used in the Verrazano-Narrows Bridge further reduced the need for splicing wires on the catwalk. Spinning operations at the Hirado Bridge proved so successful that Japanese engineers estimated spinning time and labor in a 1,000-m span could be reduced by 20% and 40%, respectively. The new controlled-tension method, as it would come to be called, would revolutionize spinning worldwide.

Two subsequent truss-stiffened suspension bridges brought further refinements in prefabricated parallel wire strand cables—the 216-m span Fukase (1979) and the 238-m span Ohdo Dam (1983) bridges. These spans would provide a deceptively diminutive prelude to the completion of the first of a plethora of suspension bridges in the world's most ambitious bridge-building program.

14.3 LINKING HONSHU AND SHIKOKU

Separating the islands of Honshu and Shikoku is the island-studded Seto Inland Sea—the Seto Naikai. Its 400-km length contains a profundity of scenic, ecological, and historic treasures. Shikoku lagged behind the explosive growth that transformed the main Japanese island of Honshu following World War II; in the mid-1980s, it still relied on ferries hampered by rough seas, high winds, fog, and immense volumes of shipping to transport 110,000 people daily between the two islands.

Stirrings for a bridge can be traced to 1889. A bridge was advocated in 1916 to a committee of the National Diet, and in 1939, rail tunnels were proposed beneath the Akashi, Naruto, and Hoyo straits to link Honshu and Kyushu to Shikoku. The following year, Chujiro Haraguchi, an engineer with the Ministry of the Interior, suggested a bridge across the Naruto Straits from Shikoku to Awaji Island, with a ferry across the 4-km Akashi Kaikyo (strait) completing the connection to the Honshu city of Kobe. Although inspired by the newly completed Golden Gate Bridge, World War II killed this initiative. The first systematic proposal then came with the inclusion of a Honshu-Shikoku link in the 1953 national rail plan, with bridges across the Akashi and Naruto straits via Awaji. Studies began in April 1955 with the Japan National Railways (JNR) adding a tunnel beneath the Akashi Straits as an option. The imperative for a link was underscored by the sinking of a JNR ferry in September 1954; then in May 1955, calamity struck when two ferries collided in fog with the loss of 168 lives, many of them children.

By 1955, Haraguchi was mayor of Kobe and advocating a road bridge across the Akashi Kaikyo. Road crossings also attracted the interest of the Japanese Ministry of Construction, which in April 1959 began studies of five routes. In 1960, the JNR decided the Awaji crossing should combine road and rail, and a second rail crossing mid-way down the Seto Inland Sea where Honshu and Shikoku are about 10-km apart was also examined. The Japan Society of Civil Engineers (JSCE) assumed control of the road/rail bridge studies in January 1962, whereas the Ministry of Construction continued to examine road-only spans ranging from 650 to 1,500 m, as did the JNR of similarly sized road/rail bridges. The five road routes under study included the 81-km long Kobe-Naruto link, three short routes in the central portion of the Inland Sea near Kojima-Sakaide, and a westerly Onomichi-Imabari link of 60 km traversing six islands. Wind tunnel testing of full-length models began at the University of Tokyo, and an 80-m tall tower was installed at the Kobe side of the Akashi Kaikyo in 1964 to collect wind data.[3] In 1965, tentative wind design specifications were issued.

Haraguchi and the city of Kobe would be the first to provide a concrete proposal. His scheme for the Kobe-Naruto route included a 3-km bridge across the Naruto Straits and a 5-km crossing of the 110-m deep Akashi Kaikyo. By 1965, the city had invested $300,000 in studies. The Akashi Kaikyo Bridge had evolved into a triumvirate of double-deck suspension spans centered by a record-breaking span of 1,300 m—just 1.5-m longer than the newly inaugurated Verrazano-Narrows Bridge. Its 650-m side spans and 2,600 m of suspended structure would also outstrip the Mackinac Bridge on both accounts. A capacious 31-m wide, 16-m deep truss would funnel six upper level auto and four lower level truck lanes and be suspended from two record 126-cm diameter cables draped over 193-m tall towers. Tower stays of Steinman vintage helped to stiffen the span. Subsidiary suspension bridges of 550 m span would share massive 50-m long anchorages with the central bridge at the ends of 275-m side spans. The total distance under cable extended almost 5 km, more than two-thirds longer than the twin suspension spans of the Transbay Bridge. Navigational clearances were later increased from 49 to 67 m in the central span, and Haraguchi even brought noted British aerodynamicist Kit Scruton to Japan for advice. Haraguchi's 1965 estimate of $217 million for the 8 km of bridge over the Akashi and Naruto straits was optimistic at best—by 1967, the Akashi Kaikyo Bridge alone was expected to cost $333 million.

Meanwhile, in 1965, the Ministry of Construction had concluded that the Akashi Kaikyo would require a single larger span of 1,500 m flanked by 750-m side spans; it also recommended just six traffic lanes on a single deck. The Ministry also urged further study of earthquake- and wind-resistant truss designs. By 1966, the Ohnaruto crossing had evolved to a 783-m suspension span flanked by 324-m side spans. Its four lanes would sit on a 25-m wide, 9-m deep truss suspended 40-m above the turbulent Naruto whirlpools from crudely fashioned 124-m tall diagonally braced towers. The Kojima-Sakaide route involved 12 km of continuous high-level bridge over 9.4 km of island-studded channel. Its centerpiece: tandem 858-m span suspension bridges over the Bisan Seto Straits near Shikoku. A massive central anchorage would link their 329-m side spans, and diagonally braced towers would stand up to 156-m tall, and. A 24-m wide, 13-m deep truss would support four upper level lanes and a single rail line on its lower deck, all clearing the straits by more than 70 m. Twin 360-m cantilever spans and a 688-m suspension span with 184-m side spans over the Shimotsui Seto abutting Honshu would round out this ambitious link.

By 1967, the first deep-water borings at the Akashi Kaikyo had been made. Foundation conditions—on which Haraguchi had largely speculated—would largely determine the span's feasibility, even more than its size. A shallower but longer alignment was also investigated that same year involving a string of three bridges centered by a 1,515-m span suspension bridge. Its 557.5-m side spans would be flanked by suspension bridges with spans of 669 and 870 m off Kobe and Awaji, respectively. Other unconventional ideas emerged in the late 1960s; one involved a five-span continuous suspension bridge of 5,000 m between anchorages, with spans of 500, 1,250, 1,500, 1,250, and 500 m supported by four 220-m tall towers. Another involved a hypothetical four-span bridge extending 3,220 m centered by twin 1,150-m spans; a Bunyanesque A-framed tower in the middle of the straits provided rigidity, but it would have also constituted a major navigational hazard.

The serpentine four-lane, road-only Onomichi-Imabari route had by 1967 evolved to include 10 large spans. Seven of these were suspension bridges, with truss-stiffened spans of 608 m at Innoshima, a 428-m span to Ikuchijima, an 842-m span over the Tatara Straits, and a 340-m span to Ohshima. Most spectacular was the final island-hopping leap to Honshu across the 5-km wide, 100-m deep Kurushima Straits involving end-on-end suspension bridges with main spans of 970, 628, and 1,150 m. A marked asymmetry was reflected in shipping clearances ranging from 35 to 67 m and many combinations of suspended and unsuspended side spans. Unusually, two of the tower piers of the first two bridges also served as anchorages for one other, and a sharp curve between the last two bridges placed one anchorage astride its side span.

In May 1967, the JSCE reported all five routes to be feasible. Two years later, after 9 years and $7.2 million of study, three routes were endorsed—the Kobe-Naruto, Kojima-Sakaide, and Onomichi-Imabari—each terminating on Shikoku about 80-km apart. Then, in July 1970, the government approved the simultaneous construction of all three routes and created the Honshu-Shikoku Bridge Authority (HSBA) to handle the enormous logistical and technical challenges of designing and building 30 km of large, open-sea bridges. The HSBA's research program and an ambitious construction schedule was approved in December 1970. All three routes would be completed by 1985 at a cost of $1.5 billion. The Kobe-Naruto would consume $700 million and 14 years of work, with the Akashi Kaikyo Bridge alone costing $523 million, more than the entire $509 million cost of the Kojima-Sakaide route; the Onomichi-Imabari route would cost another $355 million. Some 865,000 tonnes of steel were thought to be required to build the bridges. The Kobe-Naruto and Kojima-Sakaide routes would carry twin tracks of Japan's famed Bullet trains—the Shinkansen—on a lower deck. Moreover, the Kojima-Sakaide bridges would support two additional local rail lines; this and its four-lane (expandable to six) upper deck would generate the heaviest loads ever supported by a suspension bridge. Its main spans were also lengthened; the Bisan Seto bridges to 990 and 1,100 m, the Shimotsui Seto to 920 m, and the twin cantilever spans to 504 m. Six Onomichi-Imabari spans would now be suspension bridges, with the Kurushima trio evolving to spans of 860, 550, and 1,000 m.

The HSBA research program was equally daunting. Sixty large foundations would be sunk to depths of up to 50 m, many in open-sea conditions. Underwater concreting, seabed excavation, and platforms to permit work in fast tidal currents were improved. Fecund fishing grounds compounded design and construction restrictions. Most demanding was the need for earthquake resistance; the Akashi Kaikyo Bridge would be the most exposed, but all spans would have to withstand earthquakes of 8 on the

Richter scale centered 150-km away. As part of the research effort, ground vibrations recorded from actual earthquakes were used to oscillate bridge models. Winds of up to 191 km/hr recorded in the Inland Sea had also to be weathered. Models of entire bridges were subjected to wind tunnel tests, and enormous 1/10 scale sectional models were even built and rotated 360° in a giant outdoor frame exposed to the elements.

The Shinkansen posed yet another problem; the design had to allow for the simultaneous passage of two 1,400-tonne trains traveling at 160 km/hr. This made steel fatigue a major concern. Higher-strength steels pioneered in Osaka's Minato Bridge (a 510-m span cantilever completed in 1974) would reduce weight, but their fabrication would require great care to avoid "blowholes" from minute welding imperfections that could generate cracks under even low stresses. Excessive truss deflection posed another dilemma. The 1,100-m span Minami (South) Bisan Seto Bridge could sag up to 5 m if the truss was hinged at the towers; such deflections could also trigger oscillations. A continuous truss between anchorages was found to be sufficiently rigid, but this in turn brought other complications in the form of wider towers and truss strengthening to counter high stresses near the towers.

A major challenge was to maintain a level passage for trains between flexible bridge decks and the rigid approach viaducts. In response, engineers developed an ingenious 15-m long transition girder to cope with up to 1.5 m of expansion and contraction; high-speed Shinkansen trains would pound a test installation for 5 years before its adoption. Steel battledecks that would reduce weight in long continuous spans were also investigated, as were improved spinning and prefabricated parallel wire strand cabling methods for large spans in exposed winds. A corrosive marine environment generated efforts to reduce the amount of exposed steel in the spans themselves, and steel would be coated during fabrication—and not after on-site assembly—to help improve corrosion resistance.

A final issue concerned the fact that portions of all three routes traversed Seto Inland Sea National Park. The turbulent Naruto whirlpools and Mount Washu at the Honshu end of the Shimotsui Seto Bridge were particularly sensitive areas, both visually and ecologically. Concern was such that an agency was even created specifically to influence the appearance of these bridges.

Research was paralleled by the emerging designs of the spans themselves. In 1970, the Akashi Kaikyo Bridge was proposed as a 1,523-m span flanked by 761-m side spans. By 1972, two challengers had emerged—a 1,580-m span with 790-m side spans and an ambitious 1,880-m span flanked by side spans of 940 m. This longer span placed the anchorages on land and reduced foundation depths at the towers; in the smaller span, water depths alone at the tower piers reached 55 m. In 1973, a compromise was reached in the form of a record 1,780-m span with 890-m side spans. A leviathan 18-m deep truss would be aerodynamically stable in winds of up to 227 km/hr and support six upper-level lanes and twin Shinkansen lines below; its two 263-m tall towers would support two pairs of record 107-cm diameter cables. In total, the bridge would consume 236,000 tonnes of structural steel and 90,000 tonnes of cable wire. Huge tower piers would be sunk to depths of up to 67 m; these and the larger anchorages would together consume 1.3 million m^3 of concrete and 130,000 tonnes of steel.

As designs matured, costs rose, from the $1.5 billion estimated in 1970 to $2.6 billion in 1972. Yet the project's economics remained positive, and the HSBA remained committed to a November 1973 construction start on all three routes. Its request for $130 million to initiate work, however, came under darkening economic skies. The government had clamped down on expenditures to quell inflation but, in October 1973, endorsed the HSBA's ambitious 15-year construction schedule. Then the Arab oil embargo struck. Its impact was immediate; on November 20, just 5 days before groundbreaking ceremonies were to be held, the government postponed construction.

Studies continued, but it was not until August 1975 that the HSBA was permitted to proceed with any construction at all. Only the Kojima-Sakaide route and three other bridges on the other two routes would be initially built; the crown jewel, the Akashi Kaikyo Bridge, was put on hold. Work on the Ohmishima Bridge, a 279-m span half-through steel arch on the Onomichi-Imabari route, began in late 1975. It ranked as Japan's longest arch and the 14th worldwide when it opened in May 1979. The Ohnaruto Bridge was begun in July 1976, and work on the $142 million Innoshima Bridge (now a larger 770-m span) commenced in January 1977. Then, in April 1977, the government accelerated construction of the Kojima-Sakaide route, on the condition that objections to the crossing's location in Seto Inland Sea National Park be resolved. These concerns were, with some controversy, overcome, and groundbreaking ceremonies in October 1978 launched a massive construction effort lasting almost a decade to build more than 12 km of bridge.

The $285 million Innoshima Bridge was the first of the project's suspension bridges to be completed. It opened in December 1983, more than a year behind schedule because of two interruptions totaling 19 months; one of a year following tower completion, and another that delayed truss erection. The bridge's 770-m main and 250-m side spans are conventional in appearance, but its deep 9-m truss features a hefty depth-to-span ratio of 1:86. Other aerodynamic features include a 2-m wide median of open steel grating and 1.75-m gaps between the road deck and the truss edge; these help provide a critical velocity just shy of 200 km/hr. The Innoshima Bridge would introduce several features that would become standard in later Honshu-Shikoku spans: radar-absorbing materials to reduce interference with ship navigation, the bolting of the truss during erection, the use of a steel (but uncomposite) battledeck, and motorized maintenance scaffolds for the main cables. One feature repeated on no other Honshu-Shikoku bridge is a footpath tucked onto the lower chord of the truss.

Nippon Steel also improved prefabricated parallel wire strand methods in the Innoshima Bridge. Field tests in 1977 of a 127-wire strand saw the bridge become the first to use this now-standard size when cable erection began in late 1980. A slightly larger diameter wire wrought further efficiencies such that the 62-cm diameter main cables each contain just 11,557 wires. Each diagonally braced 145-m tall tower weighs just 2,835 tonnes and is of simple and graceful outline; the tower legs feature a simple three-box rectangular cross-section. The lower diagonal brace clears the roadway by almost three times the Kanmon's 5-m clearance, opening up views considerably, and the cable saddles are hidden in short extensions of the tower legs. The triangular anchorages resemble Othmar Ammann's efforts; horizontal grooves in their walls provide texture and reduce their apparent mass, no small feat considering one anchorage required the removal of 200,000 m^3 of soil.

In January 1979, a second suspension span on the Onomichi-Imabari route—the Ohshima Bridge—was approved for construction. Unsuspended 140-m side spans and a 145-m span steel box girder bridge flank its 560-m main span. Originally planned with a 7-m deep truss, the Ohshima evolved to a slender 2.2-m deep orthotropic box girder for reasons of economy and aesthetics, and it became the first such span on Japanese soil. Its slender 1:255 depth-to-span ratio constituted a radical departure from Japanese practice. The suspended box girder is heavier per unit deck area than the Severn and Namhae bridges—problems with the Severn Bridge had by now emerged, and it was felt a heavier box would be stronger and would compensate for the lower damping of the span's vertical suspenders. Nevertheless, the Ohshima's main cables are each comprised of just slightly more than 6,600 wires of 127-wire prefabricated parallel wire strands. Also notable are the elegant 97-m tall towers, whose upper portal flares gracefully into cubical tower legs with rounded edges. Construction began in March 1981, and the bridge was opened in January 1988 at a cost of $310 million. Only two of the span's four lanes were initially opened, a reality emphasized by the abrupt drop of half the deck cross-section over the unsuspended side spans.

The difficulties of inaccessibility and environmental exposure were epitomized by the arduous conditions encountered in the 1,300-m wide, 90-m deep Naruto Straits, where construction on the six-lane, 876-m span Ohnaruto Bridge began in July 1976. Tidal currents range up to 18 km/hr, and the area is especially prone to typhoons and winter winds; 288-km/hr gales and 11-m waves have been recorded. The straits form part of Seto Inland Sea National Park and are famed not only for their scenic value, but for large whirlpools created by the interaction of the strong tidal currents with a shallow and complex sea bottom. This phenomenon has been venerated for centuries and could not be disturbed, nor could nearby rich fishing grounds. The rugged, pencil-thin Tozaki peninsula was also out of bounds for the span's approaches. To complicate matters, the passage of up to 700 boats daily, mostly during slack tide, would coincide with peaks in sea-borne construction traffic.

Tidal currents and steep slopes had rendered caisson and cofferdam foundations unsuitable for the towers, but in any event it was necessary to reduce the mass of the footings themselves to avoid disruption of the whirlpools. The solution was to set the tower pedestals well above high tide in a method used in Japan previously, but never at this scale. Two clusters of nine columns—each cluster comprising 8 columns of 4-m diameter surrounding a central 7-m diameter column—protrude 6-m above water to support a huge 9-m deep, 70- by 25-m pier cap that in turn supports the tower legs. Water flows *beneath* the pier cap around the columns. Environmental restrictions were so tight that excavation-muddied seawater was purified onshore before release back into the straits. Other difficulties abounded. Submerged reefs compounded wave action and complicated the navigation of work barges. On one occasion, the work platforms shifted from forces generated by the drilling of the tower shafts, and in September 1979, a typhoon battered the incomplete foundations with winds of 176 km/hr.

Elongated 330-m side spans placed the anchorages—70-m long and up to 44-m wide—onshore, yet even here the splash zone could reach a height of 55 m. Cofferdams were used, although a steep slope on Shikoku required special attention to prevent slippage of the foundation. A 93-m backstay span off the Tozaki peninsula pushed the Awaji anchorage away from the whirlpools; protruding multiple columns also support its nearby cable bent. Together, the massive trapezoidal anchorages consumed 230,000 m^3 of concrete, although great care was expended to reduce their apparent bulk by adding texture and stepping to their sides.

A measure of the difficulties was reflected in the 4 years it took to build the five major foundations of the Ohnaruto Bridge; construction of the work platforms alone required 2 years and devoured almost 11,000 tonnes of steel. Tower erection began in January 1980. The supply of steel of course depended on barge transport, and the severe conditions saw the program soon slip behind schedule. Then, as the towers ascended, oscillations not only made bridgeworkers nauseous but raised concerns about the security of the creeper cranes and even the towers themselves. A restraining rope added to the Awaji tower then, too, began to oscillate. Engineers were forced back to the wind tunnel, and a second rope solved the problem. The towers, finally finished in August 1981, benefited from the contributions of architects and industrial designers, a first in the design of Japanese suspension bridge towers. The process spawned six alternative designs, including portal-braced towers and an unusual variation of diagonally braced towers with no top portal strut. Diagonal bracing was ultimately adopted for its structural efficiency and reduced level of steel exposure; Japan's high earthquake and wind loads encouraged its use in these early spans. A triple box cross-section comprised the towers' slender legs.

Spinning had been contemplated for the record size, 97-cm diameter main cables. An enlarged wire diameter then helped to produce a smaller 84-cm thick cable of 19,558 wires assembled from 127-wire prefabricated parallel wire strands. The 1,722-m long cables were placed by Nippon Steel, ending in 1982, and were then the longest and largest of their type. Another innovation was the draping of pilot wires for the catwalks under high tension to prevent their immersion in water. Truss erection then began, a year behind the April 1982 target. Tidal currents made the lifting of truss units impossible, and small members were instead barged to the towers, lifted to deck level, assembled into larger panels, transported over previously erected truss, and then placed. All was not achieved without tense moments. The periodic temporary hinges, as used in the Kanmon and Innoshima bridges, made the incomplete truss aerodynamically unstable and had to be abandoned; temporary adjustments to suspender lengths maintained a straight profile as the truss inched out over the straits, with the cables subjected to huge—if temporary—distortions. On another occasion, unbalanced forces over the Awaji cable bent caused the cables to slide, threatening to deform the stiffening truss; two strands were later added to each cable to connect the bent to its anchorage soon after the span opened.

No Honshu-Shikoku span is more exposed to wind than the Ohnaruto Bridge, and its critical velocity of 317 km/hr is the highest of all the project's bridges. Open steel grating forms its central median, and the steel battledeck is separated from the truss edge by air gaps. A unique vertical fin—or stabilizer—extending below the deck along the median also helps to modify airflow and reduce torsional motion. The truss design had undergone much alteration, first with the January 1979 decision to include

Shinkansen lines, which initially led to the use of a continuous truss. By the time of detailed design, the poor economics of the Shinkansen led to its removal from the opening configuration, and designers reverted to tower-mounted expansion joints. A single track would be planned for (a second line could also be added, but only one train would be permitted on the bridge at a time), and a transition rail girder could be added at the towers. The Ohnaruto Bridge would thus become the first large suspension span structurally ready to accept rail traffic.

The wind and rail demands produced a herculean truss 12.5-m deep and 34-m wide. Its 1:70 depth-to-span ratio equaled the Kanmon Bridge's as the largest of any major span of the post-Tacoma Narrows Bridge era, and the effect is emphasized by a relatively low clearance of 41 m and a flat vertical arc. In stark contrast stand the slender towers, perched atop the immense "floating" pier caps. Their deep top brace appears almost as a portal strut and is unique among diagonally braced towers, as is the placement of the cable saddles inside the legs rather than above the top brace. The Ohnaruto is an impressive bridge if somewhat contradictory in appearance; its gray-white patina exudes brilliance against the rugged and turbulent open sea setting. Technical achievements also extended to 1,005 m of high-level viaduct that flanks the Tozaki promontory; its seven steel box girder spans were placed in only seven lifts of up to 2,410 tonnes from huge floating cranes. Here, flaps and skirts streamline winds to prevent overturning.

The Ohnaruto Bridge opened in June 1985 after 9 years of difficult work at a cost of $440 million, well beyond the 7-year schedule and 1975 estimate of $337 million. Labor laws prohibited work in winds greater than 55 km/hr, an almost constant condition in winter, and down time consumed more than half the schedule. The span is heavily instrumented for traffic, wind, temperature, and even earthquake, but it is not entirely complete, for only four of its six lanes are open. Pier stubs tucked beneath the Tozaki Viaduct will eventually support a deep deck truss for the Shinkansen, and only one of the two parallel road box girders of the Tozaki Viaduct is in place. In 2000, a glass-floored promenade extending from the Naruto anchorage over the side span and 130 m into the main span was opened, providing dramatic vistas of the famed whirlpools; wind tunnel tests verified the inconsequential aerodynamic effects of this unusual feature on the stability of the truss.

14.4 THE SETO OHASHI

With the completion of the Ohnaruto Bridge, the limelight now shone on the Kojima-Sakaide route (now officially the Seto Ohashi), a gargantuan effort involving 12 km of continuous high-level bridge. Its three suspension, two cable-stayed, and single cantilever spans would individually impress, and the Bisan Seto spans would be the first tandem suspension bridges since the Transbay. Yet the 1977 decision to accelerate the link's construction through Seto Inland Sea National Park and the area's rich fishing grounds was contentious. Environmental scrutiny was rigorous, and the project was approved in spite of one agency's objections to it. The 940-m span Shimotsui Seto Bridge, which launches the route from the shores of Honshu, exemplifies these concerns. Although the smallest of the link's three suspension spans, it is perhaps the most technically innovative (*see* Fig. 14-1).

Priority had been assigned the mammoth 5-year task of building the Bisan Seto foundations, and so work on the Shimotsui Seto Bridge did not begin in earnest until the spring of 1981. As for all spans, basic standards and designs were developed by the HSBA, aided by various committees comprising government, academia, and the private sector; countless engineers in the private sector followed up with detailed designs. The Hitsuishijima Island tower stands atop a spread footing, as does its huge 68- by 54-m, 48-m tall gravity anchorage. A pneumatic caisson supports the Honshu tower on reclaimed land. It was essential that the Honshu end of the Shimotsui Seto Bridge be anchored without removing a good chunk of Mount Washu, a vantage point visited by more than 1 million tourists annually. Large gravity anchorages were out of the question, and as early as 1974, it had been decided to use tunnel anchorages. The final configuration involved 77-m long tunnels up to 13-m wide that anchored the cables via a compact system of high-tensile steel rods prestressed into concrete-encased anchor frames.

Fig. 14-1. Shimotsui-Seto Bridge
PHOTOGRAPH COURTESY OF THE HONSHU-SHIKOKU BRIDGE AUTHORITY.

Cabling methods were also part of the Mt. Washu equation. Prefabricated parallel wire methods required more than four times the number of strands as in situ spinning—and thus more space—to anchor the 24,888-wire main cables. Spun strands could be larger and would in consequence reduce the anchorage cross-section. In fact, the 552-wire strand size in each 44-strand, 93-cm diameter cable constitutes a record. Nippon Steel and Kobe Steel began cable spinning in June 1985 and further improved the controlled-tension methods introduced in the Hirado Bridge. Wire tension was set at just 25% of free hang, and spinning wheels were tilted to drop four dead wires in the strand while the four live wires were placed alongside on rollers. On its return trip, the wheel would pick up the former live wires and place them in the strand. Six months after operations began, the cables hung complete.

With the two 146- and 149-m towers dominating views from Mount Washu, aesthetics were a major preoccupation. Hybrid diagonally/portal-braced towers were mooted in the early 1970s, and some 26 other alternatives were then evaluated. From this, 20 variants of a portal-braced design were devised and even reviewed by tourists at the mountain. A distinctive inward flaring of the upper portions of the tower legs was adopted that also helps to channel the cable loads down their centers. The 7- by 9-m thick legs are the largest in the Honshu-Shikoku project save those astride the Akashi Kaikyo, yet in cross-section they comprise just three cells. Impressive loads also spawned an enormous truss of a design used in the Bisan Seto spans as well (*see* Fig. 7-2). Its 13-m depth is surpassed only in the Akashi Kaikyo Bridge, and a 1:72 depth-to-span ratio is just shy of that in the Ohnaruto Bridge. Aerodynamic measures occupy almost a half the 30-m truss width, with flanking air gaps of more than 6 m on the upper road level. Open steel grating is used on both upper road and lower rail decks over the central portion of the main span, in a 3.5-m wide upper deck median and in 2.5-m wide swaths at the edges. Diagonal and vertical suspenders are also found at midspan, and are used on all Seto Ohashi suspension spans to quell torsional motion. A relatively low 31-m vertical clearance almost equals the width of the truss, and this tends to emphasize its great size. Erection of the truss, which commenced in April 1986, was completed a year later.

Most unusual are the Shimotsui Seto Bridge's 230-m side spans. Unsuspended and suspended side spans were initially compared. But with both side spans over land, engineers were able to reduce both cable size and truss stresses from rail loads in a unique—if somewhat visually cumbersome—way. The main span is cantilevered 130 m into the side spans without any support from the main cables (save the use of temporary suspenders during erection). These cantilevered side spans terminate at massive four-column piers accommodating rail expansion joints, and the remaining 100 m to the anchorages is supported by concrete viaduct. Only in Norway and the Japanese-designed Matadi Bridge in Africa (discussed later in this chapter) had side span cantilevering been previously used.[4]

One other unique feature of the Shimotsui Seto Bridge is virtually unseen. The 1973 plans had called for a massive cut through the flanks of Mount Washu as the difficulties of shallow tunneling were thought to be insurmountable. Park authorities forced further study after the 1973 postponement, but when work began in 1978, the HSBA remained unmoved, notwithstanding its earlier decision to use tunnel anchorages. Park authorities rebelled and dictated that a tunnel be built unless technically impossible. Newly developed methods then led to a 1980 decision to bore four tightly clustered tubes under only 30 m of heavily weathered granite. None of the tunnels exceed 230 m in length, but the need to bore sequentially—and from the center out to reduce blast noise—would make the project a 6-year undertaking.

The massive job of building the four tower and three anchorage piers of the 990- and 1,100-m span Bisan Seto bridges would similarly require 6 years—and them some. Copious maritime traffic, strong tidal currents with 3-m tides, typhoons, the span's location in the national park, and nearby fishing grounds all posed major constraints. Foundation research had begun as early as 1965, followed by large-scale land and underwater tests at the Kojima (Shikoku) end of the bridge in the early 1970s. Huge caissons would rest at depths of up to 50 m, but a thick layer of weathered rock made sinking through overburden impractical. Moreover, deep water rendered cofferdams unsuitable. Engineers instead devised the "laying-down caisson method," in which huge pre-assembled steel caissons would be dropped on exposed, levelled bedrock shorn of overburden and then concreted.

Explosives first blasted away hundreds of thousands of tonnes of rock, including a small island that made way for the Pier 4A double anchorage. Blasting had not only to avoid damage to nearby Shikoku industries, but to the underlying bedrock that would support the caissons. Pier 7A, the Minami Bisan Seto's Shikoku anchorage was situated just 300-m offshore but would rest at a depth of 50 m. Even so, a huge grab dredger brought up 540,000 m³ of loosened rock—more than twice the amount dredged from the other five open-sea Bisan Seto piers—in a 3-year period. Large drills aboard a huge self-elevating platform then smoothed bedrock with variations of no more than 1 cm over areas as large as 59 by 75 m. Enormous prefabricated caissons—the 55-m tall Pier 7A anchorage caisson shell weighed 17,800 tonnes—were dropped into place beginning in October 1980 by huge floating cranes. Shore-based lasers helped to reduce alignment errors to less than 1 cm. Thick rubber seals overcame any remaining bedrock undulations. Prodigious amounts of crushed stone were then poured into the caissons in a variation of Prepakt intrusion concreting used in the Mackinac Bridge. Cement ships, sand and wastewater barges, and computer-controlled mortar plants swarmed around the piers. More than 220,000 m³ of aggregate and half that volume in mortar were placed into Pier 7A in a 2-year period, exceeding the amount poured into all other five open-sea piers. Teething problems were experienced, but the method ultimately cut the concreting schedule in half and saw the piers completed in February 1985. Work was not without incident; the self-elevating platform was sidelined for almost 3 months when its legs were damaged at the south tower, cracks in the base of the central anchorage delayed work for a month, and a small cargo ship rammed into the same pier's work platform after its navigator fell asleep. The Hitsuishijima and Iwakurojima cable-stayed spans would also use this foundation technique. These 420-m spans had been adopted in March 1980 as a replacement for the ungainly cantilevers of the 1973 scheme. This followed a May 1978 ultimatum from environmental authorities that cable-stayed bridges be built unless technically impossible. These two spans would establish type records for capacity and strength.

By the mid-1980s, work was in full swing on the Seto Ohashi, Ohnaruto, and Ohshima bridges. Annual expenditures of $600 million in 1982 and $850 million the following year were without precedent. Almost 1,500 contractors—including all of Japan's largest—tackled 499 separate contracts. The Seto Ohashi foundations also required 10 separate contractors, and its 10 steel towers required six. At the peak of construction, 5,000 workers were on-site. Costs were also measured in terms of superlatives—by 1982, the price tag for the Seto Ohashi Bridge had risen to an estimated $5 billion.

The diagonally braced towers of the Kita (North) Bisan Seto Bridge began to rise in mid-1983, followed in early 1984 by those of the Minami Bisan Seto Bridge. Although outwardly similar, all four towers are slightly different, to the point that the 194-m Minami north tower is 19-m taller than its lowest counterpart. The huge tower shafts are comprised of just three cells in cross-section and were erected with creeper derricks, although the southern-most tower used a 220-m tall self-supporting crane—then the world's tallest. Nippon Steel and Kobe Steel then strung more than 1,000 prefabricated parallel wire strands of 127 wires each over the two bridges. The 29,718-wire main cables of the Kita Bisan Seto Bridge—almost 1 m in diameter—were placed first. Equipment was then moved to the Minami Bisan Seto Bridge where, in spite of a severe winter, erection efficiency was actually improved. Its 1,648-m long main cables, with 34,417 wires and a diameter of 106 cm, were the largest erected to that time. In fact, the 50,000 tonnes of wire that went into all three Seto Ohashi suspension spans required 5 years to manufacture. Curiously, wire diameters in each span are slightly different.

Pier 4A, which anchors the more than 1,000 strands in the Bisan Seto spans, was unique. The strands of each bridge intersect one another before being fixed at opposite ends of the anchorage; in this way, one span helps to anchor the other. Such an intricate web in confining quarters was possible only with prefabricated parallel wire strands. The 255,000-m^3 behemoth—79-m long, 62-m wide, and 86-m high—features a distinctive angular shaping that offered superior cost and aesthetics over a simpler rectangular form. The Pier 7A anchorage also consumed immense quantities of concrete—200,000 m^3 above water alone—to bring its total volume to 500,000 m^3. It weighs a mind-boggling 700,000 tonnes. At both anchorages, floating cranes placed huge prefabricated anchor frames complete with anchor rods and splay saddles. All splay chambers are also dehumidified to avoid cable corrosion from the Inland Sea's salt-laden air.

Truss erection began with the lifting of huge truss blocks by floating crane at the towers, beginning in the Kita Bisan Seto Bridge in May 1986 and the Minami Bisan Seto Bridge that August. The remainder of the truss was erected from small assemblies delivered over already-erected truss from the towers. By March 1987, the Kita Bisan Seto truss was closed. Two months earlier, work on the Minami Bisan Seto Bridge had been brought to an abrupt halt by an alarming accident when the traveling scaffold supporting a large safety net ran off the end of the incomplete trusses. Its 65-m plunge into the sea took eight bridgeworkers with it but miraculously only one was killed. Truss assembly on the bridge was halted for 2 months, and criminal charges were eventually brought against two employees of the contractors. In spite of this tragic incident, crews closed the Minami Bisan Seto Bridge truss that July—on schedule. Although identical to those of the Shimotsui Seto Bridge, the Bisan Seto trusses appear somewhat shallower on account of the longer spans and higher navigational clearances, but still feature large depth-to-span ratios—1:76 for the Kita bridge and 1:85 for the Minami. In both spans, the truss is continuous between anchorages and, in the Minami Bisan Seto Bridge, was the longest anywhere save the Tagus. Large pendulum-shaped steel links allow for the considerable longitudinal movement of the truss through the towers.

The Bisan Seto bridges create an impression of great strength (*see* Fig. 14-2). Their short 274-m side spans enhance the dominance of the deep rigid truss over the main spans. The towers' robust appearance stems from strong legs; a relatively high placement of the truss and of the diagonal bracing connecting the tower legs also emphasizes this effect. The asymmetry of the two spans is not overly noticeable, and in fact the crossing is imbued with an overriding sense of order. These results vindicate the whittling down of 125 tower groupings to 26 and then five configurations before adoption of the final design. Some 45 color schemes were also scrutinized by lay people and experts alike before a light gray—applied to the entire crossing—was selected. Navigation safety also figured prominently. Steel plates embedded in the upward-sloping horizontal bars that provide texture in the anchorage faces were coated with thin sheets of rubber and ferrite to absorb radar waves, as was bridge steel. Unobtrusive double-walled collapsible steel fenders and inflated rubber cushions protected the Bisan Seto towers,[5] and all spans remain heavily instrumented to track their behavior in wind and earthquake. Some 17 km of welds and 2,000 tiny

Fig. 14-2. Kita (*left*) and Minami (*right*) Bisan Seto bridges
PHOTOGRAPH COURTESY OF THE HONSHU-SHIKOKU BRIDGE AUTHORITY.

blowholes will require careful observation to monitor possible long-term fatigue effects induced by the 6 million train crossings expected in the 100-year life of the Seto Ohashi. Train speeds are limited to 120 km/hr, and only one train is permitted at any one time (although a load test with a 1,000 tonne train comprising 11 locomotives saw the suspension spans deflect to the 80 cm predicted). The Shinkansen is also absent; only two standard lines currently operate.

Opened in April 1988 at a staggering cost of $9.2 billion, the Seto Ohashi ranks among the world's great engineering and construction management achievements. Its 14 km of high-level bridge easily unseated the Transbay as the world's greatest bridge complex. A 1987 completion date was missed by less than a year, even though 20% of working time was lost to poor weather. Some 3.6 million m^3 of concrete—more than 1 million of this in the Bisan Seto bridges—went into the crossing, as did more than 700,000 tonnes of steel. Four million bolts went into the stiffening girders alone. Individual bridges also ranked highly. The 1,100- and 990-m Bisan Seto spans ranked 5th and 11th longest in the world upon opening; the Minami span surpassed Lisbon's Tagus Bridge as the world's longest and strongest combined road/rail span. Its towers stand 32 mm farther apart at their summits than at their bases on account of the earth's curvature, and in 216-km/hr winds, its truss is designed to flex horizontally by up to 7 m. Sadly, the human cost of 17 lives to build the Seto Ohashi—six in the last year of construction alone—was also high.

The Seto Ohashi is not universally admired. Japanese sculptor Masayakuki Narare condemned it as "too imitative, too glitzy, too expensive, too ill-conceived" (Billington, 1990). Critics bemoaned the consumption of Shikoku's placid rural areas by hotels, tourist attractions, and industry; at the same time, the bridge was feared to divert growth from as far away as Hiroshima. A huge parking lot and visitor center on former forest on Yoshima at the Bisan Seto's north terminus was a symptom of this phenomenon (although 100,000 trees and shrubs were planted along the entire route). Ecologists feared the bridge would disrupt formerly isolated wildlife on Honshu and Shikoku; indeed, in 1994, illumination levels were reduced to minimize effects on migrating birds.

On the other hand, engineering historian David Billington has lauded the span's "delicate tracery through the wide reaches of the inland sea." The stunning panorama of the bridge from Mount Washu—dominated by the Honshu tower of the Shimotsui Seto Bridge—is indeed impressive, and the Mount Washu tunnels propel the motorist into a stunning spectacle of bridge, sea, and sky. Yet the almost Egyptian scale and angularity of some parts of the crossing serve to visually overwhelm their immediate surroundings. No matter, the opening celebrations attracted people in the hundreds of thousands, and 6.5 million people flocked to the crossing in its first 6 months. A competition on future bridge forms attracted 651 entries from 41 nations, and to prepare for the opening Shikoku boasted 20 new hotels, a bridge exposition, and the world's first permanent bridge museum.

As the euphoria of the summer-long celebrations ended, the bitter pill of high tolls took hold. Crossing the Seto Ohashi, for all its convenience, was not for the penurious. With $50 automobile tolls, expected traffic volumes failed to materialize. Truckers faced with $150 tolls remained loyal to the ferries, combining mandatory rest periods with impressive views of the bridge they were circumventing. Improved expressway connections did improve the picture, and by 1995, upwards of 150 trains crossed the structure daily. Yet the Shinkansen line remains in the far distant future, if ever to be built at all. Ironically, the uncertainty associated with the Shinkansen would help secure a future for the largest bridge of all, the Akashi Kaikyo.

14.5 THE AKASHI KAIKYO BRIDGE—A GIANT REBORN

By 1988, the cost of all three Honshu-Shikoku routes was pegged at a staggering $23 billion, and work on the largest, most expensive span of them all—the Akashi Kaikyo—had yet to begin. Research and design work on this mammoth 1,780-m span 60-km north of the Ohnaruto Bridge continued in the absence of a definitive timetable for construction. Yet by the time the Seto Ohashi was completed, the Akashi Kaikyo had undergone a profound transformation.

The first rumblings of change emerged in 1978 when the economics of a second Shinkansen line less than 150 km east of the Seto Ohashi were questioned. Lines on the Ohnaruto and Seto Ohashi spans were subsequently deferred, making an Akashi line unlikely. A second concern involved the great depths of the Akashi Kaikyo's tower foundations. Local pressure to get on with the long-proposed span also escalated. In 1981, the government suggested the HSBA investigate a road-only bridge. Reduced loads, strength, and weight could make a longer span possible using existing technology. The towers would stand in shallower water farther away from the 1,500-m wide navigation channel, an important consideration given a nearby bend in the channel and a daily volume of 1,400 large ships and countless small fishing boats. The anchorages could also be placed just offshore in shallow water, although dense development would preclude their placement on land. Over the next 4 years, spans ranging from 1,780 to 2,100 m were examined. A road-only bridge of 1,900 to 2,100-m span was thought to be up to 25% less expensive than the 1,780-m road/rail span, with increased superstructure costs more than offset by lower foundation costs. In August 1985, the HSBA recommended a longer road-only span, and in December, construction of the long-debated crossing was finally given the go-ahead.

The rejuvenated bridge was a monster. Its 1,990-m main span was more than 40% longer than the Humber Bridge; even its 960-m side spans would rank as the world's 14th longest single spans (*see* Fig. 17-1). A total suspended length of 3,910 m would easily surpass the Mackinac Bridge and even outdistance the twin Bisan Seto spans by 500 m. Some 209,000 tonnes of superstructure steel—although 20% less than in the 1973 scheme—would exceed the Verrazano-Narrows Bridge's 142,000 tonnes by a huge margin. Other records would be shattered; the 333-m tall towers would exceed those at the Golden Gate Bridge by more than 100 m and rank among the tallest structures of any kind. The 14-m deep and 35.5-m wide truss would be the largest ever built. Four 84-cm diameter main cables, each containing 21,463 wires, would support it. No single bridge approached this one in magnitude, or its associated design and construction challenges of deep water, poor soils, high wind, and earthquake risk.

With the new design, a 13-year construction period was now reduced to a decade. Formal groundbreaking occurred in April 1986, but site investigations and detailed design dominated the next 2 years. Full construction began in May 1988, a month after the Seto Ohashi opened. The four main foundations would take 5 years to build, and the superstructure would take another 5 years before a planned 1998 opening.

The tower piers remained such a formidable challenge that they were still considered key to the span's feasibility. A precipitous underwater slope of granite saw only the Awaji anchorage rest on hard rock; this left poor footings for the remaining piers. Water up to 50-m deep, tidal currents of up to 16 km/hr, and abundant shipping would further complicate design and construction. The placement of the Awaji tower in sandstone and the Honshu tower in gravel and sand required new methods of seismic analysis to enable the bridge to resist 8.5 Richter earthquakes centered 150-km away. Initial proposals for the tower piers called for huge rectangular steel caissons to be built downward by successive concreting of excavated layers to a depth of 70 m. Model tests then produced the alarming finding that scour could threaten their stability. Further research revealed that shallower footings were possible, and this generated fundamental change. With the Kobe pier now 60-m deep and its Awaji twin resting at a depth of 57 m, the laying-down caisson method developed for the Seto Ohashi was adopted. The use of circular steel caissons would also reduce the potential for scour. At 80-m in diameter and 70-m high, the Kobe tower caisson and its slightly smaller Awaji counterpart would be the largest bridge caissons built to that time.

Yet months into excavations, fundamental questions concerning the superstructure remained. A major dilemma involved the main cables. Prefabricated parallel wire strands were preferred, and the successful manufacture of a 4,085-m long test strand by Nippon Steel and Kobe Steel suggested they were feasible. A bigger issue involved the number of cables. Replacing the planned four cables with two larger ones would simplify construction and reduce the number of suspender and deck connections by half. To do so required stronger cable wire, but in the 80 years since the 1909 completion of the Manhattan Bridge, wire strength had increased by less than 10%. An increase in an already minute percentage of silicon offered promise, partly by reducing the loss of strength normally imposed by galvanizing. By 1989, wire with a tensile strength of about 180 kg/mm^2 was perfected, representing a strength increase of more than 10%. Still, more was required. Here, mathematics and the enormous size of the bridge came into play. Because live loads would constitute just 8% of the total load carried by the cables, it was felt that the factor of safety to which the bridge was designed could be reduced from 2.5 to 2.2 times the maximum axial stress experienced by the cables.[6] This, combined with the stronger cable wire, made two cables possible. Record-breaking they would be; each 112-cm diameter, 36,830-wire cable would weigh 25,250 tonnes. Manufacture of the cable wire alone would take 3 years.

The stronger wire had also permitted a lower cable sag, and this had the benefit of making shorter towers possible. By 1990, their height had been pared down to 297 m, but they would still rank as the tallest suspension bridge towers ever built and stand just 3 m shy of the Eiffel Tower (*see* Fig. 13-3). Five unusual motifs were investigated, including anvil-shaped bracing, two massive diagonal braces without a top strut, a solid face pierced with large vertical ovals, and a deep gothic strut capping three diagonal braces. In the end, four diagonal above-deck braces were adopted for each 23,000-tonne tower—one less than in the initially planned 333-m tall towers—with one diagonal brace below the truss. The 6.6-m wide legs are almost 15-m thick along the line of the span and stand more than 46 m apart at their bases; they taper to 10 m and stand more than 35 m apart at their apexes. One other issue: wind tunnel tests suggested vortex shedding off the towers could induce oscillation both before and after the span's opening. To quell such instability, a cruciform leg shaping was augmented by the placement of 10 tuned mass dampers—each weighing 10 tonnes—at varying levels within each tower leg.

The suspended structure posed perhaps the most vexing problem. The scale of the Akashi Kaikyo Bridge presented a completely new problem in terms of wind design, one that criteria developed for the other Honshu-Shikoku spans could not address. Flutter was a potentially serious problem, as the long span made the suspended structure very flexible; horizontal displacements of up to 30 m at the static design wind speed of 216 km/hr would induce a twist of up to 3°. Compounding this challenge was a high required flutter critical velocity of 280 km/hr (averaged over 10 minutes, as in all Honshu-Shikoku spans) on account of the exposed wind environment. Many truss designs were investigated, including one with a system of curved bottom lateral bracing. A 7-m deep truss with nonstructural fairings met the flutter velocity requirement, but it was prone to lower-speed vortex-induced torsional oscillation of surprisingly large amplitudes. Box girders, which would reduce wind loads by as much as two-thirds and simplify maintenance, were also studied. A trapezoidal box with a small central air gap and vertical median stabilizer met the flutter critical velocity, but sufficient torsional strength was achieved only by devouring more steel than a truss; wind tunnel tests also revealed severe vortex-induced oscillation that could induce long-term fatigue. Interestingly, one unusual box girder arrangement did appear to satisfy the wind criteria. A 7.4-m box depth was proposed for torsional strength on either side of the towers and near the anchorages, whereas a shallower 4.6-m box depth was adopted for aerodynamic profile and weight reduction for the central two-thirds of the main span and central third of the side spans. With the thicker boxes providing residual stiffness to the thinner boxes, this unusual configuration actually became the sole challenger to a truss. Its flutter velocity was acceptable, as was its slightly lower steel weight.

A deep truss, however, was a formidable competitor. It presented no stability problems, even in wind speeds greatly exceeding the required flutter velocity. Moreover, its erection posed no obstacles to navigation. By contrast, a box girder would be aerodynamically susceptible during erection and require lifting in congested sea lanes, a method rejected by maritime authorities. It was this latter factor in particular that swung the pendulum toward the record-size 14-m deep, 35.5-m wide truss (*see* Fig.7-2). Yet a measure of the Akashi Kaikyo's large span is found in a relatively slender depth-to-span ratio of 1:142, in part achieved with the aid of high-tensile steel and a steel (but not composite) orthotropic deck supporting the six-lane roadway to reduce weight. A 2.5-m wide air gap separates the deck from the truss edge, and the deck features 2.5-m wide swaths of open steel grating along its extremities and a 3.5-m swath in the median. At center span, inclined suspenders connect several adjacent panel points in lieu of a large midspan cable clamp to help minimize longitudinal motion.

A concern that the enormous size of the bridge might render sectional models inadequate to describe some of its aerodynamic behavior led the HSBA to commission the world's largest boundary layer wind tunnel. Opened in 1991 in Tsukuba City near Tokyo, this facility easily swallowed a 1/100 scale, 40-m long aerolastic model of the entire bridge. Tests performed here led to the adoption of a vertical fin dropping more than 2 m below the deck along the centerline of the main span. This fin—as in the Ohnaruto Bridge—acts with the median grating to equalize air pressures above and below the deck, thus reducing the potential for torsional motion. Roadway curbs are also perforated; a solid curb was found to reduce the aerodynamic stability of the truss by a surprisingly large amount. The final design was found impervious to flutter in winds exceeding 331 km/hr.

During this period, foundation work acquired full momentum. Dredging at the tower piers began in May 1988. A 15- to 20-m deep layer of material—almost 460,000 m^3 at the Awaji tower—covered bedrock, but could be removed only at slack tides. In consequence, this operation required more than a year to complete, with much of the spoil going to Osaka's nearby open-sea Kansai Airport, then under construction. In March 1989, the 15,300-tonne Kobe tower caisson was guided into place; its 14,000-tonne Awaji counterpart was deposited in June. Copious amounts of rock were immediately dropped to prevent scour, and a 3-day continuous underwater concrete pour secured each caisson to bedrock. A protracted period of concreting then began with the Kobe pier in October 1989 and the Awaji caisson 3 months later. A new concrete with improved consistency permitted longer, continuous underwater pours, and in this foundation construction diverged from the Prepakt method used in the Seto Ohashi. The Kobe caisson consumed more than 265,000 m^3 of concrete—the Awaji pier slightly less—in a year-long operation hampered by poor weather, high waves, and strong winds. Placement of a top slab up to 20-m thick saw both piers completed in March 1992, 4 years after work began. Engineers would consider the method suitable for depths of 100 m or more.

In late 1989, land filling for the anchorages was completed. Both anchorages measure about 85 by 63 m and stand 52-m tall, yet radically different foundations support them. The Kobe anchorage sits on what was then the largest bridge foundation ever built.[7] Work began with the excavation of a 2.2-m thick, 85-m diameter circular slurry wall to a depth of 73 m. Almost 325,000 m^3 of material within its perimeter was then excavated in an 11-month period to a depth of 63 m—enough to swallow four of Rome's Pantheons. More than 230,000 m^3 of concrete was then roller-compacted to re-fill this enormous hole over the course of another 22 months. In September 1992, the foundation was complete. The shallower, stepped rectangular Awaji foundation rests on solid granite up to 21-m deep and was completed after 21 months of work ending in December 1992. Four 2,000-tonne anchor frames, complete with eyebars, were then transported by floating crane to both anchorages and encased in another 140,000 m^3 of above-ground concrete. A new smooth-flowing concrete penetrated this maze of steel without any need for vibration or compacting. Unusually, the back end of the Kobe anchorage sits unsupported behind its circular footing, thanks to the huge horizontal pull of the cables; the gargantuan size of the anchorages is such that one even contains an eight-story maintenance building undetectable from the outside. Both anchorages stood ready for cable work in April 1994. In total, the foundations of the Akashi Kaikyo Bridge would consume 40% of the construction budget.

Tower erection began in the spring of 1992. The two towers together consumed more than 49,000 tonnes of steel whose fabrication required a period of 3 years. In spite of their size, each tower leg is built from just seven boxes in cross-section, with each tier placed in just three lifts by an erection crane clambering up a parallel self-elevating shaft. This unusual system not only saved time and money but also helped to stabilize the towers. As this aerial ballet began, an ingenious self-elevating observation platform near the Kobe anchorage began to whisk viewers to heights of more than 100 m to observe the proceedings. Yet the towers' ascent also threatened television interference throughout a larger area around the bridge; to forestall this calamity, the HSBA had earlier paid for some 37,000 cable television connections. By January 1993, the Kobe tower was finished, and completion of its Awaji twin followed in April. Some 700,000 bolts hold each tower together.

The first linking of both shores by a pilot rope that November generated much attention. A helicopter undertook the task, tidal currents and shipping ruling out a floating crane. Of most interest was the rope itself—a lightweight aramid fiber about a tenth the weight of an equivalent steel wire—made the lift possible. Stabilizing catwalk storm cables were also eschewed because of the long span; only triangular wire frames slung beneath the catwalk at the cross-braces served to minimize vibration from passing bridgeworkers. Amazingly, the 6 months of catwalk erection and outfitting actually exceeded the time it took to place the 300,000 km of cable wire—enough to encircle the globe seven and a half times and almost a third more than spun into the four cables of the Verrazano-Narrows Bridge. The first 87-tonne prefabricated parallel wire strand was draped in June 1994. In spite of an accident that stopped work for 2 weeks, all 580 strands were placed by December with an efficiency more than twice that achieved in Denmark's East Bridge.

It was during compaction of the main cables that the Akashi Kaikyo Bridge was severely tested. In the early morning of January 17, 1995, a 7.2 Richter-scale earthquake struck nearby Kobe, wreaking horrible devastation. Some 6,400 people were killed and more than a quarter of a million buildings damaged. The earthquake caused no casualties on the bridge itself. But initial relief that the bridge remained standing was tempered by concern over its health. Work was immediately halted. Engineers had known of a fault traversing the line of the bridge, but with the epicenter of the earthquake less than 1 km from the Awaji tower, a previously unknown 14-km deep fault had now opened up beneath the bridge. Frantic inspections dominated the following month. Concern over the tower foundations was particularly high given the soft nature of the supporting soils. Yet a combination of good design and luck with the timing of the earthquake saw the bridge sustain virtually no damage—the absence of the truss had permitted the bridge to flex, and the cables had helped to stabilize the towers.

The earthquake, however, had wrought a most unusual change; because of crustal movement, the towers now stood 80-cm farther apart than before. The Awaji side span was also 30-cm longer, increasing the overall distance between anchorages by 1.1 m. The Kobe tower also stood 30-cm west—and the Awaji tower 20-cm east—of the span's centerline, and the latter tower stood 20-cm lower than before. Incredulous engineers at first questioned the findings, but other measurements would confirm them. Although the towers and cables could support the small increase in stresses, the main cables now hung 1.3-m higher at midspan than before. All suspenders had been fabricated, so to resolve the problem engineers modified several girders to lengthen the truss and let it rise to meet them. The earthquake had also imposed other problems. Some of 200 workers compacting the cables had been injured (although not on the bridge), and others aided in relief efforts. An HSBA field office was also damaged. Luckily, truss members would arrive by sea, so transport remained unaffected. Work on cable compaction resumed a month after the earthquake.

The erection of more than 85,000 tonnes of high-tensile truss steel began in June 1995, a month behind schedule. Floating cranes first lifted huge 84-m long truss blocks at the towers, followed by even larger 114-m long blocks at the anchorages. The lifting of all 4 km of truss in large blocks would have saved several months but was only feasible during slack tides, a period when much shipping darted through the straits; the collision risk—as with the box girder—was too high for maritime authorities to accept. Instead, partially assembled panels were moved out from each tower toward midspan on previously erected truss, and from the anchorages to the towers. Some 40% of time was lost to winds exceeding 36 km/hr, but in September 1996, the truss was in place.

A final 18 months of work included the first-ever dehumidification of main cables. Minor corrosion in the cables of other Honshu-Shikoku spans had spawned this effort, in which dry, desalted air is injected into the interstices between cable wires, which typically constitute 20% of a cable's cross-section. Rubber was also laid over the traditional steel wrapping. Advances extended to the suspenders as well, where European-style pin connections were used for the first time in a Japanese suspension bridge. More durable parallel wires protected by polyethylene tubes were also used in place of helical strands. This innovation did have unintended consequences—a number of braces guarding suspender pairs against low-speed vortex shedding excitation were soon destroyed by wake-induced flutter in high winds from the upstream suspender; to solve the problem, helical ropes were later wound around the suspenders to disrupt the wake pattern.[8]

After a decade of work, the Akashi Kaikyo Bridge opened in April 1998, a decade less 5 days after the Seto Ohashi and a month ahead of schedule; its $3.8 billion cost was on budget. Hundreds of thousands of people participated in the opening celebrations. The statistics associated with this bridge are staggering. Its 1991 m main span is 80% longer than Japan's previous longest span, the 1,100-m Minami Bisan Seto (*see* Fig. 14-3). Its superstructure consumed 193,000 tonnes of steel, and the bridge as a whole devoured 1.4 million m^3 of concrete. Its towers stand 93-mm farther apart at their summits than at their bases because of the earth's curvature; each transmits a load of 100,000 tonnes to its foundation. The cables carry a tensile force of 120,000 tonnes, much more than in any other suspension bridge. The truss can safely sway laterally by up to 30 m at the design wind speed, and its midspan elevation can vary by up to 13 m just from the expansion and contraction of the main cables. Truss expansion of up to 2.7 m is distributed across huge expansion joints at the towers and anchorages. For shipping safety, the trusses are coated with a radar-absorbent coating, and the anchorage panels deflect radar waves. The tower piers' massive size renders them impervious to even the largest of vessels; their collapsible steel fenders actually serve more to protect ship hulls by reducing impact pressures.

Fig. 14-3. Akashi Kaikyo Bridge
PHOTOGRAPH COURTESY OF R.A. DORTON.

The immense job of building the bridge also involved dozens of consortia. Up to five companies combined to build each tower pier, anchorage, and tower. Nippon Steel and Kobe Steel joined forces for the cables, whereas the truss involved four consortia. Perhaps most remarkable is the fact that not one worker was killed during construction. For tourists, an observation gallery extends 150 m into the Kobe side span truss; its anchorage sits within a park on the newly reclaimed land and beside a Museum of Bridge Technology, where the 40-m wind tunnel model of the bridge now resides. As with the Seto Ohashi, plans were laid for the transformation of Awaji from a relative backwater to resort destination, a future not lauded by all.

The sheer size of the Akashi Kaikyo Bridge made aesthetics an important part of its design. With more than 100 hectares of exposed steel, the bridge sports a green-gray hue to blend with the urban landscape of Kobe, the sea, and the sky. The diagonally braced towers are perhaps the finest of their genre; slightly inclined legs and the hiding of the cable saddles within the tower legs greatly enhance their appearance. Exceptionally long side spans emphasize the enormous width of the strait; concrete panels of varying hue visually break up the enormous mass of the anchorages, whose polyhedron shape softens a basic rectangular form. The main cables are also lit in combinations of red, green, and blue that vary with the time of day, month, and season. Finally, the Kobe entrance to the span offers great visual drama when motorists find themselves propelled from a 3.3-km, $636 million tunnel almost immediately onto the bridge.

Although some have suggested that the Akashi Kaikyo's truss embodies a philosophy of a previous era, its size and innovation stretched the bounds of truss technology and many other fields of suspension bridge engineering. Ironically, the Akashi Kaikyo Bridge may be the first—and last—ultralong span truss-stiffened suspension bridge; longer spans will likely require new concepts that challenge even current box girder orthodoxy. Japanese engineers may, in fact, pioneer them.

14.6 THE KURUSHIMA TRIUMVIRATE

The route that launched the Honshu-Shikoku project in 1975 also brought the quarter-century effort to a close with the 1999 opening of five bridges on the Onomichi-Imabari route. Three of these final spans were suspension bridges. The four-lane route—freed of rail loads—would broach new frontiers in suspension and cable-stayed design.[9]

The Ikuchi Bridge—standing just south of the Innoshima Bridge—illustrates this evolution. Poor footings saw an initially planned 250-m span concrete box girder jettisoned in favor of a record cable-stayed span of 490 m. Work began in 1986, and the bridge was opened in December 1991.[10] A second span then catapulted Japanese cable-stayed technology to a new plane altogether. In 1973, the Tatara Bridge was envisaged as an 890-m span suspension bridge sporting a 10-m deep truss suspended between 140-m tall towers. Yet advances in bridge technology had by late 1989 convinced the HSBA to adopt a record 890-m span cable-stayed bridge. This new design would avoid a massive gash from an anchorage in the flanks of Ikuchi Island, part of Seto Inland Sea National Park. A 2.7-m deep steel box girder draped between 230-m twin-spired steel towers also replaced the truss of the suspension bridge. Construction began in late 1992, and the $605 million bridge—exceeding the main span of France's Pont de Normandie by 34 m—was opened in May 1999.

The final link to Shikoku crosses the 4-km wide Kurushima Straits. The 1973 plan had featured tandem 860- and 550-m span suspension bridges sharing a common anchorage, followed by a sharp curve to launch a 1,000-m span to Shikoku. Trusses up to 11-m deep were to be suspended between towers up to 161-m tall. This plan had languished in the wake of the 1973 postponement, but by 1987, the Kurushima Bridge was seen as a way to stimulate an ailing regional economy. With the completion of the Onomichi-Imabari route approved for century's end in June 1988, the Kurushima Bridge received a final go-ahead in December of the same year.

Yet the 1973 alignment was flawed. Narrow, crowded, and circuitous navigation channels—two of them international shipping lanes funneling 1,000 vessels per day—harbored tidal currents of up to 15 km/hr and provided a fecund ground for marine accidents. Environmental and scenic concerns were also paramount given the span's presence in the national park. Although not much had been settled by the time of the April 1988 groundbreaking, fundamental change soon emerged. The crossing's east abutment was moved almost 2 km north to remove the awkward kink of the 1973 plan, and one island was avoided altogether. Nor was a triple suspension bridge assured; options included two 1,030-m spans anchored by a 600-m suspension or cable-stayed span, continuous twin spans of 1,030 m suspended between three towers, and a single 2,000-m suspension span. The latter configuration was the most expensive, whereas the continuous twin span bridge required a large rigid central tower for stiffness against deck deflections. The three-span option prevailed and, with two common anchorages, the Kurushima would become the world's first triumvirate suspension bridge. The Kurushima would also be the largest Japanese suspension bridge to use box girders. As built, its 4,105 m of suspended structure would exceed even the total suspended length of the Akashi Kaikyo Bridge, and its two larger spans would individually rank 10th and 12th longest in the world (*see* Fig. 14-4).

The elements of the crossing are diverse but visually well-coordinated. A 600-m span rises sharply toward the 1,020-m span Second Kurushima Bridge, a markedly asymmetrical span thanks to suspended and unsuspended side spans and the footing of one tower high up the flanks of Umashima. The 1,030-m span Third Kurushima Bridge follows the 1973 alignment, its Shikoku side span curving out from beneath the main cables before their anchorage.[11] A fine line is etched across the straits, with shipping clearances rising from 36 m in the 600-m span First Kurushima Bridge to 65

Fig. 14-4. The three Kurushima bridges
PHOTOGRAPH COURTESY OF THE HONSHU-SHIKOKU BRIDGE AUTHORITY.

m over the international shipping channels in the two larger spans. Achieving this harmony required a smoothing out of the line between the six tower tops, which stand from 116- to 184-m tall; in consequence, one tower in the First Kurushima Bridge is 30-m taller than structurally necessary. The slender, graceful towers feature multiple portal struts that neatly dovetail into cruciform-shaped, gently raked legs.

Wind tunnel tests served to refine the box girder design. The region's complex topography saw a model of an area encompassing a 9-km radius around the bridge used in wind tunnel testing, which also benefited from a 1991 typhoon that provided valuable, if alarming, evidence of gusts up to 151 km/hr. Box girder depths ranging from 2.5 to 4.3 m were studied, as were different edge shapings. Several unusual variants also emerged, including a triangular-shaped box with no horizontal bottom and a configuration of two parallel split boxes connected over a narrow gap by open trusses. This second option coped well with strong winds, and, in the 1,030-m span Third Kurushima Bridge, would be 1 m shallower than a single box. Yet a 4.3-m deep single box was found to be stable at the required 263-km/hr flutter velocity. It also used narrower foundations and towers, and would be easier to maintain. In consequence, a single box was adopted with a 2.2-m depth over the 600-m span First Kurushima Bridge and a 4-m depth over the two larger spans. As in Denmark's East Bridge, internal cross-trusses replaced solid bulkheads, and vertical suspenders would support the box girder (*see* Fig. 8-3).

Excavations began in September 1990. The success of the tower caissons at the Akashi Kaikyo saw this method used to foot the only open-sea tower, in the Second Kurushima Bridge. Here, a central 2-m diameter pile bored into rock stabilized the 34-m high circular caisson in powerful tides until it was firmly entrenched on the sea bottom. Great care was devoted to the appearance of the four anchorages. Both flanking anchorages are tucked out of sight behind intervening ridges, the Shikoku cables diving down 63-m long shafts into rock as the crossing's sole tunnel anchorage. The crossing's two shared anchorages are 64-m tall and use the cross-stranding approach of the Bisan Seto bridges; that between the First and Second Kurushima spans is asymmetrically shaped to visually emphasize the larger 1,020-m span. Also unusual is a bus stop perched atop the anchorage between the Second and Third Kurushima bridges, accommodated by a flaring of the box girder in the adjacent unsuspended side spans.

Six years were required to complete all nine major foundations, and for half this period the floating concrete plant used at the Akashi Kaikyo toiled in the straits. Tower erection, which began in June 1994, required precise scheduling as only two tower cranes were used to build three towers each, but consecutively. Tuned mass dampers also helped to stabilize the towers but, unlike those at the Akashi Kaikyo Bridge, were used only during construction. The stronger cable wire and cable dehumidification pioneered in the Akashi Kaikyo Bridge was also used, as was a new interlocking cable wrapping wire. Nippon Steel and Kobe Steel began stringing prefabricated parallel wire strands for the main cables in 1996. The largest number of strands are found in the 1,030-m span Third Kurushima Bridge, whose 65-cm diameter cables each consumed 102 strands of 127 wires. In total, 19,400 tonnes of cable wire had been draped into the three bridges when the cables were finished at the end of 1997. By this time, box girder erection was well underway on the First

Kurushima Bridge. The 36-m long segments were lifted from a self-propelled barge into all three spans and weighed up to 540 tonnes each. Sadly, as with the Golden Gate Bridge, tragedy struck late in the program. In June 1998, crews were lowering a 50-tonne erection platform from beneath the finished unsuspended north side span of the Second Kurushima Bridge when three restraining cables slipped from their clamps. The platform tipped sharply, hurtling eight bridge workers down in a 60-m plunge to the ground. Seven of them died in what was the worst Japanese bridge construction accident in decades. Work at the anchorage was stopped for more than 2 months while investigators sought to identify the cause. The system—widely used in Japan—was subsequently condemned, and cranes dismantled the remaining sections.

In spite of this tragedy, the Kurushima Bridge opened on schedule in May 1999 at a cost of $2.9 billion. Its construction required no fewer than 15 joint ventures involving most of Japan's largest contractors. The simultaneous opening of the Tatara Bridge and the entire Onomichi-Imabari route marked an end to almost a quarter century of continuous construction on the $27 billion Honshu-Shikoku project, the largest and most expensive bridge undertaking ever. Its 10 suspension bridges have catapulted Japan to the forefront of this technology, while generating significant advances in construction techniques and management that were necessary to bring such an enormous project to completion.

14.7 BEYOND HONSHU AND SHIKOKU

The Honshu Shikoku project so dominated Japanese bridge building in the 1980s and 1990s that it overshadowed other notable suspension bridges built both at home and abroad. One such bridge—the 520-m span Matadi—spans the Congo River 100 km from the Atlantic Ocean and links the inland Congolese capital of Kinshasa and nearby mineral deposits to Atlantic ports. Japanese contractor IHI had developed design options for this road/rail span between 1971 and 1976, and the American firm Steinman, Boynton, Gronquist & Birdsall also designed what would have been a record 500-m span truss-stiffened cable-stayed bridge. Japanese funding in 1978 ensured the use of a Japanese design, although ironically, belt tightening after the 1973 oil crisis removed the original impetus for the crossing—the rail link.

The Congo River funnels ocean-going vessels, and this dictated both a 53-m navigational clearance and a single bank-to-bank span. Arch and cantilever bridges were considered too expensive and a rail cable-stayed bridge, in spite of research by the HSBA, appeared risky. Dozens of HSBA engineers contributed to the design of the Matadi Bridge, and the adoption of rail suspension bridges in seismically active Japan (a concern here, too) led to a 520-m suspension span with a two-lane—expandable to a tight four—upper deck and provision for a lower-deck rail line.

The rugged site greatly influenced the bridge's design. The base of one tower—footed just below the truss—stands high up a steep slope 17-m above its counterpart on the opposite shore. Simple fabrication, transport, and on-site assembly also dictated the use of canted tower legs comprising single hollow boxes; simple portal struts at the tower tops and below the truss brace them together. Each 47-cm diameter main cable comprises 54 prefabricated parallel wire strands of 127 wires, with two additional strands draped over each side span. In all, 13,000 tonnes of Japanese-fabricated steel and cable wire were delivered to the site in just six shiploads. The attention to logistics paid dividends, for soon after work began in 1979, the (then) Zaire government sought to accelerate the pace. IHI responded, completing the bridge in April 1983, 14 months ahead of the 66-month schedule.

The Matadi Bridge, despite its relatively modest span, holds several distinctions. The cantilevering of its side spans all 91 m to the anchorages without any intermediate support extended the approach used in the Shimotsui Seto Bridge. The bridge is only one of three truss-stiffened suspension spans designed and built with a fully composite orthotropic deck (the others being the Emmerich and A. Murray MacKay bridges). It remains the longest span in Africa and, for a brief 2 years until the 1985 opening of the Ohnaruto Bridge, was the world's longest rail suspension span—the Tagus, although longer, was initially designed for lighter rail transit. The Matadi Bridge was also the first large suspension bridge to be built of weathering steel. Its 9-m deep truss—a function of the rail requirement—produced a depth-to-span ratio of 1:58, the deepest of any large post-Tacoma Narrows Bridge span. Its towers are somewhat utilitarian in appearance, but the inventiveness of the Matadi's builders in overcoming the span's isolation earned the bridge a coveted Tanaka Award, Japan's recognition of civil engineering excellence.

One truly innovative span is found in the port area of Osaka, Japan's second-largest city. The 300-m span Konohana Bridge was the world's first large monocable suspension bridge and its longest extant self-anchored span when opened in 1990. Soft soils had precluded the use of gravity anchorages, and self-anchorage was favored for aesthetic reasons as well. The bridge, however, embodies major differences from Leonhardt's variant of the monocable concept. Suspenders hang vertically along the deck median, instead of radiating to the deck edges. They are also inclined longitudinally, as in the Severn Bridge, to create a truss action that stiffens the bridge and helps damp out aerodynamic motion. Other peculiarities include an A-framed steel tower supported on large concrete pedestals, and a box girder that flares out to accommodate approach ramps over one side span. Sophisticated computer analysis made the bridge possible. Behavior under wind during construction and with traffic—torsional resistance against asymmetric loads being critical—required the testing of two large models. A 3.2-m deep, four-lane orthotropic steel box girder was ultimately adopted with an unusual trapezoidal shaping. Its deck also extends in the form of cantilevered wings to support flanking sidewalks. Suspender fatigue was also a crucial area of research because it was one of the afflictions then plaguing the Severn Bridge.

Construction began in the early 1980s. Piles extend up to 50-m deep below the tower piers and drop even further beneath some of the approach spans. As is common in self-anchored spans, the suspended structure over the main and 120-m side spans was placed on temporary supports in advance of main cable assembly. Here, a floating crane placed the box girder in just five sections weighing up to 2,700 tonnes each. The towers were then placed, each in a single lift, atop the concrete pedestals; a generous 1:6 sag ratio reduces horizontal forces in the anchorages, a typical requirement of self-anchored spans. Last to be installed was the main cable, its 30 prefabricated parallel wire strands each containing an almost-record 184 wires. Birdcaging—the problem Bethlehem Steel encountered in the 1960s—plagued early strand laying, but these problems were eventually overcome. This unusual bridge has earned many plaudits and stands as an icon for Osaka, in spite of Billington's (1990) criticism—recalling Tunnard and Pushkarev's disapproval of the Throgs Neck Bridge—over its lack of symmetry with the channel it spans.[12]

Steel box girders soon found their way into larger Japanese suspension bridges. The 720-m span Hakucho Bridge across Muroran Harbor on the northern island of Hokkaido is one such span; it opened in 1998 after a protracted 14 years of construction. Its multiply braced portal towers (in place of initially considered diagonally braced pylons) accentuate a well-proportioned and graceful structure over the main and 330-m side spans. Deep rock presented imposing foundation challenges, especially for the north tower. An access bridge to the pier site was built in 1985, but not until 1988 was an artificial island ready to drive an outer steel cofferdam almost 67 m in diameter and an inner 37-m diameter concrete wall, the latter to the prodigious depth of 103 m. The area between the walls served as a work platform from which to dig out a 73-m deep central well. A foundation slab was then poured to support a cellular foundation filled with a special slurry that exerted just slightly more than 5% of the pressure typical of sand fill. This foundation was finished in June 1992. The walls of its southern twin were driven to a depth of 67 m; its foundation slab bears at -57 m.

As with so many Japanese spans, the towers of the Hakucho Bridge are of unequal height—they stand 136- and 139-m tall. A cruciform shaping of the tower legs, aided by the temporary use of tuned mass dampers during construction, served to minimize oscillation from vortex shedding. Floating cranes lifted large sections of pre-assembled tower into place, followed in 1994 by Nippon Steel's installation of the 52 prefabricated parallel wire strands of 127 wires each in each 47-cm diameter cable. As in the Akashi Kaikyo Bridge, suspenders are wrapped in polyethylene tubes, and the Hakucho's cable wrapping actually pioneered the special S-shaped, self-locking wires later used in the Kurushima Bridge. Several box girder designs had been considered, including a bathtub form similar to Denmark's East Bridge and another resembling that used in the Konohona Bridge. In the end, an 18-m wide steel box girder just 2.5-m deep was adopted with aerodynamic fascias that—like Hong Kong's Tsing Ma Bridge—are nonstructural. One concern—and a first for Japan—was to ensure the railings shed snow and maintained aerodynamic stability; a mock-up of the adopted design was even field-tested. The box girder was in place by the end of 1995, but not until the summer of 1998 was the bridge opened; its construction period exceeded even that of the Akashi Kaikyo Bridge. No matter, the Hakucho Bridge is one of the most attractive suspension spans ever built; the smooth transition between side, backstay, and approach spans and their integration with the unobtrusive recessed anchorages are exceptional. One peculiarity, however, is the pronounced and confined arching of the main span over its central portion in a manner reminiscent of Steinman's Deer Isle Bridge.

A contemporary of the Hakucho Bridge is the slightly larger 750-m span Akinada Bridge east of Hiroshima. Built between 1992 and 2000, this span features a 19-m wide box girder supporting two lanes. The span's narrowness prompted wind tunnel tests on a full aerolastic model, but the box girder's 2.4-m depth is, at least to the eye, of almost alarmingly slender proportions. The 128-m tall towers of this graceful span combine the upper portal flares of the Ohshima Bridge with the Kurushima Bridge's multiple braces. Side spans of 170 and 255 m respond to the site's rugged topography; one tower stands on a midchannel outcrop, and the other rests on 18 pipes anchored into steeply sloping bedrock, as in the Ohnaruto Bridge. Self-propelled barges aided in box lifting, and the main cables use the dehumidification system introduced in the Akashi Kaikyo Bridge.

The Akinada Bridge—whose construction was accident-free—is the first in a 12-bridge link to connect Hiroshima to the Onomichi-Imabari route of the Honshu-Shikoku bridges. Plans for a second suspension bridge on this link call for a span of 540 m with unsuspended side spans. The towers are to be braced by four portal struts. Most unusual is the proposed suspended structure, comprising 2-m deep streamlined trapezoidal box fairings connected by 13 m wide, 1.6 m deep cross-frames. Open steel grating is proposed to form the two-lane road deck, a first for Japan. Detailed design of the Toyoshima Bridge began in 1999, and construction should commence in a few years.

One notable truss-stiffened suspension bridge is the 570-m span Rainbow Bridge, completed across Tokyo's inner harbor in 1994. The choice of span type was not simple—no part of the bridge could exceed 156 m in height because of its proximity to Tokyo's major domestic airport at Haneda, yet the suspended structure had to clear the water by 50 m. Steel arches and trusses were ruled out, as was a record 530-m span box girder and a cable-stayed span with 200-m tall towers. A suspension bridge proved the only realistic option, in spite of difficulties in footing anchorages in soft soils.

The span's 1987 construction debut offered rarely used, but impressive, technology. Pre-assembled anchorage caissons measuring 70 by 45 m almost doubled the size record for pneumatic caissons, whereas the 49- by 25-m tower caissons ranked among the top five in size. All used the pressurized air dome method used in the Tagus Bridge, with remote-controlled digging equipment sinking the deeper offshore anchorage to a 46-m depth. Pre-assembly also figured prominently in superstructure erection. The 650-tonne anchor frames were placed by floating crane, which also hoisted each 126-m tall tower into place in just three lifts, the largest upper lift including already installed cable saddles. A total of 127 prefabricated parallel wire strands of 127 wires each were then placed in each main cable by Nippon Steel and Kobe Steel; an additional 3 strands were added over each 114-m side span cable.

The bridge's prominence as a gateway to Tokyo's inner harbor brought aesthetics to the fore. A committee of citizens and experts held considerable sway, selecting one of five stylings generated by designers. Three portal struts were adopted for the towers—one at their apex, another beneath the truss, and a third buried within the truss. A planned park beneath the mainland side span also prompted careful tower and anchorage detailing; corners are rounded, the towers are devoid of exposed joints, and strakes were splashed across the anchorages to break up their mass and mimic the splayed, but hidden, cable strands. The span's 9-m deep, 29-m wide Warren truss supports four lanes on an upper level that are jettisoned from the mainland anchorage via dramatic, curved approaches. A lower deck carries four more lanes and a rapid transit line. Walkways and a viewing gallery also provide views of the port area. Most dramatic is the use of multicolored floodlighting that varies with the time of day and season; the system is 40% solar powered, and artists are even encouraged to manipulate the lighting controls. All the visual prosthetics with which the span is imbued came at a price, absorbing some 20% of its $1.1 billion cost.

One recent bridge made possible through Japanese government financing is in Semipalatinsk, the capital of Kazakhstan. The Irtysh River Bridge is a 750-m span box girder that combines features of the two spans across the Bosporus at Istanbul. Built by Japanese construction giant and second Bosporus Bridge veteran IHI, the six-lane span sports a 3-m deep, 35-m wide box girder strongly resembling that of the first Bosporus Bridge, including cantilevered sidewalks. The almost 100-m tall towers are look-alikes to those of the second Bosporus Bridge, but the box girder hangs just 27-m over the water and actually spans the tip of a midchannel island. With land-based towers, the 168-m side spans are unsuspended. Construction, which began in early 1998, required special techniques to cope with temperature extremes of -50° to 50° C, among the largest in the world. Spinning of the two 8,192 wire main cables into strands of 512 wires began in the spring of 1999 after a winter shut-down in which temperatures plunged to -40° C. The main cables were completed in November 1999 before the onset of yet another severe winter, and lifting of the box girder commenced the following spring. The bridge's opening in October 2000—a full year ahead of schedule—was a remarkable achievement given the harsh climate.

For all their record-shattering advances in so many fields of suspension bridge design and construction, Japanese engineers have been accused of overly conservative engineering by some Western counterparts, an approach reflected in the heavy, trussed appearance of many of Japan's suspension bridges. Yet Japanese engineers point to the Severn Bridge as an example of innovation without rigorous advance testing; gradual introduction of new technologies into bridges of increasing span is the preferred approach. Moreover, Japanese testing techniques are anything but conservative—decks and towers have been deliberately pounded into vibration to verify design calculations and obtain measurements of structural damping; such tests are rare elsewhere. Recent spans have reversed the reputation for ponderous design, and the subtle cultural motifs woven into tower designs often express greater cultural meaning than their Western counterparts. When one considers the strength necessary to resist Japan's typhoons and earthquakes, such bridges as the Seto Ohashi, Akashi Kaikyo, and Kurushima become all the more impressive.[13] Moreover, few Western spans have received the attention to aesthetics lavished upon Tokyo's Rainbow Bridge.

Noted Japanese bridge engineer Manabu Ito once observed that, "Japanese engineers are rather conservative" (O'Neill, 1993). This seeming concurrence with Western criticism in fact celebrates an approach that produced more large suspension bridges in the last 2 decades of the 20th century than any other nation on earth, in environments that would challenge the most experienced bridge-builders. A quarter century ago, Japanese engineers had virtually no experience in the genre. Japan's bridges have been expensive, and there is a recognition that costs must diminish while maintaining the typically high quality of construction. But the evidence confirms Ito's rejoinder to concerns over Japan's conservative engineering approach that, "when you can afford to be so, why not?"

ENDNOTES

1. In 1989, an arch superimposed over the suspension system saw the towers and cables of the Miyoshi Bridge removed.
2. A Vierendeel truss uses large vertical members in place of diagonals.
3. A 234-km/hr gust the year of the tower's installation was never approached again in 20 years of observation.
4. The simply supported side spans of Budapest's 290-m span Elizabeth Bridge (1965) were also structurally continuous with the main span.
5. These fenders have since been replaced by the steel versions developed for the Akashi Kaikyo Bridge. They received their first test in March 2001 when a small oil tanker with a slumbering crew collided head-on with a Kita Bisan Seto Bridge tower pier. Only minor damage to ship and pier resulted.
6. Compared to this is the 16% design live load figure of the Innoshima Bridge.
7. In 2000, 90-m diameter caissons were sunk for Greece's cable-stayed Corinth crossing.
8. Vortex-induced suspender vibration also occurred in the Kurushima and Akinada bridges, both of which use parallel wire suspenders. The cause was attributed to the round shape and smooth surface of the polyethylene pipes.
9. The route is the only one of the three to permit cyclists and pedestrians.
10. The Ikuchi Bridge held this record for just 2 weeks—Norway's 530-m span Skarsundet Bridge opened later that month.
11. Subsequent modifications reduced the Shikoku side span from 370 to 280 m and substituted tunnels for its gravity anchorages.
12. The Konohana Bridge inspired a similar but unbuilt scheme in east London, England, over the River Thames; this span of 340 to 380 m was one of several considered in a 1993 competition driven by aesthetics (Billington, 1990).
13. The Kurushima and Tatara bridges successfully weathered a major earthquake measuring 6.7 on the Richter scale in March 2001. Seismic motion sensors on the First (600-m span) Kurushima Bridge verified design calculations for the span's natural vibratory period. Moreover, the motion-damping triangular suspenders connecting the box girder to the main cables at center span broke—as planned—to prevent damage to the main cable in case of faster, non-aerodynamically-induced vibrations from earthquakes. Minor repairs saw the installation repaired by June, with no disruptions to traffic save the temporary closure of the span immediately following the earthquake.

Chapter 15

CHINA'S METEORIC RISE

The rapid growth of China throughout the past decade has produced a veritable revolution in suspension bridge design and construction that may even outpace the remarkable ascendance of Japanese engineers. Many design and construction approaches have been developed indigenously, but Chinese engineers have availed themselves of British and other expertise to aid in a phenomenally ambitious program of suspension bridge building to help serve this nation of more than 1 billion people.

15.1 A SECOND REVOLUTION

The economic renaissance of China from reforms launched in 1977 not only stimulated demand for large bridges but exposed previously isolated Chinese engineers to Western bridge technology, materials, and construction methods. China completed its first modern cable-stayed bridge in 1975 and, by 1993, boasted (if briefly) the world's longest of the type—Shanghai's 602-m span Yangpu Bridge. Although China was early progenitor of suspension bridges centuries ago, it only embraced the form in a major way in the late 1980s. Communist "production brigades" had built many small bridges, with several flexible spans of less than 200 m found in mountainous southwest China. Two of these bridges were completed as early as 1951 in Sichuan Province; the larger span over the Tatu River is a 130-m truss-stiffened span with diagonally braced towers adorned with state propaganda. The high-level Feishienkuan Bridge features a span of just 67 m, but its towers are an unusual hybrid of concrete legs braced with steel diagonals. In 1958, a 120-m truss-stiffened suspension bridge was completed across the precipitous Yachih River gorge in Kweichow. Two unstiffened spans completed in Sichuan Province in 1966 and 1967 comprised a second 150-m span—the Oepian Bridge—across the Tatu River and a 124-m span on the road to Tibet, respectively. The single arched portal struts of the latter span's concrete towers were again emblazoned with slogans. A 176-m span completed across the Yalung River in 1966, again in Sichuan Province, was stiffened by a truss and was suspended between concrete towers braced—unusually—with concrete diagonals. Work brigades then spanned the Yellow (Huang) River in 1970 with the 133-m span Pingpao Bridge in Kansu; it featured portal-braced concrete towers. The following year, the 100-m span Shiamen Fenho River Bridge was completed. Its steel towers and stiffening trusses were built from preassembled units. One curiosity is the Chaoyang Bridge, completed near the city of Chongqing, Sichuan Province in 1969. Its 186-m span is supported by two pairs of main cables describing different arcs from which hangs a slender welded and bolted steel box girder whose top is capped not by steel but by a concrete slab.

The daring 500-m span Dazi Bridge became China's longest single span in late 1984. Its orthotropic deck caps a through truss that just clears the water of this shallow but untamed Tibetan river. Wild topography shaped the unusual design; sheer cliffs anchor one end of the parallel wire main cables at a height several times that of the single tower situated on the flats of the opposite shore. A similar arrangement is found in the Jinhu Bridge, a 284-m span set amid the spectacular rocky spires of Fujian Province. This 1989 span features a slender filigree truss. Towers were then dispensed with altogether in the Meixihe Bridge, a 205-m span completed in a mountainous region of Sichuan Province in 1990. The main cables are anchored in steep slopes abutting both ends of the span; the complete absence of towers is a little unnerving. This bridge also features inclined suspenders of the Severn Bridge type supporting a lithe steel truss. One conventional steel truss suspension bridge did emerge in 1987—the North Bridge in Liaoning Province—a 132-m span flanked by 40-m suspended side spans hung from squat steel towers.

Concrete has figured prominently in Chinese bridges as it is labor-intensive and inexpensive. The material was first used as a suspended structure in the Qileharen'eerqisihe Bridge, a small 108-m span completed in 1992 in the Xinjiang Autonomous Region. A slender concrete box girder with a strong yellow fascia is complemented by simple portal-braced concrete towers, yellow-wrapped cables, and the free passage of the continuous girder above the tower's lower portal strut that all make for a handsome bridge.

These efforts were soon eclipsed by the 452-m span Shantou Bay Bridge. This unique span is situated along China's exposed South China Sea coast, about 300 km east of Hong Kong in Guangdong Province. It is the world's longest and most capacious prestressed concrete box girder suspension bridge and the first of note since Canada's Hudson Hope Bridge of 1964. Including 154-m side spans, the box girder is suspended over a length of 760 m. The six-lane suspension bridge was designed by the Major Bridge Engineering Bureau of the Ministry of Railways, and was chosen over cable-stayed and prestressed concrete girder spans, the unusual choice of concrete for the box girder reflecting a need for mass against strong winds. It was also economical, even though larger cables were required. The three-celled, 25-m wide box girder is just 2.2-m deep at its centerline; its bottom curves gracefully to slightly less than 1 m at its edges.

The environment here is challenging, with a tidal range of more than 4 m, typhoons, and the potential for earthquakes of up to 8 on the Richter scale. Although beneficial against wind, the mass and inertia of the box girder were problematic in terms of earthquake response. Girder movement had to be carefully guided and the bridge designed to absorb energy, rather than resist motion that could overstress the bridge, especially the towers. Longitudinal sway was of particular concern. Restraining cables between the towers and the girder provide a first line of damping defense for moderate tremors. In more severe earthquakes, these cables are designed to snap, allowing energy to be absorbed by steel and rubber buffers in the box girder near the towers and at the abutments. The buffers near the towers transmit even higher seismic forces to "antipunching" shear walls at the abutments, and these walls are even designed to break under the most severe forces. Lateral sway was another concern and is accommodated by a transition box extending 4 m on either side of the towers. Connections between the suspended and transition boxes will withstand moderate earthquakes but under larger forces will break, allowing the main span boxes to swing freely across the line of the bridge without hitting or overstressing the towers. The longitudinal restraint cables between the towers and the box girder are also designed to restrict lateral movement.

The 95-m tall portal framed towers, with two above-deck struts and one below, are also built of concrete; in an unusual design treatment, the box girder hangs well above the lower strut. Each main cable comprises 100 prefabricated parallel wire strands of 91 wires each, and reflects the remarkable speed at which Chinese engineers have developed the method. The box girder was cast in 5.7-m long segments each weighing 170 tonnes. All 121 segments were lifted into the span and temporarily connected before being concreted together. Six other units were poured in place at the towers, the backstay bents, and the anchorages. Four to five segments were cast per month, but 4 months in a storage yard were then required to cure the segments as a means to minimize concrete shrinkage on the bridge. The design was approved in February 1992, deck casting began in April 1993, and the bridge was opened in December 1995. With an overall length of 2.4 km, the Shantou Bay Bridge provides a key link in China's coastal expressway and ranks as one of the most innovative suspension bridges anywhere.

By the time the Shantou Bay Bridge opened, three large suspension bridges were under construction, all of indigenous design. Two almost doubled the Shantou Bay Bridge in span; the third—incredibly—would rank fourth longest in the world. The first to open was the Boca Tigris Bridge across the mouth of the Pearl River delta north of Hong Kong. This 3.6-km long crossing also features a 270-m span concrete box girder bridge, one of the world's longest. The main east channel, however, required a span of at least 300 m and a 60-m vertical clearance; in the end, the Ministry of Communications' Highway Planning and Design Institute adopted an 888-m steel box girder suspension span.

Piles footed in poor soils support most of the bridge. The east tower stands in 16 m of water, with each leg supported by 16 piles bored up to 33-m deep; a pier cap serves as a base for the tower legs. Bored piles also support one leg of the west tower. The anchorages are dissimilar, too, and work on the east anchorage was actually delayed for a year when excavations uncovered the remains of an ancient Chinese fortress that in 1839 unsuccessfully repelled the British fleet during the First Opium War. Its west counterpart sits on a 60-m diameter, 10-m deep circular concrete block situated on an artificial island. Concrete has been the preferred tower material in China's large suspension spans, and the Boca Tigris Bridge is no exception. Its 154-m tall concrete towers feature three portal struts—two above-deck—connecting slightly inclined legs. Prefabricated parallel wire strands were used for a second time in the 69-cm diameter cables, each containing 110 larger strands of 127 wires each. To help perfect techniques for use in other more remote spans, engineers reeled all strands at a temporary factory erected on the span's approaches. A 3-m deep orthotropic steel box girder supports six lanes over a 35.6-m width and was designed to withstand severe typhoon exposure. Unsuspended side spans are supported on concrete viaducts 180 and 350 m long.

The entire Boca Tigris link was opened in July 1997, with the suspension bridge costing $112 million. The span is remarkably transparent, its litheness derived from its thin box girder and the sparse towers. The high placement of the intermediate tower portal is unusual (but typical of Chinese practice), as is a dense forest of piers extending from the suspension span's west tower over unsuspended side spans and a long over-water viaduct. The Boca Tigris Bridge was a major step for Chinese engineers, and clearly demonstrated their ability to use modern design and construction methods.

A second major effort is the Xiling Bridge, a slightly larger span of 900 m built across the Yangtze River to aid in the construction of the huge and controversial Three Gorges Dam. The bridge sits just 4.5-km downstream of the dam and, in a single-bound, spans the main river, an approach to a shipping lock, and a diversion channel. After assessing cable-stayed and arch spans, the Major Bridge Engineering Bureau in Wuhan adopted a suspension bridge. The dam's demanding construction schedule required that the bridge be built in just 32 months; it would also carry 54-tonne trucks, about twice the maximum individual vehicle loads accommodated on most Chinese bridges.

Large 2.2-m diameter piles extending to depths of up to 50 m support the span's 121-m tall concrete towers. Unusually, each tower leg rests on an independent foundation. Unsuspended prestressed concrete side spans meet land about halfway to the anchorages, which are set back 225 and 255 m from the towers. Again, prefabricated parallel wire strands were used in the 57-cm diameter, 1,500-m long cables, but each cable used only 91 wires in each of its 110 strands. Unusually, the edges of the four-lane orthotropic steel box girder drop so precipitously that the 21-m wide and 3-m deep box appears almost bluff in cross-section. The towers are similar to those of the Boca Tigris Bridge, but the strut placement is visually more pleasing and the top portal strut flares gracefully into the tower legs as in Japan's Ohshima Bridge (although slogans etched into the middle portal strut reduce the artistic gain). The $60 million span—then China's longest—was opened in August 1996, a remarkable 4 months ahead of the punishing 32-month schedule.

A third suspension bridge built in the 1990s crosses the Yangtze River midway between Nanjing and Shanghai. Here, the river narrows to 1,400 m, its smallest width in the 400 km between the two cities. The Jiangyin Bridge would not only replace the massive 1960s Nanjing truss as China's most easterly Yangtze River crossing, but as the world's fourth longest span, it would become a new symbol of Chinese prowess in bridge engineering.

Government approval to proceed with the crossing came in 1991, before a bridge type had even been selected. A 1,385-m span suspension bridge eventually won out against a cable-stayed span, and British firm Mott MacDonald was retained in 1994 to help prepare detailed plans with Chinese engineers from design institutes in Beijing, Nanjing, and Shanghai. The Jiangyin Bridge would unseat Hong Kong's recently completed Tsing Ma Bridge by a mere 8 m.

Foundation work began in late 1994. The south shore foundations benefited from hard granite, notwithstanding a slope beneath the south tower that required the drilling of 24 piles each with a diameter of 3 m up to depths of 40-m. Under the north shore, rock lay at considerably greater depths of up to 120 m. Here, a massive 69- by 51-m anchorage caisson—the largest ever sunk in China—was sunk to a depth of 58 m in weak soils, a 20-month operation that necessitated the removal of some 200,000 m^3 of soil. When the caisson was finally in place, extensive but unplanned grouting beneath its base was then required to alleviate fears (well-founded, as it turns out) about insufficient ground contact. The north tower was simpler and rests on 30 piles each of 2-m diameter drilled into rock through up to 96 m of sand and alluvium.

The Jiangyin Bridge's side spans are unsuspended prestressed concrete viaducts, and these were placed even before the catwalks for erecting the main cables were installed. The bridge's 336-m north side span is flanked by almost 1,000 m of approach viaduct. The south side span meets land just 163 m back of the towers; its anchorage, completely buried, sits 309-m back of the tower. The 197-m tall concrete towers bear the now-ubiquitous Chinese three-portal strut motif, but are better proportioned than those of the Boca Tigris Bridge. Typhoon concerns were not as prominent as in the Tsing Ma Bridge, but they still warranted investigation for the six-lane steel box girder. A large wind tunnel opened in 1994 at Shanghai's Tongji University aided in the crafting of a 32.5-m wide box girder with a depth of just 3-m; its 1:462 depth-to-span ratio is unsurpassed for slenderness in a large span. With cantilevered walkways extending its width to 37 m, the Jiangyin's box girder very much resembles the first Bosporus Bridge design of a quarter century earlier.

Foundations, towers, and side and approach spans were Chinese-built, but British financial assistance for cable and deck work would, notwithstanding the delicate politics of Hong Kong, pay dividends for British industry. Through subsidiary Cleveland Bridge & Engineering, Trafalgar House had invested 4 speculative years in the project and even offered—unsuccessfully—to take complete responsibility for it. The contractor was viewed as a heavy favorite when superstructure bids were called in 1996, even to the point that the presence of rumored competitors Nippon Steel and Storebælt East Bridge contractor Coinfra was suggested to be a mere formality. Whereas Coinfra had 2 more years of East Bridge work ahead of it, the newly renamed Kværner Cleveland Bridge could transfer a now-experienced team winding down work on the Tsing Ma Bridge.[1] In November 1996—2 months after Chinese contractors topped out the towers—Kværner Cleveland Bridge took the $160 million contract, its efforts aided by the offer of an $89 million British government construction loan to Jiangsu Province, owner of the bridge. The company would undertake detailed engineering, fabricate 18,000 tonnes of box girder steel in Great Britain, assemble the cables and box girder units in China, and open the bridge by October 1, 1999 to help mark a half century of Chinese communist rule. To speed work, digital design drawings were transferred via satellite link from Great Britain to China.

Great Britain's Bridon Ropes supplied wire for the two 86-cm diameter (and almost 90 cm in the side spans) cables, as it had for the Shantou, Xiling, and Tsing Ma bridges. Cleveland Bridge & Engineering's performance at the Tsing Ma Bridge might have suggested the use of in situ spinning, but a change from the prefabricated parallel wire strand cables specified by the Chinese—developed with the aid of Japanese long-span specialist Chodai—would have required a complete redesign of the 169-strand cables and the cable saddles. As such, the first prefabricated 127-wire strand was draped in the summer of 1998 by a Chinese company under subcontract. Work did not start well. The schedule called for 24 strands to be laid per week, but the first 24 strands took no fewer than 7 weeks to install; what was planned as a 2-hour strand draping operation became a 3- to 4-day effort because of technical, communication, and productivity problems. Translators eventually bridged the English/Chinese language gap, and cabling rates literally skyrocketed once technical problems were ironed out and productivity incentives were provided. Thirty strands were typically laid per week and in the second-last week of operations, 47 strands were laid. After 5 months of work all cable wire was in place, with two-thirds of all strands installed in the final 2 months. Unfortunately, efficiency problems would also dog the early stages of assembling the 44 main span box girder sections—each 32-m long and weighing 450 tonnes. It took 5 months to build the first 8 boxes, but dramatic productivity improvements saw the remaining 36 boxes assembled in just 6.5 months. The boxes were lifted into the span beginning in February 1999 at an impressive average of one per day (*see* Fig. 15-1).

The $200 million Jiangyin Bridge was opened on schedule in September 1999 and, like the Tsing Ma Bridge, is heavily instrumented; lasers are even used to monitor box girder movement. As with other Chinese spans, the towers are embellished with slogans. The box girder—dehumidified over the main span—and main cables are painted a light blue, and the box is dramatically lit on its underside.

Fig. 15-1. Jiangyin Bridge, lifting of box girder segments
PHOTOGRAPH COURTESY OF MOTT MACDONALD.

The Jiangyin Bridge is clearly the most impressive and aesthetically pleasing of the large Chinese suspension bridges.

The opening of the Jiangyin Bridge did not signal an end to Chinese aspirations. Just 2 months later, in December 1999, the Xiamen Haichang Bridge, a 648-m span with 230-m side spans, was opened along the South China Sea coast in Fujian Province about 500-km east of Hong Kong. This six-lane span features a slender box girder that, as in Denmark's East Bridge, is continuous through portal-braced concrete towers that provide no vertical support; the bridge is the third longest span of this type. Another similarity to the East Bridge is the use of open-framed triangular anchorages. The Fengdu Bridge is a 560-m span box girder suspension bridge that opened in December 2000, and Hubei Province's Yichang Bridge, again spanning the Yangtze River, is to open at the end of 2001. This impressive 960-m span also features a steel box girder.

More is to come. Groundbreaking on yet another huge span across the Yangtze River about midway between the Jiangyin Bridge and the city of Nanjing was held in October 2000. Planning for the six-lane Runyang Yangtze River Bridge began in 1992; approval to proceed was obtained in 1998, and the span's design was fixed in February 2000. The crossing involves two large bridges across a distance of 23 km, using Siyezhou Island as a stepping-stone. The 1,490-m span of the larger southern crossing will place this suspension bridge as the world's third longest upon its planned opening in October 2005. In outline, the bridge strongly resembles the Jiangyin Bridge, in its towers, box girder main span, and unsuspended side spans. The smaller northern crossing involves a cable-stayed span of 406 m. Construction of an even larger suspension bridge across the Jiaozhou Wan on China's northeast coast near Qingdao may begin in late 2001. A span of either 1,652 or 1,800 m is contemplated, which would make it only second to Japan's Akashi Kaikyo Bridge in length.

China's ambitious plans for expanding and improving its vast transportation network will undoubtedly produce more large suspension bridges in the near future. Its remarkably rapid ascendancy in the genre of long span suspension bridges has made it well-positioned to take on this task.

ENDNOTES

1. Norwegian construction giant Kværner had taken over Trafalgar House. But as Kværner finished the bridge, Cleveland Bridge & Engineering was sold off as an independent entity.

Chapter 16

NEW BRIDGES—THE NEAR FUTURE

At the end of the 20th century, a hiatus of sorts was reached after a frenetic decade of activity that saw five of the world's 10 longest suspension spans completed between 1997 and 1999. Other smaller spans have entered construction, and more large bridges remain in the planning stages. America is now beginning a resurgence with several new long-span suspension bridges entering construction. China's huge transport program will produce several sizable spans, as could Southeast Asia's, and the consolidation of Europe's transport network may see a focus for new suspension spans shift from northern to southern Europe, where the world's longest—and long-discussed—span across Italy's Messina Strait may soon be built. These myriad projects continue to challenge existing convention in suspension bridge design.

16.1 SPANS ON THE CUSP

Turkey's Bosporus Strait had been an obvious choice for the first permanent connection between Europe and Asia; the more than 75 million vehicles crossing its two spans annually by 1992 proved the logic of this choice. A third span—as discussed in Chapter 12—remains a possibility. Yet other, even larger spans could make Turkey home to four of the world's 15 longest suspension bridges. The first involves a 3-km crossing of the Gulf of Izmit, an eastern arm of the Sea of Marmara about 50-km east of Istanbul. A crossing here would considerably shorten connections between Istanbul to points south, where long ferry waits are avoided only by a long detour on a tortuous coastal road. One speculative proposal developed by Turkish contractors in the early 1990s involved a Dischinger-like cable-stayed/suspension hybrid of record-setting 2,000-m span; a cable-stayed concrete box girder extended to the main span's quarter points whereupon a steel box would be suspended conventionally over the central half of the main span. Tekfen Muhendfisklik, the Turkish firm that designed the span, claimed a cost as low as $400 million. In 1995, the government selected six consortia to bid on a bridge; the winner would not just build the span, but finance and operate it as well. Delays associated with the complex package emerged, but in May 1997 a consortium headed by Kværner Cleveland Bridge, with the Japanese contractor IHI and the Turkish contractor ENKA, beat out two French competitors (one of which proposed a COWI-designed suspension bridge of 1,576-m span) for the prize, subject to negotiations on price. The combination of Cleveland Bridge & Engineering's experience with the original Bosporus Bridge and IHI's work on the second Bosporus span proved a key factor in the selection.

The 1,688-m main span—flanked by 560-m side spans—will oust Denmark's East Bridge as the world's second longest. Its 263-m tall steel towers feature three above-deck portal struts and stand on preassembled caissons sunk to a depth of 50 m in 35 m of water. The towers are unusual, for their concrete piers rise 25 m above water and are capped by 10-m tall pedestals for ship protection before tower steel is grasped. The 4.25-m deep box girder is of British vintage, complete with cantilevered walkways that make it similar to the original Bosporus Bridge, save for the use of vertical suspenders. One anchorage is buried in a flanking hillside, and an open triangular anchorage inspired by Denmark's East Bridge marks the other end of the span, followed by 1.3 km of approach viaduct. Each 18,648-wire main cable is to be spun from 37 strands of 504 wires—by coincidence, the exact same numbers as in the East Bridge. In October 1997, a 27-year, $1.8 billion concession was signed for this six-lane, 3.3-km bridge, subject to the availability of financing. Hyder Consulting (formerly Acer Freeman Fox) was mooted as the candidate to prepare the detailed design, having beaten Mott MacDonald in June 1998. An ambitious 49-month construction period would allow for the bridge's opening in 2004, but in August 1999, the calamitous earthquake struck and devastated the Izmit area. The disaster not only underscored the challenge of earthquake resistance, but it has helped to postpone a construction start on this huge undertaking.

Another enormous Turkish bridge is planned over the Dardenelles, where the Sea of Marmara flows into the Aegean Sea. It is here that Xerces is said to have lashed 676 ships together to form a pontoon bridge in 480 B.C. in his war with Greece. A crossing near Chanakkale had been mooted by the Turkish government since 1976, and by 1978, the firm of Freeman Fox & Partners was working up a proposal. Priority was then assigned to the second Bosporus Bridge, and not until 1992 did a newly elected government launch new studies. A desire to avoid foundation work in strong currents produced a 1,510-m span, one that was required to withstand earthquakes of up to 7.5 on the Richter scale. The Turkish-developed preliminary design called for a six-lane box girder just 2.7-m deep resembling that used in Denmark's East Bridge—but wider, to include footpaths. Flanking the main span were 230-m side spans suspended from 167-m tall steel towers with multiple portal braces. Design of the estimated $340 million bridge was expected to be completed by 1995.

This concept was then abandoned. British engineer William Brown, as head of a new design team, introduced the split box concept he had proposed unsuccessfully for the second Bosporus Bridge. Brown had examined a single box with a curved underbelly and split boxes with symmetric or asymmetric cross-sections. In the end, a 1,440-m span was selected, with each three-lane roadway supported on separate boxes connected by grillage and periodic steel braces, much in the manner of Brown's design for Italy's Messina Strait Bridge described later in this chapter. In cross-section, each box was shaped as a teardrop—thicker toward the inside edge—and floated through the towers without any connection save for restraints on lateral movement. The unusual and elegant towers featured deep fluted portal struts above and below the roadway bracing inclined tower legs, with the top strut placed well below the tower apex in the manner of many cable-stayed bridges. The span would be built under the same build, finance, and operate model as the Izmit Bridge at a 1998 cost of $600 million, but this dramatic scheme has yet to proceed further.

Hong Kong, however, will likely build the first split box suspension bridge. In May 1995, 2 years before the Tsing Ma Bridge was opened, a suspension bridge to the west end of the Lantau Link from the New Territories—near the route rejected in 1978—resurfaced. The Tsing Lung Bridge would provide a second access to the new airport at Chek Lap Kok and eventually form part of a direct link from the New Territories to Hong Kong Island. A span of about 1,300 m would make landfall on the east tip of Lantau Island and pass over the approaches to the Kap Shui Mun Bridge. Yet the span's 196-m tall towers were a concern for airport authorities. The lopping off of 16 m of tower suggested a span some 180-m less to reduce otherwise high cable stresses, but this would place the towers in the busy shipping channels. Engineers retained the 1,300-m span and suspended it from 176- and 184-m tall towers resembling those of the Tsing Ma Bridge. A slender 4-m deep box girder would support six lanes of traffic.

Still, issues remained. Should the bridge carry six or eight lanes? What technologies would make it possible? In early 1998, the firm of Mott Connell (Mott MacDonald's Hong Kong office) was retained to examine such questions. Many staff involved in the Tsing Ma Bridge contributed. In the end, a lengthened span of 1,418 m was adopted that would make the Tsing Lung Bridge the world's third longest single span. Its towers would stand just 176- and 169-m tall, providing an unusually low sag ratio of 1:15. British firm Bridon Ropes had overcome the issue of great tensile forces in the main cables by launching a wire even stronger than that developed for the Akashi Kaikyo Bridge; high-quality wire and improved wire drawing made this 1999 advance possible.

The firm Maunsell Asia introduced more significant change when it began detailed design of the span in early 2000. A key factor behind the firm's selection was the proposal, developed with the aid of William Brown, to reduce the box depth to just 2 m by using Brown's revolutionary split box concept. Japanese firm Chodai–a key player in the design of the Akashi Kaikyo Bridge–was involved, while Danish firm Dissing & Weitling of East Bridge fame would provide architectural input for the overall crossing. Louvered windscreens will be accommodated if necessary, and the unsuspended side spans may become suspended for structural reasons; this could serve to lengthen the main span to 1,458 m. Either way, the New Territories will sport tunnel anchorages, and Lantau Island will support a gravity anchorage. Construction of this $1.1 billion bridge should begin in early 2002, with completion scheduled for 2007. This third generation of box girder will represent the most significant development in suspended structures since the Severn Bridge.

The Tsing Lung Bridge will give Hong Kong four world-class bridges within 5 km of one another. Incredibly, more are planned. In 1997, design of a Tsing Yi Island-Kowloon link via Stonecutters Island began. A planned cargo terminal on Tsing Yi and poor soils on Stonecutters Island hindered the anchoring of a span expected to exceed 1,000 m in length. An initial concept featured a 1,100-m span Dischinger hybrid that reduced the size of the anchorages; one variation was even self-anchored to avoid the need for anchorages altogether. The span's visibility from Hong Kong's business district then prompted the launch of an international design competition for a "signature" bridge in late 1999. Among the 22 cable-stayed, two suspension, and three hybrid forms submitted, a record 1,018-m span cable-stayed bridge took the prize. Plans call for a 2003 construction start, with completion of the estimated $645 million crossing in 2007. It too will feature a split box suspended structure.

Another imminent span is the huge 27-km Lingdingyang crossing over the mouth of China's Pearl River just west of Hong Kong. Unlike the Hong Kong spans, the design of this enormous bridge is a purely Chinese undertaking promoted heavily by the nearby city of Zhuhai. Studies by Beijing's Highway Planning and Design Institute began in the mid-1990s. Initial plans envisaged four bridges connected by causeways. The West Lingding Channel would be crossed by a 920-m span suspension bridge flanked by 311-m side spans; the East Channel would feature a 1,408-m suspension span flanked by 408-m side spans. Two cable-stayed bridges would also span between 300 and 320 m. Steel box girders with a width of 31 m would support the crossing's six lanes; the 160- and 210-m tall towers of the two suspension bridges again reflected Chinese practice in the use of highly placed intermediate portal struts. Towers and anchorages of the smaller suspension span would rest on piles driven deep into the river bottom. Zhuhai officials sought an initial blessing from Beijing in early 1997, hoping the $2 billion cost would be partially met through foreign investment. A flurry of competing schemes then emerged; Portuguese officials pitched a 38-km, $1 billion bridge linking Lantau Island directly with Macau to Beijing officials that spring. Hong Kong business interests also proposed a $1.8 billion, 27-km variant that would branch off Lantau to both Macau and Zhuhai. In December 1997, Beijing offered to fund 15% of the Lingdingyang route, and this allowed detailed studies to commence.

Typhoon exposure demands extremely high flutter critical velocities of 306 km/hr. To generate a broad spectrum of solutions, a competition involving the various Chinese bridge design institutes was held. The larger of the two suspension bridges—spanning from 1,400 to 1,450 m—generated the most interest. One 1,400-m span devised by Tongji University's Bridge Design Institute involved a Dischinger cable-stayed/suspension hybrid with a concrete box girder near the towers and a steel box over the suspended portion of the main span; both box girders were just 3.8-m deep but almost 37-m wide. A second variant involved a suspension bridge with split boxes. For the smaller 956-m span, suspension and cable-stayed options were explored. All suspension bridges will likely use prefabricated parallel wire strand cables. Preliminary construction on this fantastic project is said to have begun.

China is also contemplating other major spans within the realm of present technology. One span of at least 1,200-m across the Yangtze River just north of Shanghai may be built in the next 20 years. South of the city, several long spans to a deep-water port have been investigated by Tongji University, including the crossing of one channel by a 900-m span cable-stayed or 1,650-m span suspension bridge.

The South American country of Chile will also soon craft a huge long-span suspension bridge that will erase the 40-year reign of Venezuela's Angostura Bridge as the continent's longest span. The nation has recently undertaken to connect Chiloe Island off the Pacific coast across the Chacao Channel 1,000-km south of the capital, Santiago. In 2000, Danish firm COWI and the Chilean firm Icuatro began feasibility studies for this 2.3-km wide crossing. Water depths of up to 80 m are mitigated by a midchannel outcrop that suggested a two-span continuous box girder suspension bridge with spans of about 1,000 and 1,300 m centered by a large A-framed central tower. A cable-stayed span was also studied as were two end-on-end conventional suspension bridges. Seismicity is a major concern; the outcrop was actually an island before the devastating Great Valdivia earthquake of 1960. At 8.5 on the Richter scale, it remains one of the strongest ever recorded. Its epicenter was less than 200 km from the bridge site and caused displacements of up to 16 m on either side of the fault; it also generated the largest tsunami on record. Uncertainty over the outcrop's suitability for footings also saw a single suspension span approaching 2,200 m studied. By mid-2001, a final design was adopted comprising a unique two-span suspension bridge with spans of 870 and 1,270 m flanking a rigid central A-framed tower. Design for severe seismic events, as well as for strong winds and currents, make this a challenging project. The current schedule calls for construction tenders to be invited in late 2001 with construction to begin in 2002.

Northern Greece will soon be the site of a considerably smaller but remarkable span without towers from the British firm Ove Arup, aided by architect Wilkinson Eyre. Their winning 1999 concept for the four-lane Metsovitikos Bridge in Greece's rugged Pindos Mountains places the anchorages on the mountainsides well back of a 565-m main span. The two 800-m long, 55-cm diameter main cables are inclined transversely at an angle of 27° so that the suspenders cant in to the deck, which hangs a breathtaking 150 m above the valley floor. This feature was selected largely for aesthetic reasons, but it does enhance the stability of parallel 2.5-m deep steel box girders, which are also noteworthy. The two-lane boxes are braced over a 3.1-m wide central air gap to create a 23-m wide deck that in some respects resembles a split box; the use of an orthotropic instead of a composite concrete deck was estimated to save $5 million. Other design challenges include the region's seismicity, which prompted Ove Arup's use of tough new California design standards. Moreover, one of the four independent cable anchorages had to be moved downhill to avoid an old landslide; it will be gravity-anchored, as will one other anchorage because of the topography. The other two anchorages will be tunneled into the mountainsides and tied to rock by friction and end bearings. With the cable inclination, the anchorages at each end of the span will be separated by more than three times the width of the suspended structure. Tunnels at the immediate ends of the span will emphasize the drama of this spectacular crossing. Construction of the $60 million span is scheduled to begin in 2001, with completion expected by 2004. The 19 strands of 450 wires in each main cable will be spun in situ; the cables will then be temporarily tied together to achieve an inward cant until the 35 deck sections are lifted into place and do the job permanently.

Midspan suspension bridges have also experienced a renaissance of sorts, and many are quite innovative. In Elche, Spain stands a unique 164.5-m "half-span" suspension bridge completed in 2001. Its 24-m wide suspended structure comprises an enclosed steel grid capped with a composite concrete deck and is no more than 0.9 m deep. Two main cables diverge from a single central pylon outward to the deck edges at the other end of the "half" main span. Anchorage behind the tower is in the center of a traffic circle, and the carefully crafted concrete tower even sports a small observation platform.

Innovation also permeates an unusual South Korean self-anchored suspension span of 300 m opened in November 2000. The Yongjong Grand Bridge is the world's first double-deck self-anchored span, is the first to carry rail and road traffic and, with side spans of 125 m exceeding those of the Konohana by a mere 5 m, is the world's longest overall extant self-anchored span. It is also the world's first sizable vehicular suspension bridge with outwardly canted main cables. The $365 million suspension bridge forms the centerpiece of a $670 million, 4.4-km oversea bridge to Inchon International Airport some 60 km from Seoul. The proximity of the new airport ruled out a cable-stayed bridge because of the need for tall pylons, and a small gravity-anchored suspension bridge was uneconomic and would take longer to build. Aesthetics and symbolism were also important determinants of span selection; in outline, the selected design—produced by Yooshin Corporation with the assistance of Japanese consultant Chodai—resembles a Korean Kiwa house.

Six road lanes occupy the upper deck, and below, four more lanes cradle two (as yet unbuilt) high-speed rail lines. The upper deck is actually supported by a 3-m deep box girder with a massive 41-m width—including small nonstructural triangular fairings that house various utilities. The box girder suspends a 35-m wide lower deck from a 12-m deep Warren truss—without vertical members. To help prevent uplift, approach span trusses protrude into the side spans, which are also secured to their end piers. Diamond-shaped towers 104-m tall are more typical of cable-stayed spans and feature an impressive width of more than 48 m at deck level. Work began in late 1995 with enormous tower foundation challenges. Huge 47- by 17-m prefabricated pneumatic caissons were set to depth of up to 45 m and then remotely excavated by operators using personal computers. The hybrid box girder/truss was built first using temporary piers, with all six 75-m long sections each weighing up to 6,000 tonnes being placed in August 1999. Each tower was erected in just three lifts by floating crane. Strands of 480 wires were then spun into each 6,720-wire, 47-cm diameter cable in an operation ending in late 1999. The two cables—with suspenders attached—were then flared from a 3-m separation at the tower tops to a 30-m distance at the edges of the deck at midspan, and to an almost 35-m gap at the ends of the side spans. Special rotating cable bands allowed the suspenders to incline out to the deck edges as the main cables were raised; with a length of 4 m, the clamps are believed to be the longest ever used. Dehumidification is used extensively, in the box girder, in the hollow box truss members, and, as in recent Japanese spans, in the main cables.

A larger, more conventional suspension bridge is due to open in September 2002 across Pusan Bay in the south of the country. The Kwang Ahn Grand Bridge features a span of 500 m and 200-m side spans. A 9.85-m deep and 18-m wide truss—whose aerodynamic stability was verified at the University of Western Ontario's Boundary Layer Wind Tunnel—supports eight lanes over two levels on uncomposite orthotropic decks. The $620 million bridge is dominated by lengthy approach spans to avoid the congested urban area of Pusan and stretches a total of 7.4 km; the suspension bridge alone will consume $196 million of the total cost.

Construction, which began in 1994, made extensive use of pre-assembly. Long approach truss spans were raised in large sections of up to 3,000 tonnes by floating crane, each 105 m tall tower was erected by floating crane in just three lifts, and truss sections were lifted into the suspension bridge in lengths of 18 m. Aerial spinning was used to assemble the 37 strands in the 60-cm diameter main cables which, like those of the Yongjong Grand Bridge, are dehumidified. The visibility of the bridge from central Pusan required special attention to details. The portal braces of the towers, for example, are semi-circular in outline and the upper braces, unusually, sport a reverse curvature of their upper edge.

Variations of monocable designs have also emerged. The 300-m span Chavanon Viaduct, completed in 1999 in France's Massif Central, is only the second vehicular span after Japan's Konohana Bridge to be suspended from its median. A pair of 3,721-wire main cables spaced just 1.1-m apart are draped over 75-m tall inverted V concrete towers whose legs taper elegantly toward their bases. A 3-m deep, 22-m wide (but not orthotropic) steel box girder floats through the towers without any connection whatsoever; torsional restraint is provided at the abutments framing stubby 30-m unsuspended side spans, and the anchorages are completely buried beneath the approaches. Unusually, the 10-m long steel box sections were pushed out into the span by the incremental launching of successive boxes from assembly yards at the span's extremities. Permanent vertical suspenders then replaced the triangulated suspenders used in launching once the box girder was in place. Great attention to aesthetics has created this stunning and unusual structure from noted French cable-stayed bridge designer Jean Muller International.

In 1997, the Danish firm COWI proposed an audacious 1,100-m span suspension bridge between Panay and Guimaras Islands in the Philippines. Twin cables flaring out to the deck edges at midspan from A-framed towers suspended a slender 3-m deep, 16-m wide box girder in a manner similar to that of Korea's Yongjong Grand Bridge. A true monocable suspension bridge was one of two finalists in a mid-1990s competition to span Lake Geneva near Geneva, Switzerland. Its four-lane span of 950 m featured inclined suspenders—as in Osaka's Konohana Bridge—and gracefully tapered A-framed towers. Unusually, the main cable dipped *into* the 4-m deep triangular box girder (with a concrete deck for weight) over the central 200 m of the main span; the cable itself was of varying diameter to accommodate changing axial stresses over the length of the main span, and in this respect recalled an approach of 19th century British engineer James Dredge in thc 1830s and 1840s. The bridge was also partially self-anchored and behaved structurally somewhere between a cable-stayed and a suspension bridge. This controversial crossing remains unbuilt, as is a contentious third crossing of Scotland's Firth of Forth a few hundred meters west of the 1964 road bridge. An early-1990s study generated concepts for a 1,200- to 1,400-m span suspension bridge, a record-setting 900-m cable-stayed span, a four-span cable-stayed bridge, and two end-on-end suspension bridges. Most intriguing was a Dischinger hybrid that allowed deck assembly to proceed simultaneously on five fronts, reducing construction time by more than a year. The 1997 election of a Labour government scuttled the scheme, but the idea has been recently revived.

A new generation of large suspension bridges is also emerging after a quarter century hiatus in America. Since the 1973 completion of the William Preston Lane, Jr. Memorial Bridge, only one suspension span of note has been built, a small but interesting span of 191 m over the Mississippi River in Minneapolis, Minnesota. The Hennepin Avenue Bridge opened in 1990. Its parallel twin decks are supported by three-post towers in a configuration repeated only in Cologne's 1990s Rodenkirchen twinning. Its use of a concrete box girder suspended structure is also unusual. But construction presented a difficult aerodynamics problem. The bridge would be built in halves, the first half standing just 1.3 m from the steel arch its twin would replace. Taut strip model tests during 1986 and 1987 at Canada's National Research Council wind tunnel in Ottawa generated surprising findings when vortex shedding was found to be less with the arch than with the single *or* twin spans in place. Interestingly, once the gap between the deck slab and the supporting box girders was sealed, torsional motion was found to disappear in winds of up to 225 km/hr.

Three large American suspension bridges reached the agenda in the late 1990s, although public controversy has surrounded two of them. The least acrimonious of the three is a new span on the U. S. west coast over the Carquinez Straits. This bridge will replace David Steinman's salt-ravaged and congested 1927 cantilever, whose 1958 twin will remain. A number of alternatives were investigated beginning in 1989, when three consultants were engaged to undertake separate feasibility studies for concrete box girder, steel box girder, and cable-stayed bridges. In 1992 further studies examined multiple span cantilever truss, concrete arch, and cable-stayed spans, as well as a suspension bridge. Although studies had by 1996 suggested a cable-stayed bridge to be slightly less expensive than a suspension bridge (construction bids would later suggest much less so), the latter option avoided a midchannel pier and featured superior seismic response. This second factor was critical, as one fault lies just 1 km from the bridge. A four-lane, 728-m span suspension bridge won out, bringing Steinman's 1920s still-borne dream of a suspension bridge for this crossing to fruition. Coincidentally, the firm that now carries his name would design the new bridge.[1]

The span will introduce two features to a large American suspension bridge—an orthotropic steel box girder and concrete towers. The 131-m tall towers will feature portal braces, one below deck and another at their apex. Their foundations pose a major challenge. Only six 3-m diameter steel piles driven to depths of 47 m, with "rock sockets" extending another 45 m, will support each tower leg. One anchorage will be buried high in a hillside, and the other will be tucked beneath backstay spans; this arrangement has spawned asymmetrical side spans of 147 and 181 m. Of note is the 29-m wide, 3-m deep box girder supporting four lanes and a pedestrian/cycle path. Extensive wind tunnel tests of sectional and full aerolastic models were required, largely on account of the temporary proximity of the other two spans and the remaining 1958 span alone—British experience with the Runcorn-Widnes and Tamar bridges of the 1950s would suggest caution. The sectional model tests did in fact reveal the need to carefully detail the span's railings to avoid negative effects on aerodynamic stability. Construction began in March 2000 on the $188 million bridge. Newly independent Cleveland Bridge & Engineering will spin the 51-cm diameter, 8,584-wire cables and erect all 12,700 tonnes of box girder. Boxes will be fabricated along the banks of the Yangtze River in China and shipped across the Pacific Ocean in 48-m long sections. In spite of environmental restrictions limiting work on one tower pier to 3 months of the year, completion is expected in 2005, making the new Carquinez Bridge the first major American suspension bridge in 3 decades.

California's Transbay Bridge would appear an unlikely candidate for a new suspension bridge, but it is, thanks to the 1989 Loma Prieta earthquake. The dropping of a section of upper deck roadway on the East Bay crossing led to a 1990 estimate of $250 million to bring this portion of the Transbay up to current earthquake standards. In 1997, the twin West Bay suspension bridges were slated for an expensive seismic upgrade, but by this time the East Bay prognosis had become dire. The 427-m span cantilever bridge was perilously susceptible, and engineers concluded that the cost to rehabilitate the entire East Bay crossing would exceed $900 million. Total replacement, costing just $100 million more, was appealing, and the state developed a design for multiple concrete box girder spans. This plan immediately aroused the aesthetic sensibilities of many Oakland residents who demanded a span as notable as San Francisco's Golden Gate Bridge. In February 1997, the state allowed Bay Area governments to determine the form of the crossing, stipulating that Bay Area residents foot the bill for any additional costs. So began a tumultuous process that would produce an equally controversial structure.

A March 1997 proposal call for concept ideas was a rare event in American bridge design, but the allure of such a prestigious design prize attracted 16 entries from leading engineering firms around the world. Most proposals involved single-tower cable-stayed bridges, but several involved single tower suspension spans. One proposal featured a self-anchored span centered by a single narrow A-framed pylon situated in the median and supported on rock just off Yerba Buena Island. The concept of a single tower suspension bridge was intriguing, not the least because it would help isolate the bridge from earthquakes. Still, the relative merits of suspension and cable stayed spans were unclear. San Francisco-based T. Y. Lin International, with Moffatt & Nichol, secured the coveted design award for the now-$1.2 billion structure in January 1998. Its first task was to carry designs for 490-m cable-stayed and 565-m span suspension spans to 30% completion for comparative purposes. The suspension bridge, which also involved Weidlinger Associates in the design of the cable and deck system, included towers with one, two, and three columns.

In May 1998 the project's advisory panel—including world-renowned engineers and architects—endorsed the use of a single pylon in the median, but not the span type. A cable-stayed bridge appeared to have an edge given its lower cost, simpler construction, and superior seismic strength. Some pundits felt that the roughly equal spans of the suspension bridge made little sense in terms of the channel to be bridged, but a subsequent shortening of the suspension bridge backspan saw the panel make a surprise endorsement of the suspension bridge. This decision was controversial in light of the higher cost of the selected option and the fact that barely half the panel's 35 members were present to vote.

By the end of 1998, a single-towered self-anchored suspension bridge had been given the nod, and this allowed detailed design to commence. A 385-m main span will make the bridge the longest self-anchored span ever built.[2] The use of a single pylon in the roadway median combined with only a "half" main span and a single 180-m shoreside side span is also highly unusual, having been only recently pursued on a much smaller scale in Spain's Elche Bridge. Added to the challenge of aesthetics is an overriding need for seismic stability; serviceability is required in earthquakes of up to 7.5 on the Richter scale on the nearby Hayward Fault and 8.0 on the notorious San Andreas Fault. Still, perhaps the biggest challenge will be the foundations. With bedrock up to 137 m deep, rock-based foundations are out of the question. Some 200 piles will instead be driven up to 100-m deep through mud, clay, and finally sand, and this alone is estimated to devour some 40% of the total cost of the entire crossing. The single 162-m tall suspension bridge tower will comprise four closely spaced trapezoidal columns positioned between two box girders each supporting 5 lanes of traffic. Struts connecting the four legs will break—by design—in a severe earthquake. The box girders—separated by 10 m in a proto-split box arrangement—are braced by struts every 30 m; the boxes are not connected to the tower but are restrained only at the abutments. Suspenders are connected to the outer edge of each box and will hang at an angle of 20° from two 78-cm diameter main cables diverging from the top of the single tower to the outer deck edges. Unusually, anchorage will occur only at the span's east end. For aesthetic reasons, the cables at the west end of the span will be looped around the box girder and return eastward in a manner similar to that used by Vandepitte in his small Belgian concrete suspension bridges; a strong pier and tie-down cables will help prevent uplift. All these design anomalies will make spinning difficult, and prefabricated parallel wire strand cables are preferred.

The development of the design has not been devoid of additional controversy. The proportions of the suspension bridge have come under more criticism, as has the alleged blandness of the steel box girder approach spans dropping down to Oakland. In early 1999, the alignment itself came under attack from the U.S. Navy, the city of San Francisco, and heritage advocates because its west approach would stand astride a historic part of Yerba Buena Island. Also contentious have been the rising costs of the bridge. The 1997 estimate of $1.2 billion soon rose to $1.5 billion, abetted by the cantilevering of a bicycle lane on the south side girder and a redesign of the inner lane on each box girder for possible conversion to light rail transit. In the spring of 2001, Caltrans, the state agency sponsoring the span, revealed the troubling news that costs could rise to over $2.4 billion. And this was not all; by summer, an audit of planned seismic upgrades of major state-owned bridges in the Bay Area revealed the cost could reach almost $2.9 billion. The design, alignment and—so far—cost issues have weathered these storms, and construction of the $700 million, 2.4-km long approach viaduct is slated to begin in late 2001. Construction tenders for the suspension bridge are planned for August 2002. This unusual span will be one of the most distinctive—if not inexpensive—suspension bridges ever upon its planned opening in 2005.

One other contemplated span with a controversial pedigree is a twin for the 1950 Tacoma Narrows Bridge. In a 1980s assessment, the firm of Arvid Grant Associates found the 1950 span to be in excellent condition thanks to meticulous maintenance, but the firm predicted that in catering for future traffic needs a parallel bridge would be more than twice as expensive as the addition of lanes to the existing span. In the early 1990s, many options were examined, including the addition of a third tower leg and cable in the manner of Germany's expanded Rodenkirchen Bridge and a $330 million parallel bridge of identical appearance to the 1950 span. However, the placement of the 1950 bridge on the National Register of Historic Places in 1992 (including the underwater remains of the first span, touted as the largest human-created reef in the world) would require a twin span to be of slightly different design.

The crossing's form was only one of several obstacles. Opposition emerged from residents of the bucolic Kitsap Peninsula. Moreover, it was also not clear how any bridge expansion would be paid for. State legislation passed in 1991 opened the way for private finance, and in mid-1994, three consortia bid to span the Narrows. In August, United Infrastructure—led by construction giant Bechtel Corporation—was tapped to negotiate a contract to design and build a toll bridge, as well as expanded approach highways.

Preliminary design work undertaken by Parsons Transportation Group subsidiary Steinman generated parallel cable-stayed and suspension bridges, as well as a lower deck on the existing span. All options would convert the upper deck on the original bridge to three one-way lanes, and in the process eliminate its open steel deck slots. To expand the original bridge, Steinman proposed three lower-deck lanes supported on deeper bottom lateral bracing. From there, the approach resembled its recent strengthening of Lisbon's Tagus Bridge. Two new 8,700-wire, 51-cm diameter cables would be spun in situ and would equal the original cables in size. Suspenders would be attached to the bottom chords of the truss and tower extensions would support the new cables. The 1950 bridge had not been designed to accommodate a lower deck, and, unable to ram the lower roadway through the anchorages, Steinman engineers instead snaked it out one side of the truss halfway over the side spans. A truss half as deep as the original would extend from here to the anchorages. Following diversion from a large cable bent, the new cables would be directed into new anchorages hived onto the sides of the existing monoliths. To strengthen the towers, Steinman proposed to graft new legs onto the outsides of the originals, bolting old and new into a single structural unit once the truss was revamped.

An alternative scheme in the form of a parallel suspension bridge 60 m to the south would exceed the original's 853-m span by one meter but feature unequal side spans of 366 and 426 m. A 7.7-m deep, 22-m wide orthotropic truss continuous between anchorages would support three lanes of traffic. In future, the truss could cradle four additional lanes or two transit tracks on its lower deck with the addition of stronger bottom lateral bracing and a second pair of 4,100-wire cables in a Tagus-like makeover. These auxiliary cables would hang about 2.5-m above the initial 7,100-wire cables via extensions to the 155-m tall reinforced concrete towers. Rounding out the field was box girder cable-stayed bridge whose 854-m span would stand just 2 m shy of France's record-shattering Pont de Normandie.

In the end, the parallel suspension bridge won out on account of its lower cost, higher capacity, and greater traffic flexibility. A second deck on the existing bridge would disrupt traffic and might run afoul of preservation objectives—particularly the enlarged towers and the altered side spans. Moreover, the bridge was almost 50 years old, and, notwithstanding its excellent condition, a good portion of its service life was behind it. A cable-stayed span was feared to visually overwhelm the original bridge and was rejected. The portal-braced towers of the new suspension bridge will be supported by open-dredged caissons up to 70-m deep, similar to those used in the original span; their construction will require care, as the concrete blocks used to stabilize the caissons of the 1940 span remain littered on the Narrows bottom.

As a resolution on design matters neared, political opposition to privately tolled facilities in the state intensified. By late 1998, the bridge constituted the only state transportation initiative eligible for private financing. Design/construct contracts with Bechtel/Kiewit Pacific were signed in June 1999, but additional controversy over escalating costs of the entire project—which by early 2000 had reached $650 million—and new legal challenges have delayed a planned spring 2001 construction start. The planned 2004 completion date now seems unlikely.

16.2 THE ALLURE OF THE MESSINA STRAIT

One giant suspension bridge for which definitive plans exist, and which may be on the cusp of construction, is the crossing of Italy's fabled Messina Strait. This deep and turbulent channel separates the toe of the Italian mainland from the island of Sicily. It has tantalized engineers for decades. A suspension bridge across the 3-km wide gap will propel the genre to another plane; the challenges over the past decade have been primarily financial and political, not engineering. The crossing has tapped the imaginations of many leading long-span bridge engineers, and the currently proposed structural concept is as epoch-making as its enormous span, one far longer than anything yet built.

Ancient literature describes the perilous crossing between Scylla and Charbydis, where violent storms, strong reversible tidal currents of up to 18 km/hr, and treacherous whirlpools all prevail. Of perhaps greater notoriety is the region's restless geology; a 1908 earthquake killed an estimated 100,000 people. Water depths approach 200 m, bedrock lies more than 300-m below the seabed, and the strait funnels copious volumes of shipping. Such obstacles have combined to provide perhaps the greatest challenge yet encountered in bridge engineering.

Mythology has it that Consul Cecil Metellus built a bridge in 251 B.C. during the battle of Palermo. A floating bridge was attempted by the Spanish viceroy in Sicily in 1701, but it immediately succumbed to a severe storm. The idea of a Messina Strait bridge reemerged in the 1850s, and an 1867 article in the British journal *Engineering* proposed a suspension bridge, citing the success of John Roebling's span at Niagara. In the 1870s, rudimentary sketches depicting a bridge were published and a rail tunnel was also proposed. But the idea of spanning the strait at this time was illusory—many competent engineers even doubted the soundness of Roebling's Brooklyn Bridge, then entering construction.

Regular ferry service began in 1899, but not until 1950 was the first serious step taken when the Italian Steel Institute retained David Steinman to design a road and two-track rail bridge. Emerging from a long design hiatus, Steinman proposed a double-deck suspension bridge with a record 1,524-m span. Although never built, the design is of interest. Even its 732-m side spans would eclipse all but three of the world's then longest single spans, and its 2,987 m of suspended span would exceed that in the Mackinac Bridge by 731 m. Huge main tower piers would descend 122 m, more than 50% deeper than the record foundations of the Transbay Bridge. The $60 million bridge would consume almost 70,000 tonnes of structural steel, more than 380,000 m^3 of concrete in its piers and anchorages, and 42,000 tonnes of cable wire, four times the amount at Mackinac.

Rail loads, wind, and seismicity prompted Steinman to provide a high degree of torsional stiffness. To avoid what he argued would require a 1:40 depth-to-span ratio and a 38-m deep conventional truss, Steinman proposed a Florianopolis truss, using the main cable as its top chord. This form would significantly reduce dead loads, and the deepening of the truss to 50 m at the quarter points of the main span maximized rigidity against wind and traffic at the point of maximum torsion and bending. Steinman also extended the concept to the side spans, whereas in the 1926 Ponte Hercilio he had not. The 13-m deep truss was fortified with two systems of lateral bracing, with the lower level stringers supporting two rail lines. Initially only a third of the 27.5-m wide truss system would support an upper-level two-lane roadway, separated from the truss edge by wide air gaps. This measure would later find its way into the Mackinac Bridge, but in the Messina Strait Bridge it was much more prominent. Finally, Steinman used tower stays—a source of much bitter debate with Othmar Ammann in the aftermath of the Tacoma Narrows Bridge collapse—radiating to points along the main cable. Although Steinman (1953) considered the stay and truss combination to produce "a strikingly beautiful artistic composition, with harmony, balance and symmetry," today the proposal appears heavy.[3] Still, the Messina Strait Bridge's towers, anchorages, and main-to-side span proportions were in many ways prototypes for the more refined designs that later emerged at Mackinac.

Italy was then mired in post-war reconstruction, and a bridge at that time was an unaffordable luxury. Steinman dabbled in the project for the next 5 years, but eventually abandoned it. Then in 1959, a huge $200 million span with six upper-deck lanes and a four-track lower level with two lanes for military traffic emerged, but only briefly. A more serious effort was launched in 1967 when the Italian government retained the American firm Sverdrup & Parcel to examine several schemes—including Steinman's—with a view to developing a design and pursuing construction. A consortium of Italian firms even retained popular French undersea explorer Jacques Yves Cousteau to film the seabed. Although studies suggested a bridge might devour up to $800 million, Gruppo Ponte di Messina Spa (GPM)—a consortium formed in 1955 by 17 state and private firms to promote a crossing—pressed on. Sverdrup & Parcel (a member of GPM along with U.S. Steel) prepared new plans and, benefiting from its review of the Steinman design, proposed a suspension bridge with an even longer span of 1,676 m. By June 1968, a government decision on the bridge appeared imminent. Final plans of a selected design could be ready by the spring of 1969. Unfortunately, financing disagreements ground the project to a halt; design approval was postponed, as it turned out, indefinitely.

In July 1969, Italian state road and rail agencies gave it another try. The response to its call for designs of a combined road and rail crossing was effusive; by December, 144 submissions ranging from the conventional to the fantastic had been received from around the world. Included were some 40 proposals for suspension bridges. Capacity and seismic concerns favored a bridge, an orientation reflected in the six concepts that were awarded top honors in December 1970—three suspension spans, a cable-tension bridge, and a floating tunnel. Sverdrup & Parcel, with American contractor Morrison-Knudsen, proposed a 1,600-m span suspension bridge flanked by 770-m side spans. A deep truss supported four upper-level lanes and dual lower-level rail lines, and its record 255-m tall towers stood atop huge rock islands dumped into 125 m of water. The scheme's $670 million cost, too, was outsized. Another finalist used rock islands to support the towers of a 1,300-m main span flanked by 850-m side spans.

The U.S. firm of E. Lionel Pavlo produced another winning design for an Italian consortium. Its unconventional design featured twin continuous spans of 1,360 m separated by a large A-framed tower standing on a midchannel artificial island. Its cables were continuous over the two main and 465-m side spans, the latter spans extending from two conventional shoreside towers resembling those of the Delaware

Memorial Bridge. A truss of Lindenthalian proportions supported an eight-lane upper deck, with two rail lines and space for an additional six vehicular lanes on its lower level. All this came at an estimated cost of $780 million. The same consortium also offered an ambitious single-span suspension bridge with a 1,842-m main and 906-m side spans suspended between two 285-m tall towers. Interestingly, portal-braced towers dominated the winning designs, whereas diagonally braced towers at that time dominated proposals for Japan's equally earthquake-prone environment. The competition generated other ingenious designs. One 1,530-m span suspension bridge featured towers more than 150-m thick along the line of the span to provide rigid cantilever support from below portions of the main and side spans. Cable-stayed designs were offered, too. A record-*trebling* cable-stayed box girder span of 1,300 m was developed by Fritz Leonhardt for the Italian consortium Gruppo Lambertini, whereas another scheme involved three 1,000-m cable-stayed spans.

The need for deep-water piers burdened all six winning schemes, save one. Professor Sergio Musmeci of the University of Rome combined cable-stayed and suspension systems in an ingenious cable-tension hybrid that spanned the strait in a single leap of 3,000 m. Musmeci's scheme replaced the main suspension towers by hanging the suspension cables from single large cable stays extending from 603-m tall land-based pylons. These stays merged with the main cables 500 m into the span and 297 m above the straits—where the towers would normally stand—to form a larger main cable over the main span. To compensate for poor lateral wind stability, a horizontal cable and suspender system fanned out from each shore to link with the main cable suspension system, stabilizing a slender 40-m wide steel plate road deck and a lower-deck twin-rail line. In concept, the scheme resembled an updated version of Joseph Strauss's ungainly 1921 proposal for a hybrid suspension/cantilever bridge across the Golden Gate. Musmeci's finely spun web was the most spectacular, if technically uncertain, means to span the Messina Strait. Noted Italian architect-engineer Pier Luigi Nervi also proposed a single 3,000-m span using multilevel trapezoidal girders; in some ways, this heavily embellished design recalled Gustav Lindenthal's monumental Hudson River Bridge of 1921.

Construction of a winning design was expected to take 5 years, with completion hoped for in 1977. Momentum increased in 1971 when the government enacted legislation allowing creation of a public-private agency to manage the project. Yet the bridge was excluded from the government's 1971 to 1975 national development plan, and the agency itself would not be created until 1981. No single winning scheme was selected, and, as had frustrated early progress of the Mackinac and Humber bridges, no mechanism existed to fund the crossing's construction.

Studies during the 1970s assessed the single- and double-span winning entrants, as well as a bored tunnel plunging up to 210-m below the strait. One 1973 concept featured two 1,750-m span continuous suspension bridges with unsuspended side spans of 1,000 m on the mainland and 850 m on Sicily. In this scheme, two main cables supported a truss hung from inclined suspenders. Although the two flanking 230-m towers were footed on land, one option for footing the central A-framed tower involved an artificial island built from 20 million m^3 of fill from Sicily's Mt. Etna volcano. By 1978, GPM had also proposed a porous latticed underwater pier to reduce water forces during an earthquake.

A second GPM scheme involved a single span bridge, building on the work of Musmeci and a similar scheme proposed by D. Danieli in the mid-1970s; Danieli's 3,000-m span combined several cable-stayed systems radiating out from points along the main cables, but required a deep mid-span anchorage. Freeman Fox & Partners devised the basics of an updated single span beginning in 1976 under William Brown. Further massaging produced a bridge dwarfing all predecessors. The strait would be crossed by a single suspension span of 3,300 m. Four 1-m diameter cables—in pairs—would hang from 380-m tall shore-based reinforced concrete towers; almost 50% of the cables' strength would be devoted to supporting their own weight. Freeman Fox & Partners' record 1,410-m span Humber Bridge—more than 3 years from completion—would be eclipsed by a span more than twice as long. Unsuspended side spans of 1,000 m on Sicily and 1,150 m on the mainland would flank the main span.

Brown, aided by aerodynamicist D.E. Walshe, devised a unique concept for the suspended structure. The six-lane roadway consisted of a flat "perforated" plate with parallel slots supported on 50-m wide cross-frames between the twin sets of main cables. In cross-section, the area of openness equaled that occupied by deck, thus minimizing uplift. Although the deck itself was virtually devoid of torsional stiffness, the great distance between the main cables provided it, enhanced by the use of inclined and transverse suspenders. A two-track lower deck was hung from a central post and was braced by tensioned cables extending to the edges of the upper cross-frames. Overall flutter susceptibility was reduced and horizontal stiffness was increased—avoiding problems associated with the Musmeci and Danieli concepts.

The concept was, in fact, an evolutionary step that would later emerge in the form of the split box. Its $1 billion cost and construction period of just 5 or 6 years proved to be the most favorable of the three bridge and tunnel options, largely because deep-water piers were avoided. The design also offered superior seismic, meteorologic, and maintenance performance. Finally, ship collision risk was eliminated altogether, a factor of increasing concern when Sweden's Tjörn and Florida's Sunshine Skyway bridges were brought down within months of one another in 1980.

The twin-span alternative had involved a single deck cradling twin rail lines between two sets of three lanes on a 38-m wide trapezoidal box girder. Freeman Fox & Partners' unique twin parallel-deck concept was then modified for use in the twin-span scheme. A reduced cable-to-cable distance of 40 m reflected the greater stability of smaller spans, and only two cables were necessary on account of the smaller spans.

Other interesting concepts emerged during this period. In 1979, F. Bianchi di Castelbianco produced a bizarre twist to Musmeci's 1970 concept. Castelbianco's 450-m tall and 200-m wide latticed towers would support a 3,000-m span suspension system festooned with inclined suspenders and stays. Most unusual was the splitting of the deck to a 300-m separation over a 700-m distance near the towers—only over the central 1,600 m of main span was a single suspended structure used. German engineer Max Herzog also proposed an elegant 3,500-m span suspension bridge. In this scheme, a flat plate just 3.6-m deep supported six lanes, with two rail lines in its median. Inclined suspenders were hung from two pairs of main cables draped over 405-m tall towers. Herzog later (in 1983) speculated on the possibility of aluminum to reduce deck weight. Finally, the Lambertini-Leonhardt partnership enlarged its 1970 cable-stayed span to a remarkable 1,750 m, but its deep-water tower piers remained a disadvantage.

Government response in late 1978 to the GPM findings was encouraging but non-committal. GPM considered Freeman Fox & Partners' single span feasible and requested continued studies. By March 1981, however, inflation and design changes had escalated the bridge's cost to $2 billion, still less expensive than the $3.14 billion for a tandem suspension bridge and a $3.8 billion bored tunnel. Three months later, the government finally implemented the 1971 legislation and created the state-owned Societa Stretto di Messina (SdM)—funded by state-owned industrial, railroad, and highway agencies and the regions of Calabria and Sicily—to undertake the project. However, not until December 1985 was SdM actually authorized to initiate design studies. Moreover, in technical terms, this was almost a step backward, for SdM was also required to re-examine floating and bored tunnels in addition to bridges.

In June 1986, SdM concluded that although all concepts were technically feasible, a suspension bridge used proven technology, was safer to operate, was cheaper and faster to build, and was less expensive to maintain. By this time, the concept had evolved to a single 3,320-m span, 60-m wide road and rail bridge. Its 400-m tall towers and four 1.6-m thick main cables would be record setting, as would the span's 630,000 tonnes of superstructure steel and its now $3.1 billion price tag. A decision on the crossing type was expected that summer, followed by 3 years of design and site preparation; design work alone was expected to cost up to $150 million, with 10 years of construction to be launched in 1989. At this time, Leonhardt offered an audacious 2,000-m cable-stayed span, claiming the design would cost 60% less than the suspension bridge. He further questioned the ability to model the aerodynamic behavior of the enormous 3,320-m span suspension bridge, and argued that the strait's high winds would hamper cable erection. His unsolicited scheme went nowhere.

A decision on whether to go over or under the Strait finally came in April 1988 when the government and SdM's sponsoring agencies endorsed a bridge. SdM was authorized to develop preliminary designs for the twin- and single-span suspension bridges. The twin bridge featured two 1,800-m spans centered by a 240-m tall, 50,000-tonne, A-framed steel tower supported by a 270,000-m^3 pier standing in 153 m of water. Partially suspended side spans of 450 and 500 m would direct the four 85-cm diameter main cables into anchorages that would each devour up to 250,000 m^3 of concrete. The single span—now 3,360 m—was hung from 390-m tall steel towers, each weighing an astounding 70,000 tonnes. Each of the four main cables would be a record 130 cm in diameter. Two huge 400,000-m^3 concrete anchorages would restrain them.

In May 1990, SdM presented the two concepts to the government. By year's end, the single span had been adopted, and designers were instructed to prepare detailed preliminary designs for the end of 1992. Brown, now of Brown Beech & Associates, led the team crafting the overall design. The overriding design philosophy was to minimize forces on the bridge at their source—for both flutter and seismic movement, and much attention was given to devising a rigid yet light suspended structure. Traditional box girders were found to require stiffening to the point that a 3,300-m span would require several magnitudes of strength more than the Humber Bridge. Brown's team also varied suspender patterns—rather than using a strengthened single box—to achieve flutter resistance; one concept alternated groupings of vertical and transverse suspenders from the towers to the quarter points of the main span.

Then a breakthrough occurred. As early as 1989, Brown had revisited the twin-box concept he proposed unsuccessfully for the second Bosporus Bridge. Aerodynamic studies—principally at Milan Polytechnic under Professor Giorgio Diana—solidified evidence in support of the concept. A three-box configuration emerged involving a twin-track central box flanked by outer boxes carrying three road lanes each. Wind tunnel testing of 1:30 scale sectional models at Turin's Pinin Farina wind tunnel refined the box shaping, their separation, and the frequency of cross-bracing. Experts from Italy and around the world in government, academia, and private sector design fields were extensively consulted. A 200-year specified design life was a challenge given the need to resist earthquakes of more than 7.1 on the Richter scale— stronger than the infamous 1908 quake. More than 100 faults riddle the span's environs and an active fault even lies beneath the strait; four other faults transect the Calabria side span.

In December 1992, SdM submitted its design (*see* Fig. 16-1). A slightly reduced span of 3,300 m would triple the longest road/rail span—Japan's Minami Bisan Seto Bridge—and be two-thirds again as long as the record-setting Akashi Kaikyo Bridge, then under construction (*see* Fig. 17-1). In contrast, only 180 m of each side span would be suspended; on Sicily, an 800-m viaduct would curve away from under the main cables, which would be anchored 960-m back of the Sicily tower. The Calabria side span cables would be anchored 810-m back of the tower high up on a hill through which the road/rail lines tunneled.

Each of the four 1.24-m diameter paired main cables would contain a record 44,352 wires—slightly more over the side spans—over a 5.3-km length. Their 166,600-tonne mass was more than double the weight of the bridge's suspended structure and more than three times heavier than the cables of the Akashi Kaikyo Bridge; here, the exponential rise in cable weight in comparison to suspended structure weight in very long spans was clearly demonstrated. Each cable would be spun in situ into 88 strands of 504 wires each, the same strand size used in Denmark's East Bridge. The separation of the paired main cables by a gap of just 50 cm would reduce drag but would also require new, compact cable wrapping machines.

Other records would fall. Each 376-m tall steel tower weighed in at an unprecedented 54,100 tonnes. Enormous canted legs shaped as ovoids for aerodynamic reasons would stand 78 m apart at their bases and 52 m apart at their apexes. Three portal braces above, and one below, the deck, would connect the tower legs. In spite of the towers' immense size, only 22 tiers involving three lifts per tier would be required to build them.

Apart from its size, perhaps the most revolutionary aspect of the design was the slender 2.25-m deep trimaran box girder, which would clear the strait by 70 m. The 12.75-m wide outside boxes would each support three lanes and flank an 8-m wide central rail box; all boxes were separated by 8.25-m air gaps. Transverse steel beams connecting the boxes every 30 m would deepen to 4.5 m at the central box. Footwalks and windscreens extended the overall width to more than 60 m. With service lanes and traffic shoulders, no less than 12 lanes would be spread over a deck area of 22 hectares. For all that, its 70,500 tonnes of steel is more than 15% lighter than the

Fig. 16-1. Proposed Messina Strait Bridge
PHOTOGRAPH COURTESY OF STRETTO DI MESSINA S.P.A.

trusses of the Akashi Kaikyo Bridge. This economy of weight required only a marginal increase in cable size over the Akashi Kaikyo Bridge, in spite of the significantly higher live loads the Messina Strait Bridge must carry and a relatively taut 1:11 sag-to-span ratio (although Messina's overall suspended length of 3,660 m is less than in the Akashi Kaikyo Bridge).

Competing demands influenced the design of the suspended structure. Inclined suspenders and cable stays were investigated to augment stiffness, but refinements to the box shaping rendered such measures unnecessary. Curved underbellies, whose ease of fabrication had been proven on a 450-m approach viaduct at the second Bosporus Bridge, were one tool. The outer boxes would be teardrop-shaped to smooth wind flow (*see* Fig. 8-3), but traffic safety created a requirement for windscreens, which tended to increase drag. By 1991, Diana's team at Milan Polytechnic had devised a series of windscreen slats that not only reduced wind speeds on the deck by two-thirds but, by acting as aerofoils, significantly increased the flutter critical velocity of the split boxes to more than 288 km/hr. At the design flutter speed of 216 km/hr, the suspended structure would sway just 10 m off the centerline, one-third the specification of the Akashi Kaikyo Bridge; even in 100 km/hr winds, sway would amount to a meager 2 m. This rigidity was in great measure also attributable to the combined mass and stiffness of the cables—amounting to 56 tonnes for every meter of span—and the bridge's great live load capacity.

Enormous expansion movements and the weight of two 400-m long trains on the bridge at any one time posed additional challenges. The outer road boxes are hinged at the towers, but the rail box is continuous through the towers to minimize bending; only at the first side span piers is rail expansion of up to 3.4 m taken up over the record 3,660 m of continuous span. Other than the restraint of transverse motion, the box girders float through the towers as in Scandinavia's East and Höga Kusten bridges. The box girders and towers would also be dehumidified to avoid any need for interior painting. Each tower leg would sit on huge bell-shaped plinths up to 55 m in diameter and 20-m thick, supported at depths of up to 60 m. Softer soils beneath the Sicily tower would require grouting with a concrete volume of 10 times the 86,000-m^3 in the foundation itself. Moreover, each anchorage would consume up to 328,000 m^3 of concrete in addition to large amounts of stone ballast. Incredibly, only the housings for the Calabria anchorage cable saddles and the upper quarter of the Sicilian anchorage would protrude above ground level.

In 1994, the bridge was estimated to cost $2 billion. Towers and box girders would consume almost half this cost, cables and suspenders just less than a third. Numerous approach road and rail viaducts and tunnels added another $1.1 billion to the project's total cost. These daunting financial obstacles were then compounded by political upheaval caused by national corruption scandals that in 1993 threatened to scuttle the project. SdM's hopes to call for construction tenders by year's end and to open the span by 2000 appeared dim. Engineering debate also emerged because of the need for heavy trucks to travel on the inside lanes of the outer boxes on account of their lower torsional stiffness in comparison to conventional box girders;[4] this issue would be alleviated by the reversal of lane directions across the bridge.

In spite of the uncertainties, rail agencies had endorsed the design by July 1994. Aerodynamic studies continued, aided by a year of detailed measurements of the Humber Bridge's behavior to aid in mathematical modeling. Sectional models and a 1:1250 scale full aerolastic model of the bridge were also tested in the new boundary layer wind tunnel at the Danish Maritime Institute; full aerolastic models have also been tested at London, Ontario's Boundary Layer Wind Tunnel. Then in November 1995, SdM's road sponsors endorsed the concept. Yet the total cost—now $4.5 billion—remained a huge obstacle. Rumors suggested either imminent construction or the total demise of the project, but in October 1998, the government approved the preliminary design. Legislation enacted in mid-1998 had also opened the door for private financing, but the scheme's profitability remained questionable as the road and rail ferries still operated at a loss. Nevertheless, some form of joint public/private financing will likely prevail. SdM has advocated that the government fund only 40%

of the cost, and in late 1998 Japanese construction giant Mitsubishi Heavy Industries reportedly expressed an interest in the project. Yet even the most recent work has been dogged by controversy. In early 2000, Parsons Transportation Group subsidiary Steinman won a contract to review the design as a precursor to a final decision on the crossing. This work was then delayed by 6 months of legal challenges—unsuccessful, as it turns out—from a number of losing contenders, including T.Y. Lin International and COWI, the latter finishing second behind Steinman.

A final go-ahead may be imminent with a reported May 2001 pledge of private investment to cover half the $4.7 billion project cost. A formal launch of the project—when it comes—will initiate 2.5 years of detailed design and then 6.5 years of construction, considerably less time than at the Akashi Kaikyo Bridge, thanks in large measure to the span's land-based foundations.

The Messina Strait Bridge represents a second generation of stiffened suspended box structure. This new form may see its first expression in Hong Kong's Tsing Lung Bridge, planned for completion in 2007. Designers of Tokyo Bay's 2,250-m span have also expressed interest in the concept, although this bridge has also generated new directions in the design of suspended structures (*see* Chapter 17). But the Messina Strait Bridge may also represent a limit for conventional suspension systems. To make truly long spans anywhere near economically feasible, many researchers have concluded that changes in cable systems and materials are required. It is here that the suspension bridge has almost come full circle, bringing forth in modern form the imaginative explorations that characterized its modern conceptual beginnings in the early 19th century.

ENDNOTES

1. Steinman had first proposed a 488-m span suspension bridge flanked by 244-m side spans in 1923. The span was subsequently increased to 594 m to assuage the War Department, but it lost out to the 1927 cantilever bridge which features tandem 351-m spans.
2. It will be the first to exceed Germany's 314-m span Cologne-Mulheim Bridge, opened in 1929 but destroyed in World War II.
3. Interestingly, a box girder suspension bridge has yet to carry rail loads—Hong Kong's Tsing Ma Bridge is structurally a truss, as is Korea's Yongjong Grand Bridge.
4. A 1994 paper describing the preliminary design (Leto, 1994) sparked the controversy.

Chapter 17

ULTRALONG-SPAN SUSPENSION BRIDGES AND AERODYNAMIC STABILITY

John Roebling, in his 1867 report on the Cincinnati-Covington Bridge, wrote:

> It is in large spans where the principle of suspension will forever stand without a rival, no matter what novelties may turn up in the future. Within the whole range of engineering, no system of construction presents such favorable conditions as are offered by the suspension system. . . . I know that I am exposing myself to criticism, but I also know that the future history of engineering will do me full justice (Steinman, 1946b).

For more than 130 years, this view has remained valid. In the period since the Tacoma Narrows Bridge collapse, advances in design, materials, and construction methods have made possible suspension spans heretofore the domain of dreamers. Computers perform calculations of great complexity with improved theories at light-speed. The simplifications and approximations required to avoid prolonged and tedious calculations in applying the Deflection Theory are things of the past; calculations of unerringly full accuracy are today handled easily by personal computers, and finite-element models have made the treatment of nonlinearity much easier. Aerodynamic and seismic analysis has advanced considerably, as have wind tunnel testing techniques. Improved truss designs, box girders, and fabrication and assembly advances have all advanced the state of the art. High-strength steels permit greater spans carrying greater loads and have brought the railroad back to the long-span suspension bridge. Construction techniques have wrought tremendous advances, some injected from seemingly unrelated fields such as deep-sea oil platform technology. Finally, new stronger, lightweight materials such as carbon composites are now beginning to infiltrate bridge construction.

James Finley had prophesized spans of more than 450 m; Claude L.M.H. Navier envisaged spans twice that length. Roebling would postulate spans of more than 1,000 m, and Gustav Lindenthal considered rail suspension spans of more than 1,800 m possible. His protégé Othmar Ammann, upon completing the George Washington Bridge, suggested (1933) that spans exceeding 3,000 m "could be built with perfect safety;" in the 1950s, David Steinman would express much the same view.

In spite of these predictions, the longest suspension span increased by just 10% between 1937 and 1982. Only with the recent completion of the Akashi Kaikyo Bridge has there been a significant advance in span length, but it is not anywhere near—in percentage terms—that achieved by the George Washington Bridge in 1931 (*see* Fig. 17-1). Nonetheless, since the early 1970s, as the Akashi Kaikyo and Messina Strait crossings acquired momentum, the theoretical limits of span length and supporting design concepts and construction technologies have come under increased scrutiny. Much of this research has focused on cable and deck-stiffening systems.

In 1973, Anthony Flint, who would later indict the Severn Bridge's live load capacity, published an assessment of theoretical span limits. Dead loads figured prominently as a limiting factor, particularly the main cables, whose weight increased out of proportion to span length in very long spans to the point that they could carry little other than their own weight. He suggested that a 1,524-m span would require twice the steel of a 1,000-m span, but that a doubling of the span to 3,050 m would increase this factor to seven times the steel—and 12 times the cable weight—of a 1,000-m span. Flint favored box girders, suggesting that the total weight of main cable in the Verrazano-Narrows Bridge could support a six-lane box girder of 2,286-m span. He further asserted that the total steel weight per unit length of the Verrazano-Narrows Bridge was twice that required to support a six-lane box girder of more than 3,200-m span. Flint also suggested that in spans exceeding 1,980 m, a three-level deck—with rail lines stacked atop one another beneath an upper road deck—would minimize torsional stresses from asymmetric rail loads.

Myriad long-span bridge concepts continue to emerge. A competition to mark the 1988 opening of the Seto Ohashi generated many ideas, including a suspension bridge with split decks supported by a single main cable and triangulated suspenders, cable-stayed bridges, and even a bridge using lighter-than-air gas to reduce weights. Serious research, however, has focused on suspension bridges in light of the expected demand for spans of more than 2,000 m. Advances in structural systems, construction methods, and materials are all under investigation. Yet issues of aerodynamic stability and dramatically rising deck and cable weights have eliminated the extrapolation of traditional solutions.

So just how far can suspension bridges span? Several key factors require consideration. Weight is but one issue. Horizontal deflections from static wind forces are another. Aerodynamic stability—particularly for flutter—is a major challenge because of increased torsional flexibility over very long spans. The contribution of a suspended structure to overall stiffness has been calculated to drop precipitously from 47% in a 1,000-m span restrained at the abutments to just 2% in a 5,000-m span (Astiz, 1996). Engineers of the Akashi Kaikyo Bridge concluded that the deep torsionally stiff box girders required in ultralong spans would weigh—and cost—more than a truss. Other researchers have concluded that single box girders will not be structurally or economically feasible for spans exceeding 2,000 m (Diana *et al*, 1998; Richardson, 1984). Yet ultralong truss spans present their own complex and uncertain aerodynamics. To overcome the stability problem, research has focused on three areas: torsionally stiff structural systems, aerodynamically designed flutter-free profiles, and the enhancement of damping through the use of active or passive control systems. New materials may also offer promise to overcome the weight issue. Many ideas are highly speculative, whereas others have been devised for application in specific projects. The research has generated somewhat contradictory findings, with the result that no definitive guidelines have yet been established.

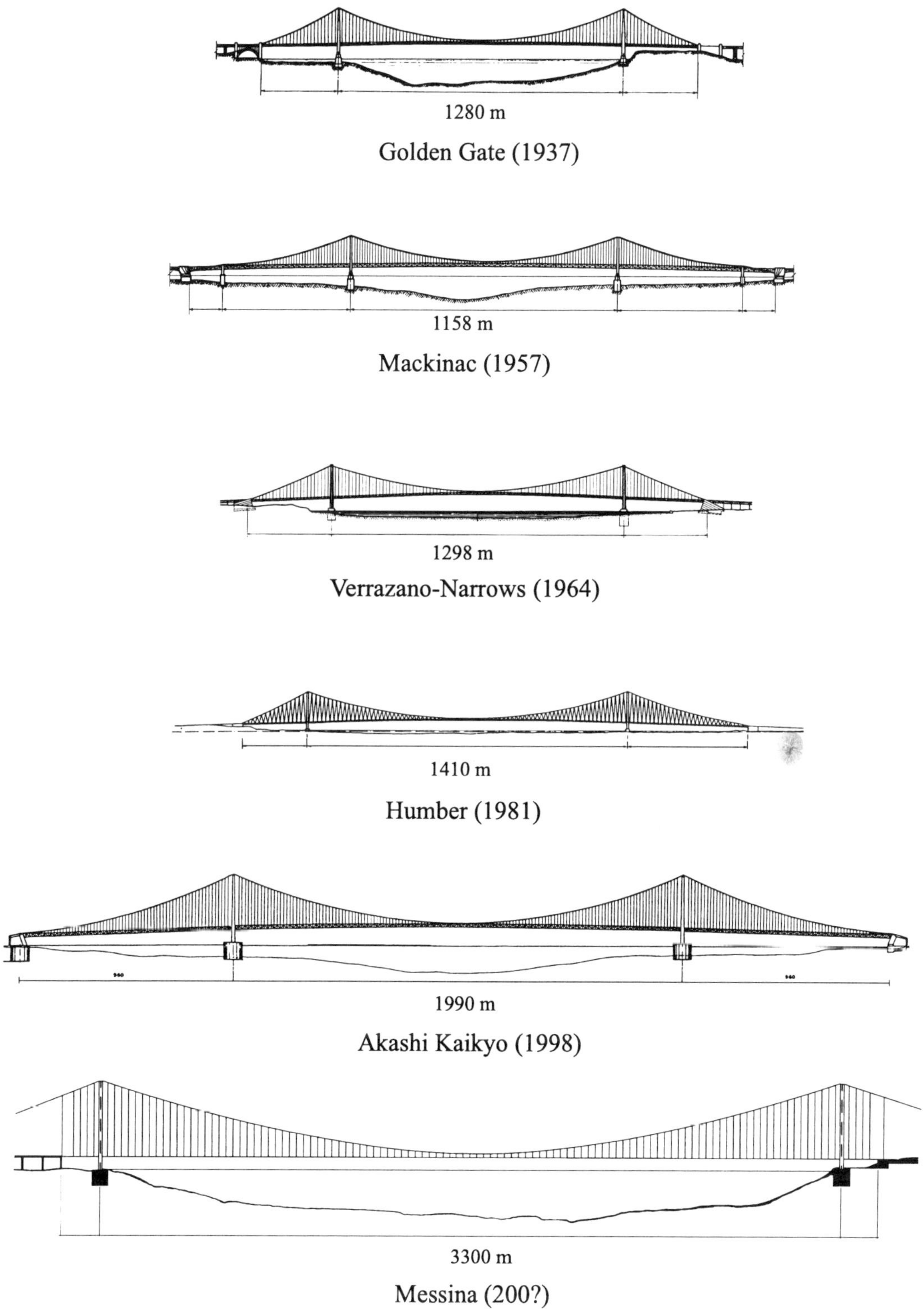

Fig. 17-1. Growth in suspended span length
MODIFIED FROM ORIGINAL DRAWINGS COURTESY OF THE MACKINAC BRIDGE AUTHORITY, HUMBER BRIDGE BOARD, HONSHU-SHIKOKU BRIDGE AUTHORITY, AND STRETTO DI MESSINA S.P.A.

17.1 ACHIEVING ULTRALONG-SPAN STABILITY

Much research has focused on extending the range of torsionally stiff box and cable systems, both independently and as integrated systems. British engineer Oleg Kerensky was possibly right on two fronts in his 1959 view that although analysis had for decades concentrated stiffening efforts on trusses and girders, for "really long spans this [cable/suspended structure integration] may be the correct solution." The computer, and improved design theory, has made such analysis—and a host of imaginative integrated systems—possible.

In addition to stiffness, air gaps and heavier weight have been suggested. As early as 1954, Yves Rocard (1957) suggested that ballast along a span's centerline could "offer new possibilities for certain types of large bridges not otherwise feasible." The injection of concrete between double-walled box girders has also been proposed. Conversely, lighter weight has also been advocated. Studies of a hypothetical 3,000-m span published in 1983 by Niels Gimsing and A.B. Sørensen of the Technical University of Denmark suggested that the truss developed for the 1978 Storebælt East Bridge proposal could have been 25% lighter—and would have reduced cable weight by 15%—had it been in the form of a truss with three widely separated roadways above the narrow lower rail deck. They also suggested glass fiber cables could reduce cable weight by an incredible four-fifths. Gimsing's work later that decade shed further light on reduced cable weights. A hybrid "cable net" with superimposed cable-stay and suspension systems in various combinations was, for a 2,000-m span, estimated to reduce cable weight over that in a suspension bridge by almost half, with only a small premium in steel for taller towers. This configuration also reduced structural steel needs by more than a third in comparison to a 2,000-m pure cable-stayed span. Even accounting for the anchorages (required to support about half the total dead load), a cable net structure was seen as advantageous, if unproven.

Swiss engineer René Walther has suggested a very long-span Dischinger arrangement incorporating a concrete box girder supported by the stays and a steel box girder by the suspension system. Wind tunnel tests on full-length aerolastic models of 2,000- to 5,000-m spans for a six-lane box girder with two rail tracks on its bottom web at the Danish Maritime Institute suggested critical velocities at least 40% greater than with a conventional suspension system. Walther predicted a slashing of superstructure costs of up to one half, with savings emanating mainly from dramatic reductions in cable weight. Japanese engineers have also investigated the use of Dischinger-type systems. In 1994, David Billington and Swiss engineer Christian Menn offered a speculative and unusual 2,500-m span hybrid system. To augment aerodynamic stability, inclined stays fanned out from large transverse brackets mounted at a 45° angle on the A-framed towers to various points over the six-lane, two-track main span. The two engineers also suggested advanced composites for spans of increasing length using this concept—first in the stay cables, then suspension cables, and then in the deck itself. The concept was also considered suitable for shorter narrow spans and in pipeline bridges.

Japanese researchers have been especially prolific. The relative merits of trusses versus box girders have received considerable attention. Studies in the early 1990s suggested 2,500-m truss spans would be possible by enlarging the size of steel members, even with the penalty of increased drag. A lower center of gravity in trussed spans was also found to increase critical flutter velocities, which had otherwise rapidly declined in spans exceeding 2,000 m, by up to 15%. Japanese researchers have also examined monocable suspension bridges and other concepts. Suspenders hung vertically and transversely across a 36-m wide, 7-m deep box girder were found to dramatically increase critical flutter velocity in a 3,000-m span. Studies have also built on the unbuilt but unique variable-depth box girder concept devised for the Akashi Kaikyo Bridge, with greater box stiffness over just one-quarter of the main and side spans near the towers. Other studies of spans of up to 3,000 m have suggested the feasibility of 36-m wide box girders with depths of up to 10 m.

One unusual Japanese concept is the "gravity-stiffened girder," in which large frames cradling two stacked box girders extend far above the upper box to grasp the suspenders. Its low center of gravity provided similar stiffening benefits revealed by research on suspended trusses. More intriguing is the temporary addition of mass to enhance stiffness in periods of high wind. Trucks, even water, would be placed on the bridge in advance of severe weather. Amazingly, calculations have suggested the lower amount of required built-in stiffness would avoid up to 70,000 tonnes of cable and deck steel, reducing the weight of a 3,000-m span by 15% to 20%. The idea is not as bizarre as it seems, as seen in the use of water pallets to help stabilize the box girders of the Humber Bridge during their early stages of erection.

Recognizing the rapid decay of torsional strength and critical flutter velocities in ultralong-span box girder suspension bridges, Japanese researchers have proposed strengthened cable systems. Stays extending from the middle of portal struts in the towers to the outer edges of main and side spans near the towers were found effective in a 2,500-m span. Periodic cross-stays extending from one cable to the opposite deck edge—recalling Steinman's remedies in the Deer Isle Bridge—in concert with ties between the main cables at midspan have also been found to enhance torsional rigidity; tests on a hypothetical 2,500-m box girder span revealed a dramatic increase in critical flutter velocity from 36 to 216 km/hr. Other engineers have examined variations of Walther's hybrid concrete/steel concept, augmented by horizontal wind cables and A-framed towers.

The high-strength cable wire developed for the Akashi Kaikyo Bridge is considered suitable for spans of up to 2,400 m. But the disproportionate increase in cable weight over very long spans requires their strength be devoted increasingly to their own support and of little else. Ergo, more cables: one Japansese study postulated the use of four 1-m diameter cables in a 3,000-m span supporting a 37-m wide, 6.5-m deep box girder. A 4,000-m span was even considered possible with eight cables.

Other approaches abound. Spanish engineer Miguel Astiz has proposed a variety of one- to four-cable systems combined with various patterns of framed and crossed suspenders to enhance torsional stiffness. These systems would allow for the retention of a shallow, lightweight box girder, but such box girders still need to be very wide. Engineers at COWI have devised an enclosed elliptical tube for spans of 3,500 to 5000 m. The tube is suspended from two tightly clustered cables draped over A-framed towers, and its edges resemble those of an aircraft wing. The "Space Web" concept launched in 1997 by British engineer Gordon Rose comprises a series of suspension systems inclined from the vertical to the horizontal that are strung from "towers" shaped as large transverse arches. Rose has claimed the system would eliminate the need for stiff decks and even make horizontally curved decks possible, and Great Britain's Steel Construction Institute has sought international patents for the concept (Matthews, 1995). Whether it is the greatest single suspension bridge advance since the box girder, as touted by its promoters, remains to be seen.

Following his early work with William Brown, British aerodynamicist J.R. Richardson continued to develop the split box concept as a means to raise critical flutter velocities. In 1988, he modified the ovoid deck housing of a failed 1986 bid to bridge the English Channel with multiple 2,000-m single box spans. Tests of a flatter, elliptical shape in the NMI wind tunnel revealed less drag and significantly reduced vortex shedding; torsional resistance was found to be excellent. Richardson then applied these findings to the split box in the form of two pear-shaped housings connected by upper and lower cross-beams. The suspenders would be attached to the upper cross-beams just inside of the enclosed housings, and tower legs would extend through the open gap between the boxes. Brown has, of course, since proposed the split box in proposed spans at Messina, the Dardenelles, and at Bali; the concept will be used most imminently in Hong Kong's Tsing Lung Bridge. He suggests its use in spans as small as 500 m could reduce costs by up to one-third over conventional truss or box girder designs.

The split box was also combined with stiffer cable systems in a 1996 study that included French engineers Michel Virlogeux—of Pont de Normandie fame—and Jacques Mathivat, an authority on cable-stayed bridges. These two engineers assessed a 3,000-m span Dischinger hybrid incorporating inclined suspenders of the Severn Bridge type over the central half of the main span. The 70-m wide split box was comprised of a 10-m wide rail box centering two 15-m wide road boxes edged with nonstructural aerodynamic fairings; cross-trusses bridged the intervening 15-m gaps between boxes. The span, supported by 450-m tall A-framed concrete towers, featured unsuspended 1,000-m cable-stayed side spans anchored in large earth-filled concrete housings. Lightweight cables made from synthetic materials and an aluminum suspended structure were considered, but their economics were uncertain and so steel was assumed. A variation of this split box system was proposed by Jörg Schlaich and Walther in the 1988 design competition to replace New York's Williamsburg Bridge. Walther also devised a hypothetical variant for Denmark's 1,624-m span East Bridge, with stays superimposed over a suspension system in the side spans and the first 332 m of the main span; a similar design for a 1,271-m span rail crossing of the Vierwaldstättensee in Switzerland has also been proposed. In these latter two examples, the stays support concrete boxes whereas steel is found over the majority of the main span that is suspended from the suspension cables.

Japanese engineers have also explored the split box. One mid-1980s study involved a 1,900-m span with two 20-m wide box girders separated by a large 80-m gap. Each box was just 3-m deep, and the entire suspended structure was suspended from four cables draped over towers comprising four legs arranged in pairs. The concept was thought to generate substantial savings over a single 37-m wide, 6.5-m deep box girder suspended from two 2.5-m diameter main cables. Each box could be erected independently of the other, and fewer cross-braces would also be required, as each box featured independent aerodynamic stability over several hundreds of meters.

More recent Japanese research has suggested that split boxes can reduce weights by as much as 30% over trusses. One proposal that converted a single box to a twin box over the central half of the main span—where flutter is considered to be most problematic—was discovered to increase critical velocity by at least 50% over a single box alone. Side- to main span proportions have also been found to affect the flutter susceptibility of the main span. One concept currently under consideration features split boxes connected by open steel grating in the median that also serves as part of the driving surface; a small vertical stabilizer descending from the median—as was first used in the trusses of the Ohnaruto and the Akashi Kaikyo bridges—serves to enhance torsional rigidity. Wind vanes placed above and below the box edges have also been suggested. Other investigations have sought to optimize the number of cables, the stiffness and frequency of cross-bracing, suspender systems, and the distance between box girders. Spans of 2,500 to 5,000 m have been examined. Continuous multiple span suspension bridges with flexible—in lieu of rigid—towers to reduce foundation footprints have also been studied.

The Danish firm COWI has devised two means to enhance damping, which such investigators as Alan Davenport suggest offers untapped potential. With *passive* damping, the firm has sought to counter motion by way of eccentric mass damping. This system involves passive masses that move across the inside of a box girder to counteract the effects of varying wind conditions. Another form of passive damping uses movable tuned mass dampers buried on the downwind side of the box girder to increase its critical velocity throughout a range of wind conditions.

For *active* damping, Klaus Ostenfeld of COWI has devised computer controlled flaps attached to the sides of ultralong-span box girders that respond continuously to changing winds. These flaps are intended to modify airflow around the suspended structure and thus reduce flutter potential. Active control has been used in aircraft wings since the 1970s and, in effect, would transform the box girder into a form of highly controlled airfoil. In the mid-1990s, COWI patented two variants it asserted would make bridges of any span aerodynamically stable. Flaps cantilevered below and just beyond the edges of a box girder shaped as a flat ellipse would intercept natural winds in advance of any upstream flow modification (that is, turbulence) created by the box itself. Airflow would then be channeled to damp out vertical or torsional motion. Computers using information obtained from deck-mounted sensors would control a continuous series of 5- to 10-m long flaps that would individually respond to varying wind conditions along the length of the span. A variation using rounded moldings that swivel around the edges of an ellipse-shaped box girder was found to create less drag and to also offer a cleaner appearance; the elliptical box shape also provided enclosure and all-weather protection for traffic. Tests have shown active controls can increase the critical velocity of even a flat steel plate by 50%, and one British study suggests that reductions in box depths of up to a third are possible.

One concern with active control is the failure of power systems in severe weather, when such control is most needed. In response, hybrid active/passive systems have been proposed. One Japanese concept features an enclosed pendulum whose motion under box torsion changes the position—via connecting cables—of external control surfaces below and outside the edges of the box girder. These control surfaces then modify the wind flow up- and downstream of the box to restore it to equilibrium. In

1999, British engineers patented a purely mechanical system of flaps activated by slight changes in angle of the deck. Active control may also help to stabilize bridges during construction. One issue is the aerodynamic interaction of moveable flaps with the windscreens considered necessary for ultralong spans. Research into passive and active control continues.

As discussed, a major problem associated with ultralong spans is cable weight. Japan's Manabu Ito has calculated that the steel cables of a 4,000-m span could support themselves and nothing more.[1] Advanced composite materials—involving carbon, aramid, or glass fibers extruded in a polymer resin such as epoxy—or fiber-reinforced plastics may offer the promise of cables with a strength of up to twice those of the Akashi Kaikyo Bridge at a fifth the weight. Carbon fiber composites have gained widespread application since the early 1960s, particularly in aircraft. Such materials cannot corrode, have excellent fatigue resistance, expand and contract less than steel, can be easy to transport, and are suitable for automated production. A rod of fiber-reinforced plastic is stronger than a similarly sized rod of steel yet is one-quarter the weight. As early as 1986, Swiss engineer Urs Meier predicted these new materials might become economical for suspension bridges with spans exceeding 3,900 m; by 1995, others were suggesting spans of 5,000 m might be feasible. Some Japanese engineers believe that carbon or aramid fibers could reduce cable weight by 60% to 70% in spans of up to 3,000 m.

At present, manufacturing costs remain prohibitive, and longevity in service is unproven. One long-term prospect may be to extrude carbon fiber into strands similar to prefabricated parallel wire strand cables. For the immediate future, composite materials have been suggested for nonstructural aerodynamic fairings and windscreens. They have been used in a number of small experimental bridges, most notably in the deck and cables of a 1992 cable-stayed footbridge in Aberfeldy, Scotland. The first fiber-reinforced suspension bridge was completed in 1999 in Kentucky. Although the towers and cables are steel, the deck of this 32-, 64-, and 32-m pedestrian span is made of fiber-reinforced polymers. A smaller 48-m span British suspension footbridge completed in 2000 near Cornwall features a glass fiber-reinforced deck just 40-cm deep. The firm of Flint & Neill in association with Wilkinson Eyre Architects designed this highway overpass, which is also distinguished by its backward-leaning towers.

Aluminum has also been touted for its potential use in suspension bridges—in fact, David Steinman had proposed its use very early on in a 1934 proposal to upgrade the Brooklyn Bridge. In spite of higher material costs, aluminum is light in weight and is simple to install; it also resists corrosion and requires minimal maintenance. A 1996 pilot rehabilitation of the one-lane 97-m span Corbin suspension bridge in rural Pennsylvania used this material to skirt the problems of weak foundations. Its use of an aluminum deck weighing 50% less than a traditional steel and concrete deck allowed the span's live load capacity to be increased almost three-fold. All 22 aluminum deck panels were lifted into place using only a lightweight crane. Notwithstanding such impressive results, aluminum has yet to be considered a serious contender for steel, or for the even newer composite materials, which hold even greater promise for long-span bridges.

17.2 A PLETHORA OF FANTASTIC SCHEMES

Much research has been of a purely theoretical nature, without any specific application in mind. Yet many schemes previously considered to be beyond the realm of engineering or economic feasibility are now being explored, placing new design concepts under the gaze of practitioners. Although many of these projects remain speculative in economic terms, they have focused more sharply the challenges that lie ahead for ultralong spans. Asia is home to many of them, but the grandest of all lies across the Straits of Gibraltar.

British engineer William Brown has devised an ingenious plan for a 2,100-m span split box suspension bridge between the Indonesian islands of Bali and Java that can metamorphose with traffic increases. Two parallel curved-bottom boxes separated by Brown's customary grillage would support an initially built two-lane deck. The boxes are supported in the middle of long cross-beams, which in turn are suspended from two main cables at their extremities. Two additional boxes can be added on either side of the existing boxes to provide a total of six lanes, augmented by two more cables placed outside the originals. The towers feature the same motif as in Brown's Dardenelles design, but with three narrow above-deck struts; the lower strut, unusually, would just clear the roadway.

Another design for the same crossing emerged in 1997 for a Dischinger-type hybrid of 2,300-m span with an effective suspended length of 1,600 m. A notable feature of this scheme—from British engineering firm Flint & Neill with Danish architects Dissing & Weitling—is its split box girder, the two boxes being separated by a 16-m wide gap formed from a trussed connecting grid. The striking 300-m tall towers of this four-lane span are formed from two elongated isosceles triangles that meet only at deck level; here, the twin boxes squeeze through openings in each pylon. Four suspension cables are used, two per box. Concrete would be used over the side spans and 450 m into the main span near the towers where stays provide most of the support. One attraction of this hybrid solution is that construction can occur simultaneously at four locations over the main span alone. One major challenge—a channel that is believed to widen by 3 mm per year because of plate tectonic movement. This visually stunning proposal, along with a 6-km bridge between Java and Madura Island, was in 1997 given priority by the Indonesian government; both projects outranked a fantastic $21 billion, 27-km bridge with a 3,300-m span across the Sunda Strait between Java and Sumatra. Indonesia's economic troubles have since delayed progress on any of these ambitious bridges, but this island nation will inevitably provide several venues for ultralong spans.

The 1998 completion of the Akashi Kaikyo Bridge may not signal an end to Japan's large bridge-building program. In fact, myriad huge spans are now under study. The long-discussed plan to span the 12-km entrance to Tokyo Bay is one of them. In 1971, a 1,500- to 2,000-m span was contemplated, but not until 1991 was a stimulus provided when the government created a joint academic/industry committee to report on technologies required for post-Akashi Kaikyo Bridge strait crossings. The committee's August 1993 report launched a research program that by 1998 saw five ultralong suspension bridges under consideration. The two most feasible bridges include a 2,250-m span for the Tokyo Bay crossing and, in a second Kansai-Shikoku link across the 11-km Kitan Straits, cable-stayed spans of 450 and 750 m and a suspension span of at least 2,100 m and possibly 2,500 m. As currently proposed, the Tokyo Bay crossing would use a box girder suspended structure, two main cables, and towers somewhat reminiscent of those at the Kurushima Straits. Nagoya's 20-km Bay of Ise may be crossed by a 2,100-m span suspension bridge and—for its west half—twin continuous 1,500-m span suspension bridges or two cable-stayed spans of 1,000 and 850 m. At Shikoku's west end, a 15-km link across the 190-m deep Hoyo Strait to Kyushu could involve either a four-span suspension bridge with two continuous 3,000-m central spans and a 2,100-m span, or three Kurushima-style spans, including two of 2,000 m and one of 2,800 m. A 7-km string of three bridges on Kyushu's west coast and a second crossing of the Kanmon Strait are also mooted. Such bridges will demand a quantum leap in technology to deal with greater seismic, wind, depth, and geological challenges than posed by the Honshu-Shikoku spans. Typhoons, for example, could push the suspended structure of a 3,000-m span up to 53 m off its centerline. Government, university, and industry continue joint research into increased wire strength, cable-restraining systems, girder designs, earthquake vibration damping, and lightweight deep-sea piers.

Yet even larger spans have been proposed in Japan. A road crossing of the Tsugaru Strait between Honshu and Hokkaido—site of the record-length rail tunnel—could encounter 22-m high waves, 270-m water depths, up to 290-km/hr winds, and extreme seismic exposure. A 1994 plan for a 19-km east crossing called for five continuous suspension bridges of 2,000-m span, whereas a 1998 scheme envisaged two continuous spans of 4,000 m flanked by 2,000-m side spans; in the latter concept, conventional 2,000-m spans with 1,000-m side spans would land the crossing on each shore. The use of a cable-stayed/suspension concept similar to that devised by Sergio Musmeci in 1969 for the Messina Strait was thought able to reduce the effective main span of the central bridge to 2,600 m. Three main cables would suspend a 50-m wide split box girder just 3-m deep, and 10-m deep transverse triangular trusses would connect the twin boxes. Foundations for the 450-m tall towers would use deep-sea oil platform technology. The lure of Russian natural gas and oil spawned an even more fantastic proposal in 1994 to cross the La Perouse Strait from the northern Japanese island of Hokkaido to the Russian island of Sakhalin. This $75 *billion* road, rail, and pipeline bridge would involve multiple 3,000-m spans with 1,500-m side spans over a 42-km distance, and second 7-km crossing over the shallower Tatar Strait from the north end of Sakhalin to the Russian mainland. Although the 70-m pier depths of the La Perouse Strait are within the bounds of existing technology, the superstructure would pose formidable obstacles.

Clearly, the Tsugaru and La Perouse crossings are decades away, but the future of the less ambitious projects is equally uncertain. The failure of Japan's extensive public works program to help end the country's decade-long economic stagnation has forced a change in policy. Moreover, past investments have proven more burdensome than planned. In 2001, for example, a plan was floated to provide more than $660 million in interest-free loans annually for the next decade to help pay for the Honshu-Shikoku project. Toll revenue from these bridges, which carry 70% of the traffic predicted, can't cover interest on the loans that helped to fund their construction.

In China, several fantastic schemes are under study. A 20-km crossing of the

Qiongzhou Strait to Hainan Island faces the challenges of 100-m depths, typhoons, earthquakes, strong tides, and chaotic foundation conditions; here, multiple suspension bridges of 2,000- or 3,000-m span, cable-stayed bridges of 1,000-m span, and bridge-tunnels have been examined. One option for the deep direct line involves a seven-span suspension bridge comprising four continuous central spans of 2,000 m flanked by end spans of 1,000 m. A more circuitous but shallower route has lead engineers to propose two tandem suspension bridges arranged as in the Transbay and Bisan Seto spans. Each bridge would feature a main span of 2,500 m and 1,250-m side spans. This longer crossing would also require a cable-stayed bridge of two central 1,000-m spans flanked by 500-m end spans. Even more audacious is a 115-km crossing of the Bohai Sea Strait between Shandong and Liadon peninsulas; the 42-km wide north channel alone would require causeways and up to seven long bridges.

One crossing in particular has generated an enormous amount of applied research extending almost 2 decades—the Straits of Gibraltar between Europe and Africa. Although a bridge is not likely to be built for many years, if ever, this fabled connection has helped to define technologies that could propel the suspension bridge to a new scale and form. The crossing has a long—and as yet unfinished—history. A tunnel was proposed as early as 1869, and again in 1908, 1918, and 1927; submerged tunnels were put forth in 1919 and in 1928. A Spanish commission studied the project from 1931 to 1936, and in 1956, Spaniard Alfonso Peña Boeuf (who had devised several concepts for spanning the Tagus at Lisbon) proposed a suspension bridge. A 1964 Moroccan study examined several solutions, as did a France-sponsored study that same year. In 1968, a bridge of seven 3,050-m spans was proposed. The Spanish Public Works Ministry then undertook a 3-year study ending in 1972.

Most schemes were sketchy, as little was known of the straits' chaotic geology. Conditions are severe, with high seismic risk, gales (in 1960, 225-km/hr winds struck nearby Tangiers), complex currents of up to 9 km/hr, and waves up to 11 m high. The straits also funnel 50,000 ships per year—more than two times the number plying Denmark's Storebælt—including large oil tankers. The work of a Spanish body established in 1972 to identify information gaps led to a 1979 agreement with Morocco to sketch out a concept, and by late 1980, the firm of Freeman Fox & Partners was at work. A 14-km direct line harbors an 800-m deep gorge, although at midchannel, an undersea mountain rises to a depth of about 450 m. To the west, a subterranean ridge reduces depths to a maximum of 350 m, but the bridge length is doubled to 28 km. Bored tunnels, bridges, submerged bridges, and floating bridges and/or tunnels were examined. A 1.5 *billion* m³ causeway was even proposed, flanked by 1,000- to 2,000-m suspension spans over the coastal shallows. But tunnels fared poorly—ventilation issues nixed a 33-km road tunnel, and a rail tunnel would stretch a prohibitive 50 km and might take as long as 14 years to build.

Of the bridge concepts, an 11-span suspension bridge of 2,000-m spans over the longer but shallower west alignment was preferred. Steel towers would stand in up to 250 m of water on piers derived from deep-sea oil platform technology. A flat 34.5-m wide split box separated by a wide air gap was envisaged; each two-lane box was flanked by a collapsible skirt and was hung via inclined suspenders from two pairs of main cables. Construction was thought to require up to 10 years.

In 1983, American-based engineer T.Y. Lin examined the question and, in March

1984, urged the 14-km direct route be tackled.[2] Lin proposed two 5,000-m spans centered by a pier on the midchannel ridge with flanking side spans of 2,500 m. Three 600-m tall towers would stand in water up to 450-m deep supported by foundations using North Sea oil platform technology. Unusually, the tower legs branched out up to 201 m apart as they rose, allowing transversely inclined suspenders to aid stability. The main cables would be braced together and almost converge at midspan. As an alternative, Lin also proposed extremely wide conventional towers that retained the canted suspension system.

Five more years of study began in 1985. COWIconsult was retained to lead the work, aided by Carlos Fernández Casada, a leading Spanish engineer, and Ben C. Gerwick, the U.S. foundation specialist firm (and currently part of COWI). Several concepts for the western alignment were examined: continuous multiple spans stabilized by rigid A-framed towers with shore-based anchorages, the use of shared anchorages as in the Transbay and Bisan Seto bridges, and multiple but staggered cable systems. In this last scenario, one set of main cables would be anchored at the tower/deck connection whereas another set would be draped over the top of the same tower. A Dischinger hybrid was also proposed, as was a variation of Musmeci's 1969 stayed suspension bridge for the Messina Strait. Although more expensive, a continuous multiple span road and rail bridge of 2,000- to 3,000-m spans on the west alignment avoided deep water anchorages; its 300- to 400-m tall A-framed towers also afforded superior protection against ship collision.

Many suspended structures were examined for the 2,000-m span option. Engineers modified the 1978 Storebælt design in a truss with highly canted sides supporting a single rail line on a narrow lower deck. A conventional box girder was examined, as were several split box variants supported from two to four cables; one even replaced the box girders with trusses. Wind tunnel tests at the NTH in Trondheim, Norway revealed the split boxes to possess a critical flutter velocity of up to 234 km/hr, 61 km/hr higher than with a traditional box. Increased gap widths were also found to increase critical velocity. Most intriguing was an enclosed elliptical girder which offered complete weather protection, 30% of the drag of a box girder in spite of an 11-m depth, and superior torsional resistance. One drawback, however, was its great weight. Efforts were also made to enhance rigidity via the main cables. One solution bundled the main cables to one another as far as the quarter-points of the main span, whereupon they diverged over its central portion. In another concept, the cables could be anchored in massive enclosures near the top of A-framed towers to prevent collapse of the entire bridge should one span fail.

Three types of oil platform technology formed the basis for foundation design. North Sea concrete gravity pier platforms had by 1988 reached 220-m depths, and a 350-m platform was planned for Norway's Troll field. A 330-m deep steel jacket in the Gulf of Mexico was under design, but this type's lighter weight made it vulnerable to ship collision. Tetrapods—large, partially buoyant steel tubes—were untested; here, engineers proposed four tubes set 150-m apart on the sea floor that would converge at deck level to support a conventional tower. Concrete gravity piers were favored in terms of cost—$1.45 billion for the piers of the 3,000-m span variant—and for their superior strength against ship collision, a major consideration given the $100 million of protection required for each pier.

The most economical span length was found to hover somewhere between 1,800 and 2,500 m. A series of 2,000-m road spans would cost $3.8 billion—$4.8 billion with rail—and a road/rail bridge of 3,000-m spans would cost $5.4 billion. In the end, COWIconsult recommended 10 continuous spans of 2,000-m, the ensemble flanked by 1,000-m end spans. Although the firm did not rule out a twin 5,000-m span with 2,000-m end spans on the shorter direct line, it felt such a concept must await advances in lightweight composite materials for the cables and suspended structure.

Unsolicited proposals during this period also demonstrated great imagination. In 1986, Swiss engineer Meier proposed a hypothetical 8,400-m road/rail cable-stayed span with carbon fiber cable stays and a flat plate deck of carbon fiber just 1.8-m deep. A 1987 Moroccan proposal of Victorian visage used huge tower-mounted trussed frames extending 650 m into 2,625-m spans to support suspension cables as in Musmeci's Messina Strait concept. In 1990, a Norwegian engineer suggested a 350-m deep "submerged tube bridge," and German engineer Jörg Schlaich proposed floating cable-stayed spans. The following year, Lin updated his 5,000-m span bridge on the direct line by also adapting Musmeci's Messina Strait concept; cable stays extending from 900-m tall towers supported an inclined frame protruding 1,000 m into each span to support 3,000-m main spans. One other unsolicited 1994 scheme for 5,000-m spans proposed a "tensioned cable network" formed from taut horizontal and vertical cables describing an elliptical tube narrowing at center span.

In 1989, Spain and Morocco agreed to 5 years of additional study. On the direct east route, COWIconsult devised three 3,550-m continuous spans anchored by 1,500-m side spans; an alternative with two spans of 5,000 m and 646-m tall A-framed towers was also mooted. Suspended structures included a closed elliptical tube housing two decks, or a split box of two braced girders separated by a 31-m gap. Two closely spaced cables suspended the elliptical tube from its outer edges, save for several hundred meters near midspan where the girder was suspended from its center; horizontal wind cables could also be used to aid stability. The wire strength achieved for the Akashi Kaikyo Bridge permitted a main cable diameter of just 116 cm, the 38,481 wires in each cable to be built from a total of 303 prefabricated parallel wire strands comprising 127 wires each. The main challenge would involve the use of continuous cables extending over some 13,650 m of suspension bridge.

Two configurations of the split box for the three 3,550-m span variants were elaborated. One, a 55-m wide arrangement comprised two 13.5-m wide, 2.5-m deep rectangular boxes separated by a 28-m air gap. Nonstructural fiber-reinforced plastic fairings were fitted on the outer edges of each box, and 3.5-m high perforated windscreens would reduce winds on the two-lane decks. Cross-girders connected the boxes every 26 m with additional bracing near the towers and center span, and the main cables would be bolted to the deck at midspan. A second variant added a central, 7-m wide twin-track rail box that would, with two 16-m air gaps, extend the overall width of the suspended structure to 66 m. Wind tunnel testing in the spring of 1994 suggested high performance, with a twin-box flutter critical velocity of 280 km/hr exceeded by a fantastic 360-km/hr figure for the trimaran box scheme.

Engineers estimated the longer 28-km westerly bridge to cost $11 billion, and the 14-km scheme, $14 billion. Either alternative would take a decade to build, and Gerwick estimated each pier would cost as much as Denmark's East Bridge. Torsional stability and vertical stiffness remained concerns, especially if lightweight composites were used in the suspended structure or main cables. Triangulated cable systems, midspan deck/cable connections, and hydraulic, tuned mass, or viscous damping systems to counter deck motion were investigated, as were reduced cable safety factors.

In spite of the 14-year, $49 million research program, the bridge received a fatal blow in 1995 when both governments decided to focus on a tunnel on the westerly alignment. Although the design investigations have contributed greatly to an understanding of ultralong-span bridge design, the diversity of concepts associated with this daunting crossing suggests that there remains a lack of firm direction for tackling such long and expensive spans. Research around the world continues into the development of economical, serviceable, and aerodynamically stable spans exceeding 2,000 m. Many more ideas will emerge before one or more forms emerge as an accepted standard, as did the truss-stiffed suspension bridge in the 19th century and—more recently—spans stiffened by box girders. The split box may, in fact, represent the next evolutionary step towards the ultralong suspension bridge.

ENDNOTES

1. Norwegian engineers have strung a high-voltage hydro line over a single 4,888-m span across the Sogne Fjord.
2. Lin was no stranger to ambitious plans; since 1959, he had advocated an 88-km long "Intercontinental Peace Bridge" across the Bering Strait between Siberia and Alaska.

Chapter 18

THE LEGACY OF THE TACOMA NARROWS BRIDGE

An understanding of the aerodynamic behavior of suspension bridges and the means by which to devise and test alternative designs has evolved considerably since the discipline first emerged in the wake of the Tacoma Narrows Bridge collapse. The work of F.B. Farquharson was, of course, groundbreaking and served to launch the science. Allan Larsen, a bridge aerodynamics specialist with the Danish firm COWI, has (1998) described the most enduring legacy of the initial investigations to be the discovery that sectional models could simulate the aerodynamic response of full aerolastic models (this is at least true for smooth wind flow). Although consensus pervades many areas of bridge aerodynamics, it remains a subject littered with varying findings and opinions from those immersed in it. In fact, not until the early 1960s was a broad consensus even reached on the aerodynamic phenomena being studied. Many of the mechanisms involved—flutter, buffeting, turbulence, galloping, vortex shedding—still stir debate. These phenomena are held to play differing combinations of roles by different researchers, and more than two vibration modes may occur simultaneously in large spans or those under construction. Some confusion over terminology even persists. Moreover, the relative roles of full aerolastic models, sectional models, computer simulations, and mathematical analysis continue to evoke debate.

18.1 THE EVOLVING SCIENCE OF BRIDGE AERODYNAMICS

Frederich Bleich's mathematical description of flutter established a baseline for many early investigators—such as Norwegian Arne Selberg—who simplified and/or modified the theory of thin airfoils for use in bridges. Yet R.A. Frazer and Kit Scruton in their 1952 summary of the Severn Bridge studies considered primarily mathematical approaches difficult and prone to inaccuracy. American George Vincent expressed much the same view in 1958 (Kerensky, 1959), as did D.E. Walshe (1967), a key figure in the investigations of the Severn Bridge's box girder. Noted British suspension bridge theorist Sir Alfred Pugsley (1968) concluded that, "designers will for long be forced to turn to empirical methods [in this sense, methods derived from observed performance, such as those of Othmar Ammann, David Steinman, and Selberg] for preliminary design and thereafter, in all important cases, to resort to sectional models and other oscillation tests to check the stability of the resulting bridge." The testing of full aerolastic models has in fact become increasingly common with the enormous size of contemporary spans. However, several engineers, most notably the late American Robert Scanlan (1914-2001), have made significant mathematical advances that may soon challenge Pugsley's assessment.

Several researchers contributed significantly to mathematical and wind tunnel methods in the 1950s and 1960s after the work of Farquharson, Bleich, Vincent, and others involved in the early investigations. Most of this work occurred outside the United States, as the focus of suspension bridge building shifted to Europe and then Japan. Scruton's work under Frazer on the early Severn Bridge studies confirmed the validity of sectional model testing in lieu of more expensive full aerolastic models in smooth flow. Scruton also launched the first International Conference on Wind Effects on Buildings and Structures in 1963 at Great Britain's National Physical Laboratory (now the National Maritime Institute) in Teddington; this conference is today recognized as a seminal event in the field.[1] Frazer also contributed to suspension bridge wind engineering through his work in flutter instability and vortex shedding, as did colleague Walshe.

Scandinavian engineers were—and remain—particularly active. S.O. Asplund—designer of Sweden's Älvsborg Bridge—investigated vortex shedding and flutter and devised practical rules for avoiding flutter through stiffness. Asplund also developed analytical methods to aid in suspension bridge design generally, as did Norwegian Selberg. Through extensive testing of sectional models from 1954 to 1958, Selberg, using airfoil theory, developed simplified approximation equations for flutter. He also developed simplified diagrams and formulae to aid in the determination of flutter velocities and oscillation characteristics. This was useful for intermediate-span suspension bridges for which even sectional model testing might be too expensive. The method, quite simple in comparison to Bleich's, was also suitable for the preliminary design of larger spans to narrow down truss alternatives in advance of model testing (Selberg, 1961). Selberg's measure of flutter efficiency, which compared the critical velocity of a suspended structure with that of a flat plate with the same mass and natural frequencies, proved useful for assessing streamlined bridge sections, where wind disturbance at the leading edge was minimized. In the early 1960s, Yves Rocard of France would produce similar formulae, as would Danish engineer A.G. Frandsen and—later that decade—German engineers Kurt Klöppel and F. Thiele.[2] Selberg and Klöppel both devised factors to modify flat plate flutter speeds when dealing with bluff forms, but these factors were based on wind tunnel testing and not mathematical theory. It is somewhat ironic that for all the efforts to modify flat plate theory for trusses, the theory would become more relevant in its pure form with the advent of streamlined box girders, which feature less separation of airflow. In fact, the Lillebælt Bridge is what attracted Frandsen's attention to the use of flat plate theory, and Bleich's original flutter work using Theodore Theodorsen's method can actually provide reasonably accurate results for suspended box girder analysis.

Selberg was also known for unorthodox—if effective—research methods. In 1949 and 1950, his pioneering studies of the effects of structural damping on vertical and torsional motion involved up to 25 young adults jumping in unison to the natural frequency of various spans from 70 to 160 m to initiate oscillation; Selberg (1950) described the method as "very successful." He would later use even more unusual—and dramatic—techniques to assess damping. Participants at the 1966 Lisbon International Symposium on Suspension Bridges were treated to film footage taken the previous year in which Selberg (1966) induced torsional motion in a 229-m span suspension bridge; he did it by firing rocket engines in opposite directions on either side of the truss! Selberg's work, as with Ammann's and Steinman's stiffness indices, provided useful preliminary indicators of expected aerodynamic behavior that, although not always precise, enabled engineers to further develop their designs.

An early but relatively unknown contributor was Japanese engineer Atsushi Hirai, who in 1942 sought to explain the Tacoma Narrows Bridge collapse in terms of lateral and torsional buckling under static wind forces, a concern for long, slender spans (Theodore von Kármán had also recognized lateral buckling as a concern in his letter to *ENR* immediately following the Tacoma Narrows Bridge collapse). Hirai's work, which remained virtually unknown because of Japan's isolation during World War II, entered the mainstream only with its English-language publication in 1956. It has been considered to offer the first theoretical explanation of the Tacoma Narrows Bridge collapse and was later expanded to include the role of aerodynamic torsion (Hirai *et al*, 1966). Hirai and I. Okauchi's sectional model testing of the Wakato Bridge in the late 1950s would also pioneer the study of aerodynamic forces on bridge models, which are discussed later in this chapter.

The effects of wind turbulence on suspended structures and critical flutter velocities have become recognized as a key determinant of aerodynamic instability. Farquharson (1946) had early on recognized that turbulence could explain how instability in the wind tunnel was not always reflected in the field. Charles Andrew (1945), who led the design of the second Tacoma Narrows Bridge, also considered the "constant wind in the [wind] tunnel" to exaggerate stiffness requirements. Perhaps Ammann's wariness of wind tunnels emanated from the fact that turbulence was then not being accounted for.

In 1955, Danish engineer Martin Jensen (for whom the wind tunnel at the Danish Maritime Institute is named) established the basic principles of turbulence in the so-called "boundary layer," in which wind shear from ground effect, the influence of variable topography (natural and cultural), and friction from air viscosity are manifested in turbulence. Alan Davenport would be the first to properly simulate and codify turbulence in the wind tunnel. As suspected by his predecessors, Davenport demonstrated that this natural characteristic of wind—greatly influenced by adjacent terrain (unlike smooth flows in the wind tunnel)—disrupts the cyclical production of vortices, thus reducing their regularity and intensity of impulse. This in turn modifies the suspended structure response and in general was found to reduce the need to design for the higher critical flutter velocities indicated in smooth flow.

As an example, tests in turbulent flow at the University of Western Ontario Boundary Layer Wind Tunnel in Canada on New York's World Trade Center produced spectacular results, in which an unacceptably high 12.2-m sway in smooth flow was reduced to mere centimeters. Even the simulation of moderate turbulence on a bluff body was found to significantly reduce motion otherwise experienced in smooth flow. Davenport's (1966) study of the Golden Gate Bridge using sectional models actually predicted its critical velocity to be 50% higher in turbulent than in smooth flow. His studies of the A. Murray MacKay Bridge in the late 1960s would confirm the effects of turbulence, as would studies in the 1970s on Vancouver's Lions Gate Bridge by Canadians Peter Irwin and Robert Wardlaw. Davenport *et al* (1971) would later suggest on the basis of wind tunnel tests that turbulence might have also reduced the incidence of vortex-induced vertical motion over the Tacoma Narrows Bridge's short life.

Turbulence can be considered a key factor in determining a structure's flutter critical velocity (although it can incite vertical and torsional motion in bluff structures in even low wind speeds through buffeting). After the resolution of flutter concerns, vortex shedding and buffeting need to be addressed to avoid steel—and motorist—fatigue. The treatment of all phenomena, of course, applies both during construction and after a span's completion (Scanlan, 1979). The advent of computers made the study of turbulence—which introduces all manner of complex variations in wind speed, duration, and direction—possible. The simulation of wind variability at a specific site also allowed engineers to tailor a design to its particular wind environment. This testing is best done with full-length aerolastic rather than sectional models to account for the influence of turbulence across the entire span. The acceptance of this approach is reflected in the many wind tunnels around the world that today generate turbulent flows. Nonetheless, caution is called for; recent research has revealed similar flutter critical velocities in smooth and turbulent flow under certain circumstances (Cai *et al*, 1999).

Scanlan suggested that as bridge models for the calculation of static loads have been replaced by computer simulation, the reliance on "empiricism" (that is, wind tunnel tests on full aerolastic models) for aerodynamic assessment is—and should be—diminishing. There are, however, fundamental disagreements over this direction; for example, sectional models cannot simulate turbulence over the entire length of a suspended span. Nonetheless, the goals—to determine how suspended structures receive and respond to aerodynamic forces—are the same.

A significant tool in this quest involves flutter—or aerodynamic—derivatives. These dimensionless (that is, they do not represent a unit of measurement) mathematical expressions have been explored by several researchers, most notably Scanlan, and can be used to determine critical wind speeds for aerodynamic instability. Here, theory has diverged from the work of Bleich, Rocard, Selberg, and others whose basis was airfoil flutter. Bleich's approach in particular was complex, and in the discussion of Bleich's 1948 ASCE *Transactions* paper, Pugsley suggested that knowing the magnitude and rates of displacement from model experiments could simplify analysis enormously. This approach now provides key parameters to aid bridge designers. Flutter derivatives measure the ratio of energy input from external wind forces to the response—or motion—of the suspended structure. They distinguish between aerodynamic cause and dynamic effect, and provide a means by which to obtain the magnitude of dynamic forces and aerodynamic damping for various bridge cross-sections. Flutter derivatives can thus provide an overall assessment of stability—the net addition of aerodynamic damping (which in a stable bridge is positive and in an unstable bridge negative) and structural damping (which is always positive). If negative aerodynamic damping exceeds structural damping, as measured by the in-phase motion of the suspended structure, catastrophic oscillation characterized by an exponential growth in amplitude can occur. It is interesting to note in this regard that a bridge with an inferior aerodynamic shape can be stable if strong in torsion, which separates torsional and vertical frequencies.

Scanlan (1990) stated that flutter derivatives were in fact computed as early as 1950 for the torsional motion of the Tacoma Narrows Bridge by von Kármán and Louis Dunn; Pugsley's criticism of Bleich's approach had also in effect advocated their use. Hirai and Okauchi's work with the Wakato Bridge sought to determine flutter derivatives through experiment, and at the 1966 Lisbon Symposium, flutter derivatives on a variety of girder and truss sections were examined by Japanese researchers Nobuhiro Ukeguchi, Hiroshi Sakata, and Harumitsu Nishitani. The first comprehensive exploration of flutter derivatives came from A. Sabzevari and Scanlan in 1967 and 1968, and then from Scanlan and J.J. Tomko in a seminal 1971 paper. These early efforts involving Scanlan considered vertical and torsional motion and the effects of torsion on lift and vice versa. More recently, the effects of drag (that is, lateral displacement)—which can increase tilt—have been introduced; this analysis involves the solving of an 18(!) variable equation.

Scanlan (1990) also cited the use of sectional models for determining flutter derivatives—as opposed to the purely mathematical derivation of airfoil derivatives—as a clear indication of the fundamental differences between the aerodynamic behavior of bluff decks and airfoils. There are two main approaches by which to determine these terms. The *forced vibration method* measures and compares forces from vibration both in still air and in wind at the supports of a fixed model, whereas the *free vibration method* interpolates aerodynamic forces indirectly by measuring the motion of a model hung elastically. Japanese researchers were the first to experiment with the forced vibration method (Ukeguchi *et al*, 1966) and have continued to focus on this approach, whereas Scanlan pursued the free vibration method (Scanlan and Sabzevari, 1967; Scanlan and Tomko, 1971). Both methods have advantages and disadvantages, but recent developments in the free vibration method have seen its use increase in the United States, Denmark, Norway, and Japan. Improvements—most notably by Canada's Guy Larose—have also maintained the appeal of the forced vibration method.

Most flutter derivative work has assumed a smooth wind flow using non-dynamically scaled sectional models. These models do not account for varying vibration amplitudes and aerodynamic forces across the length of a span, nor do they account for the effects of turbulence. This has led to some criticism of efforts to create a catalogue of flutter derivatives for generic bridge sections. Moreover, injecting the effects of turbulence into the derivation of flutter derivatives has been difficult. Their determination even in smooth flow is time-consuming, and there is a case to be made for the wind tunnel testing of dynamically scaled full aerolastic models in turbulent flow, which takes significantly less time than the testing of many differently shaped sectional models and accounts for a specific wind environment.

On the other hand, Scanlan touted the inexpensive nature of sectional models. Even Davenport, with others, has suggested promise for their use in turbulent flow; in a study of the Bronx-Whitestone Bridge, sectional models in turbulent flow generated results that "bore a marked resemblance to that of the taut strip model simulation" (Davenport *et al*, 1979). Scanlan also suggested a need to better understand the theory and characteristics of turbulent flows, which are inherently complex; the accuracy of their simulation in the wind tunnel, he argues, is difficult. Moreover, Scanlan and Joseph Vellozzi (1980) cautioned that actual wind can also be surprisingly smooth, which, although rare, could raise concerns over critical flutter velocities derived from studies under turbulent flow. In any event, Scanlan (1983) suggested that the *aerodynamic admittance*—a factor devised by Davenport to approximate wind-induced force fluctuations using wind velocity as a proxy—is of critical importance to the determination of flutter derivatives (the shape of the suspended structure will influence how the forces are interpreted).[3] However, Davenport *et al* (1971) have suggested that taut strip models—which Davenport largely developed—can avoid the need for this complex calculation. Although the debate continues, flutter derivatives have become useful tools for comparing the aerodynamic behavior of different cross-sections of suspended structure.

Scanlan (1990) suggested a need for research in several areas: the development of a flutter derivative catalogue for standard deck cross-sections to aid in assessing the behavior of preliminary designs; research into flutter derivatives in turbulent wind for more complex flutter patterns; the study of wind turbulence generated by the suspended structure itself as a contributor to oscillation; and the development of damping devices where inherently aerodynamic solutions are not possible. The retrofitting of existing bluff-shaped spans, such as the Deer Isle Bridge, falls into this latter category. With more advanced analytical techniques and computing power, COWI's Larsen (1998) has suggested that mathematical models have now matured to the point of providing useful design information. COWI, in fact, has helped devise a computer simulation model in conjunction with the Technical University of Denmark that has generated close approximations to wind tunnel tests on sectional models.

The monitoring of completed bridges to correlate predictions with reality has long been considered an area requiring research; this gap may be narrowed with extensive instrumentation on such recent spans as Hong Kong's Tsing Ma and China's Jiangyin bridges. Davenport (1996) has written that structural damping has not been "fully exploited in long span bridge design and remains one of the challenges for the future;" this long-standing research need was recognized by all of the early researchers, including Farquharson (1949) and Steinman (1947b). Given the fact that welding diminishes structural damping, as compared to riveting and bolting (witness the early concerns with the Severn Bridge's design), there is a need to better understand its potential. Larsen (1999) states that measurements on actual bridges are still required. Unfortunately, full-scale testing is an expensive and difficult proposition and modeling to simulate structural damping remains crude. Japanese engineers have perhaps most vigorously pursued large-scale tests on many of its newly completed bridges.

The debate continues, and many approaches are still under development to assess and improve the aerodynamic stability of suspension bridges. There is little doubt, however, that a combination of theoretical and wind tunnel analysis will jointly contribute to the understanding of the aerodynamic behavior of the ultralong spans now contemplated.

18.2 THE PROBLEMATIC PLATE GIRDER

The 50th anniversary of the Tacoma Narrows Bridge collapse was perhaps most distinguished by a paucity of attention marking its passing. Various community events held in the Tacoma area celebrated the event as a novelty. The engineering press published articles recounting the collapse. In Great Britain, more sobering reflection was evident, primarily through two engineering society lectures marking the death earlier that year of British aerodynamicist Kit Scruton and assessing the state of suspension bridge aerodynamics after 50 years.

The specifics of the Tacoma Narrows Bridge collapse have been considered to be largely of academic and historical interest; even the initial investigations recognized that the inherent instability of torsionally weak plate girders rendered them unlikely candidates for dynamic stability. The subsequent research focus on truss designs and the pursuit of means to avoid coupled flutter saw plate girders receive scant attention; they have never re-emerged in a span anywhere near the size of the original Tacoma Narrows Bridge. Research continues to focus on truss, box girder, or newly emerging forms of suspended structures. This state of affairs prompted American Daniel Dicker to write in a 1966 ASCE Normal Medal-winning paper on the aerodynamics of H-shaped sections that, "It is somewhat disquieting to realize that in the 25 yr since the failure of the Tacoma Narrows Bridge, the theoretical problems associated with the collapse have never been truly solved." COWI bridge aerodynamicist Larsen (1999) would echo these sentiments a quarter century later. One other reason for the dearth of attention was the apparent complexity of the task. As a line of enquiry, Scanlan long argued (Scanlan and Tomko, 1971; Scanlan, 1990) against the relevance of airfoil theory to bluff (or blunt) suspended structures.[4] Yet whereas predictive mathematical models for airfoil-like flat plates are well-developed, those for bluff decks remained elusive for some time (Buckland and Wardlaw, 1972; Wyatt, 1992).

Larsen has, however, gone on to pursue an explanation for the onset of the final and fatal torsional motion of the Tacoma Narrows Bridge. Moreover, he suggests that rudimentary principles of fluid mechanics help provide it. Billah and Scanlan (1991) had, of course, implicated vortex shedding over the leading edge girder as a precondition for the onset of flutter (von Kármán and Dunn had as early as 1952 considered leading edge vortices to play some undefined role in the torsional motion of the Tacoma Narrows Bridge), but Larsen (2000a) suggests the Billah and Scanlan approach "failed to connect the vortex pattern to the shift of apparent section damping from positive to negative, which signifies the onset of torsion instability."

It is here where even the benign vertical motion of the Tacoma Narrows Bridge, long ascribed to vortex shedding off the leeward girder, has been challenged. Research published in 1986 by Y. Nakamura and M. Nakashima at Kyushu University in connection with the Honshu-Shikoku crossings appeared to dispel a connection between downwind vortex shedding and the vertical oscillation of the Tacoma Narrows Bridge. They found that a so-called splitter plate—a horizontal plate placed downwind of the girder that broke up the trailing vortices—had no effect on the vertical motion of the H-shaped cross-section. They posited instead that vortices observed to spin off the *leading* edge girder were responsible for the vertical oscillation, not the downwind vortex trail.

Larsen has taken these lines of enquiry a step further to determine whether leading edge vortex creation could be responsible for the onset of torsional motion in the Tacoma Narrows Bridge. The continual and rapid measurement of such parameters as motion and forces now possible with COWI's advanced computer simulation model made the analysis possible. Larsen's explanation emanates from an understanding of the formation of vortices at the girder's leading edge and their movement across the deck cross-section. These vortices are generated when the wind's angle of attack changes from the horizontal; their spacing across the top and bottom of the deck also varies with wind velocity. Larsen's premise is that wind speeds generating vortices alternately across the upper and lower sides of the deck with a spacing equal to one-half its width or greater signify the onset of torsional instability for a structure with very low levels of structural damping.

The explanation is as follows (*see* Fig. 18-1): the bluff windward girder creates a leading edge vortex, say, on the top side of the deck, and a previously created vortex is positioned beneath the deck centerline. The newly generated upper vortex—which rotates in a clockwise direction—creates, as Billah and Scanlan concluded, a positive suction force as it begins to move across the cross-section in much the same way that uplift is created on a tennis ball hit with a back spin; at the same time, the bottom vortex, rotating with an opposite spin, moves across the leeward half of the cross-section. The upper vortex lifts the windward side of the deck, and the lower vortex pulls down the trailing side; the forces (and the motion) are mutually reinforcing and the cross-section twists, with maximum torsion occurring when the top and bottom vortices are one-quarter and three-quarters across the cross-section, respectively. As the upper vortex approaches the median and the lower vortex approaches the leeward edge, the lift forces and the deck motion—which is now reversed in an effort to restore an equilibrium (that is, horizontal) position—oppose one another. When the upper vortex passes over the median and the lower vortex leaves the leeward edge, there are—momentarily—no unbalanced forces at work, and the girder moves through the horizontal. As the vortices move across the second half of the cycle (that is, the upper vortex travels over the leeward half of the girder and a newly generated lower vortex moves beneath the front half of the girder cross-section), the scenario is repeated, with the motion reversed. The cross-section thus oscillates in a self-sustaining motion; a critical velocity has been reached.

Larsen then examined vortex production whose spacing is greater than half the width of the deck cross-section—that is, generated by a wind speed greater than the critical velocity. In this case, the sympathetic lift forces on top and bottom halves of the deck imposed by the vortices are more in sync with the deck motion for more than half the cycle of oscillation. This generates extra energy that can be absorbed only through ever-higher amplitudes of torsional motion—a catastrophic situation. Next, Larsen examined a vortex spacing of less than half the girder width—that is, at wind speeds less than the critical velocity; the presence of three vortices over much of an oscillation cycle (two below, one above, and vice versa) means that forces tend to cancel one another out over most of the oscillation cycle. The girder is stable because of aerodynamic damping.

Larsen was able to confirm these phenomena with computer simulations and tests on a sectional model of the Tacoma Narrows Bridge in a water flow tunnel. The

A counter clockwise vortex is formed under the deck as the bridge twists with the windward edge downwards, thus exerting a downward force (black arrow) acting on the upwind deck edge. An instant later the vortex has drifted to a middle position and the downward pull now acts close to the deck center line. Still later, the upwind part of the deck twists upwards and the counter clockwise vortex has drifted to the leeward, now exerting a downward force on this portion of the deck. A new clockwise vortex and associated force (outline arrow) is created upwind and the process is repeated. The vortex creation and drift process will cause instability at wind speeds at which the vortices travel faster across the deck than the time period of the twisting motion.

Fig. 18-1. Torsional excitation mechanism—Tacoma Narrows Bridge
ORIGINAL DRAWINGS AND TEXT COURTESY OF ALLAN LARSEN, COWI ENGINEERS AND PLANNERS A/S.

results were remarkably consistent. In the water tunnel, Larsen used a splitter plate similar to that used by Nakamura and Nakashima to disrupt the vortex stream and found no change in torsional motion. He concluded the motion was being “excited” by vortex production off the leading edge girder, and not by vortices in the downstream wake. Larsen also studied the venting of the upwind side girder, which both the Advisory Board on the Investigation of Suspension Bridges and Steinman had suggested might have reduced stability. To the contrary, Larsen found that vents had a beneficial effect by disrupting the generation of vortices off the girders' leading edge. Moreover, a streamlined section superimposed over the H-section was found to double what he found to be a 34-km/hr critical velocity to one of 65 km/hr. Larsen (2000b) still considered this higher critical velocity insufficient to render the Tacoma Narrows Bridge safe "through the first winter."

Whether Larsen's work will be accepted as the definitive explanation of the Tacoma Narrows Bridge collapse remains to be seen. Larsen (Personal communication) has since uncovered some intriguing history in the form of F. Stüssi and J. Ackeret’s 1941 article on H-sections. This article, which had caught the attention of Farquharson, also revealed the creation of leading edge vortices in tests comparing fixed and elastically sprung sectional models of the Tacoma Narrows Bridge in a water tunnel; in the latter case, the effects were particularly pronounced. Stüssi and Ackeret recognized the significance of this leading edge vortex creation in torsional instability, but the inability to measure changing forces in real time made it impossible for them to verify their role. It is not clear why Farquharson did not focus on this finding but instead highlighted the role of deck slots obtained through subsequent correspondence with Stüssi and Ackeret. Interestingly, similar tests using a fixed model at the Massachusetts Institute of Technology in the summer of 1941 were also performed under the Advisory Board’s auspices. Larson (Personal communication) speculates that several factors may explain Farquharson’s avoidance of this line of enquiry: an excessive (greater than 70°) angle of twist in the Stüssi and Ackeret studies, the lack of information regarding the model characteristics, and communications made more difficult by World War II. Had Farquharson pursued these findings, it is conceivable that a whole new research direction might have opened up for the Advisory Board.

Larsen’s findings may in fact have more than just passing historical or academic interest, for plate girders remain a relevant subject of study. Several contemporaries of the Tacoma Narrows Bridge have remained aerodynamically active, and engineers continue to seek means to quell their unwanted motion. Both of Steinman's Thousand Island spans received an elaborate make over in 1976 and 1977 with the addition of rigid braces connecting the plate girders and two main cables at the quarter-points of the main span and at the halfway mark over the side spans. The bridges also acquired additional cable stays of Steinman vintage over the main span; the side spans, previously devoid of these appurtenances, now received them. Most intriguing was the addition of the very tower stays that Steinman objected to so strenuously in his war of words with Ammann after the Tacoma Narrows Bridge collapse. Two stays now connect the top of each tower leg to each of the main and side spans.

Steinman's Deer Isle Bridge has been more problematic. Its elaborate system of stays, cross-stays, and braces proved only partially effective; some studies suggest critical flutter velocities were raised from 40 to just 82 km/hr, much less than the required 175 km/hr (Cai *et al*, 1999). Oscillation inflicted minor damage on several occasions, and in one May 1978 episode, the bridge was even closed. With the bridge having been modified four times without success, engineers festooned it with electronic sensors in 1981. This led to yet another defense when wedge-shaped steel fascias were clipped on the plate girder sides in 1994. The span now resembles a streamlined box girder, albeit without its torsional rigidity. Selberg's Fyksesund Bridge in Norway also received a major overhaul during the late 1970s when its suspended structure was beefed up with a shallow truss to augment the ground-to-deck stays installed in 1945.

Ammann's Bronx-Whitestone Bridge has similarly defied efforts to suppress aerodynamic motion. In November 1968, 112-km/hr winds incited vertical undulation felt by motorists crossing the span. The tower stays and superimposed truss of the 1940s have never fully done the job, because torsional rigidity was not significantly enhanced. With more sophisticated means of assessing the bridge's behavior, the situation warranted re-examination; there was also concern over long-term fatigue from the more than 100,000 cycles of vibration that could occur over the bridge's life. By 1984, the findings suggested a need for remedial work to reduce vertical oscillations expected to peak at about 65 km/hr and to eliminate the threat of torsional vibrations in higher winds. If the midspan deck/cable connection were to break in extreme winds—an albeit remote possibility—engineers feared the span could even suffer the fate of the Tacoma Narrows Bridge. Moreover, it has been speculated that the bridge could succumb were New York's strongest winds—which habitually strike from the south more parallel to the span—to strike the bridge at right angles.

Strengthening the steel connecting the two plate girders was deemed too expensive and not fully effective; streamlined fascias were also thought too expensive. In the end, it was decided to strengthen the midspan deck/cable tie and—for the first time in the suspended structure of a suspension bridge—to wedge counterweights in the form of tuned mass dampers into the plate girder. Four 22-tonne weights placed beneath the roadway at center span on the inside walls of the plate girder were attached to springs and brakes. Activated by winds exceeding 56 km/hr, the dampers rode back and forth along a 73-m length of main span in a motion tuned to—and counteracting—the plate girders' natural twisting frequency. The system constituted less than 0.5% of the bridge weight, but it was designed to dissipate all energy generated by the suspended structure from wind. The system, completely hidden, was installed in the late 1980s at a cost of $2 million.

This remedy would not mark the end of the story. In 1999, Weidlinger Associates, under contract to the Triborough Bridge and Tunnel Authority, proposed the most comprehensive and radical surgery yet. To improve what was determined to be an unacceptably low 80-km/hr flutter instability, the firm plans to completely remove the truss superimposed in 1940, move the lateral bracing to the bottom of the plate girders, and replace the concrete deck with a lighter steel orthotropic deck forming a top lateral system. The overhauled suspended structure, devised with the aid of wind tunnel tests at the University of Western Ontario's Boundary Layer Wind Tunnel, will in effect act as a torsionally stiff open box. Most significant is a reversal of a decision made in the 1980s; wedge-shaped aerodynamic fascias built from fiber-reinforced polymers will be added to the girder edges in the same manner as they have been in Steinman's Deer Isle Bridge. The tuned mass dampers will also be removed. When work begins in 2001, more than 6 decades after the Tacoma Narrows Bridge collapse, this dramatic retrofit should restore the Bronx-Whitestone Bridge closely to its original, and much vaunted, appearance.

18.3 THE LEGACY OF THE TACOMA NARROWS BRIDGE

The Tacoma Narrows Bridge collapse continues to fascinate the public and remind engineers of the need to think beyond contemporary practice. Engineers of the era that produced the Tacoma Narrows Bridge were well versed in the consideration of static wind forces, and the liberation offered by the Deflection Theory created suspension bridges of increasing flexibility and slenderness. Unfortunately, the confidence of engineers verged on an atmosphere of hubris that, in the end, created fallible designs.

Suspension bridges remain the most aerodynamically vulnerable of contemporary bridge forms. Could such a disastrous oversight ever occur again in an engineering world of greater knowledge and vastly more sophisticated analytical, wind tunnel, and computing tools? An understanding of engineering history would not be remiss. John Roebling's success was founded on the recognition of earlier engineering failures and an intuitive grasp of how to avoid them, if that of a later generation of engineers in the early decades of the 20th century was not. The bridge collapses of the 19th century were too far in the past, and it was not easy for succeeding generations of engineers to appreciate the lessons these failures offered for contemporary bridge design.

In 1977, P.G. Sibly and A.C. Walker examined an interesting pattern of major bridge failures extending from 1847 that occurred, on average, at 30-year intervals. Included were the Dee (1847), the Tay (1879), the Quebec (1907), the Tacoma Narrows (1940), and the Milford Haven (1970) bridges. Although Sibly and Walker admitted the pattern could be coincidental, they speculated that the use of new technology and the loss of knowledge of prior experiences—including failures—could have influenced this trend. A loss of communications between generations of engineers, as in the Tacoma Narrows Bridge, was evident in all these collapses. Henry Petroski (1994) extrapolated this trend to suggest that daring span leaps in cable-stayed bridges could, conceivably, around the year 2000, perpetuate this possible trend of failure. Whether this comes to pass is speculation indeed. The recent, well-publicized, and surprisingly perplexing vibration of London's 330-m long Millennium Bridge under its May 2000 opening day crowds is pause for thought. Its main cables feature an extremely low cable sag, hang well beyond the edges of the suspended pedestrian walk, and actually drop below the level of the thin 50-cm deep deck at the mid-point of the 144-m main span. The bridge, designed by engineering firm Ove Arup, with architect Norman Foster and sculptor Anthony Caro, embodied a deliberately leading-edge approach to suspension bridge design, with results that were not predicted.

The tools, design techniques, and suspended bridge forms—truss, streamlined box girders, split boxes, cable-stiffening systems, and others—now emerging in the design of ultralong-span bridges would astonish the likes of Thomas Telford and Roebling, and perhaps even Steinman, Ammann, and Gilbert Roberts. Dischinger-inspired hybrid suspension/cable-stayed forms, and even pure cable-stayed bridges, may challenge the long-held hegemony of the suspension bridge. Whatever this future holds, the vivid and haunting imagery of the Tacoma Narrows Bridge collapse will be indelibly etched in the minds of long-span bridge engineers for generations to come. The Tacoma Narrows Bridge will forever influence the practice of their craft.

ENDNOTES

1. This conference, now called the International Conference on Wind Engineering, is held every 4 years. The second conference (in 1967) was held in Ottawa, Canada; the 11th is planned for Lubbock, Texas in 2003.
2. Klöppel had designed one of the four finalists in the 1959-60 competition to span the Tagus at Lisbon and would later perform the independent design check on Freeman Fox & Partners' design for the first Bosporus Bridge at Istanbul.
3. Davenport (1996) has since suggested that the measurement of wind pressure, rather than wind velocity, is more appropriate, as it correlates more directly to wind loading across the length of a span. Scanlan (1990) suggested that the relationship of the aerodynamic admittance and flutter derivatives may lead to what he calls a "signature buffeting theory."
4. Interestingly, Scanlan emerged out of the field of aeronautics, as had Scruton. During World War II, Scanlan became Chief of Aerolasticity at the American aircraft manufacturer Republic Aviation and later wrote *Aircraft Vibration and Flutter*, considered a major contribution to the early development of that field. Scanlan later went on to teach at Princeton and Johns Hopkins universities. His book *Wind Effects on Structures* (with Emil Simiu) is also regarded as a classic.

REFERENCES

(full listings are contained in the Bibliography)

Chapter 1

page

2 Hopkins 1970, 176.
Peters 1987, 30.
Ammann 1923, 1073.

3 Petroski 1994, 130.
Ammann 1923, 1073.

5 Peters 1987, 201.

8 Petroski 1994, 131.
Buonopane and Billington 1993, 958.

9 Steinman and Watson 1957, 217 (There is not one good...).
Ibid., 221 (Bridges of half a...).
Billington 1979, 681.
Henry and Jerome 1965, 20.
Plowden 1974, 76.

10 Ibid., 80, 117.
Steinman and Watson 1957, 223 (no competent builder...).
Hopkins 1970, 221

11 Steinman and Watson 1957, 234

12 Billington 1977, 1656
Plowden 1974, 136
Kemp 1993, 24
Steinman and Watson 1957, 210.

Chapter 2

15 Reier 1977, 29.
Ammann 1923, 1073.

16 Moisseiff 1937, 1342.
Steinman 1934, 1134.

17 Moisseiff 1939, 47.
Condit 1961, 131.

18 Ammann 1929, 800.

22 Billington 1977, 1655.

23 Ammann 1933, 38 (very shallow...).
Ammann 1953, 235 (excessive...).

24 Ammann 1933, 41 (The system...).
Ibid., 43 (bridge as a whole...).
Doig and Billington 1994, 552-53.
Buonopane and Billington 1993, 972.
Steinman 1929a, 12.
Condit 1961, 139.

25 Le Corbusier 1947, 75.
Petroski 1994, 153.

26 Plowden 1974, 242.

27 Seim 1987, 47.

28 Halprin 1966, 48.
Strauss 1938, 109.
Moisseiff 1939, 207.

Chapter 3

30 Leonhardt 1982, 287.

31 Leonhardt 1966, 126.
Moisseiff 1939, 48.

32 Petroski 1994, 145-46.

33 Mock 1949, 60.

35 Ratigan 1959, 244-45.

37 Anonymous 1940b, 58.
Ammann 1939a, 218 (It is now...).

39 Ibid., 218 (It was the aim...).
Ibid., 219 (architectural...).
Embury 1938, 87.

40 Ammann 1939b, 626.

Chapter 4

42 Bd. of Eng. 1941, App. II-1 (the towers...).
Ibid., App. II-2 (at great cost).
Ibid., App. II-2 (may practically...).
Ibid., App. IV-6 (We have full...).
Ibid., App. IV-6 (It might seem...).

43 Ibid., App. IV-7 (In a long narrow...).
Ibid., App. IV-7 (This feature...).
Ibid., App. IV-8 (We believe...).
Ibid., App. IV-3 (In view of...).

44 Ibid., App. IV-3 (reasonably...).
Ibid., App. IV-4 (Condron).
Ibid., App. IV-4 (To obtain a...).
Ibid., App. IV-4 (rather...).
Ibid., App. IV-5 (advisable to...).
Ibid., App. IV-5 (is probably...).

45 National Youth Admin. 1940, 2.
47 Farquharson 1949, 41.
Andrew 1949, 956.
Farquharson 1940, 445.
Hopkins 1970, 231-32.
49 Bd. of Eng. 1941, App. V-3 (in the customary…).
Ibid., App. V-14 (alternately…).
Ibid., App. V-10 (a series of…).
50 Ibid., App. V-10 (while checking…).
51 Ibid., App. V-10-11 (on reaching…).
Ibid., App. V-11 (began to shift…).
53 Robson 1994, 187.
54 von Kármán 1940, 40.
55 Farquharson 1949, 13.
Russell 1841 (in Farquharson 1949, 84).
56 Finch 1941, 75.
Rocard 1957, 95-96.
Larsen 1999, 56-57.
57 Podolny and Scalzi 1986, 287.
Steinman and Watson 1957, 364.
Finch 1941, 78.
Kranakis 1997, 151.
58 Billington 1979, 681.
Farquharson 1949, 14.
59 Carriere et al 1966, 356.
Bd. of Eng. 1941, 71.
Ammann 1953, 236.
Ibid., 236.
60 Ratigan 1959, 248.
61 Steinman 1941a, 36.
Ammann 1941, 36 (It must be…).
Ibid., 37 (such a complex problem).
62 Ibid., 37 (It would be of…).
Ibid., 37 (constant watching).
Steinman 1941b, 60 (a sincere…).
Ibid., 60 (bring main span …).
Ibid., 60 (be obvious…).
Ibid., 60 (admission).
Ibid., 61 (by his own criteria…).
Ibid., 61 (I am not advertising…).
63 Ammann 1946, 103.
Anon. 1941, 79.
64 Bd. of Eng. 1941, 64 (That under…).
Ibid., 137 (an extraordinarily…).
Ibid., 132 (It must not be…).
Steinman 1946a, 21.
65 Bd. of Eng. 1941, 89 (greatly overestimated…).
Ibid., 73 (give no clue…).
Ibid., Concl. 2 (The excessive…).
Anon. 1941, 77, 79.
66 Steinman 1941c, 59 (valuable contribution…).
Ibid., 59-60 (the body of the…).
Ibid., 60 (In the case of…).
Ibid., 61 (of the Whitestone…).
Ibid., 61 (I submit that…).
Ibid., 61 (leaves many questions…).
67 Bd. of Eng. 1941, App. VI-1 (even more complex…).
von Kármán and Edson 1967, 213.
Bd. of Eng. 1941, 115 (The mechanism…).
Ibid., 130 (It is very improbable…).
68 Ibid., 131 (It seems that…).
Ibid., Sum. of Concl. 3 (There is…).
Ibid., 129 (explained satisfactorily…).
Ibid., 103 (will be amplified…).
Ibid., Sum. of Concl.13 (Further…).
von Kármán and Edson 1967, 214.
Ibid., 213 (the structural…).
69 Farquharson 1941, 37.
Steinman 1945c, 504.
Petroski 1994, 144.
Steinman and Watson 1957, 366.
70 Rubin 1985, 35.

Chapter 5

74 Steinman 1941d, 51.
75 Farquharson 1952a, 113 (interesting suggestions).
Ibid., 113 (The effect of opening…).
Maney 1941, 64.
76 Modjeski & Masters 1941, 99.
77 Hadley 1943, 317 (How wind…).
Ibid., 317 (the extraordinary…).
78 Andrew 1949, 955.
Farquharson 1952a, 11.
79 Farquharson 1952b, 147 (a rapid…).
Farquharson 1952a, 42.
Ibid., 87.
Farquharson 1952b, 148.
80 Vincent 1965, 489.
Vincent 1951, 34.
Smith and Vincent 1950, 56.
81 Bleich *et al* 1950, 1.
Scanlan and Sabzevari 1967, 4.
Davenport 1966, 85.
Farquharson 1954, 12.
82 Andrew 1945, 49.

von Kármán and Dunn 1952, 59.
84 Balog 1945, 523.
Rocard 1957, 185.
Steinman 1954b, 69.
Steinman 1945d, 558.
Vincent 1952, 23 (in full...).
Ibid., 36 (more research is needed).
85 Rocard 1957, 185.
Steinman 1946a, 22 (assured...).
Ibid., 22 (helpful in resisting...).
Steinman 1946b, 67.
86 Steinman 1946c, 462 (There is no...).
Ibid., 462 (The provision of...).
Ibid., 462 (It would be...).
Farquharson 1947a, 90 (a dangerous oversimplification).
Ibid., 90 (it would be most...).
Steinman 1947a, 20.
87 Steinman 1947b, 79.
Steinman 1946a, 23.
Steinman 1947a, 20.
Steinman 1947c, 80 (of assured...).
Farquharson 1947b, 44.
Steinman 1954a, 410.
Steinman 1947c, 80 (the public...).
88 Farquharson 1950, 1242.
Steinman 1950, 1245.
Vincent 1954, 86.
Ammann 1953, 231.
89 Ibid., 240 (inadequately explored...).
Ibid., 245 (more unfavorable).
Ibid., 238 (Greater caution...).
91 Billah and Scanlan 1991, 122.
Scanlan and Vellozzi 1980, 252.
92 Billah and Scanlan 1991, 119.
93 Larsen 1999, 60.
Farquharson 1947b, 44.
94 Anon. 1990, 57.
Frandsen 1966, 610.
Ito and Nakamura 1982, 41.
Ratigan 1959, 254.
Richardson 1984, 731.
Davenport *et al* 1969, 41-42.

Chapter 6
96 Birdsall 1944, 69.
98 Mock 1949, 63.
Andrew 1949, 959 (without...one...).
Ibid., 961 (set up a condition...).
104 Anon 1969, 61.
107 Tunnard and Pushkarev 1963, 240.
108 Scanlan and Vellozzi 1980, 259.
Balog 1945, 527.

Chapter 7
111 Modjeski & Masters 1940, 12.
112 Rubin 1985, 9.
113 Ibid., 35.
120 Brumer and Hanson 1956, 16.
122 Tunnard and Pushkarev 1963, 235.
123 Steinman 1929b, 51 (Some designers...).
Ibid., 53 (subterfuge and evasion).
Ibid., 51 (No attempt...).
Ibid., 53 (not by ornamentations...).
Ibid., 51 (peal out anthems).
Ibid., 51 (symbol of Liberty...).
124 Pfeiffer 1978, unknown.
126 Cohen and Stahl 1978, 1030.
129 Furnas *et al* 1966, 121.
Anon. 1979, 109.
Leonhardt 1982, 290.
Stahl 1979, 5.
130 Plowden 1974, 242.
Condit 1961, 138.

Chapter 8
133 Troitsky 1967, 38.
Westhofen 1890, 1.
143 Frazer 1951, 271.
Walshe 1967, 94.
145 Kerensky 1967, 132.
147 Shirley-Smith 1967, 184.
154 Parsons 1988, 4/2.
Heywood and Birchall 1987, 22.
155 Anon. 1964b, 88 (The great ...).
Ibid., 88 (Are the British...).
Ammann 1964, 10 (must be...).
Ibid., 10 (I am confident...).
Gronquist 1964, 22.
156 Roberts 1964, 8.
157 Brumer 1965, 47.
Chang and Cohen 1979, 32.
158 Anon. 1998, 4.
Crosthwaite 1969, 548.

Chapter 9
161 Leonhardt 1982, 292.
162 Leonhardt 1966, 137.
165 Vandepitte 1966, 555.
166 Cain 1975, 736.
169 Schlaich 1990, 307.

Chapter 10
193 Scanlan and Vellozzi 1980, 250.
200 Minshall 1966, 413.
Leonhardt 1966, 145.

Chapter 11
205 Ostenfeld and Jønson 1966, 155.
209 Cohen *et al* 1980, 30, 33.

Chapter 12
222 O'Neill 1982a, 13 (no new knowledge…).
Ibid., 14 (important).
223 O'Niel 1982b, 13.
Anon. 1983a, 4 (plucked out of …).
Ibid., 4 (a freak extreme loading).
Brown 1983, 12.
Anon. 1983c, 13.
224 Middelboe and Ferguson 1984, 5.
Anon. 1984, 17.
225 Heywood and Birchall 1987, 23.
Boyd 1987, 9.
Brown 1983, 12.
227 Anon. 1978, 17.
Anon. 1983d, 11.
228 Anon. 1983b, 27.
229 Richardson 1981, 28.
Heywood and Birchall 1987, 24.
Anon. 1986, 21.
230 Barrett 1985, 17 (foolish).
Ibid., 17 (stick in the gullet).
Heywood and Birchall 1987, 24.
Byrd 1986, 30.
232 Heywood 1988, 36.
233 Anon. 1987, 5.
Goltz and Reina 1988, 11.
237 Homberg 1990, 49.

Chapter 13
243 Balog 1943, 876.
259 Bolton 1995, 7.
265 Frankland 1934, 105.
Selberg 1950, 184, 191.

Chapter 14
286 Billington 1990, 50 (too imitative…).
Ibid., 50 (delicate tracery…).
301 O'Neill 1993, 33 (Japanese engineers…).
Ibid., 33 (when you can…).
302 Billington 1990, 52-53.

Chapter 16
323 Steinman 1953, 6.

Chapter 17
331 Steinman 1946b, 66.
Ammann 1933, 37.
332 Astiz 1996, 1060.
Diana *et al* 1998, 132.
Richardson 1984, 730.
334 Kerensky 1959, 200.
Rocard 1957, 171.
336 Matthews 1995, A12.

Chapter 18
345 Larsen 1998, 73.
Kerensky 1959, 203.
Walshe 1967, 92.
Pugsley 1968, 132.
346 Selberg 1950, 183.
Selberg 1966, 365.
347 Farquharson 1946, 345.
Andrew 1945, 49.
Davenport 1966, 87.
Davenport *et al* 1971, 1217.
348 Scanlan 1979, 8.
Cai *et al* 1999, 182.
349 Scanlan 1990, 65.
Ibid., 64.
Davenport *et al* 1979, 814.
Scanlan and Vellozzi 1980, 262.
350 Scanlan 1983, 2832-33.
Davenport *et al* 1971, 1217.
Scanlan 1990, 69-70.
Larsen 1998, 74.
Davenport 1996, 1037.
Farquharson 1949, 5.
Steinman 1947b, 79.
Larsen 1999, 57.
351 Dicker 1966, 56.
Larsen 1999, 56.
Scanlan and Tomko 1971, 1733.
Scanlan 1990, 64.
Buckland and Wardlaw 1972, 12.
Wyatt 1992, 320.
von Kármán and Dunn 1952, 59.
Larsen 2000a, 244.
354 Larsen 2000b, 46.
356 Petroski 1994, 176.
357 Davenport 1996, 1034.
Scanlan 1990, 70

BIBLIOGRAPHY

The preparation of this book has involved the consultation of hundreds of individual sources of information, ranging from books, journal articles, and conference proceedings to technical literature from bridge agencies, engineering consultants, contractors, and others.

Only a partial listing of source material is possible for this bibliography. In consequence, the bibliography focuses on journals and technical reports consulted in the course of research.

The bibliography begins with a list journals consulted and then identifies specific items, either references cited in the text or publications having special value for source information.

JOURNALS

The journals are listed alphabetically; the language of non-English language journals is listed in parentheses, as is the country of origin when an identically titled journal published in another nation exists:

AASHTO Quarterly, *Acier Stahl-Steel*, *Aluminium* (in German), *Annual Report of Roads* (Japan), *APWA Reporter*, *Der Bauingeneiur* (in German), *Bridge Design and Engineering*, *Bridge Engineering* (in Japanese), *Bridge Update*, *Building in Steel*, *Bygningsstatiske Meddelelser* (in Norwegian), *Cement Beton* (in French), *Civil Engineering* (Great Britain), *Civil Engineering and Public Works Review*, *Civil Engineering ASCE*, *Civil Engineering in Japan*, *Civil Engineer International*, *Concrete Construction*, *Concrete International*, *Concrete Journal* (in Japanese), *Construction Today*, *Costruzioni metalliche* (in Italian), *COWINYT*, *Der Stahlbau* (in German), *Die Bautechnik* (in German), *The Economist*, *The Engineer* (Great Britain), *Engineering* (Great Britain), *Engineering and Contract Record*, *Engineering Journal*, *Engineering Structures*, *Engineering News-Record (ENR)*, *Le Freyssinet Magazine*, *Génie Civil* (in French), *Heavy Construction News*, *The Highway Engineer*, *History and Technology*, *History of Technology*, *Hong Kong Engineer*, *IABSE Bulletin*, *IABSE Structures*, *IHI Engineering Review*, *Ingénieurs et Architectes Suisses* (in French), *International Construction*, *International Highways*, *Japanese Railway Engineering*, *Journal for the Society for Industrial Archeology*, *Journal of Bridge Engineering ASCE*, *Journal of the Construction Division ASCE*, *Journal of the Franklin Institute*, *Journal of the Structural Division ASCE*, *Journal of Wind Engineering and Industrial Aerodynamics*, *Kobelco Technology Review*, *Mitsubishi Technical Review*, *Nature*, *New Civil Engineer*, *New Scientist*, *News from Storebælt*, *Newsletter on Long-Span Bridges*, *Nippon Steel Technical Report Overseas*, *Pacific Builder and Engineer*, *PCI Journal*, *Popular Science*, *Proceedings ASCE*, *Proceedings ICE*, *Public Innovation Abroad*, *Public Works Review*, *The Rhodesian Engineer*, *Roads and Engineering Construction*, *Roads and Streets*, *Roads in Japan*, *Routes*, *Rural and Urban Roads*, *Scientific American*, *The Scientific Monthly*, *Steel Structures in Japan*, *The Structural Engineer*, *Structural Engineering International IABSE*, *Structural Engineering Review*, *Technology Review*, *Transactions ASCE*, *Transportation Research Record*, *Travaux* (in French), *The Wheel Extended*, *World Construction*, *World Highways*.

SOURCE INFORMATION

Addis, W. 1994. Design revolutions in the history of tension structures. *Structural Engineering Review* 6: 1-10.

Alexander, D. 1990. Wind power. *New Civil Engineer,* 1 November: 26-29.

American Society of Civil Engineers. 1979. *Long Span Suspension Bridges: History and Performance*. Sessions No. 70 and 78, Preprint 3590. Boston: ASCE Convention and Exposition, April 2-6.

Ammann, O. H. 1923. Possibilities of the modern suspension bridge for moderate spans. *Engineering News-Record (ENR),* June 21: 1072-1078.

———. 1929. Suspension bridges revised (a review of David Steinman's *A Treatise on Suspension Bridges*). *ENR,* May 16: 800.

———. 1933. George Washington Bridge: General conception and development of design. *Proceedings of the American Society of Civil Engineers* 59, *Transactions of the American Society of Civil Engineers* 97, No. 8, Part 2, October: 1-65. Discussion by G. Lindenthal. October: 422-425.

———. 1939a. Planning and design of Bronx-Whitestone Bridge. *Civil Engineering—ASCE,* April: 217-220.

———. 1939b. Discussion of *Preliminary Design of Suspension Bridges*, by Shortridge Hardesty and Harold E. Wessman. *Transactions of the American Society of Civil Engineers* 104, October: 624-626.

———. 1941. Letter to the editor. *ENR,* February 27: 36-37.

———. 1946. Additional stiffening of Bronx-Whitestone Bridge. *Civil Engineering—ASCE,* March: 101-103.

———. 1953. Present status of design of suspension bridges with respect to dynamic wind action. *Journal of the Boston Society of Civil Engineers* 40: 231-255.

———. 1964. Letter to the editor. *ENR,* July 2: 10.

Andrew, C. E. 1945. Design of a suspension superstructure to replace the former Tacoma Narrows Bridge—Part 1. *Pacific Builder and Engineer,* October: 43-49.

———. 1949. Unusual design problems—Second Tacoma Narrows Bridge. *Transactions of the American Society of Civil Engineers* 114: 955-978.

Anonymous. 1940a. Stays and brakes check oscillation of Whitestone Bridge. *ENR,* December 5: 54-56.

———. 1940b. Two recent bridges stabilized by cable stays. *ENR,* December 5: 56-58.

———. 1941. Why the Tacoma Narrows Bridge failed. *ENR,* May 8: 75-79.

———. 1964a. Advances in bridge design, editorial. *ENR,* May 7: 92.

———. 1964b. On suspension bridges, editorial. *ENR,* May 28: 88.

———. 1965. Can the old spinning wheel retire? *ENR,* November 25: 15-16.

———. 1969. Stiffening systems vs. aerodynamics. *Civil Engineering—ASCE* June: 61.

———. 1978. Freeman Fox wins Bosporus job. *ENR,* January 12: 17.

———. 1979. Designer's innovations span industry spectrum *ENR,* April 12: 106,109.

———. 1983a. Designer irate over Severn assessment. *New Civil Engineer,* 19 May: 4.

———. 1983b. Bridge-deck twists go straight. *ENR,* July 14: 26-27.

———. 1983c. New Severn Bridge scare. *ENR,* November 10: 13.

———. 1983d. Big bout about bridge bids. *ENR,* November 24: 11.

———. 1984. Severn strengthening set. *ENR,* March 1: 17.

———. 1986. Yen for steel wins Bosporus contract. *New Civil Engineer,* 29 May: 21.

———. 1987. Third Bosporus crossing seems doubtful. *ENR,* November 19: 5.

———. 1990. Threat of war brings prosperity. *ENR,* January 4: 57.
———. 1998. Severn Bridge listed. *New Civil Engineer*, 4 June: 4-5.
Aoki, M. 1984. Marshal Mobutu Sese Seko Bridge (Matadi Bridge). *Civil Engineering in Japan /'84*, JSCE: 58-71.
Appleton, J. 1990. To please or not to please? *Landscape Design,* October: 16.
Association Française Pour la Construction (AFPC). 1994. *International Conference A.I.P.C.—F.I.P.: Cable-Stayed and Suspension Bridges.* Deauville, France: AFPC, October 12-15.
Astiz, M. A. 1996. Wind related behaviour of alternative suspension systems. *15th Congress, International Association for Bridge and Structural Engineering.* Copenhagen: International Association for Bridge and Structural Engineering: 1079-1090.
Balog, L. 1943. Letter to the editor. *ENR,* December 16: 74-75.
———. 1945. Discussion. *Transactions of the American Society of Civil Engineers* 110: 522-532.
Barrett, N. 1985. Bosporus brings new loans pressure. *New Civil Engineer,* 23 May: 17
Billah, K. Y., and R. H. Scanlan. 1991. Resonance, Tacoma Narrows Bridge failure, and undergraduate physics textbooks. *American Journal of Physics* 59(2), February: 118-124.
Billington, D. P. 1977. History and esthetics in suspension bridges. *Journal of the Structural Division*, ASCE, August 1977: 1655-1672. With discussion (January 1978: 246-249; February 1978: 378-380; March 1978: 619; April 1978: 732-734; June 1978: 1027-1035; July 1978: 1174-1176) and closure by Billington (March 1979: 671-687).
———. 1990. Creative connections: Bridges as art. *Civil Engineering—ASCE,* March: 50-53.
Birdsall, B. 1944. Construction record set on Alcan Suspension Bridge. *ENR,* January 13: 64-69.
Bleich, F. 1948. Dynamic instability of truss-stiffened suspension bridges under wind action. *Proceedings of the American Society of Civil Engineers* 74, October: 1269-1314.
Bleich, F. M., C. B. McCullough, C. B. Rosecrans, and G. S. Vincent. 1950. *The Mathematical Theory of Vibration in Suspension Bridges*. Washington, D.C.: Bureau of Public Roads, U.S. Department of Commerce, Government Printing Office.
Board of Engineers. 1941. The failure of the Tacoma Narrows Bridge. *The Failure of the Tacoma Narrows Bridge, A Reprint of Original Reports.* Bulletin of the Agricultural and Mechanical College of Texas, Fourth Series, Vol. 15, No. 1, Bulletin No. 78, January 1, 1944.
Bolton, A. 1995. Great Belt dampers ease drivers fears. *Civil Engineer International*, April: 7.
Boyd, E. L. 1987. Letter to the editor. *ENR,* July 23: 9.
British Bridge Builders. 1981. *Bridging the Humber*. York: Cerialis Press.
Brown, W.C. 1980. Long-span suspension bridges: A British approach. *Long-Span Bridges, O.H. Ammann Centennial Conference.* Annals of the New York Academy of Sciences, Vol. 352: 1-26.
———. 1983. Letter to the editor. *ENR,* September 29: 11-12.
Brown, W.C., M.F. Parsons, and H.S.G. Knox. 1976. *Bosporus Bridge, Design and Construction*, London: Institution of Civil Engineers.
Brumer, M. 1965. U.S. vs. British suspension bridge design. *ENR,* September 23: 44-47.
Brumer, M., and C. W. Hanson. 1956. Development and design of the Walt Whitman Bridge. *Journal of Structural Division*, ST 4, Paper 1019, ASCE, July:1-39.
Buckland, P. G., and R. L. Wardlaw. 1972. Some aerodynamic considerations in bridge design. *Engineering Journal,* April: 10-16.

Buonopane, S. G., and D. P. Billington. 1993. Theory and history of suspension bridge design from 1823 to 1940. *Journal of Structural Engineering.* ASCE, Vol. 119, No. 3, March: 954-977.

Burden, A. 1991. Modern Japanese suspension bridge design. *Proceedings Institution of Civil Engineers* (90), Part 1, February: 157-177.

Byrd, T. 1986. Brown builds Bosporus Two. *New Civil Engineer,* 11 December: 30-32.

Cai, C. S., P. Ablrecht, and H. R. Bosch. 1999. Flutter and buffeting analysis II: Luling and Deer Isle bridges. *Journal of Bridge Engineering,* ASCE, August: 181-188.

Cain, J. F. 1975. Fifty-year development—Construction of steel pipeline suspension bridges. *Journal of the Construction Division*, ASCE, Vol. 101, CO4, December: 733-749.

Carriere, R., F. Huguet, and H. Bachelart. 1966. Les ponts suspendus modernes. *International Symposium on Suspension Bridges*, Lisbon: 347-360.

Chang, F-K., and E. Cohen. 1979. Overview on new and current trends for long span bridge structures. *Long Span Suspension Bridges: History and Performance.* ASCE Preprint 3590: 1-51.

Chapin, F. D. 1941. Letter to the editor. *ENR,* May 22: 42-43.

Chow, C. K., R. Hodges, and C. K. Lau. 1982. The planning and design of the Lantau fixed crossing. *Hong Kong Engineer*, March: 13-21.

Cohen, E., and B. Birdsall, eds. 1984. *Long-Span Bridges, O.H. Ammann Centennial Conference.* Annals of the New York Academy of Sciences, Vol. 352.

Cohen, E., A. M. Custen, and F. L. Stahl. 1980. Long-span suspension bridges: The American approach. In Cohen, E., and B. Birdsall, eds. *Long-Span Bridges, O. H. Ammann Centennial Conference.* Annals of the New York Academy of Sciences, Vol. 352: 27-39.

Cohen, E., and F. L. Stahl. 1978. Discussion. *Journal of the Structural Division*, ASCE, ST6, June: 1027-1030.

Condit, C. W. 1961. *American Building Art—The Twentieth Century.* New York: Oxford University Press.

Crossthwaite, C. D. 1969. Discussion on Severn Bridge. *Proceedings Institution of Civil Engineers,* April: 548-549.

Davenport, A. G. 1962. Buffetting of a suspension bridge by storm winds. *Journal of Structural Division, Proc. ASCE*, ST3, June: 233-268.

———. 1966. The action of wind on suspension bridges. *International Symposium on Suspension Bridges*, Lisbon: 79-100.

———. 1996. The vulnerability of long span bridges to wind. *15th Congress, International Association for Bridge and Structural Engineering.* Copenhagen: International Association for Bridge and Structural Engineering: 1031-1042.

Davenport, A. G., N. Isyumov, D. J. Fader, and C. F. P. Bowen. 1969. *A Study of Wind Action on a Suspension Bridge During Erection and on Completion.* BLWT-3-69, Boundary Layer Wind Tunnel, University of Western Ontario.

Davenport, A. G., N. Isyumov, and T. Miyata. 1971. The experimental determination of the behaviour of suspension bridges in turbulent wind. *Proceedings Third International Conference on Wind Effects on Buildings and Structures*, Tokyo: 1207-1219.

Davenport, A. G., N. Isyumov, H. Rothman, and H. Tanaka. 1979. Wind induced response of suspension bridges – wind tunnel model and full scale observations. *Proceedings 5th International Conference on Wind Engineering*, Fort Collins: 807-824.

Delcamp, A. 1960. The Tancarville suspension bridge (France). *Acier Stahl-Steel*, April, No. 4: 145-157.

Dicker, D. 1966. Aerodynamic stability of H sections. *Journal of Engineering Mechanics Division, Proc. ASCE*, E 2, April: 1-23.

Diana, G., S. Bruni, A. Collina, and A. Zasso. 1998. Aerodynamic challenges in super long span bridges design. In Larsen, A., and S. Esdahl, eds. *Proc. International Symposium on Advances in Bridge Aerodynamics*. 10-13 May, 1998, Copenhagen. Rotterdam: Balkema: 131-143.

Doig, J., W., and D. P. Billington. 1994. Ammann's first bridge: A study in engineering, politics, and entrepreneurial behavior. *Technology and Culture* 35: 537-570.

Dorton, R. A. 1968. Design of the Narrows suspension bridge at Halifax. *Engineering Journal* (Canada), December: 15-26.

Dudra, J. 1966. Design and construction of Hudson Hope Bridge. *PCI Journal*, April: 52-72.

Durkee, J.L. 1966. Advancements in suspension bridge cable construction. *International Symposium on Suspension Bridges*, Lisbon: 425-449.

Dwyer, J.D. 1975. Fifty-year development: Construction of steel suspension bridges. *Journal of the Construction Division*, ASCE, Vol. 101, No. CO1, March: 105-125.

Embury II, A. 1938. Esthetics of bridge anchorages. *Civil Engineering—ASCE,* February: 85-89.

Farquharson, F. B. 1940. A dynamic model of the Tacoma Narrows Bridge. *Civil Engineering—ASCE,* July: 445-447.

———. 1941. Letter to the editor. *ENR,* July 3: 37.

———. 1946. Lessons in bridge design taught by aerodynamic studies. *Civil Engineering—ASCE,* August: 344-345.

———. 1947a. Letter to the editor. *Civil Engineering—ASCE,* February: 90.

———. 1947b. Letter to the editor. *Civil Engineering—ASCE,* July: 44.

———. 1949. Part I: Investigations prior to October, 1941. *Aerodynamic Stability of Suspension Bridges with Special Reference to the Tacoma Narrows Bridge*, Bulletin No. 116, University of Washington, Engineering Experiment Station.

———. 1950. Discussion. *Transactions of the American Society of Civil Engineers* 115: 1237-1242.

———. 1952a. Part III: The investigation of models of the original Tacoma Narrows Bridge under the action of wind. *Aerodynamic Stability of Suspension Bridges with Special Reference to the Tacoma Narrows Bridge*, Bulletin No. 116, University of Washington, Engineering Experiment Station.

———. 1952b. Model verification of the classical flutter theory as adapted to the suspension bridge. Extract from *Mémoires,* International Association for Bridge and Structural Engineering: 147-166.

———. 1954. Part IV: Model investigations which influenced the design of the new Tacoma Narrows Bridge. *Aerodynamic Stability of Suspension Bridges with Special Reference to the Tacoma Narrows Bridge*, Bulletin No. 116, University of Washington, Engineering Experiment Station.

Finch, J. K. 1941. Wind failures of suspension bridges or evolution and decay of the stiffening truss. *ENR,* March 13: 74-79.

Finzi, L. 1978. Structural problems for the Messina Strait Bridge, *Transportation Research Record* 665, September: 131-139.

Flint, A. 1973. How long a span? *Building with Steel,* May: 6-8.

Frandsen, A. G. 1966. Wind Stability of Suspension Bridges: Application of the Theory of 'Thin Airfoils'. Paper No. 43. *International Symposium on Suspension Bridges*, Lisbon: 609-627.

Frankland, F. H. 1934. *Suspension Bridges of Short Span*. New York: American Institute of Steel Construction.

Frazer, R. A. 1951. Aerodynamic oscillations of suspension bridges. *Engineering* 171, No. 4440, March 2: 270-271.

Freeman, R. 1942. Letter to the editor. *ENR,* April 23: 59.

Furnas, C. C., McCarthy, J., and the editors of Life. 1966. *The Engineer*. New York: Time, Inc.

Gabinete da Ponte sobre o Tejo. 1966. *The Salazar Bridge*. Lisbon: Ministério das Obras Públicas.

Gasparini, D. A., J. M. Spivey, S. G. Buonopane, and T. E. Boothby. 1999. Stiffening suspension bridges. In Kemp, E., ed. *Proc. International Conference on Historic Bridges to Celebrate the 150th Anniversary of the Wheeling Suspension Bridge, October 21-23, 1999*, Morgantown: West Virginia University Press: 105-115.

Gimsing, N. J. 1997. *Cable-Supported Bridges: Concept and Design,* 2nd ed. New York: John Wiley & Sons.

———. 1998. Long Span Cable Supported Bridges: Present Technology and Trends. *Symposium on Long-Span and High-Rise Structures.* Kobe, Japan: International Association for Bridge and Structural Engineering: 103-114.

———., ed. 1998. *The Storebælt Publications – East Bridge*. Copenhagen: A/S Storebæltsforbindelsen.

Gimsing, N. J., and A. B. Sørensen. 1983. *Investigations into the Possibilities of Constructing Bridges with a Free Span of 3000m*. Series R, No. 168, Department of Structural Engineering, Technical University of Denmark.

Goltz, T., and P. Reina. 1988. Levantine intrigue precedes third Bosporus bridge award. *ENR,* February 18: 11.

Gronquist, C. H. 1964. Letter to the editor. *ENR,* July 2: 10, 22.

Hadley, H. M. 1943. Letter to the editor. *ENR,* August 26: 317.

Halprin, L. 1966. *Freeways*. New York: Reinhold.

Harrison, C. O. 1941. Letter to the editor. *ENR,* September 11: 343.

Henry, D., and J. A. Jerome. 1965. *Modern British Bridges*. London: C.R. Books.

Heywood, P. 1988. Bosporus Strait is spanned quickly. *ENR,* May 12: 36-38.

Heywood, P., and J. Birchall. 1987. Design analysis doubts strength of recent bridges. *ENR,* May 28: 22-24.

Hirai, A., I. Okauchi, and T. Miyata. 1966. On the behaviour of suspension bridges Under Wind Action. Paper No. 8. *International Symposium on Suspension Bridges*, Lisbon: 249-256.

Homberg, H. 1990. Call for expressions of interest—Lantau Fixed Crossing, Tsing Ma Bridge, Hong Kong, *New Civil Engineer,* 24 May: 48-49.

Hong Kong Institution of Engineers (HKIE). 1995. *Proceedings Bridges into the 21st Century.* Hong Kong: HKIE, 2-5 October.

Honshu-Shikoku Bridge Authority. 1998. *The Akashi-Kaikyo Bridge: Design and Construction of the World's Longest Bridge*. Kobe.

Hopkins, H. J. 1970. *A Span of Bridges.* Newton Abbot: David & Charles.

International Association for Bridge and Structural Engineering. 1998. *Long-Span and High-Rise Structures.* Kobe, Japan: September 2-4.

Ito, M. 1999. Large bridge projects in the Far East. *Long-Span Bridges and Aerodynamics.* Toshio, M, Fujisawa, N, and H. Yamada, eds., Springer Verlag, Tokyo: 40-55.

Ito, M., and Y. Nakamura. 1982. Aerodynamic stability of structures in wind. *Periodica* 2/1982. International Association for Bridge and Structural Engineering May: 33-56.

Jakkula, A. A. 1941. *A History of Suspension Bridges in Bibliographical Form*. Bulletin of the Agricultural and Mechanical College of Texas, Fourth Series, Vol. 12, Engineering Experiment Station, Series No. 57, College Station, Texas.

Kamai, K, T. Maruyama, and H. Tanaka. 1992. Konohana Bridge, Japan. *Structural Engineering International* 1/92: 4-6.

Kemp, E. L. 1993. National styles in engineering: The case of the 19th-century suspension bridge. *Journal of the Society of Industrial Archeology,* 19(1): 21-36.

Kerensky, O. A. 1959. The Maitland Lecture 1959: Long span suspension bridges. *The Structural Engineer,* July: 199-219.

———. 1967. Discussion, Forth Road Bridge. *Proceedings Institution of Civil Engineers,* January: 130-133.

Kiran, A. B. 1987. Pont sur le détroit de Messine. *Ingénieurs et architectes suisses*, No. 23, 5 novembre: 417-421.

Kircher, A. 1667. *China Monumenta qua Sacris qua Profanis Illustrata.* Amsterdam: Joannem Janssonium.

Kranakis, E. 1997. *Constructing a Bridge: An Exploration of Engineering Culture, Design, and Research in Nineteenth-Century France and America.* Cambridge: MIT Press.

Krokeborg, J., ed. 1986, 1990, and 1994. *Strait Crossings*. Rotterdam: Balkema.

Laboratório Nacional de Engenharia. 1966. *International Symposium on Suspension Bridges*, Lisbon.

Larsen, A., ed. 1992. *Aerodynamics of large bridges, Proc. first international symposium on aerodynamics of large bridges.* 19-21 February, 1992, Copenhagen. Rotterdam: Balkema.

———. 1998. Advances in aeroelastic analyses of suspension and cable-stayed bridges. *Journal of Wind Engineering and Industrial Aerodynamics* 74/76: 73-90.

———. 1999. A 1999 view on aeroelastic stability—A driving force in bridge aerodynamics. In Larsen, A., G. L. Larose, and F. M. Livesey, eds. *Wind Engineering into the 21st Century: Proceedings of the 10th International Conference on Wind Engineering.* 21-24 June, 1999, Copenhagen, Denmark. Rotterdam: Balkema: 55-66.

———. 2000a. Aerodynamics of the Tacoma Narrows Bridge—60 years later. *Structural Engineering International* 4/2000: 243-248.

———. 2000b. Lessons learned. *Bridge Design & Engineering* 20, Third Quarter: 44-46.

Larsen, A., and S. Esdahl, eds. 1998. *Proc. International Symposium on Advances in Bridge Aerodynamics*. 10-13 May 1998, Copenhagen. Rotterdam: Balkema.

Latimer, M., B. Hindle, and M. Kranzberg, eds. 1984. *Bridge to the Future: A Centennial Celebration of the Brooklyn Bridge*, New York: New York Academy of Sciences.

Le Corbusier. 1947. *When the Cathedrals Were White*. London: Reynal and Hitchcock.

Leonhardt, F. 1966. New trends in the conception and design of suspension bridges. *International Symposium on Suspension Bridges*, Lisbon: 125-162.

———. 1982. *Bridges*. London: The Architectural Press.

Leto, I. V. 1994. Preliminary design of the Messina Strait Bridge. *Proceedings Institution of Civil Engineers* 102, August: 122-129.

Maney, G. A. 1941. New type suspension bridge proposed. *Engineering News-Record,* April 24: 64-65.

Matthews, R. 1995. Space-age bridge spans great lengths. *Ottawa Citizen*, April 5: A12.

McBeth, D. G. 1990. The Forth bridges. In Topping, B.V., ed. *Developments in Structural Engineering: Proceedings Forth Rail Bridge Centenary Conference.* London: E. and R. N. Spon: 3-35.

Melan, J. 1888. *Theorie der Eisernen Bogenbrücken und der Hangebrücken.* Ed 2. Leipzig.

———. 1913. *Theory of Arches and Suspension Bridges*. Translated by D.B. Steinman. Chicago: Myron C. Clark.

Middelboe, S., and H. Ferguson. 1984. Closure confusion mars Severn decision. *New Civil Engineer,* 16 February: 4-5.

Minshall, H.H. 1966. Conception, design and construction of a precast post-tensioned concrete suspension bridge. Paper No. 25. *International Symposium on Suspension Bridges*, Lisbon: 411-414.

Mock, E.B. 1949. *The Architecture of Bridges*. New York: The Museum of Modern Art.

Modjeski, R., and F. Masters. 1940. *Report on a Proposed Crossing for the Straits of Mackinac*. Lansing: Mackinac Straits Bridge Authority of Michigan.

———. 1941. Suspension bridges and wind resistance. *ENR,* October 23: 97-100.

Moisseiff, L. S. 1937. Evolution of high-strength steels used in structural engineering. *Transactions of the American Society of Civil Engineers* 102: 1333-1358.

———. 1939. Growth in suspension bridge knowledge. *ENR,* August 17: 46-49.

Moisseiff, L. S., and F. Lienhard. 1933. Suspension bridges under the action of lateral forces, with discussion. *Transactions of the American Society of Civil Engineers* 98: 1080-1141.

Mumford, L. 1963. *The Highway and the City*. New York: Harcourt Brace Jovanovich.

Muroran Road Office. 1997. *Technical Information on Hakucho Ohashi Bridge*. Muroran Development and Construction Department.

Nakamura, Y., and M. Nakashima. 1986. Vortex excitation of prisms with elongated rectangular, H, and ⎬ cross-sections. *Journal of Fluid Mechanics* 163: 149-169.

National Youth Administration. 1940. *Tacoma Narrows Bridge—Official Opening—June 30-July 4, 1940*. Tacoma: Johnson-Cox Co.

Navier, C. L. M. H. 1823. *Rapport et Mémoire sur les Ponts Suspendus*. Paris: Imprimerie Royale.

O'Neill, W. 1982a. The case against inclined hangers. *New Civil Engineer,* 15 April: 12, 14.

———. 1982b. Traffic strain takes Severn blame. *New Civil Engineer,* 1 July: 12-13.

———. 1993. Longer yet and longer. *New Scientist,* 2 October: 32-34.

Ostenfeld & Jønson. 1970. *Motorway Bridge across Lillebælt*. Publications I-XIV, Copenhagen.

Ostenfeld, C., and W. Jønson. 1966. 5.2—Discussions. *International Symposium on Suspension Bridges*, Lisbon: 155.

Ostenfeld, K. H. 1990. *Multiple span bridge concept for the fixed link across the Strait of Gibraltar*. COWIconsult Publication No. 700.

Paine, C. 1937. The world's longest span. *Annual Report Smithsonian Institute.* 1937: 565-571.

Parsons, M. F. 1988. Major suspension bridges completed since 1960. *1st Oleg Kerensky Memorial Conference*, Session 2, London: 1-6.

Peters, T. F. 1987. *Transitions in Engineering: Guillaume Henri Dufour and the Early 19th Century Cable Suspension Bridges*. Basel and Boston: Birkhäuser Verlag.

Petroski, H. 1985. *To Engineer is Human*. New York: St. Martin's Press.

———. 1994. *Design Paradigms: Case Histories of Error and Judgement in Engineering.* Cambridge and New York: Cambridge University Press.

Pfeiffer, M. B. 1978. The Verrazano-Narrows Bridge—2 men wanted to design span, but the 'gentle genius' won out. *Staten Island Advance,* September 10.

Plowden, D. 1974. *The Bridges of North America*. New York: The Viking Press.

Podolny, W. Jr., and J. B. Scalzi. 1986. *Construction and Design of Cable-Stayed Bridges*. 2nd Ed. New York: John Wiley and Sons.

Pope, T. 1811. *A Treatise on Bridge Architecture*. New York: Alexander Niven.
Port of New York Authority and Triborough Bridge & Tunnel Authority. 1955. *Joint Study of Arterial Facilities: New York – New Jersey Metropolitan Area.* New York.
Provis, W. A. 1842. Observations on the effects produced by wind on the suspension bridge over the Menai Strait, more especially as relates to the injuries sustained by the roadways during the storm of January 1839; together with brief notices of various suggestions for repairing the structure. *Transactions Institution of Civil Engineers* Vol. III.
Pugsley, A. 1968. *The Theory of Suspension Bridges*. 2nd ed. London: Edward Arnold.
Ratigan, W. 1959. *Highways Over Broad Waters: Life and Times of David B. Steinman, Bridgebuilder*. Grand Rapids: Wm. B. Eerdmans Publishing Co.
Reier, S. 1977. *The Bridges of New York*. New York: Quadrant Press.
Richardson, J. R. 1981. *The Development of the Concept of the Twin Suspension Bridge*. NMI Report 125.
———. 1984. *Influence of Aerodynamic Stability on the Design of Bridges*. 12th Congress International Association for Bridge and Structural Engineering, Vancouver: 729-736.
Roberts, G. 1964. Letter to the editor. *ENR,* August 27: 8.
Robson, R. 1994. Tacoma Narrows Bridge collapse. In Schlager, N. ed. *When Technology Fails: Significant Technological Disasters, Accidents, and Failures of the Twentieth Century*. Detroit: Gale Research: 184-190.
Rocard, Y. 1957. *Dynamic Instability: Automobiles, Aircraft, Suspension Bridges*. Translated by M. L. Meyer, London: Crosby, Lockwood and Son.
Ross, S., *et al.* 1940. *Tacoma Narrows, 1940, Construction Disasters*. New York: McGraw-Hill: 216-239.
Rubin, L. A. 1985. *Bridging the Straits.* Detroit: Wayne State University Press.
Russell, J. S. 1841. On the vibration of suspension bridges and other structures; and the means of preventing injury from this cause. *Transactions Royal Scottish Society of Arts* 1: 304-314.
Sabzevari, A., and R. H. Scanlan. 1968. Aerodynamic instability of suspension bridges. *Journal of Engineering Mechanics Division, Proc. ASCE*, EM2, April: 489-517.
Scanlan, R. H. 1979. Performance of long-span suspension bridge structures under wind loadings. *Long Span Suspension Bridges: History and Performance,* Sessions No. 70 and 78, Boston: ASCE National Convention, ASCE Preprint 3590: 1-16.
———. 1983. Aeroelastic simulation of bridges. *Journal of Structural Engineering*, ASCE, 109(12), December: 2829-2837.
———. 1990. Bridge aerolasticity: Present state and future challenges. *Journal of Wind Engineering and Industrial Aerodynamics* 36: 63-74.
Scanlan, R. H., and A. Sabzevari. 1967. Suspension Bridge Flutter Revisited. Preprint No. 468. Seattle, Washington: *ASCE National Meeting in Structural Engineering* , May 8-12.
Scanlan, R. H., and J. J. Tomko. 1971. Airfoil and bridge deck flutter derivatives. *Journal of. Engineering Mechanics Division, Proc. ASCE,* Vol. 97, No. EM 6, December: 1717-1737.
Scanlan, R. H., and J. W. Vellozzi. 1980. Catastrophic and annoying responses of long-span bridges to wind action. In Cohen, E., and B. Birdsall, eds. *Long-Span Bridges, O. H. Ammann Centennial Conference*, Annals of the New York Academy of Sciences, 352: 247-263.
Schlaich, J. 1990. Some thoughts on cable bridge design. In Topping, B.V., ed. *Developments in Structural Engineering: Proceedings Forth Rail Bridge Centenary Conference.* London: E. and R. N. Spon: 297-315.
Seguin, M. 1824. *Des Ponts en Fil de Fer*. Paris: Bachelier.

Selberg, A. 1946. *Design of Suspension Bridges*. Trondheim: Norske Videnskabers Selskabs Skrifter, No. 1.

———. 1950. Dampening effect in suspension bridges. Extract from *Mémoires*. Zurich: International Association for Bridge and Structural Engineering: 183-198.

———. 1961. Oscillation and aerodynamic stability of suspension bridges. *Acta Polytechnica Scandinavia.*

———. 1966. Aerodynamic Stability and Related Aspects of Suspension Bridges. Paper No. 20. *International Symposium on Suspension Bridges*, Lisbon: 361-366.

Seim, C. 1987. Building over the Golden Gate. *Civil Engineering—ASCE*, May: 46-49.

Shirley-Smith, H. 1964. *The World's Great Bridges*. Rev. ed. New York: Harper and Row.

———. 1967. Discussion on Forth Road Bridge. *Proceedings Institution of Civil Engineers,* January: 182-184.

Sibly, P. G., and A. C. Walker. 1977. Structural accidents and their causes. *Proceedings Institution of Civil Engineers* 62: 191-208.

Smith, F. C., and G. S. Vincent. 1950. Part II: Mathematical analyses. *Aerodynamic Stability of Suspension Bridges with Special Reference to the Tacoma Narrows Bridge*, Bulletin No. 116, University of Washington, Engineering Experiment Station.

Societé des Ingénieurs civils de France, *et al.* 1961. *Pont-Route de Tancarville – le plus grand pont suspendu d'Europe*. Pars: La Documentation Technique du Bâtiment, et des Travaux Publics.

Stahl, F. L. 1979. Letter to the editor. *ENR,* June 21: 5.

Stang, O. 1934. *Myke Hengebroker pa Norske Landeveier (Flexible Suspension Bridges for Norwegian Roads).* Oslo: Meddelelser fra Veidirektoren, No. 1.

Steinman, D. B. 1911. *Suspension Bridges and Cantilevers: Their Economic Proportions and Limiting Spans*. New York: Van Nostrand.

———. 1929a. *A Practical Treatise on Suspension Bridges: Their Design, Construction and Erection.* 2nd Ed. New York: John Wiley and Sons.

———. 1929b. *Fifty Years of Progress in Bridge Engineering*. New York: American Institute of Steel Construction.

———. 1934. A generalized deflection theory for suspension bridges. *Transactions American Society of Civil Engineers,* March: 1133-1170.

———. 1938. Construction of the Thousand Islands Bridge. *Civil Engineering—ASCE,* November: 716-719.

———. 1941a. Letter to the editor. *Engineering News-Record,* February 27: 36.

———. 1941b. Letter to the editor. *ENR,* May 8: 60-61.

———. 1941c. Letter to the editor. *ENR,* August 14: 59-61.

———. 1941d. Letter to the editor. *Engineering News-Record,* December 18: 51.

———. 1943. Rigidity and aerodynamic stability of suspension bridges. Proceedings American Society of Civil Engineers 69, November: 1361-1397.

———. 1945a. *The Builders of the Bridge*. New York: Harcourt, Brace and Co.

———. 1945b. Design of bridges against wind: I. General considerations—aerostatic stability. *Civil Engineering—ASCE,* October: 469-472.

———. 1945c. Design of bridges against wind: II. Aerodynamic instability—historical background. *Civil Engineering—ASCE,* November: 501-504.

———. 1945d. Design of bridges against wind: III. Elementary explanation of aerodynamic instability. *Civil Engineering—ASCE,* December: 558-560.

———. 1946a. Design of bridges against wind: IV. Aerodynamic instability—prevention and cure. *Civil Engineering—ASCE,* January: 20-23.

———. 1946b. Design of bridges against wind: V. Criteria for assuring aerodynamic stability. *Civil Engineering—ASCE,* February: 68-66.

———. 1946c. Letter to the editor. *Civil Engineering—ASCE,* October: 462.

———. 1947a. Simple model tests predict aerodynamic characteristics of bridges: Part I. Pressure distribution graphs. *Civil Engineering—ASCE,* January: 20-24.

———. 1947b. Simple model tests predict aerodynamic characteristics of bridges: Part II. Response curves from static lift and torque graphs. *Civil Engineering—ASCE,* February: 77-79.

———. 1947c. Use of girder-type bridge sections affirmed. *Civil Engineering—ASCE,* August: 56, 80.

———. 1947d. Wind tunnel tests yield aerodynamically stable bridge sections. *Civil Engineering—ASCE,* December: 40-43, 94.

———. 1947e. Problems of aerodynamic and hydrodynamic stability. *Proceedings Third Hydraulics Conference*, Bulletin 31, University of Iowa Studies in Engineering: 136-164.

———. 1950. Aerodynamic theory of bridge oscillations, with discussion. *Transactions American Society of Civil Engineers* 115: 1180-1260.

———. 1953. The Messina Straits Bridge to connect Sicily with Italy. *Road International* No. 11, Winter.

———. 1954a. Suspension bridges: The aerodynamic problem and its solution. *American Scientist,* July : 397-438, 460.

———. 1954b. Bridges. *Scientific American,* November: 61-71.

———. 1956. Mackinac Bridge . . . designed for complete aerodynamic stability. *Civil Engineering—ASCE,* May: 37-41.

Steinman, D. B., and C. H. Gronquist. 1959. Mackinac Bridge: Superstructure design and construction. *Civil Engineering—ASCE*, January: 48-53.

Steinman, D. B., and J. T. Nevill. 1957. *Miracle Bridge at Mackinac*. Grand Rapids: Wm. B. Eerdmans Publishing Co.

Steinman, D. B., and S. R. Watson. 1957. *Bridges and Their Builders*. 2nd ed. New York: Dover Publications.

Strauss, J. B. 1938. *The Golden Gate Bridge—Report of the Chief Engineer*. San Francisco: Golden Gate Bridge and Highway District.

Studio. G. 1966. *The Severn Bridge*. Birmingham: The Kynock Press.

Stüssi, F., and J. Ackeret. 1941. Zum einsturz der Tacoma-Hängebrucke. *Schwiezer Bauzeitung,* March: 586-589.

Talese, G. 1964. *The Bridge*. New York: Harper & Row.

Tajima, T. 1985. Construction of the Ohnaruto Bridge. *Civil Engineering in Japan*, JSCE: 105-123.

Toscano, A. M. 1979. Fattibilità di ponti a più campate per l'attraversamento dello Stretto di Messina. *Construzioni metalliche*, No. 3-1979: 123-135.

Tozöz, M. 1963. Projets de ponts et de tunnels pour le franchissement du Bosphere. *Travaux*, juin: 291-301.

Triborough Bridge & Tunnel Authority, and Ammann & Whitney. 1966. *Verrazano-Narrows Bridge: History, Financing, Design and Construction*. Reprinted from Proceedings ASCE, Journal of the Construction Division., Vol. 92, No. CO2.

Troitsky, M. S. 1967. *Orthotropic Bridges: Theory and Design*. Cleveland: The James F. Lincoln Arc Welding Foundation.

———. 1990. *Prestressed Steel Bridges: Theory and Design*. New York: Van Nostrand Reinhold.

Tunnard, C., and B. Pushkarev. 1963. *Man-Made America, Chaos or Control?* New Haven and London: Yale University Press.

Turner, C. A. P. 1940. Letter to the editor. *ENR,* December 5: 40-41.

———. 1942. Letter to the editor. *ENR,* April 9: 61.

Tyrrell, H. G. 1911. *History of Bridge Engineering.* Chicago: Published by the author.

Ukeguchi, N., H. Sakata, and H. Nishitani. 1966. An Investigation of Aeroelastic Instability of Suspension Bridges. Paper No. 11. *International Symposium on Suspension Bridges,* Lisbon: 273-28.

Uno, N., and H. Hara. 1975. Design, wind tunnel test and site observation of 'Namhae Bridge' for Ministry of Construction, the Republic of Korea. *IHI Engineering Review,* Vol. 8, No. 1, January: 100-11.

Vandepitte, D. 1966. Self-Anchored Prestressed Concrete Suspension Bridges with Parabolic Cables. Paper No. 38. *International Symposium on Suspension Bridges,* Lisbon: 549-556.

Verantius, F. 1615. *Machinae Novae.* Florence, 1615.

Vicat, L. J. 1831. Ponts suspends en fil en fer sur le Rhône. *Annales des Ponts et Chausées (Mémoires et Documents)* 1: 93-145.

Vincent, G. S. 1951. Suspension bridge vibration formulas. *ENR,* January 11: 32-34.

———. 1952. Aerodynamic stability of suspension bridges. *Proceedings—Separate No. 144A-144B ASCE,* August: 1-63.

———. 1954. Part V: Extended studies: Logarithmic decrement field damping, prototype predictions, four other bridges. *Aerodynamic Stability of Suspension Bridges with Special Reference to the Tacoma Narrows Bridge,* Bulletin No. 116, University of Washington, Engineering Experiment Station.

———. 1965. A summary of laboratory and field studies in the United States on wind effects on suspension bridges. Paper 20. *Symposium on Wind Effects on Buildings and Structures,* National Physical Laboratory: 488-513.

von Kármán, T. 1940. Letter to the editor. *ENR,* November 21: 40.

von Kármán, T., and L. G. Dunn. 1952. Chapter VII: Wind-tunnel investigations carried out at the California Institute of Technology. In Farquharson, F. B. *Aerodynamic Stability of Suspension Bridges with Special Reference to the Tacoma Narrows Bridge,* Bulletin No. 116, University of Washington, Engineering Experiment Station: 57-53.

von Kármán, T., and L. Edson. *The Wind and Beyond: Theodore von Kármán, Pioneer in Aviation and Pathfinder in Space.* Boston: Little, Brown.

Walshe, D. E. 1967. A resumé of the aerodynamic investigations for the Forth Road and the Severn bridges. *Proceedings Institution of Civil Engineers* 37, May: 87-108.

Walshe, D. E., and T. A. Wyatt. 1992. Bridge aerodynamics 50 years after Tacoma Narrows, Part II, a new discipline world-wide. *Journal of Wind Engineering and Industrial Aerodynamics* 40: 327-336.

Westhofen, W. 1890. *The Forth Bridge, Centenary Edition.* Edinburgh: Moubray House.

Whitelaw, J., ed. 1997. *Lantau Link–Tsing Ma Bridge.* Supplement to New Civil Engineer. London: Emap Construct.

Wooten, L.R. 1990. Kit Scruton: An appreciation. *Journal of Wind Engineering and Industrial Aerodynamics,* reprint.

Wyatt, T. A. 1992. Bridge aerodynamics 50 years after Tacoma Narrows, Part I, the Tacoma Narrows failure and after. *Journal of Wind Engineering and Industrial Aerodynamics* 40: 317-326.

Yamazaki, K., K. Izumi, and M. Ogihara. 1994. Rainbow Bridge, Japan. *Structural Engineering International,* November: 226-228.

Appendix

THE WORLD'S LONGEST SUSPENSION BRIDGES

The world's 150 longest suspension bridges are listed in order of span. Completion dates are noted in parentheses. Bridges under construction are noted with a "u.c"; those under active planning or under design are marked with a "u.p". Industrial pipeline/conveyor bridges are marked with an asterisk (*). Bridges no longer standing are marked with a [1]. Pedestrian suspension bridges are marked with a [2]. Twin bridges (in tandem or in parallel) are marked with a [3]. Replacement bridges are marked with a [4]. The list is current to mid-2001.

Messina Strait, Italy (u.p.)	3300 m
Akashi Kaikyo, Japan (1998)	1990 m
Izmit Bridge, Turkey (u.p.)	1688 m
Jiaozhou Wan, China (u.p.)	1652 m
East Bridge, Denmark (1998)	1624 m
Runyang Yangtse, China (u.c.)	1490 m
Tsing Lung, Hong Kong SAR (u.p.)	1418 m
Humber, England (1981)	1410 m
Lingdingyang, China (u.p.)	1400 m
Jiangyin, China (1999)	1385 m
Tsing Ma, Hong Kong SAR (1997)	1377 m
Hardanger Fjord, Norway (u.p.)	1325 m
Verrazano-Narrows, U.S. (1964)	1298 m
Golden Gate, U.S. (1937)	1280 m
Chacao Channel, Chile (u.p.)	1270 m
Höga Kusten, Sweden (1997)	1210 m
Mackinac, U.S. (1957)	1158 m
Minami Bisan Seto, Japan (1988)	1100 m
2nd Bosporus, Turkey (1988)	1090 m
1st Bosporus, Turkey (1973)	1074 m
George Washington, U.S. (1931)	1067 m
3rd Kurushima, Japan (1999)	1030 m
2nd Kurushim, Japan (1999)	1020 m
Tagus, Portugal (1966)	1013 m
Forth Road, Scotland (1964)	1006 m
Kita Bisan Seto, Japan (1988)	990 m
Severn, England-Wales (1966)	988 m
Yichang, China (2001)	960 m
Lingdingyang, China (u.p.)	956 m
Shimotsui Seto, Japan (1988)	940 m
Xiling, China (1996)	900 m

Boca Tigris, China (1997)	888 m
Ohnaruto, Japan (1985)	876 m
Chacao Channel, Chile (u.p.)	870 m
Dneiper, Russia (1964)*[1]	874 m
Tacoma Narrows U.S. (1940[1], 1950[4], u.p.[3])	853 m
Askøy, Norway (1992)	850 m
Innoshima, Japan (1983)	770 m
Akinada, Japan (2000)	750 m
Irtysh, Kazakhstan (2000)	750 m
Carquinez, U.S. (u.c.—2005)	728 m
Dneiper, Russia (1979)	720 m
Hakucho, Japan (1998)	720 m
Kanmon, Japan (1973)	712 m
Angostura, Venezuela (1967)	712 m
Transbay, U.S. (1936)[3]	704 m
Bronx-Whitestone, U.S. (1939)	701 m
Storda, Norway (2000)	677 m
Pierre Laporte, Canada (1970)	668 m
Amu-Darya, Russia (1974)*	660 m
Delaware Memorial, U.S. (1951, 1968)[3]	655 m
Mississippi River, U.S. (1955)*	655 m
Xiamen Haichang, China (1999)	648 m
Portsmouth, U.S. (1962)*	635 m
Gjemnessundet, Norway (1992)	623 m
Swiss Alps, Switzerland (1982)*	611 m
Atchafayla, U.S. (1951)*	610 m
Walt Whitman, U.S. (1957)	610 m
Tancarville, France (1959)	608 m
Lillebælt, Denmark (1970)	600 m

Osterøy, Norway (1997)	595 m
Flaming Gorge, U.S. (197?)*	585 m
Bømla, Norway (2001)	577 m
Rainbow, Japan (1994)	570 m
Metsovitikos, Greece (u.p.)	565 m
Ambassador, Canada-U.S. (1929)	564 m
Ohshima, Japan (1988)	560 m
Fengdu, China (2000)	560 m
Rainbow, Japan (1994)	570 m
Throgs Neck, U.S. (1961)	549 m
Benjamin Franklin, U.S. (1926)	533 m
Skjomen, Norway (1972)	525 m
Kvalsund, Norway (1977)	525 m
Matadi, Congo (1983)	520 m
Taylor Flats, Canada (19??)*	510 m
Emmerich, Germany (1965)	500 m
Dazi, China (1984)	500 m
Kwang Ahn Grand, South Korea (u.c.—2002)	500 m
Dalsfjorden, Norway (u.p.)	500 m
Bear Mountain, U.S. (1924)	497 m
Williamsburg, U.S. (1903)	488 m
Newport, U.S. (1969)	488 m
William. Preston Lane, Jr. II, U.S. (1951, 1973)[3]	488 m
Brooklyn, U.S. (1883)	486 m
Lions Gate, Canada (1938)	472 m
Sotra, Norway (1971)	468 m
Hirado, Japan (1977)	465 m
Mid-Hudson, U.S. (1930)	457 m
Plattsmouth, U.S. (1957)*	457 m
Vincent Thomas, U.S. (1963)	457 m
Shantou Bay, China (1995)	452 m
Manhattan, U.S. (1909)	448 m
Fraser River, Canada (19??)*	448 m
Lysefjorden, Norway (1997)	446 m
Angus. L. Macdonald, Canada (1955)	441 m
A. Murray MacKay, Canada (1970)	427 m
Triborough, U.S. (1936)	421 m
Älvsborg, Sweden (1966)	418 m
Namhae, South Korea (1973)	404 m
Similkameen, Canada (1980)*	404 m
Sioux City (1940)*	402 m
Aquitaine, France (1967)	394 m
Fraser River, Canada (19??)*	393 m
Missouri River, U.S. (1932)*	390 m
Amu-Darya, Russia (1964)*	390 m
Upper Suspension Bridge, Canada-U.S., (1869)[1]	386 m
East Bay, U.S. (u.p.)	385 m
Rodenkirchen, Germany (1941[1], 1955)	378 m
Japan pedestrian (1993)[2]	375 m
St. Johns, U.S. (1931)	368 m
Wakato, Japan (1962)	367 m
Mount Hope, U.S. (1929)	366 m
Pipeline, southwest U.S. (1947)*	366 m
Tanana, U.S. (1976)*	366 m
Bandar Sunway, Malaysia (1994)[2]	352 m
Ogdensburg, Canada-U.S. (1960)	351 m
Hercilio, Brazil (1926)	340 m
Bidwell Bar, U.S. (1965)	338 m
Chirundi, Rhodesia (1939)	320 m
Varodd, Norway (1956)	337 m
Tamar, England (1961)	335 m
Feda, Norway (u.p.)	331 m
Deer Isle, U.S. (1939)	329 m
Colorado River, U.S. (1957)*	329 m
Fraser River, Canada (19??)*	329 m
Rombak, Norway (1964)	325 m
Maysville, U.S. (1931)	323 m
Isle d'Orléans, Canada (1935)	323 m
Nærøysundet, Norway (1981)	325 m
J.A. Roebling, U.S. (1867)	322 m
Po, Italy (1950s)*[3]	320 m
Dent, U.S. (1971)	320 m
Danube, Vienna (1961)*	320 m
Lewiston, Canada-U.S. (1851)[1]	317 m
Cologne-Mulheim, Germany (1929[1], 1951)	315 m
Miampini, Mexico (1900)	314 m
Blythe, U.S. (1947)*	311 m
Wheeling, U.S. (1849)	308 m
Avalon, U.S. (19??)*	307 m
Konohana, Japan (1990)	300 m
Chavanon, France (1999)	300 m
Yongjong Grand, South Korea (2000)	300 m
Elizabeth, Hungary (1903[1], 1965[4])	290 m
Tjeldsund, Norway (1967)	290 m
Similkameen, Canada (1972)*	290 m
Grand' Mère, Canada (1929)	289 m
Cauca River, Columbia (1894)	287 m
Jinhu, China (1994)	284 m
Peace River, Canada (1943)[1]	283 m

INDEX

Entries in *italics* identify illustrative material. Completion dates are given in parentheses following the bridge name.